Smart Nanomaterials Technology

Series Editors

Azamal Husen, Wolaita Sodo University, Wolaita, Ethiopia

Mohammad Jawaid, Department of Chemical and Petroleum Engineering, United Arab Emirates University, Al Ain, United Arab Emirates

Nanotechnology is a rapidly growing scientific field and has attracted a great interest over the last few years because of its abundant applications in different fields like biology, physics and chemistry. This science deals with the production of minute particles called nanomaterials having dimensions between 1 and 100 nm which may serve as building blocks for various physical and biological systems. On the other hand, there is the class of smart materials where the material that can stimuli by external factors and results a new kind of functional properties. The combination of these two classes forms a new class of smart nanomaterials, which produces unique functional material properties and a great opportunity to larger span of application. Smart nanomaterials have been employed by researchers to use it effectively in agricultural production, soil improvement, disease management, energy and environment, medical science, pharmaceuticals, engineering, food, animal husbandry and forestry sectors.

This book series in Smart Nanomaterials Technology (SNT) publishes cutting edge research monographs and edited volumes by top researchers and professionals across the globe. This book series aims to comprehensively cover topics in the fabrication, synthesis and application of these materials for applications in the following fields:

- Energy Systems—Renewable energy, energy storage (supercapacitors and electrochemical cells), hydrogen storage, photocatalytic water splitting for hydrogen production
- Biomedical—controlled release of drugs, treatment of various diseases, biosensors,
- Agricultural—agricultural production, soil improvement, disease management, animal feed, egg, milk and meat production/processing
- Forestry—wood preservation, protection, disease management
- Environment—wastewater treatment, separation of hazardous contaminants from wastewater, indoor air filters.

Jen-Tsung Chen
Editor

Nano-Delivering for Plant Genetic Engineering

Springer

Editor
Jen-Tsung Chen
Department of Life Sciences
National University of Kaohsiung
Kaohsiung, Taiwan

ISSN 3004-8273 ISSN 3004-8281 (electronic)
Smart Nanomaterials Technology
ISBN 978-981-95-4164-5 ISBN 978-981-95-4165-2 (eBook)
https://doi.org/10.1007/978-981-95-4165-2

This Springer imprint is published by the registered company Springer Nature Singapore Pte Ltd.
The registered company address is: 152 Beach Road, #21-01/04 Gateway East, Singapore 189721, Singapore

Preface

Plant gene transfer is a fundamental biotechnology for obtaining genetically modified plants, which leads to crop improvement and enhanced agricultural food production. For decades, certain protocols have been developed to achieve successful plant genetic engineering events; however, there are still some limitations, particularly in the way to enhance their efficiency and precision.

This book summarizes the knowledge of smart nanoparticles in advancing plant gene transfer, focusing on their capacity for penetrating plant cells and delivering genetic engineering cargo, chiefly nucleic acids, and eventually to enhance the efficiency and precision of plant genetic engineering. The benefits and limitations of nano-biotechnology on plant gene transfer have been comprehensively discussed in each subtopic covering carbon, metallic, silicon, and bioinspired nanoparticles.

This book presents the edge knowledge of nanocarriers in plant genome editing, including their technical advancements and recent applications, particularly including the part of CRISPR/Cas9 technology.

From fundamentals to practical applications of nano-biotechnology, this book can serve as an ideal reference for readers in the fields of plant genetics, plant physiology, plant biotechnology, plant breeding, and materials science.

The editor appreciates contributors' innovative chapters and the publisher's instruction and assistance.

Kaohsiung, Taiwan — Jen-Tsung Chen

Contents

Contributors

Banzeer Ahsan Abbasi Department of Botany, Rawalpindi Women University, Rawalpindi, Pakistan

Hajra Abid Department of Botany, University of Poonch Rawalakot, Rawalakot, Pakistan

Emmanuel Aguirre-García Divisi6n de Investigaci6n y Posgrado, Facultad de Ingeniería, Universidad Aut6noma de Querétaro, Santiago de Querétaro, Querétaro, Mexico

Khawaja Shafique Ahmad Department of Botany, University of Poonch Rawalakot, Rawalakot, Pakistan

Zubaria Tul Ain Department of Plant Sciences, Faculty of Biological Sciences, Quaid-i-Azam University Islamabad, Islamabad, Pakistan

Mariam Akhtar Agricultural Biotechnology Division, National Institute for Biotechnology and Genetic Engineering (NIBGE), Constituent College of Pakistan Institute of Engineering and Applied Sciences (PIEAS), Faisalabad, Pakistan

Ayesha Akram Agricultural Biotechnology Division, National Institute for Biotechnology and Genetic Engineering (NIBGE), Constituent College of Pakistan Institute of Engineering and Applied Sciences (PIEAS), Faisalabad, Pakistan

Imran Amin Agricultural Biotechnology Division, National Institute for Biotechnology and Genetic Engineering (NIBGE), Constituent College of Pakistan Institute of Engineering and Applied Sciences (PIEAS), Faisalabad, Pakistan

Muhammad Anas Department of Plant Sciences, Faculty of Biological Sciences, Quaid-i-Azam University Islamabad, Islamabad, Pakistan

Majid Anwar Department of Pharmacology, Multan University of Science and Technology, Multan, Pakistan

Muhammad Arif Agricultural Biotechnology Division, National Institute for Biotechnology and Genetic Engineering (NIBGE), Constituent College of Pakistan Institute of Engineering and Applied Sciences (PIEAS), Faisalabad, Pakistan

Sanam Ashraf Department of Plant Sciences, Faculty of Biological Sciences, Quaid-i-Azam University Islamabad, Islamabad, Pakistan

Muhammad Asif Agricultural Biotechnology Division, National Institute for Biotechnology and Genetic Engineering (NIBGE), Constituent College of Pakistan Institute of Engineering and Applied Sciences (PIEAS), Faisalabad, Pakistan

Loulaq Banday Department of Biotechnology, University of Kashmir, Srinagar, J&K, India

Piyush Bania Palli Siksha Bhavana (Institute of Agriculture), Visva-Bharati, Sriniketan, West Bengal, India

Madhumita Barooah Department of Agricultural Biotechnology, Assam Agricultural University, Jorhat, Assam, India

Sana Basharat School of Breeding and Multiplication (Sanya Institute of Breeding and Multiplication), College of Tropical Agriculture and Forestry, Hainan University, Sanya, China

Amitesh Bhattacharyya Biotechnology Unit, Crop Improvement Division, ICAR-Central Research Institute for Jute and Allied Fibres, Barrackpore, West Bengal, India

Sudipta Sankar Bora Silchar Regional Forensic Science Laboratory, Ghungur, Silchar, Assam Under the Directorate of Forensic Science, Kahilipara, Guwahati, Assam, India

Directorate of Forensic Science, Kahilipara, Guwahati, Assam, India

Madhurjya Protim Borah Department of Biosciences and Bioengineering, Indian Institute of Technology Jammu, Jammu, Jammu and Kashmir, India

José Antonio Cervantes-Chávez Laboratorio de Microbiología Molecular, Cuerpo Académico Biotecnología y Sanidad Vegetal, Facultad de Ciencias Naturales, Universidad Autónoma de Querétaro, Santiago de Querétaro, Querétaro, Mexico

Dyuti Ballav Chongder Palli Siksha Bhavana (Institute of Agriculture), Visva-Bharati, Sriniketan, West Bengal, India

Fung Swrangshee Dalmari Department of Botany, Bodoland University, Deborgaon, Assam, India

Ankan Das Palli Siksha Bhavana (Institute of Agriculture), Visva-Bharati, Sriniketan, West Bengal, India

Anirneeta De School of Agriculture, Swami Vivekananda University, Barrackpore, West Bengal, India

Sandip Debnath Department of Genetics and Plant Breeding, Palli Siksha Bhavana (Institute of Agriculture), Visva-Bharati, Sriniketan, West Bengal, India

Tabasum Mohi Ud Din Department of Botany, Amar Singh College, Srinagar, J&K, India

Mohsen Mohamed Elsharkawy Agricultural Botany Department, Faculty of Agriculture, Kafrelsheikh University, Kafr Elsheikh, Egypt

Karen Esquivel Escalante División de Investigación y Posgrado, Facultad de Ingeniería, Universidad Autónoma de Querétaro, Santiago de Querétaro, Querétaro, Mexico

Alexandro Escobar-Ortíz División de Investigación y Posgrado, Facultad de Ingeniería, Universidad Autónoma de Querétaro, Santiago de Querétaro, Querétaro, Mexico

Asifa Farooqi Department of Environmental Sciences, Faculty of Biological Sciences, Quaid-i-Azam University Islamabad, Islamabad, Pakistan

Sana Fatima Department of Botany, The Government Sadiq College Women University, Bahawalpur, Bahawalpur, Pakistan

Ana A. Feregrino-Pérez División de Investigación y Posgrado, Facultad de Ingeniería, Universidad Autónoma de Querétaro, Santiago de Querétaro, Querétaro, Mexico

Bertha Isabel Carvajal Gámez Laboratorio de Proteogenómica y Diagnóstico Molecular. Cuerpo Académico Salud Animal y Microbiología Ambiental, Facultad de Ciencias Naturales, Universidad Autónoma de Querétaro, Santiago de Querétaro, Querétaro, Mexico

Prakash I. Gangashetty International Crops Research Institute for the Semi-Arid Tropics (ICRISAT), Hyderabad, India

Rahul Gogoi Department of Agricultural Biotechnology, Assam Agricultural University, Jorhat, Assam, India

Jude W. Grosser Citrus Research and Education Center, University of Florida/IFAS, Lake Alfred, FL, USA

Mariana Yalid Hernández-Ramírez División de Investigación y Posgrado, Facultad de Ingeniería, Universidad Autónoma de Querétaro, Santiago de Querétaro, Querétaro, Mexico

Ishfaq Majid Hurrah Plant Biotechnology Division, CSIR-Indian Institute of Integrative Medicine, Sanat Nagar, Srinagar, J&K, India

Sajjad Hussain Department of Botany, University of Poonch Rawalakot, Rawalakot, Pakistan

Shumaila Ijaz Department of Burn and Plastic Surgery, Shenzhen Institute of Translational Medicine, The First Affiliated Hospital of Shenzhen University, School of Biomedical Engineering, Shenzhen University Medical School, Shenzhen University, Shenzhen, China

Muhammad Ikram Department of Botany, PMAS-Arid Agriculture University Rawalpindi, Rawalpindi, Punjab, Pakistan

Ifrah Imran Agricultural Biotechnology Division, National Institute for Biotechnology and Genetic Engineering (NIBGE), Constituent College of Pakistan Institute of Engineering and Applied Sciences (PIEAS), Faisalabad, Pakistan

Javed Iqbal Department of Botany, Bacha Khan University, Charsadda, Khyber Pakhtunkhwa, Pakistan

Rashid Iqbal Department of Agronomy, Faculty of Agriculture and Environment, The Islamia University of Bahawalpur, Bahawalpur, Pakistan

Department of Life Sciences, Western Caspian University, Baku, Azerbaijan

Marcello Iriti Department of Biomedical, Surgical and Dental Sciences, Milan State University, Milan, Italy

Anam Ishtiaq Agricultural Biotechnology Division, National Institute for Biotechnology and Genetic Engineering (NIBGE), Constituent College of Pakistan Institute of Engineering and Applied Sciences (PIEAS), Faisalabad, Pakistan

A. K. M. Aminul Islam Department of Genetics and Plant Breeding, Gazipur Agricultural University, Gazipur, Bangladesh

Shazina Jabeen Department of Agricultural, Forestry and Food Science-DISAFA, University of Turin, Turin, Italy

Daniel Mendoza Jiménez Laboratorio de Microbiología Molecular, Cuerpo Académico Biotecnología y Sanidad Vegetal, Facultad de Ciencias Naturales, Universidad Autónoma de Querétaro, Santiago de Querétaro, Querétaro, Mexico

Suresh Kadaru Medium Maturity Rice Breeding Program, International Rice Research Institute (IRRI), Los Baños, Laguna, Philippines

Muhammad Kaleem Department of Plant Sciences, Faculty of Biological Sciences, Quaid-i-Azam University Islamabad, Islamabad, Pakistan

Muhammad Faraz Khan Department of Botany, University of Poonch Rawalakot, Rawalakot, Pakistan

Rabia Kousar Department of Botany, The Government Sadiq College Women University, Bahawalpur, Bahawalpur, Pakistan

Tariq Mahmood Department of Plant Sciences, Faculty of Biological Sciences, Quaid-i-Azam University Islamabad, Islamabad, Pakistan

Sayanti Mandal Department of Chemistry and Biochemistry, Sharda School of Engineering & Sciences, Sharda University, Greater Noida, Uttar Pradesh, India

Luis Alejandro Martínez-Chávez División de Investigación y Posgrado, Facultad de Ingeniería, Universidad Autónoma de Querétaro, Santiago de Querétaro, Querétaro, Mexico

Ansar Mehmood Department of Botany, University of Poonch Rawalakot, Rawalakot, Pakistan

Azza H. Mohamed Department of Agricultural Chemistry, College of Agriculture, Mansoura University, Mansoura, Egypt

Citrus Research and Education Center, University of Florida/IFAS, Lake Alfred, FL, USA

Srija Mondal Palli Siksha Bhavana (Institute of Agriculture), Visva-Bharati, Sriniketan, West Bengal, India

Abdul Samad Mumtaz Department of Plant Sciences, Faculty of Biological Sciences, Quaid-i-Azam University Islamabad, Islamabad, Pakistan

Akhtar Munir Department of Chemistry, Quaid-i-Azam University Islamabad, Islamabad, Pakistan

Ghulam Murtaza School of Agriculture, Yunnan University, Kunming, Yunnan, China

Rubab Zahra Naqvi Agricultural Biotechnology Division, National Institute for Biotechnology and Genetic Engineering (NIBGE), Constituent College of Pakistan Institute of Engineering and Applied Sciences (PIEAS), Faisalabad, Pakistan

Syed Ali Raza Naqvi Department of Chemistry, Government College University Faisalabad, Faisalabad, Pakistan

Edgardo Ulises Esquivel Naranjo Laboratorio de Microbiología Molecular, Cuerpo Académico Biotecnología y Sanidad Vegetal, Facultad de Ciencias Naturales, Universidad Autónoma de Querétaro, Santiago de Querétaro, Querétaro, Mexico

Nor Anis Nadhirah Md. Nasir Faculty of Civil Engineering & Technology, Universiti Malaysia Perlis, Arau, Perlis, Malaysia

Hummera Nawaz Department of Botany, Division of Science and Technology, University of Education, Lahore, Pakistan

Ahmad A. Omar Citrus Research and Education Center, University of Florida/IFAS, Lake Alfred, FL, USA

Biochemistry Department, Faculty of Agriculture, Zagazig University, Zagazig, Egypt

Bodhisattya Pal Crop Improvement Division, ICAR—Central Research Institute for Jute and Allied Fibres, Barrackpore, West Bengal, India

Biswajit Pramanik Biotechnology Unit, Crop Improvement Division, ICAR-Central Research Institute for Jute and Allied Fibres, Barrackpore, West Bengal, India

Amna Rafique Agricultural Biotechnology Division, National Institute for Biotechnology and Genetic Engineering (NIBGE), Constituent College of Pakistan Institute of Engineering and Applied Sciences (PIEAS), Faisalabad, Pakistan

Abhisek Rath Department of Plant Pathology, Assam Agricultural University, Jorhat, Assam, India

Naima Razzaq Agricultural Biotechnology Division, National Institute for Biotechnology and Genetic Engineering (NIBGE), Constituent College of Pakistan Institute of Engineering and Applied Sciences (PIEAS), Faisalabad, Pakistan

Amanda K. Rico-Chávez Divisi6n de Investigaci6n y Posgrado, Facultad de Ingeniería, Universidad Aut6noma de Querétaro, Santiago de Querétaro, Querétaro, Mexico

Bristi Roy Palli Siksha Bhavana (Institute of Agriculture), Visva-Bharati, Sriniketan, West Bengal, India

Muhammad Waseem Sajjad Agricultural Biotechnology Division, National Institute for Biotechnology and Genetic Engineering (NIBGE), Constituent College of Pakistan Institute of Engineering and Applied Sciences (PIEAS), Faisalabad, Pakistan

Shreya Sarkar Biotechnology Unit, Crop Improvement Division, ICAR-Central Research Institute for Jute and Allied Fibres, Barrackpore, Barrackpore, West Bengal, India

Hridesh Harsha Sarma Department of Agronomy, Assam Agricultural University, Jorhat, Assam, India

Muhammad Muzamil Sattar Department of Plant Sciences, Faculty of Biological Sciences, Quaid-i-Azam University Islamabad, Islamabad, Pakistan

Sayantan Senapati Palli Siksha Bhavana (Institute of Agriculture), Visva-Bharati, Sriniketan, West Bengal, India

Sadhana Singh Department of Biochemistry, School of Sciences, U. P. Rajarshi Tandon Open University, Prayagraj, India

Zakir Ullah Department of Plant Sciences, Faculty of Biological Sciences, Quaid-i-Azam University Islamabad, Islamabad, Pakistan

Rooma Waqar Department of Botany, Faculty of Sciences, Mohi-ud-Din Islamic University, Nerian Sharif, Azad Kashmir, Pakistan

Muhammad Waseem School of Breeding and Multiplication (Sanya Institute of Breeding and Multiplication), College of Tropical Agriculture and Forestry, Hainan University, Sanya, China

Fang Zhiyuan Academician Team Innovation Center, Haikou, Hainan, China

Mohammad Abubakar Yasin Agricultural Biotechnology Division, National Institute for Biotechnology and Genetic Engineering (NIBGE), Constituent College of Pakistan Institute of Engineering and Applied Sciences (PIEAS), Faisalabad, Pakistan

Omar Zayed Department of Botany and Plant Sciences, University of California Riverside, Riverside, CA, USA

Nanoparticles for Advancing Plant Gene Transfer: Opportunities and Challenges

Muhammad Ikram, Azza H. Mohamed, Omar Zayed, Shazina Jabeen, Marcello Iriti, Jude W. Grosser, and Ahmad A. Omar

Abstract To ensure food security, developing innovative techniques are crucial for crop yield, productivity, and improving resilience in the changing climate conditions and increasing food demand. Conventional plant breeding methods are laborious, less effective, and limited in scope. The vector-based genetic transformation for gene delivery and CRISPR/Cas9 provides avenues to improve stress resistance, plant yield, and food quality. However, the efficacy of genetic transformation is limited in many commercially significant crops because of plant cell wall hindrance. Nanotechnology, especially green nanoparticles (NPs) facilitates efficient gene transfer due to their ability to cross the barrier of plant cell walls. NPs have been widely used for gene transfer, molecular breeding, and promoting plants' resilience to environmental stresses due to their ease of penetration, and various distinct physi-

M. Ikram · J. W. Grosser
Citrus Research and Education Center, University of Florida/IFAS, Lake Alfred, FL, USA

A. H. Mohamed
Department of Agricultural Chemistry, College of Agriculture, Mansoura University, Mansoura, Egypt

Citrus Research and Education Center, University of Florida/IFAS, Lake Alfred, FL, USA

O. Zayed
Department of Botany and Plant Sciences, University of California Riverside, Riverside, CA, USA

S. Jabeen
Department of Agricultural, Forestry and Food Science-DISAFA, University of Turin, Turin, Italy

M. Iriti
Department of Biomedical, Surgical and Dental Sciences, Milan State University, Milan, Italy

A. A. Omar (✉)
Citrus Research and Education Center, University of Florida/IFAS, Lake Alfred, FL, USA

3 Citrus Research and Education Center, University of Florida/IFAS, Lake Alfred, FL, USA
e-mail: omar71@ufl.edu

J.-T. Chen (ed.), *Nano-Delivering for Plant Genetic Engineering*, Smart Nanomaterials Technology, https://doi.org/10.1007/978-981-95-4165-2_1

cochemical properties. This chapter focuses on the potential of NPs, their application in gene delivery, and their efficient role in plant biotechnology. Despite their promising success in plant transformation, there is still a long way to go due to several challenges, including NPs optimization for gene transfer, off-target accumulation, nanotoxicity, and understanding of their impact on plant physiology. Future studies aimed at synthesizing biodegradable, biocompatible, and efficient NPs will be key in advancing plant genetic transformation and enhancing crop resilience. The integrated approach involving nanobiotechnology and traditional genetic engineering possesses a significant potential to transform crop improvement and sustainable agriculture production.

Keywords CRISPR/Cas9 · Crop improvement · Crop resilience · Genetic transformation · Green nanoparticles

1 Introduction

Changing climatic conditions, such as increased heat stress, extreme weather, soil salinity, drought stress, heavy metal toxicity, and high food demand due to overpopulation, threaten food security and agricultural productivity [1]. The mutation breeding techniques and various conventional methods have been deployed in the recent past successfully to acquire desired plant traits, crop productivity, and agricultural quality [2]. However, these breeding techniques have several limitations i.e. time-consuming, and appearance undesirable traits. In plant genetic transformation, there are various methods aimed at either silencing the expression or modifying the functions of one or more endogenous genes and the expression of one or more introduced genes [3]. Genetic transformation is considered an effective approach to enhance crop quality, yield, and resistance in response to biotic and abiotic stresses by introducing desirable characteristics. In addition, it is used to improve crop genetic and functional genomics, particularly for introducing new features and modifying existing traits [4]. Genetic transformation helps transfer transgene (gene of interest), encodes desirable traits into the plant cell, and produces plants with desired traits and characteristics [5]. Plant genetic modification techniques can be used to produce plants with high yields that are more resistant to pathogens and insects, highly tolerant to herbicides, and with improved nutritional profiling [6]. For example, rice crops have been genetically modified for vitamin A biofortification that can sustainably minimize malnutrition cost-effectively [7]. Despite various technological innovations or improvements in plant engineering, genetic transformation efficacy is extremely low in many plants or crop species of commercial interest. This is because plant cells have rigid and multilayer cell walls made up of cellulose. Therefore, protoplasts have been exploited for gene transformation due to easier DNA delivery, but the techniques or protocols for effectively maintaining and regenerating plantlets from protoplast culture have not yet been fully developed or optimized in many plant species [8, 9]. The development of new methods that facilitate the passive movement of various biomolecules across a broader spectrum of plant species is crucial to make plant genetic transformation more successful. Gene vectors play a pivotal role in transferring desired genes into a host plant cell and are crucial in genetic transformation [10]. In the last decade, the clustered regularly interspaced short palindromic repeats (CRISPR) /Crispr-associated protein (Cas) genome editing system has emerged as the most prominent and significant genetic engineering method. In plants, CRISPR-Cas gene editing showed remarkable promise for ensuring food security, and tackling abiotic and biotic stresses under changing climatic conditions; thereby contributing to food sustainability and environmental conservation [11]. Its high stability permits the efficient alteration of target DNA with convenience and high efficiency. The RNA-guided endonuclease system incorporates cells through different forms, such as RNA, ribonucleoproteins, DNA fragments, and plasmid DNA. In addition to other components, single-stranded and double-stranded oligo donors (ssODN and dsODN) are also introduced to achieve target gene replacement and insertions [12, 13]. Various applications of this technology have been shown since its discovery, including the editing of plant genomics. Despite the significant advancements and groundbreaking progress of genome editing techniques in plant genetic engineering, CRISPR applications in plant biotechnology still face many obstacles [14]. For example, the cell wall presence and need for the cell tissue culture process make it difficult to deliver these materials into plant cells [15], in contrast to other studies conducted in mammalian and bacterial cells, where these materials could be directly introduced into single cells using lipofectamine and polyethyleneimine (PEI) transformation assays [16]. Nanomaterials can potentially address key issues in CRISPR genome editing in plants by improving delivery mechanisms, overcoming species barriers, enabling germline cell transformation, and boosting gene editing effectiveness [8, 17–19]. Nanotechnology-mediated bio-transformation is quite effective; however, this technique along with commonly used methods in the development of transgenic plants could cause improvement in transformation efficacy and productivity. In recent decades, nanomaterial-mediated DNA transfer in plant cells has opened up significant possibilities for biomolecule delivery [20]. Nanomaterials are getting the attention of researchers because of their distinct size, and characteristics relative to larger and bulkier forms. Further, NPs may be synthesized using diverse materials including carbon, polymers, metals, ceramics, etc., and usually have dimensions ranging from 1–100 nm [17, 21]. Various forms of magnetic NPs have been utilized to improve and advance plant genetic transformation, especially in molecular plant breeding programs [22]. Furthermore, NPs have been explored to improve agronomic traits. In addition, NPs have also been used as sensing material, herbicides, pesticides, nano-fertilizers, agrochemical-controlled release, and nutrients for crop bio-fortification [23]. Moreover, NPs also contribute to plant adaptation to climate change by enhancing their tolerance to both biotic and abiotic stress [24]. NPs also play a prominent role in tissue culture, i.e. somatic embryogenesis, callus formation, organ development, and primary and secondary metabolite production [25]. However, the exact mechanism is still unclear, but these investigations support the idea that plant cells can absorb NPs across their cell wall. Plant genetic transformation and genome modification in plant cells are anticipated to greatly benefit from the developing field of NPs based cargo delivery in plant cells [26, 27]. Gene transfer carried out by NPs relies on their size, shape, surface modifications, ability to protect nucleic acid, their bio-compatibility with biological systems. Various forms of magnetic and carbon NPs have been deployed to develop sustainable plant genetic transformation in a sustainable manner [20]. Furthermore, in the future, NP-based gene delivery tools may also assist to carry and transform bio-macromolecules with high weight, e.g., proteins. In this current chapter, we focused the different types of NPs that are reported for genes delivery into plant cell and the challenges that occurred during their applications. Moreover, we additionally describe the progress of nanotechnology needed to accelerate and facilitate plant genome editing.

2 Conventional Methods of Plant Genetic Transformation and their Limitations

Genetic transformation is a significant approach for developing sustainable crops and advancing breeding techniques; thereby anticipated to be a significant tool for future breeding initiatives. Direct and indirect genetic transformation are two main types of plant genome transformation techniques [28]. Direct genetic transformation employs various methods including particle bombardment, microinjection, electroporation, liposomes, silicon carbide, etc., to introduce genes into plant cells [29, 30]. In contrast, indirect genetic transformation involves using organisms as vectors, such as Agrobacterium, to transfer genes into target cells to achieve plants with desired traits [31, 32]. Particle bombardment, referred to as a gene gun, is used to directly deliver foreign genes of interest into the plant genome [33, 34]. In this method, the targeted gene is coated on gold or tungsten particle surface to develop DNA-coated microcarriers which are then pushed into plant cells using high-pressure pulses through a gas acceleration tube either by a pressurized helium gas stream or electric discharge [35]. These particles with targeted gene coating gain enough momentum to penetrate inside the recipient cells rapidly and finally merge with the plant chromosomes to produce plants with desired characteristics [36]. This method has many advantages, including its wide range of targets and easy to use; however, in comparison with *Agrobacterium*-based transformation, it has a few drawbacks, including higher costs, unprotected foreign DNA, and lower transformation rate [37]. Furthermore, this method is limited to the transfer of DNA fragments under 10 kb, since larger fragments can be easily broken into pieces during bombardment or may have less affinity to adhere with metal particles, eventually leading to unpredictable DNA integration events [38]. Furthermore, the microinjection-assisted plant genetic transformation technique involves the injection of DNA into the nucleus of a plant cell using a fine microcapillary needle. In addition, the recipient cell is immobilized under a microscope in this procedure.

This process involves the utilization of two micromanipulators, a DNA solution containing a microcapillary tube used to pierce the nuclear membrane, and a holding pipette deployed to stabilize the cell. The genetic material (DNA) is injected into the plant cell cytoplasm or nucleus. After genetic transformation, the transformed cells are grown and developed into transgenic plants [39, 40]. In addition, it efficiently delivers targeted DNA to specific cells or organelle sites. Microinjection is a less popular technique as compared to other methods for plant genetic transformation. Despite having some advantages, this method can be inefficient because advanced technical skills are required to manipulate less number of cells. Moreover, the plant cell wall obstructs the microinjection method and making it challenging for microinjection pipettes to penetrate the cell [41]. Electroporation is another technique for gene transfer in which cells are placed in a solution containing the desired gene, and utilization of electric pulses to create minute pores in the cell membrane of desired cells, enabling the genetic material to pass through [42, 43]. Compared to other direct and indirect plant genetic transformation methods, electroporation is the most efficient method due to its low cost, fast application, and highly efficient and stable transformation rate [44]. The major limitation of the electroporation method is the transformation of plant cells with thick and multilayered cell walls [43]. In addition, a series of electrical pulses could destroy naked genes, eventually resulting in errors in the translation process of the subsequent product. Moreover, it could also cause cellular damage, especially to protoplast [45]. Liposome-assisted transformation involves the delivery of desired DNA into the protoplast by endocytosis across the plasma membrane [46]. Liposomes are composed of phospholipid bilayer membranes, comprising cholesterol and phospholipids [47]. The desired gene is transferred into the protoplast by the process of liposome-mediated protoplast fusion or endocytosis, followed by the combination of liposomes with DNA and incubated to produce a DNA–liquid complex, which is then combined with protoplast solution (PEG supplemented) [46]. Positively charged liposomes attract DNA and cell membranes and get attached to the surface of the protoplast. Upon fusion of liposomes and protoplast, the plasmid is then released inside the target cells [48]. Another method of genetic transformation is silicon carbide whiskers, which are needle-like micro-whiskers that range in length from 10 to 80 μm and have a diameter of roughly 0.5 μm. The whiskers create sharp edges that penetrate the cell wall and, eventually reach the cell nucleus because of their toughness and ease of cleavage [49]. In addition, this transformation is accomplished by subjecting DNA and embryonic callus to centrifugation with silicon carbide, which acts as an inert compound to DNA [50]. Silicon carbide whiskers incorporate desired DNA into target cells by creating pores on the cell membrane. [51–53]. The transformation mediated by silicon carbide whistler is easy to achieve without the involvement of any complex equipment. However, the procedure needs to be carried out very carefully because silicon carbide fibers might cause cancer. This operation could lead to cell damage and lower their regeneration ability, resulting in poor conversion efficiency [43, 54, 55]. *Agrobacterium*-based transformation, also known as the indirect genetic transformation method, is a promising approach for gene delivery and to production of genetically modified plants [56, 57]. This technique primarily uses two species of *Agrobacterium*, viz. *Agrobacterium tumefaciens* and *Agrobacterium rhizogenes* [31, 58]. These species contain plasmids that cause the induction of tumors (Ti) or hairy roots (Ri). *Agrobacterium* tumor-inducing (Ti) plasmid, a binary vector releases the T-DNA into plant cell and it becomes the part of plant genome [57]. It is easy to induce *Agrobacterium* mediated infection in dicots compared to monocots since dicots are considered as primary hosts of *Agrobacterium* [45, 59].

Various factors influence the efficacy of *Agrobacterium*-mediated transformation. However, genotype dependence is the most well-known barrier between different plant varieties, even within the species. In addition, the responses to *Agrobacterium* infection vary depending on genetic background, physical environment, and metabolic state. Furthermore, *Agrobacterium* as a vector becomes inappropriate for transferring the desired DNA due to the physical and metabolic traits of some plant species [60–62]. For instance, some particular metabolites including ethylene, γ-aminobutyric acid, and salicylic acid inhibit *Agrobacterium's* growth and gene-transformation capacity in many plant varieties. The antagonism between *Agrobacterium* and host plants could lead to severe plant disorders [63–66]. Moreover, various tissue culture conditions must be tested for each variety, such as pre-culture course, the time required for co-inoculation, the density of *Agrobacterium*, and hormone types and concentrations. In addition, Agrobacterium cannot transfer certain materials, including oligonucleotides, RNA, RNPs, etc. Furthermore, these materials are crucial for crop genome editing to produce precise, transgene-free genome-edited cultivars [67, 68]. Therefore, although *Agrobacterium*-based transformation is simple and practical in terms of application, its effectiveness needs to be increased because it is not appropriate for all plant species or applications. Figure 1 illustrates the different stages of genetic transformation and genome editing in plants.

3 Overview of CRISPR Genome Editing in Plants and Current Challenges

CRISPR/Cas system has been extensively explored in plants as a genome editing strategy. In this method, plasmids containing CRISPR/Cas reagents have been incorporated into plant cells by using different techniques to achieve either stable or transient transformation [69, 70]. Transient CRISPR/Cas expression approaches are better because they use plant machinery to temporarily express CRISPR/Cas reagents rather than chromosomal integration [71, 72]. Further, transient expression of the CRISPR/Cas complex has reportedly decreased the regulatory load by decreasing unintended mutations in bread wheat [73, 74]. However, these plasmid delivery-based expression approaches have not been extensively employed so far since these delivery techniques are not effective in the case of certain plant species. Furthermore, another alternative genome-editing method is DNA-free editing,

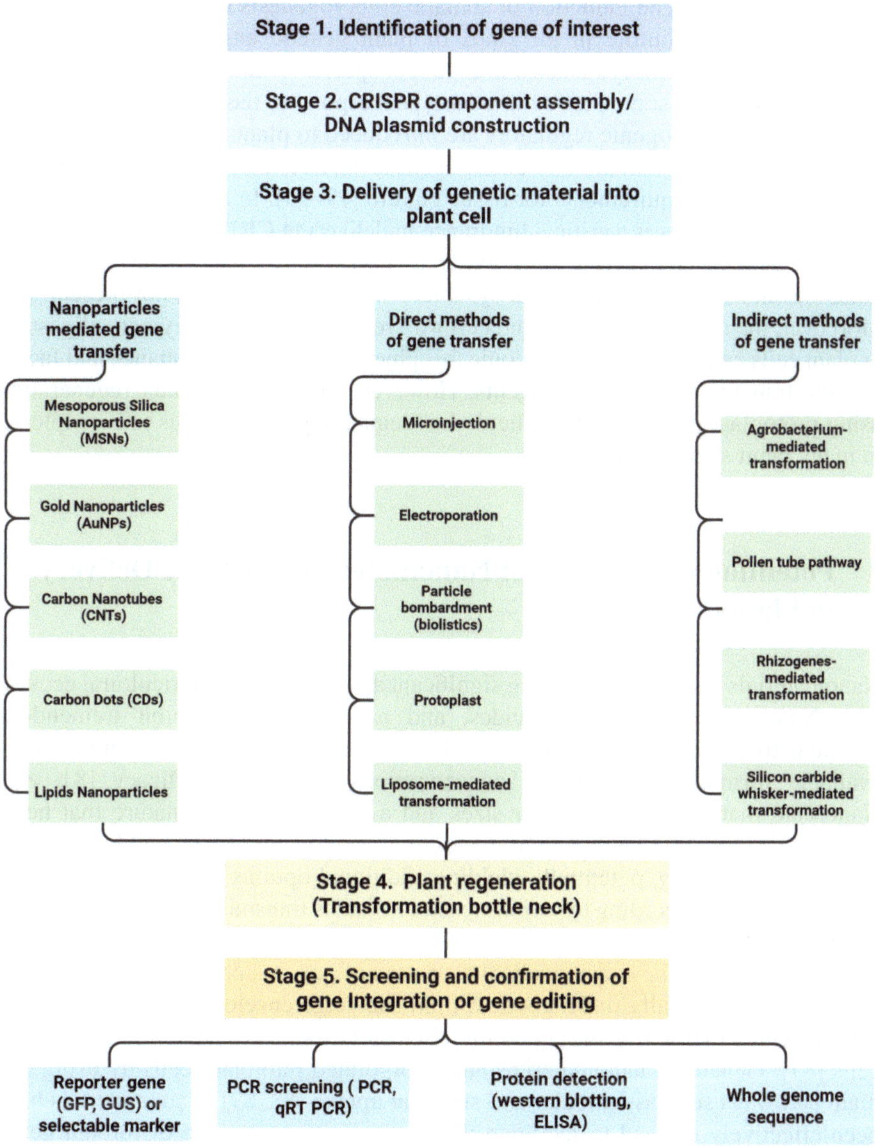

Fig. 1 Flowchart of the genetic transformation and genome editing stages in plants

which involves directly incorporating the CRISPR/Cas complex into the plant cells. Pre-assembled ribonucleoproteins are commonly incorporated through protoplast transformation successfully, but their editing efficiency remains low [75]. Moreover, the transfer of CRISPR/Cas reagents by an RNA virus was carried out which resulted in heritable mutations of numerous genes at once with 90–100% editing efficiency in infected somatic tissue [76]. However, any virus-based tools have a limited host range. The induction of somatic embryogenesis in some plants is considered a primary hurdle in all kinds of plant genetic engineering techniques. Generally, somatic embryogenesis in monocot species is laborious and technically demanding because some particular genotypes respond to tissue culture in a different way. The morphogenic regulators are introduced to plants to speed up transformation by inducing somatic cell embryogenesis and de novo meristem formation, which minimizes requirements for tissue culture [77–79]. In addition, the presence of cell walls in plants is a critical hindrance in delivering CRISPR reagents through CRISPR genetic engineering [80]. Whereas CRISPR has had numerous positive outcomes in many other biological systems where the cell wall is not present. [8]. Moreover, the use of protoplasts and cell wall removal through enzymatic digestion in plant cells can be reduced, overcome this physical barrier, and enhance and facilitate the transfer of CRISPR reagents. However, the complete plant regeneration using protoplast is technically difficult and demanding and remains to be explored in many plant species [9].

4 Potential Applications of Nanomaterials for Gene Delivery in Plants

Nanomaterials showed a promising significant advantage in the agricultural ecosystem. Nano-nutrients, nano-pesticides, and nano-fertilizers exhibit tremendous potential to enhance plant health, yield, and quality compared to commercially available formulations at lower concentrations with equal efficacy [81, 82]. Nanoscale materials have smaller sizes and a unique chemical nature that helps plant cells to easily assemble them. In addition, nanomaterials showed less toxicity and target specificity, potentially yielding additional options for future smart agriculture [82–84]. According to various research studies transmission electron microscopy confirms the delivery of nano fertilizers and nano pesticides into plant cells [85, 86]. However, the exact mechanistic action of nanomaterial delivery into plant cells has not been fully understood. A lipid exchange envelope penetration model was presented to describe the penetration of nanotubes into the lipid bilayer of plant cells [87]. Generally, nanomaterials are incorporated into plant cells by using two main pathways such as cuticular and stomatal uptake [88, 89]. Nanomaterials have been effectively utilized to facilitate the transfer and expression of foreign genes and the delivery of CRISPR/Cas reagents into plant cells [90, 91]. As mentioned above, the cell wall being a physical barrier avoids the entrance of external material into the organelles and cytoplasm of plant cells. Recently, various researchers have investigated nanomaterials as vectors to transfer and incorporate genes of interest inside the genome of plant cells [8]. Delivery of desired genes in plant cells is considered a critical challenge. Nanotechnological formulations provide valuable contributions to plant genetic engineering, enabling successful and target-specific delivery of desired material without off-site target mutations [10, 18, 45, 92].

Additionally, nanomaterials also provide controlled cargo release, targeted gene delivery, and protection of cargo from disintegration [93]. The nanomaterial-mediated DNA and protein delivery into various plant cells was carried out successfully. However, nanoparticle-assisted CRISPR/Cas genome editing remains to be investigated in plants due to various distinct physicochemical features of NPs and CRISPR reagents. Furthermore, Cas9 protein and RNP delivery during nanomaterial-assisted CRISPR genome editing in plants presents a significant difficulty due to their greater size, increased density, and little stability in ionic solutions.

However, a potential approach is to covalently link Cas9 RNPs to the nanomaterial's surface through conjugation, enabling cleavage close to the target site. Other potential methods include the use of enzymatically cleavable linkers and utilizing tissue penetrating near-infrared light to stimulate the RNPs release from nanoparticle surfaces [8]. Moreover, the delivery of nanoparticles is still not fully effective compared to the biotic delivery methods. Therefore, there is a need to enhance the effectiveness of CRISPR reagent delivery to facilitate CRISPR genome editing for real-world plant genetic applications. A range of nanomaterials including mesoporous silica NPs, carbon nanotubes, metal NPs, magnetic NPs, etc. have been explored for plant genetic engineering applications. The nanomaterial-based cargo delivery in plant cells is a rapidly evolving field, and plant genetic engineering and genome editing are anticipated to benefit from it. We reviewed the various kinds of nanomaterials currently being investigated for their potential to deliver genes to plant cells efficiently.

5 Mesoporous Silica Nanoparticles for Gene Transfers in Plant

Mesoporous silica nanoparticles (MSNs) have attracted more attention from researchers for plant growth, gene delivery, and stress alleviation. The MSNs exhibit a distinct solid framework, high thermal stability, porous structure, and the potential to link several functional groups to their surface [94]. In addition, silica particles showed high biodegradability, biocompatibility, and low toxic effects on plant and mammalian cells [95]. The silica nanoparticles contain silanol groups, providing them with high biocompatibility and stability for various agricultural and therapeutic applications [96]. The use of MSNs for drug delivery, specific site targeting, and intracellular delivery of nucleic acid on animal cells or tissues is well known but the MSN's potential for delivering biomolecules to plants is less explored. This is because larger DNA fragments are difficult for MSNs to integrate due to their smaller pores size; hence polymer coating is commonly deployed to provide a positive charge on the MSN surface. The most commonly utilized polymers for this purpose are poly (allylamine hydrochloride) (PAH), polyamidoamine dendrimers and polyethyleneimine (PEI) [20]. In tobacco plant cells (suspended), reference [97] used PEI-coated MSNs (100 ± 8.7 nm) to introduce GUS-encoding plasmid via ultrasound treatment. This study revealed the efficient transfer of GUS-encoding plasmid DNA into tobacco cells. It has been demonstrated that excessive sonoporation can negatively impact cell viability. Concludingly, these results showed the promising potential of ultrasonic treatment which provides a straightforward and economical pathway for gene delivery into tobacco-suspended cells without using complicated devices and serious concerns about safety issues. Another study [98] also demonstrated the potential role of synthesized MSNs as DNA nanocarriers for novel transient gene expression approaches in tomato plants to develop resistance against *Tuta absoluta*. Various techniques were deployed to analyze and confirm MSNs, including FE-SEM, TEM, and Zeta sizer. Using spray and injection procedures, MSNs carrying (pDNA-MSNs) were injected into the abaxial surface of the leaves and shoots of tomato plants. Various expression analysis confirms the entrance and moving of pDNA-MSNs between the cells through plasmodesmata. Moreover, they reported that the MSN-based gene transformation approach is efficient due to the reduced number of stages, saving energy, biocompatibility, and biodegradability. After 2 h of plasmid bindings to the NPs, the target plants are ready for the transformation using an MSN-based gene transfer approach but plasmid extraction is necessary in such kind of transformation. In addition, the MSNs-assisted transient gene transformation approach showed significant safety compared to *Agrobacterium*-mediated plant transformation because of their high biocompatibility and the lack of use of *Agrobacterium*. Furthermore, in addition to transferring biomolecules, the MSN-based techniques benefit developmental growth and improve plant resilience to heavy metals stress, drought, and various diseases. However, further research work is required to optimize this technique for stable and effective gene transformation. Another study [99] reported the use of optimized Au-MSNs for the delivery of both plasmid DNA and proteins to plant cells using the biolistic method. The characterization techniques revealed that Au-MSNs (10 nm) were used to deliver and release DNA plasmid and proteins simultaneously to the same cells across the cell wall upon bombardment. In addition, the authors also reported the successful release of fluorescent protein (eGFP) to plant cells, indicating the delivery of active, non-denatured proteins. A research study [100] also prepared MSNs for the delivery of DNA to examine the transient expression of candidate genes without the involvement of outside mechanical forces. In this study, roots of *Arabidopsis* seedlings were exposed to MSNs for 24 hours, and transgene expression was detected in the cortical and epidermal cells of *Arabidopsis* roots. The authors also reported that MSNs can easily move through the pores of the cell wall without any external aid due to the loosening of the microfibril network. Furthermore, another study by [101] used MSNs to designe a plant gene transformation system reulated by maize proteinase inhibitor (MPI) as an inducible promoter. The functionalized MSNs were attached to a recombinant construct (pDNA) having *cryIAb* gene under the control of (MPI), and its transformation was carried in tomato plants via injection. The real-time -PCR was deployed to confirm the integration and expression of transgene. The results of this study confirmed the successful integration of transgene in the tomato plant genome through homologous recombination.

Cai et al. [102] reported the MSN-mediated RNA interference (RNAi) delivery in plants. RNA interference is considered an efficient tool for signaling pathway modifications in plants which may enhance the new plant characteristics with high productivity. The study findings demonstrated that MSNs effectively carry siRNA into the *Nicotiana benthamiana* leaves, achieving transitory gene knockdown with molecular-level silencing efficiency of up to 98%. Furthermore, this technique doesn't harm plant leaves, and siRNA can be repeatedly delivered for long-term silencing. This study also revealed that the application of MSN-siRNA complex that targets the genes for magnesium chelatase and phytoene desaturase, white patches, and yellowing phenotypes is seen in plants. Additionally, the study also demonstrated that MSNs could mute many genes simultaneously. Based on the results, MSN-based siRNA delivery is a robust method for long-term multi-gene silencing, and it has a great deal of potential for application in plant functional genomic studies and crop improvements.

Sangwan et al. [103] reported that different-sized amine-functionalized MSNs (amine-MSNs) mediate the dsRNA delivery derived from the *AC2* (dsAC2) gene of tomato leaf curl New Delhi virus (ToLCNDV). It was demonstrated that the application of these dsRNA-loaded NPs resulted in reduced viral load. Furthermore, results also showed that amine-MSNs protect the dsRNA from nuclease, and the roots and leaves of *Nicotiana benthamiana* showed efficient uptake. In addition, the application of NPs of sizes 10 nm, 32 nm, and 66 nm triggered gene expression and demonstrated that treated plants exhibited five-, eleven- and three-fold ToLCNDV reduction in *N. benthamiana*, respectively, relative to the plants treated with naked dsRNA. The internalization of amino-MSNPs and their relative effectiveness in delivering dsRNA into plants are both size-dependent and evident in this scientific investigation. These findings would be helpful in effective topical applications to shield plant species from viruses and other infections.

According to Torney et al. [104], the MSNs can carry chemicals and DNA across the leaves and plant cells. The MSNs containing the gene and its inductor in 3 nm pores coated with Au-NPs were used in the biolistic approach to produce transgenic tobacco plants. The release of β-estradiol enabled tobacco plants to express fluorescent protein. Au-NPs were used in genetic weapon systems to cover the pores of the MSNs after they carried the gene and its inducer. When the pores were opened, the chemicals were released, which in turn caused the gene to express itself in plants. The MSNs enable the delivery of large-size biocompounds and molecules along with DNA that is not permeable to the plasma membrane of the targeted cells.

6 Metal Nanoparticles Mediated Gene Transfer in Plants

Metal nanoparticles including silver, gold, palladium, platinum, etc. showed distinct scattering and optical absorption properties, which make them the best candidate for biomedical or agricultural applications [20, 105, 106]. Silver and gold NPs are getting the attention of researchers owing to their smaller size, shape, ability to prepare, bio-conjugation, flexible surface modification, and adjustable optical properties [107]. Moreover, metal-based nanomaterials have been extensively used for various purposes such as intracellular transport of biomolecules i.e. DNA, RNA, drugs, and peptides in plant and animal cells [108, 109]. Substantial advancement has been achieved in the applications of various metal NPs in agriculture as nano-fertilizers to improve crop quality and yield [110, 111]. However, the application of NPs as a vector to deliver biomolecules in plant cells has not been explored well. Some studies reported using metal-based NPs as a capping agent and other nanomaterials as a cargo release of biomolecules. Wang et al. [112] investigated the genetic transformation potential of magnetic NPs by using GUS plasmids, red fluorescent protein, and eGFP in maize plants. This investigation revealed that 22% of transfected pollen of maize plants showed red fluorescence, and 55% of embryos displayed GUS staining during pollination, indicating that transfection was successful in the treated pollen. Vijayakumar et al. [113] used carbon bullet-functionalized gold nanoparticles (Au-NPs) as potent gene delivery vehicles through the gene gun method. In this study, Au-NPs were synthesized using fungal biomass to reduce gold ions in Au-NPs. Afterward, Au-NPs were embedded into a carbon matrix with a carrying potential of 2400 ng plasmid DNA. Interestingly, adding Au-NPs to composite could lead to enhanced surface area availability for cargo loading, efficient packaging density, and decreased nuclease degradation. The expression of GUS reporter enzymes in tobacco and rice cells indicated the successful transformation of DNA plasmid using gold nanocomposite. Additionally, this research suggested that carbon bullets loaded with nanogold potentially decrease cell damage and efficiently transform DNA plasmid in the mentioned plant cells.

Lei et al. [114] showed that AuNPs-siRNANPR1 can infiltrate leaf cells and partially silence the NPR1 gene in Arabidopsis thaliana. Compared with traditional gene-silencing approaches, the AuNP-based approach is faster, less toxic, and requires a lower dosage. This simple and rapid method offers a new tool for studying plant gene function. In another study, gold nanocrystals (Au NCs) were successfully used as a vector to deliver siRNA in plants. In this study, authors demonstrated that 1.5 μg of siRNA could be delivered into *N. benthamiana* plants through 1 μg of Au NCs, successfully silencing the GFP and ROQ1 genes [115]. These results indicated positive genetic transformation in *N. benthamiana* by using Au-NCs. In a later work, [116] studied the application of Au-NPs coated with polyethyleneimine (PEI) in *Catharanthus roseus* leaves by infiltration. In this research study, Candidatus *Liberibacter asiaticus* (CLas) a pathogen that causes huanglongbing (HLB) or citrus greening disease in citrus plants was used to infect *Catharanthus roseus* leaves. The delivery of the NPR1-GFP gene using PEI-Au-NPs resulted in the pDNA expression. This boosted the expression of pathogenesis-related genes, thereby lowering class titers.

Another study, [117] examined the efficacy of silver nanoparticles (AgNPs) as a gene carrier was investigated to substitute the gold nanomaterials for stable genetic transformation in *Nicotiana tabacum*. This comparative study to examine the efficiency of bio-transformation of AgNPs was carried out under various helium pressures (450 psi, 650 psi, 900 psi, and 1100 psi) and target distance (6 cm and 9 cm).

AgNPs at the dose of 10 mg·L^{-1} resulted in more transient GUS expressions. AgNPs at a helium pressure of 900 psi and target distance of 6 cm resulted in maximum GUS expression, while gold nanocarrier showed the highest GUS expression at a helium pressure of 650 psi and target distance of 9 cm. These results suggested that higher transformation efficiency was achieved with AgNPs than gold nanocarriers. Furthermore, the regeneration efficiency of tissues bombarded with AgNPs was reported at 70.83%, while for gold nanocarrier, it was found at 62.5%.

7 Carbon-Based Nanomaterials for Plant Gene Transfer

Carbon nanotubes (CNTs), composed of rolled-up graphene sheets, are long, thin, and cylindrical molecules [118]. CNTs have emerged as practical nano-vector-delivering tools for proteins and DNA due to their potential to enter plant cell walls with minimum injury [119, 120]. CNTs can be classified as either single-walled CNTs (SWCNTs) or multi-walled CNTs (MWCNTs) [121, 122]. According to a study by Liu et al. [123], SWCNTs can readily penetrate intact plant cell membranes as well as cell walls. They employed oxidized SWCNTs labeled with fluorescein isothiocyanate (FITC), which could penetrate intact tobacco BY2 cells. Temperature and incubation period are the factors affecting the uptake of SWCNTs by tobacco cells. Tobacco cells that were pretreated with endocytosis inhibitors demonstrated significantly reduced fluorescence intensity, indicating that endocytosis may be a primary mechanism behind the uptake of SWCNTs by tobacco plant cells.

The biological application of CNTs faces the challenge of their inability to disperse in aqueous medium. CNTs are functionalized with molecules or functional groups to increase their dispersibility through covalent and non-covalent modifications. Golestanipour et al. [124] used arginine-coated SWCNTs and cell-penetrating peptides (CPPs) engineered with particular hydrophobic and cationic domains for the delivery of DNA plasmid containing GFP gene into tobacco (*Nicotiana tabacum* var. Virginia) root cells. The solubility of SWCNTs was enhanced by functionalization with arginine and facilitated stable loading of plasmid DNA. Western blotting and fluorescence microscopy confirmed that arginine SWCNTs successfully transfer the GFP gene. This successful transformation was attributed to the SWCNT's small diameter under 3 nm smaller than the cell's pore diameter. They also tried to transport the plasmid through the peptide-plasmid complex, but this attempt failed as large-size peptides were unable to cross the cell wall pores. CNTs have an advantage over the peptide carrier as they have the ability to penetrate the cell through a needle-like mechanism that does not require endosome formation. This process enhances the likelihood of internalization of SWCNTs, which is beneficial for plant gene transfer.

Schwartz et al. [125] investigated carbon dots for siRNA delivery into the cells of *Nicotiana benthamiana* and *Solanum lycopersicum*. Low-pressure spray application of carbon dots successfully delivered the siRNA, which silenced the GFP gene in both pants' leaves. Additionally, the efficiency of carbon dots as siRNA nanocarrier was also tested by focusing on endogenic genes that encode for magnesium subunits chelates viz.; CHLH and CHLI which translate chlorophyll subunits H and I, respectively. This gene silencing could lead to distinguished phenotypes in the aforementioned plants. The authors also claimed that this is the first research study of chlorophyll-encoding genes silencing through nanomaterial-assisted siRNA delivery. A research study [126] reported that SWCNTs assisted DNA delivery into plant cells by exploiting a diffusion-based delivery method. Analysis of reporter gene expression in the leaf cells of wheat, cotton, tobacco, and arugula plants, which encodes GFP, confirmed the DNA delivery. In addition, CNTs with a size of 20 nm can easily penetrate the cell wall barrier of plants, which makes them suitable candidates for plant genetic transformation.

8 Polymeric Nanomaterials as a Gene Transfer

Polymer-mediated NPs can also be used as nano-delivery agents that can deliver genetic material such as RNA, DNA, and ribonucleoproteins to the cells. Recently, starch and cellulose have been used to produce natural polymer-based NPs with less toxicity, high biocompatibility, and biodegradability to plant cells [127]. Due to their higher stability, chemical versatility, and tunability, starch nanoparticles are abundantly employed to transfer DNA into plant cells. In another study, poly-L-lysine, a fluorescent dye, was used to modify the starch NPs surface and followed by conjugation with the pEGAD plasmid DNA. Later, through the ultrasonication method, the DNA-loaded NPs were incorporated into the cells of the *Dioscorea zingiberensis* plant. The findings of the study showed the efficient delivery of DNA inside the plant cells without degradation by starch NPs [128].

Chitosan is a natural, stable, cationic, and alkaline polysaccharide made up of β-linked d-glucosamine and N-acetyl-d-glucosamine. Chitosan binds to negatively charged DNA through its positively charged amino groups and easily degrades in plants [129, 130]. According to a study chitosan NPs loaded with plasmid DNA were efficiently delivered into the cells of *Paulownia tomentosa* and *Solanum tuberosum*. The expression of the GUS reporter gene confirmed the successful transition of plasmid DNA into the aforementioned plants [131]. Another study revealed the successful application of chitosan-based NPs for delivering DNA into *Jatropha curcas* plant cell suspension [132]. Amani et al. [133] explored the application of cationic polyamidoamine NPs by PUC18 plasmid having GUS A gene for genetic modification in *Medicago sativa* via ultrasound method. Kim et al. [134] suggested the application of DNA/RNA nanostructures for genetic transformation in plants as these delivery systems are easy, versatile, and eco-friendly and provide biological stability, as well as immune compatibility over other nanomaterials.

Additionally, a number of nucleotides can be incorporated along nucleotide NPs which may reach the nucleus by crossing the nuclear membrane. Zhang et al. [135] reported the successful delivery of siRNA using DNA NPs into *N. officinale*,

N. tabacum, and *Eruca sativa* leaf cells. In addition, Zhang et al. [136], highlighted the immense potential of DNA nanomaterials in tobacco plant cells for siRNA delivery with higher internalization. Furthermore, dendrimers composed of polyamidoamine (PAMAM) are also cationic nanostructures with low polydispersity making them viable options for effective gene transport in plants [137]. PAMAM dendrimers were also involved in the successful transfer of GFP-encoding genes into turfgrass [138].

9 Future Perspectives and Challenges

NP-mediated gene transfer methods have promising benefits and a higher potential for plant genetic transformation than traditional strategies. Interestingly, NPs are effective for gene transfer and act as a carrier for various bioactive molecules. The use of NPs for gene transformation in plant systems is still limited due to the cell wall hindrance, making NPs internalization difficult [20]. In the recent past, tremendous efforts were exercised to optimize the NPs, which provides them not only easy access to plant cell wall barriers but also to a specific organellar membrane. The advancement of plant nanobiotechnology requires precise and appropriate design analytics for more effective nanoparticle delivery vehicles. In addition, further advancement of NPs for plant gene transfer depends on the properties of NPs and various factors affecting their synthesis, such as temperature, time, pH, etc., which play pivotal roles in regulating and maintaining the nanomaterials shape and size [20]. Further studies on surface properties, which are required for target-specific transport of genetic material to the meristematic tissues, callus, and cellular organelles such as the nucleus and chloroplast, will open new avenues in genetic engineering field [91].

Most importantly, attention must be paid to various factors i.e. stability, size, shape and compactness while designing gene-delivering NPs that are responsible for the free internalization of NPs into plant cells. Along with gene delivery, it is very crucial to investigate the delivery capability of mRNA, sgRNA, and ribonucleoproteins. Recently, NPs shown promising potential in the CRISPR/Cas 9 system. However, there are still some associated challenges that exist, such as avoiding nucleic acid degradation, scaling up production, improving encapsulation methods and low transformation rate, etc. [91]. To cope with these challenges, researchers should emphasize developing innovative techniques for NPs synthesis, such as polymers, silica, and lipid-mediated NPs to improve encapsulation efficacy and limit nucleic acid deterioration. Moreover, NPs surface functionalization with various ligands can enhance uptake and target-specific delivery. Similarly, NPs coated with proteins or peptides can also protect to avoid fluid degradation in the biological system for the smooth release of desired genes into the cell cytoplasm [19].

Moreover, detailed research is necessary to examine the impact of NPs on plant physiology, biochemistry and metabolism. Various studies revealed that NPs could trigger ROS production which could cause deleterious impact on their development and growth. Keep in mind the dual nature of NPs in plants, it is crucial to alleviate their adverse effects along with beneficial aspects. The development of biocompatible, biodegradable, less toxic NPs could provide a possible solution to overcome the concerns about their toxicity. Moreover, knowledge about NPs' transport and accumulation inside plant tissues will help address the issue of off-target accumulation during gene delivery [139].

Some studies revealed that NPs work in a dose-dependent manner. Therefore, the optimization of NPs is an important factor for successful transformation. Non-optimized doses of NPs may cause cellular toxicity and off-site target gene delivery [19]. So, it is important to optimize the NPs dose on different plants to explore their effects, ensuring efficient, target-specific gene delivery while maintaining the integrity and biosafety of plant systems.

Furthermore, the binding affinity of NPs to DNA/RNA is another significant challenge that may result in the ineffective release or disintegration of genes inside the plant tissues. Different biomolecules have different binding ability to interact with NPs due to their varying size, shape, and charge. Hence, optimizing these factors may help elevate the binding affinity with biomolecules and their controlled release at target-specific sites. Concludingly, the application of nanobiotechnology in plant genetic transformation requires collaborative efforts with experts from nanotechnology, plant physiology, biochemistry, and molecular biology to ensure the appropriate use of NPs to revolutionize target-specific plant genetic transformation. Addressing the aforementioned challenges will provide pivotal potential to explore the role of nanotechnology in plant genetic transformation.

10 Conclusion

Concludingly, NPs play a pivotal role in yield improvement through genetic engineering in crops. NPs are promising in plant biotechnology due to their magnificent role as growth regulators, the release of fertilizers in a controlled manner, and efficient biomolecule nanocarriers. Various studies revealed that NPs limit traditional methods due to their tunable physico-chemical properties, making them effective agents for biomolecule delivery in plants *in vivo* and *in vitro* conditions. NPs also provide new ways for precise and effective genetic transformation and modification for target-specific gene delivery in plants. However, still, there is a big challenge to understand, how NPs affect the plant physiological characteristics, biochemical behavior, and plant metabolism.

References

1. Pradhan A, Datta A, Lal MK, Kumar M, Alam MK, Basavaraj P, Pal KK (2025) Abiotic stresses in field crops: response, impacts and management under climate change scenario, vol 8. Frontiers Media SA, Lausanne

2. Aswini M, Lenka B, Shaniware YA, Pandey P, Dash AP, Haokip SW (2023) The role of genetics and plant breeding for crop improvement: current Progress and future prospects. Int J Plant Soil Sci 35(20):190–202
3. Lalotra S, Mukherjee T, Singh A, Yadav PK, Upadhyay P, Ghazaryan K, Rajput VD (2024) Unlocking sustainable agriculture: the crucial role of plant genetic engineering. In: Sustainable agriculture: nanotechnology, biotechnology, management and food security. Walter de Gruyter, London, p 143
4. Bailey-Serres J, Parker JE, Ainsworth EA, Oldroyd GE, Schroeder JI (2019) Genetic strategies for improving crop yields. Nature 575(7781):109–118
5. Sedeek KE, Mahas A, Mahfouz M (2019) Plant genome engineering for targeted improvement of crop traits. Front Plant Sci 10:114
6. Siddique AB, Ahmed T, Khan F (2024) Techniques of plant stress management, including genetic engineering, agronomic practices, and the creation of new crop types. In: Molecular dynamics of plant stress and its management. Springer, Cham, pp 429–446
7. Thangadurai D, Soundar Raju C, Sangeetha J, Hospet R, Pandhari R (2020) Rice genetic engineering for increased amino acid and vitamin contents. In: Rice research for quality improvement: genomics and genetic engineering: volume 2: nutrient biofortification and herbicide and biotic stress resistance in rice. Springer, Cham, pp 655–667
8. Demirer GS, Silva TN, Jackson CT, Thomas JB, Ehrhardt D, Rhee SY, Mortimer JC, Landry MP (2021) Nanotechnology to advance CRISPR–Cas genetic engineering of plants. Nat Nanotechnol 16(3):243–250
9. Eeckhaut T, Lakshmanan PS, Deryckere D, Van Bockstaele E, Van Huylenbroeck J (2013) Progress in plant protoplast research. Planta 238:991–1003
10. Lv Z, Jiang R, Chen J, Chen W (2020) Nanoparticle-mediated gene transformation strategies for plant genetic engineering. Plant J 104(4):880–891
11. Ghosh SK, Chatterjee T (2024) Clustered regularly interspaced short palindromic repeats (CRISPR)-associated proteins (Cas)[CRISPR–Cas]: an emerging technique in plant disease detection and management. In: Gene editing in plants: CRISPR-Cas and its applications. Springer, Cham, pp 589–645
12. Kumar R, Shanker R, Singh P, Yadav MK, Chaudhary V, Kumar M (2024) Genome editing towards pests and disease management in agricultural crops: recent developments, challenges and future prospects. Physiol Mol Plant Pathol 134:102402
13. Wu F-H, Hsu C-T, Lin C-S (2023) Targeted insertion in Nicotiana benthamiana genomes via protoplast regeneration. In: Plant genome engineering: methods and protocols. Springer, Cham, pp 297–315
14. Wang Y, Zafar N, Ali Q, Manghwar H, Wang G, Yu L, Ding X, Ding F, Hong N, Wang G (2022a) CRISPR/Cas genome editing technologies for plant improvement against biotic and abiotic stresses: advances, limitations, and future perspectives. Cells 11(23):3928
15. Touzdjian Pinheiro Kohlrausch Távora F, de Assis Santos Diniz F, Moraes Rêgo-Machado C, Chagas Freitas N, Barbosa Monteiro Arraes F, Chumbinho Andrade E, Furtado LL, Osiro KO, Lima Sousa N, Cardoso TB (2022) CRISPR/Cas-and topical RNAi-based technologies for crop management and improvement: reviewing the risk assessment and challenges towards a more sustainable agriculture. Front Bioeng Biotechnol 10:913728
16. Zhi H, Zhou S, Pan W, Shang Y, Zeng Z, Zhang H (2022) The promising nanovectors for gene delivery in plant genome engineering. Int J Mol Sci 23(15):8501
17. Babu KH, Devarumath R, Thorat AS, Nerkar G, Purankar M, Penna S (2024) Plant genetic engineering: nanomaterials-based delivery of genetic. Innov Methods Hortic Crop Improve 1:153
18. Squire HJ, Tomatz S, Voke E, González-Grandío E, Landry M (2023) The emerging role of nanotechnology in plant genetic engineering. Nat Rev Bioeng 1(5):314–328
19. Vats S, Kumawat S, Brar J, Kaur S, Yadav K, Magar SG, Jadhav PV, Salvi P, Sonah H, Sharma S (2022) Opportunity and challenges for nanotechnology application for genome editing in plants. Plant Nano Biol 1:100001

20. Jat SK, Bhattacharya J, Sharma MK (2020) Nanomaterial based gene delivery: a promising method for plant genome engineering. J Mater Chem B 8(19):4165–4175
21. Gad MA, Li M-j, Ahmed FK, Almoammar H (2020) Nanomaterials for gene delivery and editing in plants: challenges and future perspective. In: Multifunctional hybrid nanomaterials for sustainable agri-food and ecosystems. Elsevier, Amsterdam, pp 135–153
22. Ahmar S, Mahmood T, Fiaz S, Mora-Poblete F, Shafique MS, Chattha MS, Jung K-H (2021) Advantage of nanotechnology-based genome editing system and its application in crop improvement. Front Plant Sci 12:663849
23. Zhao L, Lu L, Wang A, Zhang H, Huang M, Wu H, Xing B, Wang Z, Ji R (2020) Nano-biotechnology in agriculture: use of nanomaterials to promote plant growth and stress tolerance. J Agric Food Chem 68(7):1935–1947
24. Al-Khayri JM, Rashmi R, Surya Ulhas R, Sudheer WN, Banadka A, Nagella P, Aldaej MI, Rezk AA-S, Shehata WF, Almaghasla MI (2023) The role of nanoparticles in response of plants to abiotic stress at physiological, biochemical, and molecular levels. Plants 12(2):292
25. Prasad A, Sidhic J, Sarbadhikary P, Narayanankutty A, George S, George BP, Abrahamse H (2024) Role of metal nanoparticles in organogenesis, secondary metabolite production and genetic transformation of plants under in vitro condition: a comprehensive review. Plant Cell Tissue Organ Culture 158(2):33
26. Mujtaba M, Wang D, Carvalho LB, Oliveira JL, Espirito Santo Pereira A, Sharif R, Jogaiah S, Paidi MK, Wang L, Ali Q (2021) Nanocarrier-mediated delivery of miRNA, RNAi, and CRISPR-Cas for plant protection: current trends and future directions. ACS Agric Sci Technol 1(5):417–435
27. Zhao Y, Thenarianto C, Sevencan C, Rajappa S, Shen D, Puangpathumanond S, Yao X, Lew TTS (2024) Rational nanoparticle design for efficient biomolecule delivery in plant genetic engineering. Nanoscale 16(46):21264–21278
28. Anjanappa RB, Gruissem W (2021) Current progress and challenges in crop genetic transformation. J Plant Physiol 261:153411
29. Özkaya DE (2024) Genetic transformation: current opinion and future prospect. In: Plant physiology: from historical roots to future frontiers. Bentham Science, NewYork, p 176
30. Sharma S, Pawar S, Sagwal V, Mishra C (2023) Genetic transformation methods in plants-"a review". Int J Agric Sci 8:117–127
31. Azizi-Dargahlou S, Pouresmaeil M (2024) Agrobacterium tumefaciens-mediated plant transformation: a review. Mol Biotechnol 66(7):1563–1580
32. Rahman SU, Khan MO, Ullah R, Ahmad F, Raza G (2024) Agrobacterium-mediated transformation for the development of transgenic crops; present and future prospects. Mol Biotechnol 66(8):1836–1852
33. Rustgi S, Luo H (2020) Biolistic DNA delivery in plants. Springer, Cham
34. Sparks CA, Doherty A (2020) Genetic transformation of common wheat (Triticum aestivum L.) using biolistics. In: Biolistic DNA delivery in plants: methods and protocols. Springer, Cham, pp 229–250
35. Altpeter F, Baisakh N, Beachy R, Bock R, Capell T, Christou P, Daniell H, Datta K, Datta S, Dix PJ (2005) Particle bombardment and the genetic enhancement of crops: myths and realities. Mol Breed 15:305–327
36. Twyman RM, Christou P (2004) Plant transformation technology: particle bombardment. In: Handbook of plant biotechnology. Wiley, Hoboken
37. Keshavareddy G, Kumar A, Ramu VS (2018) Methods of plant transformation-a review. Int J Curr Microbiol Appl Sci 7(7):2656–2668
38. Anami S, Njuguna E, Coussens G, Aesaert S, Van Lijsebettens M (2013) Higher plant transformation: principles and molecular tools. Int J Dev Biol 57(6-8):483–494
39. Crossway A, Oakes JV, Irvine JM, Ward B, Knauf VC, Shewmaker C (1986) Integration of foreign DNA following microinjection of tobacco mesophyll protoplasts. Mol Gen Genet MGG 202:179–185
40. Morikawa H, Yamada Y (1985) Capillary microinjection into protoplasts and intranuclear localization of injected materials. Plant Cell Physiol 26(2):229–236

41. Lörz H, Paszkowski J, Dierks-Ventling C, Potrykus I (1981) Isolation and characterization of cytoplasts and miniprotoplasts derived from protoplasts of cultured cells. Physiol Plant 53(3):385–391
42. Ozyigit II (2020) Gene transfer to plants by electroporation: methods and applications. Mol Biol Rep 47(4):3195–3210
43. Saeed T, Shahzad A (2016) Basic principles behind genetic transformation in plants. In: Biotechnological strategies for the conservation of medicinal and ornamental climbers. Springer, Cham, pp 327–350
44. Kar S, Loganathan M, Dey K, Shinde P, Chang H-Y, Nagai M, Santra TS (2018) Single-cell electroporation: current trends, applications and future prospects. J Micromech Microeng 28(12):123002
45. Yan Y, Zhu X, Yu Y, Li C, Zhang Z, Wang F (2022) Nanotechnology strategies for plant genetic engineering. Adv Mater 34(7):2106945
46. Gad A, Rosenberg N, Altman A (1990) Liposome-mediated gene delivery into plant cells. Physiol Plant 79(1):177–183
47. Akbarzadeh A, Rezaei-Sadabady R, Davaran S, Joo SW, Zarghami N, Hanifehpour Y, Samiei M, Kouhi M, Nejati-Koshki K (2013) Liposome: classification, preparation, and applications. Nanoscale Res Lett 8:1–9
48. van Wordragen M, Shakya R, Verkerk R, Peytavis R, van Kammen A, Zabel P (1997) Liposome-mediated transfer of YAC DNA to tobacco cells. Plant Mol Biol Report 15:170–178
49. Mahmood-Ul-Hassan M-U-H, Zahid Akram ZA, Shaukat Ali SA, Ali GM, Yusuf Zafar YZ, Shah ZH, Fahad Alghabari FA (2016) Whisker-mediated transformation of peanut with chitinase gene enhances resistance to leaf spot disease
50. Mutsuddy B (1990) Electrokinetic behaviour of aqueous silicon carbide whisker suspensions. J Am Ceram Soc 73(9):2747–2749
51. Kaeppler H, Somers D, Rines H, Cockburn A (1992) Silicon carbide fiber-mediated stable transformation of plant cells. Theor Appl Genet 84:560–566
52. Kaeppler HF, Gu W, Somers DA, Rines HW, Cockburn AF (1990) Silicon carbide fiber-mediated DNA delivery into plant cells. Plant Cell Rep 9:415–418
53. Mizuno K, Takahashi W, Ohyama T, Shimada T, Tanaka O (2004) Improvement of the aluminum borate whisker-mediated method of DNA delivery into rice callus. Plant Prod Sci 7(1):45–49
54. Boffetta P, Hashim D (2017) Exposure to silicon carbide and cancer risk: a systematic review. Int Arch Occup Environ Health 90:1–12
55. Vaughan GL, Trently SA (1996) The toxicity of silicon carbide whiskers, a review. J Environ Sci Health Part A 31(8):2033–2054
56. Gelvin SB (2017) Integration of agrobacterium T-DNA into the plant genome. Annu Rev Genet 51(1):195–217
57. Shreni Agrawal ER (2022) A review: agrobacterium-mediated gene transformation to increase plant productivity. J Phytopharmacol 11:111
58. Roychowdhury D, Majumder A, Jha S (2012) Agrobacterium rhizogenes-mediated transformation in medicinal plants: prospects and challenges. In: Biotechnology for medicinal plants: micropropagation and improvement. Springer, Cham, pp 29–68
59. Koetle M, Finnie J, Balázs E, Van Staden J (2015) A review on factors affecting the agrobacterium-mediated genetic transformation in ornamental monocotyledonous geophytes. S Afr J Bot 98:37–44
60. Das D, Reddy M, Upadhyaya K, Sopory S (2002) An efficient leaf-disc culture method for the regeneration via somatic embryogenesis and transformation of grape (Vitis vinifera L.). Plant Cell Rep 20:999–1005
61. Deng W, Pu X, Goodman R, Gordon M, Nester E (1995) T-DNA genes responsible for inducing a necrotic response on grape vines. Mol Plant-Microbe Interact 8(4):538–548
62. Pu X-A, Goodman R (1992) Induction of necrogenesis by agrobacterium tumefaciens on grape explants. Physiol Mol Plant Pathol 41(4):241–254

63. Baker CJ, Mock NM, Whitaker BD, Roberts DP, Rice CP, Deahl KL, Averyanov AA (2005) Involvement of acetosyringone in plant–pathogen recognition. Biochem Biophys Res Commun 328(1):130–136
64. Ciardi JA, Tieman DM, Lund ST, Jones JB, Stall RE, Klee HJ (2000) Response to Xanthomonas campestris pv. Vesicatoria in tomato involves regulation of ethylene receptor gene expression. Plant Physiol 123(1):81–92
65. Nonaka S, Ezura H (2014) Plant–Agrobacterium interaction mediated by ethylene and super-Agrobacterium conferring efficient gene transfer. Front Plant Sci 5:681
66. Nonaka S, Someya T, Zhou S, Takayama M, Nakamura K, Ezura H (2017) An agrobacterium tumefaciens strain with gamma-aminobutyric acid transaminase activity shows an enhanced genetic transformation ability in plants. Sci Rep 7(1):42649
67. Gu X, Liu L, Zhang H (2021) Transgene-free genome editing in plants. Front Genome Edit 3:805317
68. Park J, Choi S, Park S, Yoon J, Park AY, Choe S (2019) DNA-free genome editing via ribonucleoprotein (RNP) delivery of CRISPR/Cas in lettuce. In: Plant genome editing with CRISPR systems: methods and protocols. Springer, Cham, pp 337–354
69. Kocsisova Z, Coneva V (2023) Strategies for delivery of CRISPR/Cas-mediated genome editing to obtain edited plants directly without transgene integration. Front Genome Edit 5:1209586
70. Kuluev B, Gumerova G, Mikhaylova E, Gerashchenkov G, Rozhnova N, Vershinina Z, Khyazev A, Matniyazov R, Baymiev AK, Baymiev AK (2019) Delivery of CRISPR/Cas components into higher plant cells for genome editing. Russ J Plant Physiol 66:694–706
71. Ali A, Zafar MM, Farooq Z, Ahmed SR, Ijaz A, Anwar Z, Abbas H, Tariq MS, Tariq H, Mustafa M (2023) Breakthrough in CRISPR/Cas system: current and future directions and challenges. Biotechnol J 18(8):2200642
72. Patel VK, Soni N, Prasad V, Sapre A, Dasgupta S, Bhadra B (2019) CRISPR–Cas9 system for genome engineering of photosynthetic microalgae. Mol Biotechnol 61:541–561
73. Dijkerman A (2022) Genome editing in bread wheat using CRISPR/Cas9. Stellenbosch University, Stellenbosch
74. Liang Z, Chen K, Li T, Zhang Y, Wang Y, Zhao Q, Liu J, Zhang H, Liu C, Ran Y (2017) Efficient DNA-free genome editing of bread wheat using CRISPR/Cas9 ribonucleoprotein complexes. Nat Commun 8(1):14261
75. Woo JW, Kim J, Kwon SI, Corvalán C, Cho SW, Kim H, Kim S-G, Kim S-T, Choe S, Kim J-S (2015) DNA-free genome editing in plants with preassembled CRISPR-Cas9 ribonucleoproteins. Nat Biotechnol 33(11):1162–1164
76. Ellison EE, Nagalakshmi U, Gamo ME, Huang P-j, Dinesh-Kumar S, Voytas DF (2020) Multiplexed heritable gene editing using RNA viruses and mobile single guide RNAs. Nat Plants 6(6):620–624
77. Gordon-Kamm B, Sardesai N, Arling M, Lowe K, Hoerster G, Betts S, Jones T (2019) Using morphogenic genes to improve recovery and regeneration of transgenic plants. Plants 8(2):38
78. Lowe K, Wu E, Wang N, Hoerster G, Hastings C, Cho M-J, Scelonge C, Lenderts B, Chamberlin M, Cushatt J (2016) Morphogenic regulators baby boom and Wuschel improve monocot transformation. Plant Cell 28(9):1998–2015
79. Maher MF, Nasti RA, Vollbrecht M, Starker CG, Clark MD, Voytas DF (2020) Plant gene editing through de novo induction of meristems. Nat Biotechnol 38(1):84–89
80. Zhang Y, Huang C, Xiong R (2024) Advanced materials for intracellular delivery of plant cells: strategies, mechanisms and applications. Mater Sci Eng R Rep 160:100821
81. Babu S, Singh R, Yadav D, Rathore SS, Raj R, Avasthe R, Yadav S, Das A, Yadav V, Yadav B (2022) Nanofertilizers for agricultural and environmental sustainability. Chemosphere 292:133451
82. Zohra E, Ikram M, Omar AA, Hussain M, Satti SH, Raja NI, Mashwani Z-U-R, Ehsan M (2021) Potential applications of biogenic selenium nanoparticles in alleviating biotic and abiotic stresses in plants: a comprehensive insight on the mechanistic approach and future perspectives. Green Process Synth 10(1):456–475

83. Ali SS, Al-Tohamy R, Koutra E, Moawad MS, Kornaros M, Mustafa AM, Mahmoud YA-G, Badr A, Osman ME, Elsamahy T (2021) Nanobiotechnological advancements in agriculture and food industry: applications, nanotoxicity, and future perspectives. Sci Total Environ 792:148359
84. Usman M, Farooq M, Wakeel A, Nawaz A, Cheema SA, Ur Rehman H, Ashraf I, Sanaullah M (2020) Nanotechnology in agriculture: current status, challenges and future opportunities. Sci Total Environ 721:137778
85. Adisa IO, Pullagurala VLR, Peralta-Videa JR, Dimkpa CO, Elmer WH, Gardea-Torresdey JL, White JC (2019) Recent advances in nano-enabled fertilizers and pesticides: a critical review of mechanisms of action. Environ Sci Nano 6(7):2002–2030
86. Jogaiah S, Singh HB, Fraceto LF, De Lima R (2020) Advances in nano-fertilizers and nano-pesticides in agriculture: a smart delivery system for crop improvement. Woodhead Publishing, Sawston
87. Wong MH, Misra RP, Giraldo JP, Kwak S-Y, Son Y, Landry MP, Swan JW, Blankschtein D, Strano MS (2016) Lipid exchange envelope penetration (LEEP) of nanoparticles for plant engineering: a universal localization mechanism. Nano Lett 16(2):1161–1172
88. Banerjee K, Pramanik P, Maity A, Joshi D, Wani SH, Krishnan P (2019) Methods of using nanomaterials to plant systems and their delivery to plants (mode of entry, uptake, translocation, accumulation, biotransformation and barriers). In: Advances in phytonanotechnology. Elsevier, Amsterdam, pp 123–152
89. de la Rosa G, Vázquez-Núñez E, Molina-Guerrero C, Serafín-Muñoz AH, Vera-Reyes I (2021) Interactions of nanomaterials and plants at the cellular level: current knowledge and relevant gaps. Nanotechnol Environ Eng 6:1–19
90. Mostafa M, Ahmed FK, Alghuthaymi M, Abd-Elsalam KA (2021) Inorganic smart nanoparticles: a new tool to deliver CRISPR systems into plant cells. In: CRISPR and RNAi systems. Elsevier, Amsterdam, pp 661–686
91. Rani N, Kumari K, Hooda V (2025) The role of nanoparticles in transforming plant genetic engineering: advancements, challenges and future prospects. Funct Integr Genomics 25(1):1–22
92. Cunningham FJ, Goh NS, Demirer GS, Matos JL, Landry MP (2018) Nanoparticle-mediated delivery towards advancing plant genetic engineering. Trends Biotechnol 36(9):882–897
93. Hu P, An J, Faulkner MM, Wu H, Li Z, Tian X, Giraldo JP (2020) Nanoparticle charge and size control foliar delivery efficiency to plant cells and organelles. ACS Nano 14(7):7970–7986
94. Tang F, Li L, Chen D (2012) Mesoporous silica nanoparticles: synthesis, biocompatibility and drug delivery. Adv Mater 24(12):1504–1534
95. Asefa T, Tao Z (2012) Biocompatibility of mesoporous silica nanoparticles. Chem Res Toxicol 25(11):2265–2284
96. Zhou Y, Quan G, Wu Q, Zhang X, Niu B, Wu B, Huang Y, Pan X, Wu C (2018) Mesoporous silica nanoparticles for drug and gene delivery. Acta Pharm Sin B 8(2):165–177
97. Zolghadrnasab M, Mousavi A, Farmany A, Arpanaei A (2021) Ultrasound-mediated gene delivery into suspended plant cells using polyethyleneimine-coated mesoporous silica nanoparticles. Ultrason Sonochem 73:105507
98. Hajiahmadi Z, Shirzadian-Khorramabad R, Kazemzad M, Sohani MM (2019) Enhancement of tomato resistance to Tuta absoluta using a new efficient mesoporous silica nanoparticle-mediated plant transient gene expression approach. Sci Hortic 243:367–375
99. Martin-Ortigosa S, Valenstein JS, Lin VSY, Trewyn BG, Wang K (2012) Gold functionalized mesoporous silica nanoparticle mediated protein and DNA codelivery to plant cells via the biolistic method. Adv Funct Mater 22(17):3576–3582
100. Chang F-P, Kuang L-Y, Huang C-A, Jane W-N, Hung Y, Yue-ie CH, Mou C-Y (2013) A simple plant gene delivery system using mesoporous silica nanoparticles as carriers. J Mater Chem B 1(39):5279–5287
101. Hajiahmadi Z, Shirzadian-Khorramabad R, Kazemzad M, Sohani MM, Khajehali J (2020) A novel, simple, and stable mesoporous silica nanoparticle-based gene transformation approach in Solanum lycopersicum. 3 Biotech 10(8):370

102. Cai Y, Liu Z, Wang H, Meng H, Cao Y (2024) Mesoporous silica nanoparticles mediate SiRNA delivery for long-term multi-gene silencing in intact plants. Adv Sci 11(9):2301358
103. Sangwan A, Gupta D, Singh OW, Roy A, Mukherjee SK, Mandal B, Singh N (2023) Size variations of mesoporous silica nanoparticle control uptake efficiency and delivery of AC2-derived dsRNA for protection against tomato leaf curl New Delhi virus. Plant Cell Rep 42(10):1571–1587
104. Torney F, Trewyn BG, Lin VS-Y, Wang K (2007) Mesoporous silica nanoparticles deliver DNA and chemicals into plants. Nat Nanotechnol 2(5):295–300
105. Aminzai MT, Yildirim M, Yabalak E (2024) Metallic nanoparticles unveiled: synthesis, characterization, and their environmental, medicinal, and agricultural applications. Talanta 280:126790
106. Yaqoob SB, Adnan R, Rameez Khan RM, Rashid M (2020) Gold, silver, and palladium nanoparticles: a chemical tool for biomedical applications. Front Chem 8:376
107. Singh SP, Ansari MI, Pandey B, Srivastava JK, Yadav TP, Rani H, Parveen A, Mala J, Singh AK (2020) Recent trends and advancement toward phyto-mediated fabrication of noble metallic nanomaterials: focus on silver, gold, platinum, and palladium. In: Nanomaterials and environmental biotechnology. Springer, Cham, pp 87–105
108. Han G, Ghosh P, Rotello VM (2007) Functionalized gold nanoparticles for drug delivery. Nanomedicine 2(1):113–123
109. Nishiyama N (2007) Nanomedicine: nanocarriers shape up for long life. Nat Nanotechnol 2(4):203
110. El-Saadony MT, Almoshadak AS, Shafi ME, Albaqami NM, Saad AM, El-tahan AM, Desoky E, Elnahal AS, Almakas A, Abd El-Mageed TA (2021) Vital roles of sustainable nano-fertilizers in improving plant quality and quantity-an updated review. Saudi J Biol Sci 28(12):7349–7359
111. Kekeli MA, Wang Q, Rui Y (2025) The role of Nano-fertilizers in sustainable agriculture: boosting crop yields and enhancing quality. Plants 14(4):554
112. Wang ZP, Zhang ZB, Zheng DY, Zhang TT, Li XL, Zhang C, Yu R, Wei JH, Wu ZY (2022b) Efficient and genotype independent maize transformation using pollen transfected by DNA-coated magnetic nanoparticles. J Integr Plant Biol 64(6):1145–1156
113. Vijayakumar PS, Abhilash OU, Khan BM, Prasad BL (2010) Nanogold-loaded sharp-edged carbon bullets as plant-gene carriers. Adv Funct Mater 20(15):2416–2423
114. Lei W-X, An Z-S, Zhang B-H, Wu Q, Gong W-J, Li J-M, Chen W-L (2020) Construction of gold-siRNA NPR1 nanoparticles for effective and quick silencing of NPR1 in Arabidopsis thaliana. RSC Adv 10(33):19300–19308
115. Zhang H, Cao Y, Xu D, Goh NS, Demirer GS, Cestellos-Blanco S, Chen Y, Landry MP, Yang P (2021) Gold-nanocluster-mediated delivery of siRNA to intact plant cells for efficient gene knockdown. Nano Lett 21(13):5859–5866
116. Zhang J, Lei W, Meng Y, Zhou C, Zhang B, Yuan J, Wang M, Xu D, Meng X, Chen W (2022) Expression of PEI-coated gold nanoparticles carrying exogenous gene in periwinkle mesophyll cells and its practice in huanglongbing research. Iscience 25(6):104479
117. Rajkumari N, Alex S, Soni K, Anith K, Viji M, Kiran A (2021) Silver nanoparticles for biolistic transformation in Nicotiana tabacum L. 3 Biotech 11:497
118. Eatemadi A, Daraee H, Karimkhanloo H, Kouhi M, Zarghami N, Akbarzadeh A, Abasi M, Hanifehpour Y, Joo SW (2014) Carbon nanotubes: properties, synthesis, purification, and medical applications. Nanoscale Res Lett 9:1–13
119. Safdar M, Kim W, Park S, Gwon Y, Kim Y-O, Kim J (2022) Engineering plants with carbon nanotubes: a sustainable agriculture approach. J Nanobiotechnol 20(1):275
120. Samadi S, Saharkhiz MJ, Azizi M, Samiei L, Karami A, Ghorbanpour M (2021) Single-wall carbon nano tubes (SWCNTs) penetrate thymus daenensis Celak. Plant cells and increase secondary metabolite accumulation in vitro. Ind Crop Prod 165:113424
121. Anzar N, Hasan R, Tyagi M, Yadav N, Narang J (2020) Carbon nanotube-a review on synthesis, properties and plethora of applications in the field of biomedical science. Sens Int 1:100003

122. Rashko MN, Hamad SM, Barzinjy AA, Hamad AH (2022) Mechanical properties of carbon nanotubes (CNTs): a review. Euras J Sci Eng 8(2):54–68
123. Liu Q, Chen B, Wang Q, Shi X, Xiao Z, Lin J, Fang X (2009) Carbon nanotubes as molecular transporters for walled plant cells. Nano Lett 9(3):1007–1010
124. Golestanipour A, Nikkhah M, Aalami A, Hosseinkhani S (2018) Gene delivery to tobacco root cells with single-walled carbon nanotubes and cell-penetrating fusogenic peptides. Mol Biotechnol 60:863–878
125. Schwartz SH, Hendrix B, Hoffer P, Sanders RA, Zheng W (2020) Carbon dots for efficient small interfering RNA delivery and gene silencing in plants. Plant Physiol 184(2):647–657
126. Demirer GS, Zhang H, Matos JL, Goh NS, Cunningham FJ, Sung Y, Chang R, Aditham AJ, Chio L, Cho M-J (2019) High aspect ratio nanomaterials enable delivery of functional genetic material without DNA integration in mature plants. Nat Nanotechnol 14(5):456–464
127. Chandrasekaran R, Seetharaman PK, Danaraj J, Rajiv P, Abd-Elsalam KA (2021) Polymer and lipid-based nanoparticles to deliver RNAi and CRISPR systems. In: CRISPR and RNAi systems. Elsevier, Amsterdam, pp 635–659
128. Liu J, Wang F-h, Wang L-l, Xiao S-y, Tong C-y, Tang D-y, Liu X-m (2008) Preparation of fluorescence starch-nanoparticle and its application as plant transgenic vehicle. J Cent S Univ Technol 15:768–773
129. Chouhan D, Mandal P (2021) Applications of chitosan and chitosan based metallic nanoparticles in agrosciences-a review. Int J Biol Macromol 166:1554–1569
130. Kashyap PL, Xiang X, Heiden P (2015) Chitosan nanoparticle based delivery systems for sustainable agriculture. Int J Biol Macromol 77:36–51
131. Tawfik E, Ahmed M (2022) Chitosan nanoparticles as a new technique in gene transformation into different plants tissues. Nat Resour Human Health 2:215–221
132. Wang Q, Chen JN, Zhan P, Zhang L, Kong QQ (2013) Establishment of a suspension cell system for transformation of Jatropha curcas using nanoparticles. Adv Mater Res 608:314–319
133. Amani A, Zare N, Asadi A, Zakaria RA (2018) Ultrasound-enhanced gene delivery to alfalfa cells by hPAMAM dendrimer nanoparticles. Turk J Biol 42(1):63–75
134. Kim C-J, Park J-e, Hu X, Albert SK, Park S-J (2020) Peptide-driven shape control of low-dimensional DNA nanostructures. ACS Nano 14(2):2276–2284
135. Zhang H, Demirer GS, Zhang H, Ye T, Goh NS, Aditham AJ, Cunningham FJ, Fan C, Landry MP (2019) DNA nanostructures coordinate gene silencing in mature plants. Proc Natl Acad Sci 116(15):7543–7548
136. Zhang H, Zhang H, Demirer GS, González-Grandío E, Fan C, Landry MP (2020) Engineering DNA nanostructures for siRNA delivery in plants. Nat Protoc 15(9):3064–3087
137. Dufès C, Uchegbu IF, Schätzlein AG (2005) Dendrimers in gene delivery. Adv Drug Deliv Rev 57(15):2177–2202
138. Pasupathy K, Lin S, Hu Q, Luo H, Ke PC (2008) Direct plant gene delivery with a poly (amidoamine) dendrimer. Biotechnol J 3(8):1078–1082
139. Soliman GM (2017) Nanoparticles as safe and effective delivery systems of antifungal agents: achievements and challenges. Int J Pharm 523(1):15–32

Nanoparticles for Advancing CRISPR/Cas9-Mediated Plant Genome Editing: Opportunities and Challenges

Ankan Das, Bristi Roy, Srija Mondal, Biswajit Pramanik, Sayantan Senapati, and Sandip Debnath

Abstract Nanotechnology has revolutionized CRISPR/Cas9-mediated plant genome editing by addressing limitations of conventional delivery methods, such as Agrobacterium-mediated transformation and particle bombardment, which often suffer from low efficiency, tissue damage, and challenges in editing recalcitrant plant species. Nanoparticles, encompassing metallic, polymeric, lipid-based, and hybrid systems, have emerged as advanced carriers that enhance the targeted, efficient, and stable delivery of CRISPR/Cas9 components. By protecting Cas9 ribonucleoprotein complexes from enzymatic degradation and facilitating precise in vivo genome editing, nanoparticles offer solutions to key technical barriers. Mechanisms such as magnetofection, where magnetic nanoparticles are used for targeted delivery under external magnetic fields, and the development of biodegradable systems, which reduce environmental and biological toxicity, have further expanded the scope of nanotechnology in genome editing. The integration of these advancements with speed breeding techniques enables the rapid development of climate-resilient and high-yielding crops, addressing the increasing demand for sustainable agriculture. Despite these promising developments, challenges such as nanoparticle toxicity, scalability for field applications, and regulatory hurdles remain significant. Overcoming these challenges requires interdisciplinary approaches to design next-generation nanoparticle systems that balance efficiency, biocompatibility, and cost-effectiveness. By bridging laboratory innovations with practical agricultural applications, nanotechnology-driven CRISPR/Cas9 systems hold the potential to transform plant genome editing, accelerating the development of crops capable of

A. Das · B. Roy · S. Mondal · S. Senapati
Palli Siksha Bhavana (Institute of Agriculture), Visva-Bharati, Sriniketan, West Bengal, India

B. Pramanik
Biotechnology Unit, Crop Improvement Division, ICAR – Central Research Institute for Jute and Allied Fibres, Barrackpore, West Bengal, India

S. Debnath (✉)
Department of Genetics and Plant Breeding, Palli Siksha Bhavana (Institute of Agriculture), Visva-Bharati, Sriniketan, West Bengal, India
e-mail: sandip.debnath@visva-bharati.ac.in

J.-T. Chen (ed.), *Nano-Delivering for Plant Genetic Engineering*, Smart Nanomaterials Technology, https://doi.org/10.1007/978-981-95-4165-2_2

withstanding biotic and abiotic stresses while improving global food security. This synergy of nanobiotechnology and genome editing paves the way for sustainable and rapid advancements in crop improvement.

Keywords Nanotechnology · CRISPR/Cas9 · Plant genome editing · Nanoparticle delivery systems · Speed breeding

1 Introduction

Gene editing technology, often termed "molecular scissors," enables precise DNA/RNA manipulation through tools like CRISPR-Cas9, Zinc Finger Nucleases (ZFNs), and Transcription Activator-Like Effector Nucleases (TALENs), revolutionizing plant genetic engineering [1]. Unlike traditional breeding, which relies on time-consuming backcrossing, CRISPR-Cas9 allows direct, predictable edits in elite crop varieties, accelerating trait development [2]. Originally derived from bacterial immune systems, CRISPR-Cas9 uses a single guide RNA (sgRNA) to direct the Cas9 enzyme to cut specific DNA sequences, offering unmatched precision [3, 4].

Conventional delivery methods, such as biolistics (gene guns) and *Agrobacterium*-mediated transfer, face limitations in plants. Biolistics often damages tissues and causes random DNA integration, leading to inconsistent gene expression, while *Agrobacterium* struggles with monocot transformation and risks disrupting essential genes [5, 6]. Electroporation, though rapid, can harm cells and permit unintended material uptake [6]. Viral vectors, like lentiviruses, raise safety concerns due to immune reactions and insertional mutagenesis, whereas microinjection is labor-intensive and species-restricted [1].

Non-viral nanoparticle (NP) delivery systems address these challenges. Cationic lipids, polymers, and inorganic NPs (e.g., gold nanoparticles, zeolitic imidazolate frameworks [ZIFs]) protect CRISPR components (DNA, RNA, or ribonucleoproteins [RNPs]) from degradation and enhance cellular uptake [1]. RNPs—preassembled Cas9 protein and sgRNA—minimize off-target effects and enable transient editing, avoiding prolonged Cas9 activity [7]. Gold nanoparticles, functionalized with peptides or responsive to near-infrared light, demonstrate efficient cargo delivery in animal studies, with potential adaptability to plants through surface modifications (e.g., cellulose-binding ligands) [8]. Magnetic nanoparticles (MNPs), guided by external fields, enable tissue-specific targeting; in animal models, MNP-viral hybrids boosted gene delivery five-fold, a strategy adaptable to root or vascular tissue editing in plants [5].

Nanoparticles also enhance stress resilience: nanofertilizers improve nutrient uptake efficiency by 18–29% compared to synthetic counterparts, while NPs mitigate drought and salinity stress by regulating aquaporins and ion homeostasis [5, 9, 10]. Their versatility extends to delivering DNA, RNA, and proteins—metallic NPs and carbon nanotubes (CNTs) transport nucleic acids, whereas silica and polymer NPs encapsulate proteins [6]. Hybrid approaches, such as the Magnetic

Nanoparticle-Assisted Genome Editing (MAGE) platform, combine CRISPR plasmids with magnetic core-shell nanoparticles (MNCPs) for targeted delivery, a concept applicable to crops under speed-breeding regimes [8].

By integrating nanotechnology with gene editing, plant engineering enters a new era of precision, scalability, and efficiency, paving the way for climate-resilient, high-yielding crops.

2 Nanoparticles for CRISPR/Cas9 Delivery: Mechanisms and Innovations

Nanoparticles have emerged as a versatile and effective platform for introducing CRISPR/Cas9 components inside phyto-cellular systems. They outpaced conventional delivery technologies *viz.* viral vectors, electroporation, and microinjection, which often suffer from limitations including low efficiency, tissue damage, and lack of target specificity. Engineered nanoparticles enhance the stability of biomolecules, improve cellular uptake, and facilitate precise delivery, making them increasingly indispensable for plant genome editing.

2.1 Enhancing Cas9 Stability and Efficiency

The advent of this advanced gene editing hinges on the stability and activity of its core elements, *viz.* Cas9 protein and single-guide RNA (sgRNA), commonly referred to as ribonucleoproteins (RNPs). These biomolecules are susceptible to degradation by nucleases and environmental stressors. Nanoparticles act as protective carriers, shielding these RNPs during transit and enhancing their uptake and function within plant cells.

2.1.1 Nanoparticles as Protective Carriers against Enzymatic Degradation of Cas9 RNP Complexes

Plant cells harbour various enzymes, including nucleases and proteases, which can degrade CRISPR components, especially sgRNAs, before they reach their intracellular targets. Nanoparticle encapsulation provides a crucial protective layer, ensuring the integrity and functionality of RNPs [11].

(a) Polymeric nanoparticles: Biodegradable polymers like polylactic-co-glycolic acid (PLGA), polyethyleneimine (PEI), chitosan, etc. are often employed to encapsulate Cas9 RNPs. Chitosan, in particular, offers antimicrobial benefits in addition to protection. For example, in wheat, chitosan nanoparticles successfully preserved Cas9 RNPs from degradation, facilitating the editing of drought-resistance genes [12].

(b) Lipid nanoparticles (LNPs): Mimicking the structure of biological membranes, LNPs encapsulate CRISPR components within a lipid bilayer, enhancing compatibility with plant cells while protecting against nuclease activity. In maize, LNPs were employed to deliver CRISPR/Cas9 complexes, resulting in the effective editing of genes related to drought tolerance such as ZmPL1 [13].
(c) Inorganic nanoparticles: Materials such as mesoporous silica nanoparticles (MSNs) and gold nanoparticles (AuNPs) offer excellent structural stability and cargo protection. AuNPs can bind Cas9 proteins through both covalent and non-covalent interactions, maintaining their functionality. Similarly, MSNs, owing to their high surface area and porosity, can encapsulate and protect large biomolecules.

2.1.2 Strategies for Increasing Delivery Efficiency

Numerous research groups worldwide are actively exploring diverse strategies to enhance the efficiency of nanoparticle-mediated delivery systems in plants. These approaches include modifying nanoparticle surfaces with targeting ligands, incorporating stimuli-responsive elements for controlled release, and optimising particle size and composition for better cellular uptake. Such innovations are aimed at improving the precision, stability, and bioavailability of CRISPR/Cas9 components during plant genome editing.

(a) Surface functionalisation: Modification in nanoparticle surfaces with specific ligands, antibodies, or peptides can significantly improve their interaction with plant tissues. Several recent reviews have highlighted the utility of lectin-coated nanocarriers in gene delivery platforms due to their ability to facilitate selective binding and internalisation.
(b) PEGylation: Coating nanoparticles with polyethylene glycol (PEG) minimises aggregation, enhances stability, and prolongs systemic circulation. PEGylated nanoparticles have shown superior delivery efficiency in wheat, resulting in higher genome-editing success compared to uncoated variants [14].
(c) Stimuli-responsive nanoparticles: Nanoparticles engineered to respond to specific environmental cues, such as pH or enzyme activity allow for controlled and timely release of CRISPR components. For instance, pH-sensitive polymeric nanoparticles have been designed to release their cargo within the acidic endosomal environment, ensuring effective cytoplasmic delivery [15].

2.2 Mechanisms of Action

Nanoparticles must overcome multiple structural and physiological obstacles within plant systems to successfully deliver CRISPR/Cas9 elements. Plant cell wall, with its constituents like lignin, hemicellulose, cellulose, etc., acts as a major physical barrier that limits the entry of large biomolecules. To bypass this, nanoparticles are engineered with optimised size, surface charge, and chemical functionalisation, enabling them to penetrate through cell wall pores or interact favourably with cell membrane receptors. Once inside the plant cell, these carriers must traverse additional hurdles such as endosomal entrapment and cytoplasmic degradation, for which stimuli-responsive designs, viz. pH-sensitive coatings are employed to ensure timely release of the cargo. Through these finely tuned mechanisms, nanoparticles provide a reliable and targeted approach for delivering genome-editing tools with minimal disruption to cellular function.

2.2.1 Transversing Plant Cell Walls and Membranes

Plant cell wall, formed with rigid biopolymers like cellulose, lignin, and pectin, presents a major obstacle to intracellular delivery. Nanoparticles counter this challenge through carefully tailored size, surface chemistry, and interaction dynamics.

(a) Mesoporous silica nanoparticles (MSNs): These nanoparticles feature adjustable size and porosity, which allow them to carry sizeable payloads. When functionalised with ligands such as lectins, MSNs demonstrate improved adhesion and passage through cell wall pores [16].
(b) Carbon nanomaterials (CNs): Graphene oxide and carbon nanotubes possess mechanical and electrical properties that support their penetration of the plant cell wall. PEG-functionalised carbon nanotubes successfully delivered Cas9 RNPs into mesophyll cells, leading to precise gene edits associated with salt stress tolerance [17].
(c) Lipid-based carriers: Lipid nanoparticles can fuse with plant membranes to facilitate cargo entry by mimicking the lipid bi-layer structure. Surface modification with targeting ligands further enhances their ability to bypass the cell wall and deliver payloads to specific cells.

2.2.2 Delivery Pathways

Nanoparticles utilize a combination of passive and active transport mechanisms to overcome plant cellular barriers [18]. Through these mechanisms, they can navigate the complex structure of plant tissues, including the cell wall and plasma membrane, thereby facilitating the efficient and site-specific delivery of bioactive compounds, genetic material, or agrochemicals to targeted cells or organelles within the plant system.

(a) Passive pathways: Nanoparticles smaller than 50 nm can diffuse through natural pores in the plant cell wall. The efficiency of this process is influenced by particle size, surface charge, and hydrophobicity with neutral or slightly negative charges being most favourable.

(b) Active pathways: Larger or ligand-functionalized nanoparticles often utilize receptor-mediated endocytosis for entry. For instance, auxin-functionalized nanoparticles leverage plant hormone receptors to gain access to target cells via active transport.
(c) Endosomal escape: Following internalization, nanoparticles must exit endosomes to avoid degradation. Stimuli-responsive particles, particularly those designed to react to acidic conditions, can disrupt the endosomal membrane and release their cargo into the cytoplasm.

2.2.3 Role of Nanoparticles in Achieving Precise In Vivo Genome Editing

Precision in genome editing is paramount to minimize unintended modifications. Nanoparticles aid in achieving this through targeted delivery, controlled release, and efficient nuclear localization of CRISPR elements [19].

(a) Controlled release: A critical aspect of precision genome editing lies in the timely and site-specific release of CRISPR/Cas9 components. To achieve this, researchers have developed stimuli-responsive nanoparticles that remain inert under normal physiological conditions but become active in response to specific triggers, such as acidic pH within endosomal compartments or the presence of certain plant-specific enzymes. Such smart-release systems help safeguard the integrity of Cas9 ribonucleoproteins (RNPs) during transit, minimizing premature degradation or diffusion, and ensuring that the editing machinery is released precisely where and when required. This controlled release not only improves editing efficiency but also contributes to reducing off-target effects.
(b) Targeted delivery: Achieving high specificity in plant genome editing requires precise localization of CRISPR components to the desired tissues or cell types. Nanoparticles can be surface-modified with functional ligands, such as peptides, antibodies, or carbohydrates, that selectively bind to receptors on target cells. For instance, peptide-functionalized nanoparticles have been engineered to recognize and bind to stomatal guard cells, key regulators of water-use efficiency in plants. By directing CRISPR/Cas9 complexes specifically to these cells, it becomes possible to selectively modify gene expression related to drought tolerance, without impacting surrounding tissues. This precision-driven technique elevates not only accuracy but also the overall success of gene-editing implementations in the field of crop improvement.
(c) Nuclear localization: As the CRISPR/Cas9 system operates within the nucleus, effective genome editing hinges on the ability of delivered components, particularly the Cas9-sgRNA complex, to cross the nuclear envelope. Nanoparticles designed for this purpose are often conjugated with nuclear localization signals (NLS), short peptide sequences that facilitate active transport through nuclear pores. This strategic modification ensures that once the Cas9 RNPs are released

into the cytoplasm, they are rapidly and efficiently trafficked to the nucleus. Enhancing nuclear delivery narrows down the window of off-targets along with speeding up the entire editing process, thus improving the specificity and dependability of *in vivo* gene editing.

3 Specialized Nanoparticle-Based Delivery Methods

3.1 Polymer-Based NPs

Polymer-based nanoparticles, engineered to carry CRISPR/Cas9 components, offer tailored solutions for plant genome editing by combining versatility with biocompatibility. Cationic polymers like polyethyleneimine (PEI) form stable complexes with CRISPR plasmids, enhancing delivery efficiency by protecting DNA from degradation and aiding escape from plant cell compartments (e.g., vacuoles) that break down foreign material [20]. While PEI shows high editing efficiency, its toxicity at high doses limits use—a challenge addressed by biodegradable alternatives like chitosan, a natural polymer derived from shellfish, which binds nucleic acids but often requires structural tweaks to loosen its grip on DNA for better release inside cells [21]. Innovations such as poly(β-amino ester) (PBAE) polymers overcome charge limitations, enabling the encapsulation of proteins with varied surface properties. For example, a PBAE variant (C5 polymer) achieved 77% gene knockout in mammalian cells when delivering CRISPR ribonucleoproteins (RNPs) [22], suggesting potential for plant applications where precise editing is critical. Advanced designs, like boronic acid-modified dendrimers (branched polymers), efficiently deliver proteins of diverse sizes and charges into cells while maintaining their function—successfully editing genes in multiple cell lines using Cas9 RNP complexes. Redox-sensitive polymers, such as CP/Ad-SS-GD, leverage plant cells' internal reducing environment: disulfide bonds in the polymer break upon entry, releasing CRISPR components to edit genes [23]. Light-responsive systems, like semiconducting polymer brushes (SPPF), enable the controlled release of CRISPR tools using near-infrared light [24], a feature adaptable for tissue-specific editing in plants (e.g., targeting roots or leaves). These polymers also integrate imaging capabilities, allowing real-time tracking of delivery. Hybrid systems, such as hyaluronic acid-decorated nanoparticles, tested in cancer models [25], hint at plant-specific modifications (e.g., cellulose-binding domains) to improve tissue targeting. By balancing stability in harsh plant environments (e.g., soil pH, enzymatic activity) with efficient intracellular cargo release, polymer nanoparticles minimize off-target effects and simplify workflows—key for editing crop genes to boost disease resistance, drought tolerance, or yield.

3.2 Lipid-Based NPs

Lipid nanoparticles (LNPs) have emerged as a promising tool for delivering the CRISPR/Cas9 gene-editing system in plants, leveraging their ability to protect genetic material and enhance cellular uptake. These nanoparticles, composed of biocompatible lipids, form stable complexes with negatively charged CRISPR components—such as Cas9 mRNA, guide RNA (sgRNA), or pre-assembled Cas9 protein-RNA complexes (RNPs)—through electrostatic interactions. For instance, zwitterionic amino lipids (ZAL) and cationic lipids like DOTAP, combined with helper lipids such as DOPE and cholesterol, efficiently encapsulate CRISPR elements, shielding them from enzymatic degradation and improving delivery into plant cells [26]. Studies show that lipid formulations with disulfide bonds (e.g., BAMEA-O16B) or bioreducible properties enhance the intracellular release of CRISPR cargo by breaking down in the plant cell's reducing environment, boosting editing efficiency [17]. Additionally, incorporating polyethylene glycol (PEG) in lipid coatings stabilizes nanoparticles in plant tissues, preventing aggregation and enhancing systemic delivery [27]. Challenges in delivering large RNPs—prone to denaturation in acidic conditions—have been addressed by blending ionizable lipids with permanently cationic supplements (e.g., DOTAP), enabling neutral pH encapsulation without compromising RNP integrity [28]. This approach, tested in model systems, achieved precise gene edits by facilitating endosomal escape and cytoplasmic release of CRISPR components. Innovations like lipid-hydrogel hybrids or exosome-liposome combinations further improve targeting and transfection, demonstrating the potential for editing plant genomes with minimal off-target effects. By adjusting lipid composition, these nanoparticles can be tailored to penetrate rigid plant cell walls or target specific tissues, offering a versatile, low-toxicity platform for crop improvement and functional genomics.

3.3 Inorganic-Based NPs

Inorganic nanoparticles, leveraging their unique chemical and optical properties, are emerging as precision tools for CRISPR/Cas9 delivery in plants. Gold nanoparticles, customizable in shape and size, serve as versatile carriers due to their stability and biocompatibility. For instance, gold nanoclusters modified with cell-penetrating peptides have successfully delivered CRISPR components into mammalian cells, achieving >70% gene knockdown [29]—a strategy adaptable to plant systems by coating nanoparticles with plant-specific ligands (e.g., cellulose-binding domains) to traverse rigid cell walls. Gold nanorods, responsive to near-infrared (NIR-II) light, enable the controlled release of CRISPR tools via heat generated under light exposure [30], a feature that could be harnessed for root or leaf-specific editing in plants. Zeolitic imidazolate frameworks (ZIFs)—porous, cage-like structures made of metal ions and organic linkers—outperform traditional lipid-based carriers in protecting CRISPR/Cas9 from enzymatic degradation, as shown by improved gene expression in mammalian cells. These ZIFs, with their high cargo-loading capacity, could encapsulate CRISPR ribonucleoproteins (RNPs) for efficient delivery into plant cells [31]. Magnetic iron oxide nanoparticles, when combined with viral vectors, allow spatial control of gene editing: applying localized magnetic fields directs nanoparticle-virus complexes to specific tissues, enhancing uptake while minimizing off-target effects [32]. This concept can be transferred to plants by magnetically guiding nanoparticles to roots or vascular tissues. Challenges like unintended activation in non-target areas are mitigated through material design, such as heat- or light-triggered release mechanisms. These inorganic systems, combining durability with stimuli-responsive features, pave the way for precise, high-efficiency genome editing in crops, targeting traits like drought resilience or pathogen resistance without compromising plant viability.4. Accelerating Crop Improvement with Nanotechnology-Driven CRISPR and Speed Breeding

3.4 Integrating Speed Breeding with CRISPR/Cas9

3.4.1 Importance of Combining Rapid Generation Cycles with Precise Genome Editing

The global agricultural sector faces mounting pressure to enhance crop yields, improve resistance to environmental stresses, and address the never-ending food requirements of the globally expanded population. Traditional breeding methods, while effective, are time-consuming and often require multiple generations to achieve desirable traits. CRISPR/Cas9, a reliable and targeted tool for efficient gene alteration, has revolutionized crop advancement by enabling targeted genetic modifications. However, the effectiveness of CRISPR-based breeding is often hindered by the extended time required to develop homozygous lines through conventional breeding cycles.

Speed breeding, an advanced plant propagation strategy, addresses this limitation by significantly shortening generation intervals through controlled environmental conditions. This technique, which manipulates light, temperature, and humidity to accelerate plant growth and flowering, enables up to six generations per year for some crops, compared to one or two in traditional methods. By integrating CRISPR/Cas9 with speed breeding, researchers can not only introduce precise genetic modifications but also rapidly propagate edited lines, reducing the breeding timeline from decades to just a few years.

Recent studies have demonstrated that speed breeding can effectively complement CRISPR/Cas9 applications in crops [33]. In these studies, accelerated generation cycles enabled the rapid establishment of homozygous lines carrying desired genetic edits, thereby significantly reducing breeding timelines. Moreover, integrating precise genome editing with rapid generation advancement allows researchers to stack multiple beneficial traits in a shorter time frame, enhancing crop performance under various environmental conditions.

3.4.2 Potential of Nanotechnology in Expediting CRISPR/Cas9 Delivery for Speed Breeding Programs

While CRISPR/Cas9 is a highly efficient approach, its efficiency is often constrained by the delivery method used to introduce gene-editing components into the phyto-cellular system. Agrobacterium-mediated transformation and gene gun technology can be limited by host specificity, low efficiency, and potential off-target effects. Nanotechnology offers a groundbreaking solution to these challenges by facilitating the precise and efficient delivery of CRISPR/Cas9 components.

Among the nanoparticles, CNs, MSNs, and lipid-based nanocarriers, have emerged as effective vehicles for the transport of CRISPR/Cas9 constructs into plant cells. These nanocarriers can protect genetic material from degradation, enhance cellular uptake, and promote efficient genome editing. Furthermore, they may be tailored to direct their action towards specific tissues or organelles within the plant, improving the precision of genetic modifications.

Studies have shown that mesoporous silica nanoparticles (MSNs) can effectively deliver CRISPR/Cas9 ribonucleoproteins (RNPs) into plant cells, achieving high editing efficiency with minimal off-targets [34]. Similarly, lipid nanoparticles were successfully employed for ephemeral expression of the entire constructs in various plant species, reducing the time and complexity associated with transgenic selection [35]. These advancements indicate that nanotechnology can play a crucial role in expediting genome editing and subsequent breeding cycles.

Additionally, nanoparticle-based delivery methods can be fine-tuned to minimize cellular stress and improve transformation efficiency across a wide range of crop species. This approach is particularly valuable in recalcitrant crops, where traditional transformation methods have low success rates. By integrating nanotechnology-driven CRISPR delivery systems with speed breeding, researchers can accelerate the development of improved crop varieties with enhanced traits, ultimately contributing to global food security and sustainable agriculture.

3.5 Nanotechnology and CRISPR in Genome Editing

The integration of CRISPR/Cas9 into the vast field of nanotechnology has gained substantial recognition in numerous sectors, especially the agricultural sector, with its wide applicability extending across plant biotechnology, medical research, the food industry, and livestock enhancement. This cutting-edge technology was employed in producing biofuel, pathogen diagnostics, and infection control [36]. Furthermore, it plays a significant role in gene therapy, offering solutions for treating various diseases and cancers. It facilitates gene modification, including gene pyramiding and knockouts, to introduce foreign genes. Additionally, gene activation and suppression can be studied using this system, thereby regulating gene expression. This technology has a crucial role in fostering genetic variations, aiding breeders in crop enhancement, and enabling the successful incorporation of proteins into plant genomes. Moreover, gene silencing, a pivotal aspect of crop improvement, helps in deciphering novel functionalities of respective genes [37].

Nanotechnology has been, to date, proven to be benevolent in agriculture, particularly in formulating nano-fertilizers and nano-pesticides, which enhance crop resilience against stress and protect plants from diseases. Various NPs are being employed for genetic transformation, editing, and silencing to enhance crop yield, quality, and productivity. Additionally, CRISPR/Cas9 is recognized as an innovative and highly efficient genome-editing technique.

However, genome editing does come with certain challenges. The method is ineffective for large proteins, necessitating the use of smaller proteins for more efficient genome modifications. Moreover, the CRISPR/Cas9 system can induce unintended mutations at non-target sites. It also demonstrates limited efficiency in the homology-directed repair (HDR) pathway and is less effective against viral elements. One of the biggest challenges is the regulatory and commercial barriers associated with genetically modified plants, making widespread adoption difficult. Public unawareness regarding nano-based fertilizers and pesticides further complicates the implementation of this technology. Nevertheless, current advancements in genome editing enable agricultural researchers and breeders to develop high-yield, nutritionally rich, and resilient crops, contributing to global food security.

3.5.1 Challenges in CRISPR-Based Genome Editing

Despite CRISPR/Cas9 is a streamlined and precision-driven methodology for altering genetic architecture, several obstacles must be overcome for its effective application. Concerns such as off-target effects, safety precautions, delivery methods, and overall gene editing efficiency need to be carefully addressed [38]. The integration of nanotechnology with gene therapy can help mitigate these risks through the strategic design and fabrication of nanoparticles, including liposomal, Ag-, Au-, and polymer-based NPs. Each of them exhibits distinct advantages in the CRISPR/Cas9 delivery mechanism.

Nanoparticles offer several advantages, including the ability to enable the packaging of large molecules like DNA or protein for their targeted transport with reduced immunogenic responses, making them preferable over conventional vectors constituted with foreign viruses [39]. However, exposure to nucleases and proteases may lead to the breakdown of proteins or plasmids, as the host immune system often recognizes them as foreign invaders. Encapsulating CRISPR/Cas9 within nanoparticles can help prevent degradation and enhance its efficiency [40]. Modifications in nanoparticle designs can improve the precision of CRISPR/Cas9 at target sites, thereby reducing off-target effects.

Although there are advanced biotechnological and bioinformatics tools available, researchers still face challenges in utilizing this leading-edge technology for phyto-genetic modifications. The absence of complete genome sequences for many plant species, combined with their large genome sizes, makes gene editing a highly complex task. Furthermore, the biological mechanisms governing gene expression and environmental interactions remain largely unexplored. For improved trait discovery and functional gene characterization, further research is essential. Some plant species also exhibit low transformation efficiency, slow regeneration times, and the need for extensive field trials [41].

Public acceptance always remains a significantly contentious issue both now and in the foreseeable future while working with genome editing. Additionally, nanoparticle-mediated gene delivery systems still require further refinement. The Cas9 protein, which has a molecular weight of approximately 150 kDa and a positive charge, differs from other proteins in several ways [42]. While nanoparticles help protect the CRISPR/Cas9 plasmid from degradation, their ability to efficiently penetrate the nucleus for gene editing remains crucial. Optimizing nanoparticle properties such as particle size, charge ratio, and hydrophilicity can help mitigate immune responses triggered by CRISPR/Cas9 [43]. Furthermore, increasing nanoparticle accumulation at target sites can enhance CRISPR efficiency. However, scaling up nanoparticle production remains a challenge, necessitating improvements in reproducibility and mass production techniques.

3.5.2 Advancing CRISPR-Based Genome Editing Through Nanotechnology

Early research on delivering plant biomolecules and enabling genome editing through nanomaterials primarily relied on relatively large particles, typically exceeding 100 nm in size, which required biolistic (gene gun) methods for introduction into plant cells. However, advancements in the field have shown that certain smaller nanoparticles can enter plant cells without the need for biolistic delivery. These nanocarriers have demonstrated the capability to transport DNA and RNA payloads across a wide range of plant species and tissue types. Nanoparticles-based advancement of CRISPR-generated genome modifications follows a high throughput roadmap in order to overcome the challenges posed by the tools (Table 1).

Delivery Strategies

A major hurdle in phyto-genetic modifications lies in an efficient introduction of CRISPR components into the cell. Conventional abiotically-sourced transfection techniques, *viz.* electroporation, heat shock, and lipid- or polymer-associated cellular introduction, which are effective mostly in microbes and animals, often fail in plant-system. Nanotechnology offers a transformative approach by enabling efficient gene delivery across various plant species and tissues.

Beyond precise targeting, nanoparticle-based delivery systems enhance controlled release and protection from degradation. Some nanoparticles also exhibit fluorescence, facilitating real-time imaging of CRISPR cargo inside plant cells [45].

Table 1 Strategic framework for implementation of nanotechnology to surpass CRISPR-associated hurdles in the phyto-cellular system (adapted from [44])

Challenges		Nanotechnology-based advancements
Efficient delivery	Size, higher local charge, and lower stability of Cas9 protein	• Covalent attachment of Cas9 RNPs to nanoparticles via cleavable, enzyme-sensitive, or near-infrared light-induced chemical linkers for targeted release of RNPs within plant tissues
	Donor DNA transport	• Direct delivery of donor DNA into plant cells via NPs due to their advantage of non-restricted delivery
Tissue culture and regeneration	Drawbacks in transforming germline cells in the phyto-cellular system	• Delivery of genetic material through pollen with NPs possessing greater mechanical strength, leveraging large surface apertures • Integration of NPs with physical methods such as microinjection to enable transformation of flowers and shoot apical meristem.
Species-specific constraints	Limitations in cargo delivery across a wide range of plant species	• Acquisition of NPs via mechanical entry offers wide applicability across species regardless of their genetic makeup or signaling pathways
	Challenges involved in *in vivo* sgRNA validation	• NPs deliver sgRNA into Cas-expressing transgenic plants to enable rapid, high-throughput screening of gRNA functionality directly in plant system
	Lack of appropriate genomic context with respect to PAM sequences	• Development of NP-assisted platforms for rapid screening of alternative nucleases with diverse PAM requirements to bypass conventional restraints
Low HDR efficiency	Distress in the co-delivery of Cas protein and sgRNA into the nuclear system	• Negatively charged NPs to co-stabilize Cas–sgRNA complexes and transport modified donor DNA, aiding efficient nucleus-oriented delivery of all CRISPR elements
	Cas protein and sgRNA have mismatched synthesis and stability timelines in cells	• Recruitment of time-regulated delivery systems using NPs to introduce sgRNA at the peak expression of Cas protein through phase-specific release, sequential introduction, or sgRNA stabilization strategies

However, while nanomaterials have successfully transported DNA and proteins into plant cells, direct CRISPR-mediated genome editing via nanoparticles remains unexplored owing to specialised attributes of CRISPR reagents and the demand for precise and efficient delivery.

A major hurdle is the relatively large size and charge of CRISPR-related DNA plasmids. Researchers must optimize nanoparticle chemistry to accommodate these molecular characteristics. Additionally, Cas9 proteins and ribonucleoprotein (RNP) complexes require modifications to improve their stability and target-site release. Some promising approaches involve using cleavable chemical linkers or near-infrared light-triggered release systems.

Role of Nanoparticles in Tissue Culture and Genetic Transformation

Nanoparticle-based strategies for plant germline transformation could eliminate the need for labour-intensive tissue culture regeneration, which remains a bottleneck in plant biotechnology. A promising approach is editing pollen, allowing direct inheritance of genetic modifications via pollination. However, due to the structural toughness of pollen and its fragility post-transformation, current transformation methods have had limited success. Nanotechnology, particularly carbon nanotubes with high tensile strength, could help facilitate this process.

An alternative strategy is targeting germline cells within flowers, where both male and female reproductive cells reside. Nanoparticle-mediated CRISPR delivery in flowers has the potential to edit multiple target sites, including pollen, ovules, and embryos. Another promising target is the shoot apical meristem in order to allow the generation of fully edited plants with minimal tissue culture requirements [44].

Addressing Species-Specific Limitations

A key limitation in plant genome editing is that transformation efficiency varies across species. Factors influencing this include the inability of certain species to uptake genetic cargo, issues with in vivo sgRNA substantiation, and sequence-specific PAM site requirements of CRISPR nucleases.

Nanoparticles have been shown to facilitate cargo delivery in various plant species, including *Arabidopsis*, tobacco (*Nicotiana* sp.), maize (*Zea mays*), wheat (*Triticum aestivum*), spinach (*Spinacia oleracea*), and cotton (*Gossypium* sp.) [46]. The mechanical nature of nanoparticle penetration suggests that it may be independent of plant-specific genetics or signalling pathways. Additionally, nanoparticles could be utilized for rapid sgRNA screening, allowing high-throughput identification of effective sgRNA sequences.

Enhancing Homology-Directed Repair (HDR) Efficiency

Increasing HDR efficiency remains a challenge in plant genome editing. Nanotechnology offers several potential solutions, such as the delivery of donor DNA templates via nanoparticles to facilitate precise genome modifications. Researchers have proposed the use of negatively charged nanoparticles to stabilize CRISPR complexes and shuttle template DNA into plant nuclei, a technique that has improved HDR efficiency in human cells and could be adapted for plants. Another approach is the time-controlled delivery of CRISPR components, ensuring that donor DNA and sgRNA remain available when Cas9 reaches peak expression [47]. Nanoparticle-based platforms could help delay degradation and optimize sequential cargo release. By leveraging these nanotechnology-based advancements, CRISPR genome editing could become more precise, efficient, and applicable across a diverse range of plant species, ultimately accelerating agricultural biotechnology.

4 Case Studies in Precision Editing and In-Vivo Genome Editing

4.1 Nanoparticles for In-Vivo Genome Editing

The following Table 2 highlights recent case studies demonstrating the use of various nanoparticles for real-time in-vivo genome editing in plants. These nanocarriers enable precise and efficient delivery of genetic materials, bypassing traditional transformation barriers and offering new avenues for crop improvement and functional genomics.

As illustrated in Fig. 1, the genome editing process begins with the selection of target genes using computational tools, followed by the design of sgRNAs which are then loaded into cationic lipid-assisted nanoparticles along with Cas9 under appropriate promoters. These nanoparticle complexes can be introduced into plant cells via Agrobacterium transformation, biolistic inoculation, or protoplast transformation. The delivery enables targeted double-strand breaks (DSBs) at the genome level, which are subsequently repaired through the insertion of desired gene sequences. The resulting transformed tissues are regenerated into whole plants, which are then screened and validated using molecular techniques such as PCR, sequencing, and reporter assays. This nanoparticle-assisted genome editing approach ensures high efficiency, target specificity, and reduced off-target effects, making it a promising strategy for sustainable crop improvement.

4.2 Precision Editing for Trait-Specific Improvement

Plant breeding has been more accurate thanks to certain innovative procedures that have improved the genetic engineering stream. One of these important methods is CRISPR/Cas9. With the use of this technology, it is possible to change the genome of plants without the need to introduce foreign DNA. In other words, rather than introducing foreign genes or DNA, genome modifications are carried out by precisely altering the plant genome and delivering the necessary reagents into that particular cell [59]. Several techniques, including ribonucleoprotein complexes, viral vectors, and nanoparticle-mediated delivery, can be used for this delivery [60]. Here, we're talking about delivering CRISPR components using nanoparticles, which guarantees good protection and improved CRISPR component uptake [60]. CRISPR/Cas9 technology employs precision tools to target specific stress

Table 2 Case studies of nanoparticles for the real time in-vivo genome editing

Nanoparticles	Species	Targeted cargo	Outcomes	References
Polyethyleneimine-functionalised gold nanoclusters (PEI-AuNCs)	Whole plant (any crop can be considered)	siNanRNA	siRNA can be protected from degradation and can easily bypass the plant cell wall	[48]
Chitosan-SWNT, PEGylated CS-SWNT, Carboxylated SWNTs	Chloroplast of arugula, watercress, tobacco, spinach	Plasmid DNA	CNs selectively transfer the plasmid DNA to chloroplasts without the usage of biolistic or chemical methods	[49]
Carbon dots (CDs)	*Triticum aestivum*	Plasmid DNA	CDs are successfully used to manipulate the chloroplast and nuclear genomes to induce inflorescence	[50]
LEEP-CNTs	Any crop can be desirable	Organellar membrane coated with LEEP-CNTs	Delivery of exogeneous biomolecules to organelles for gene expression without the aid of gene gun	[49, 51]
MNPs	*Cucumis* sp. and *Gossypium* sp.	DNA	DNA conjugated MNPs transfer to pollen grains show a stable expression validated by GUS staining	[52, 53]
PAMAM dendrimer NPs	*A. thaliana* roots	Double-stranded DNA for RNA interference	Development of gene silencing leading to systemic phenotypes	[54]
Gold nanoparticles	*Triticum aestivum*	Cas9 ribonucleoproteins (RNPs)	Targeted editing for disease susceptible loci and reducing regulatory hurdles	[55]
Silver NPs	*Zea mays*, *Citrullus lanatus* (watermelon), and *Cucurbita pepo* L. (pumpkin)	Direct spray on seeds	Enhance the rate of seed germination	[56, 57]
Single walled CNTs	Any desirable plant	Direct application	Direct penetration of pollen surface apertures for transformation could circumvent regeneration from tissue culture, allowing direct production of edited offspring	[58]

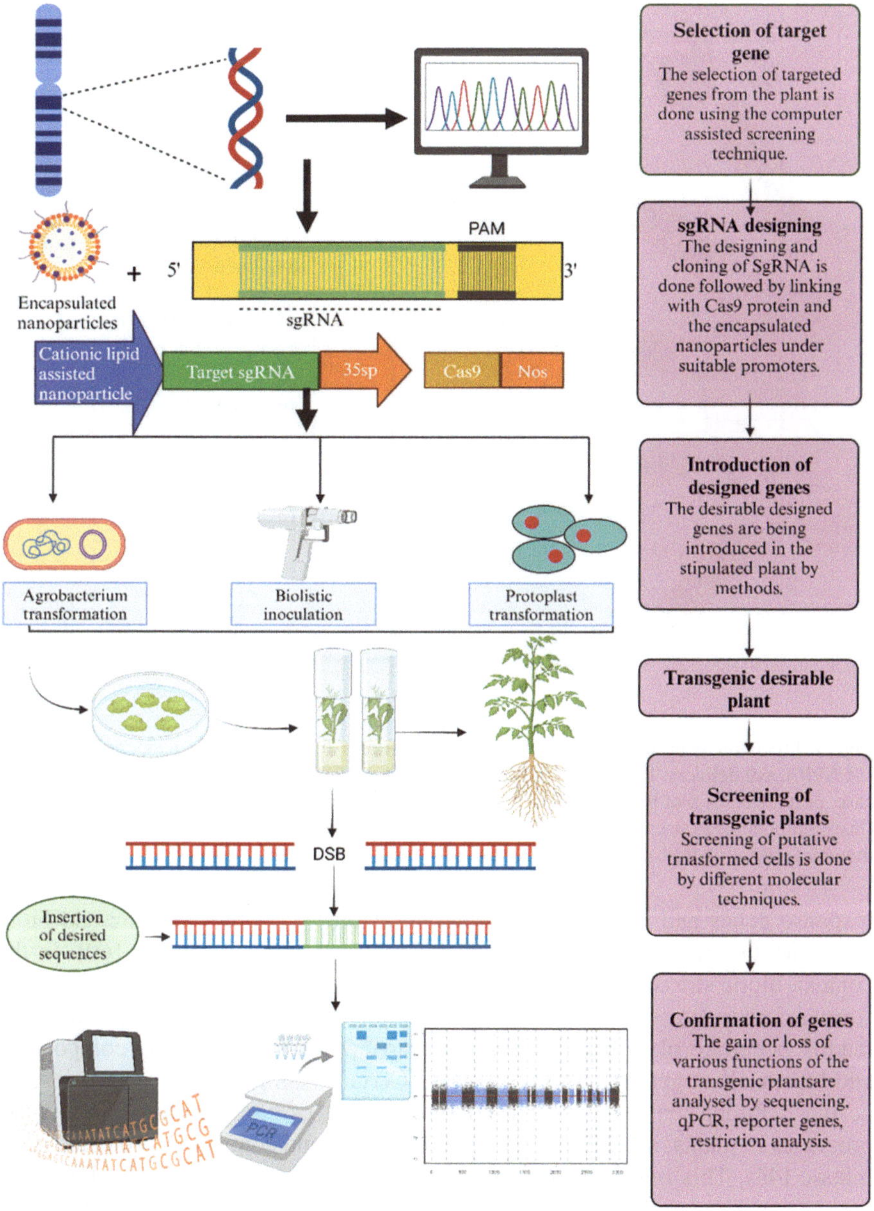

Fig. 1 Schematic representation of nanoparticle-assisted in vivo CRISPR/Cas9 genome editing workflow in plants

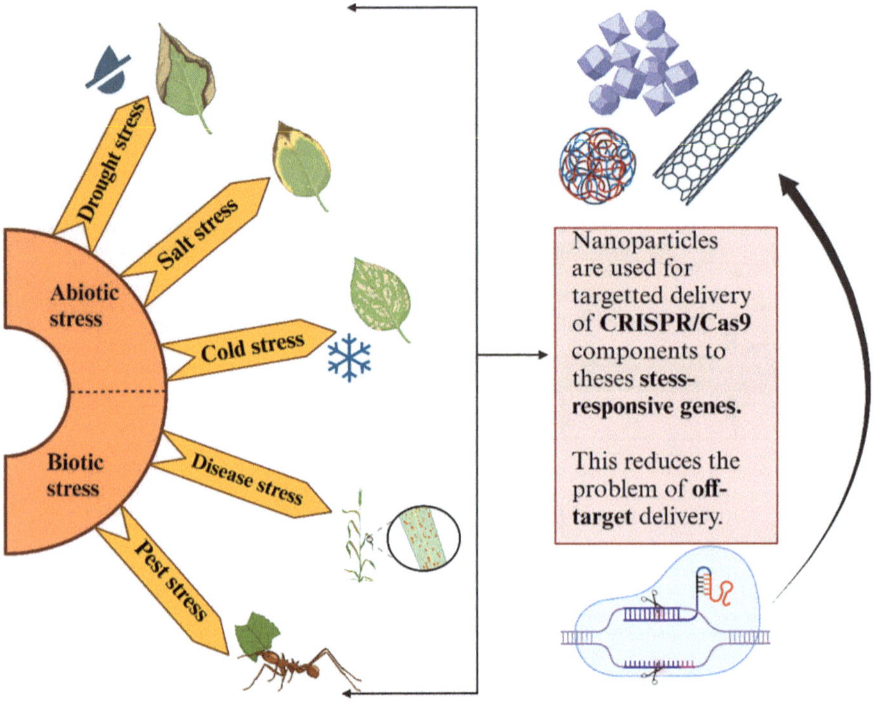

Fig. 2 Precision genome editing for trait-specific stress tolerance using nanoparticle-assisted CRISPR/Cas9 delivery. Plants face multiple abiotic (e.g., drought, salt, cold) and biotic (e.g., disease, pest) stresses that limit productivity. Nanoparticles enable the targeted delivery of CRISPR/Cas9 components to stress-responsive genes, enhancing precision and reducing off-target effects—thereby facilitating trait-specific improvement for resilient crop development

response genes and make the crop tolerant against the stress. Abiotic and biotic stress drastically impair the yield and productivity of several crops [60]. In this context, biotic stress refers to stress brought on by disease and pests, while abiotic stress refers to stress caused by drought, salinity, cold, etc. The nanoparticles can readily penetrate plant barriers, serve as carriers for CRISPR/Cas9 components, and lessen off-target effects (Fig. 2) [60]. Certain nanoparticles have intrinsic or designed fluorescent features that enable imaging of cargo delivery and release processes in plants, in addition to targeted delivery and controlled, protected cargo release [44]. This improves our understanding of cargo delivery in various plant tissues and helps to clear up any confusion surrounding the release of CRISPR/Cas9 components. This new technique makes it possible to precisely modify particular characteristics to improve vital crops and increase their quality and output (Table 3).

Table 3 Case studies in precision genome editing for yield and quality traits

Trait	Innovative technology used	Crops	Targeted genes	Application in crops	References
Yield traits improvement	CRISPR/ Cas9	Rice	*RDD1*	Raises RDD1 transcript levels, NH4 + and PO4 3− absorption, and photosynthetic activity in rice under high CO2 circumstances by inhibiting miR166-RDD1 transcript pairing	[61]
			OsHHO3	Removing this OsHHO3 repressor gene of three AMT1 genes increases rice's effectiveness in using nitrogen	[17]
		Wheat	*ARE1*	Exhibits delayed senescence, resistance to nitrogen shortage, and a markedly elevated grain yield free of growth abnormalities	[62]
			TaRPK1	Increases the productivity of wheat plants by altering the root architecture system and other yield components (such as the number of efficient tillers, grain weight, etc.)	[63]
		Barley	*GW2* and *GW2.1*	Disabling GW2 can boost yield because it has a detrimental impact on grain weight and size. Additionally, GW2.1 gene mutagenesis leads to the production of longer grains with higher protein content	[64]
Quality traits improvement		Potato	*gbssI*	Remove amylose from the production process and only produce amylopectin starch by disrupting this gene. Their freeze-thaw stability and possibly increased ethanol yields for industrial uses make them beneficial	[65]
		Rice	*SD1, Wx*	These genes can be edited to create stable lines with semi-dwarf and glutinous traits, which improve grain quality and yield	[66]
		Maize	Zmbadh2a, Zmbadh2b	When these genes were altered, 2AP was produced, which causes maize to exhibit its scent	[67]

5 Challenges and Limitations of Nanotechnology-Driven CRISPR/Cas9

5.1 Toxicity and Biocompatibility Concerns

Nanoparticles can affect physiology in a variety of ways and have long-term effects on surroundings, but they are also incredibly helpful and efficient as vectors for delivering the genetic tool CRISPR/Cas9 to the target cells. Different kinds of nanoparticles have varying effects on plant cells. In a study of ryegrass, the presence of metal oxides such as ZnO has been shown to reduce biomass, decrease root tips, and cause the collapse of root epidermal and cortical cells [68]. Nano-phytotoxicity is the term used to describe the phytotoxicity brought on by nanoparticles. There are various ways that nanoparticles might be phytotoxic. The uptake of various nanoparticles can result in oxidative stress, blockage of the plant's vascular system, and structural damage to the DNA of the plant. Additionally, the nanoparticles have a detrimental effect on plant reproductive growth [5]. Because carbon nanotubes (CNT) resemble asbestos, which can cause cancer, they are harmful to human health. CNTs have also demonstrated cytotoxicity in vitro, which includes cellular growth suppression and necrosis induction [68]. A portion of the applied nanoparticles may reach the land if the dosage is higher than what is recommended, which could contaminate the soil's surface or groundwater [68]. These have a negative impact on the environment, particularly if the nanoparticles are composed of metals, metal oxides, and other non-biodegradable substances.

There are certain actions that can help lessen the harmful effects that nanoparticles have on plants and the environment. Applying the appropriate dosage of nanoparticles is one of them. Excessive application of nanoparticles may result in their deposition and oxidative damage in cells. To avoid employing biologically hazardous nanoparticles, it is imperative that we have accurate information on the nanomaterials being employed [69]. Another issue is the nanoparticles' biodegradability. Therefore, biodegradable and environmentally friendly nanoparticles such as lipid nanoparticles (LNPs), cationic polymer nanoparticles, hybrid nanoparticles, etc., should be used in place of non-biodegradable or bio-incompatible nanoparticles of metals and metal oxides [1]. LLNs are lipid-like, biodegradable nanoparticles that effectively carry CRISPR/Cas9 payloads without posing a biological hazard. Since they are biodegradable, cationic polymers like polyethyleneimine and chitosan are frequently employed to transport various nucleic acids, including mRNA and plasmid DNA. LLNs and other CRISPR/Cas9 payloads can also be delivered with it. Compared to LLNs, it has a wider range of chemicals and is more adaptable to various structural designs for delivery [1]. We can guarantee the biosafety, biocompatibility, and biodegradability of the nanomaterials utilized for CRISPR/Cas9 delivery by taking these steps.

5.2 Delivery Efficiency in Field Conditions

CRISPR/Cas9, which is powered by nanotechnology, has shown signs of success in laboratory tests. It has been noted that certain nanoparticles may theoretically have enormous promise for efficient CRISPR/Cas9 delivery and for promoting the genetic transformation of plants [5]. Despite nanoparticles' extraordinary capabilities, there are a few important issues that still need to be fixed [5]. The first is scale-up issues, which refer to the challenge of producing these nanoparticles in large quantities without sacrificing their efficacy and quality. Poor encapsulation is another issue; in this case, the system using CRISPR/Cas9 may not be "held" or "carried" by the nanoparticles, which would decrease its effectiveness. Another issue is bioprotection, where the nanoparticles must shield CRISPR/Cas9 from being broken down by various enzymes within the plant cells. Because of the poor transfection rate, this technique might not be as effective as anticipated. This reduces the success rate of gene editing since only a tiny percentage of plant cells actually receive and employ the CRISPR/Cas9 system [5]. These are some inevitable issues that must be resolved if we hope to implement this innovative technology in actual agriculture.

The primary issue under field circumstances is the various environmental stresses, such as variations in temperature and pH, humidity, etc. The nanoparticles are affected by these environmental stresses and become unstable (Fig. 3). Numerous studies have been conducted to observe how temperature affects the nanoparticles [70]. Long-term exposure to high temperatures has been shown to induce agglomeration and the inactivation of numerous biological components, which leads to instability and decreased functionality of nanoparticles [70]. Different environmental conditions, such as ionic strength, pH, moisture content, etc., affect how the

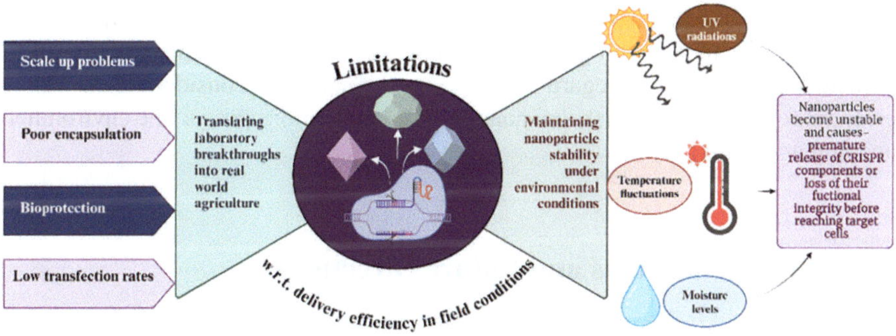

Fig. 3 Limitations affecting the delivery efficiency of CRISPR/Cas9 components in field conditions using nanoparticles. Effective translation of laboratory-based genome editing strategies into real-world agriculture faces key challenges such as poor encapsulation, bioprotection, low transfection rates, and scalability issues. Additionally, environmental factors like UV radiation, temperature fluctuations, and varying moisture levels compromise nanoparticle stability, leading to premature release or degradation of CRISPR/Cas9 components before reaching the target cells

nanoparticles behave within plant systems. These elements have an impact on the nanoparticles' rate of dissolution and transformation, which lowers their efficacy and plant bioavailability [71]. The CRISPR components may be prematurely released and lose their functional integrity before they reach the target cells when nanoparticles become unstable. To address these issues and guarantee the system's efficacy in real-world scenarios, more research on this innovative technology is required.

5.3 Regulatory and Ethical Considerations

Like other genetically modified crops, CRISPR/Cas9 raises ethical questions because it is a gene editing technology. Therefore, while GMO laws vary by nation, they would be comparable for this new technology. For the commercial use of CRISPR/Cas9 technology, several regulations may need to be revised or modified [72]. In addition, there is a considerable danger associated with CRISPR/Cas9 technology since it might result in off-target mutations and transfer features to unrelated animals, which could lead to ecological imbalance [73]. Some regulatory regulations that will support the long-term use of this innovation must be implemented in order to guarantee that these kinds of issues stay at a minimum.

The complete image of a field cannot be determined from the few applications in the lab because nanobiotechnology is a future-focused strategy that is still in its early stages of discovery and study. This raises a number of concerns in people's minds, such as whether the technology is actually successful on a bigger scale if it is safe to use, and whether it poses any threats to the environment and ecological system. Public acceptability of this technology may suffer as a result of these new concerns, which could erode public confidence in it.

It is imperative that appropriate rules and regulations be developed for the safe application of nanotechnology in agriculture, as this is a new subject that requires particular attention with regard to ethical and regulatory considerations. This is essential for reducing the risk that nanoparticle applications pose to the environment and human health [71].

6 Recommendations and Future Directions

6.1 Enhancing Research and Development

Nanoparticle-mediated genome editing promotes the extensive potential in crop improvement by overcoming technical and regulatory challenges. In the due course of time, NP-mediated genome editing procedures have taken an advanced turn by improving their efficiency. Tailored nanocarriers like carbon nanotubes, and gold NPs with tuneable surface chemistry, increase cell wall penetration and promote a smooth transfer of nucleotide sequences. It promotes specific organelle targeting (i.e., nuclei, mitochondria, chloroplast) by several mechanisms. Surfactants lower the surface tension of the nanoparticles which helps in adhesion to the organelles. However, this may denature the cargo proteins attached to the nanoparticles [74]. Thus, these denaturing properties of the nanoparticles upon specific macro biomolecules must be reduced. Advancement in this research domain will increase the conductance and uptake rates of the nanoparticles. NPs have proven achievement in provide excellent physiochemical stability and biocompatibility (mainly gold nanoparticles). CRISPR-Cas9 protein can be delivered by gold-coated nanomaterials, lipid NPs, etc. Further studies provide the confidence of utilization of more NPs like AuNP, and AuNC for the delivery of targeted genes. Recent studies depicted that in-vivo somatic genetic modification coating of RNA is possible by nanocapsule which comprises multiple layers [75]. With the amalgamation of nano-sized particles CRISPR/Cas9 has gained the advantage of feasibility, transient expression, and less immunogenicity to maintain safety and efficiency.

Moreover, the synergy of AI and Machine learning has revolutionized the usage of NPs. Many genome editing tools such as TIGER CNN, DeepCRISPR have been found to analyze any off-target risks, Cas9 variant compatibility, undesirable mutations, etc. as shown in Table 4. It provides the experimental validation of the changes caused due to the delivery of the NP-coated genes. (e.g., resistance to diseases, resistance to drought, etc.). This has improved the concept of NPs in the field of agriculture. With the advancement of technology, the phytotoxicity effect of the NPs can be discarded eventually.

6.2 Field Ready Applications

NPs in agriculture are a revolutionary approach for farmers, offering practical advantages to long-term challenges in crop improvement. By leveraging this technology, the farmers are highly benefited and will take modern turns in the near

Table 4 AI models and their respective roles CRISPR/Cas9 system

AI models	Major roles
TIGER CNN, deep CRISPR	The efficiency of gRNA prediction by incorporating insertions, deletions, etc
Support vector machines (SVM), gradient boosted trees (GBT)	Optimization of gRNA selection and the efficiency of gRNA
CRISPRon, DeepHF	Enhancement of the precision of gRNA design for specific Cas9 variants
piCRISPR	Off-target prediction algorithms, featuring chromatin accessibility and DNA methylation patterns
SkipGuide	Promotes CRISPR/Cas9-based exon skipping methods, precise gRNA selection
Plant predictor	Forecasts morphological impacts of genetic manipulation, trait-specific modifications

future. Researchers are able to deliver the CRISPR/Cas9 components precisely to the farmers. Thus, they benefit from improved crop morphology and reduction in production barriers. Tissue culture has been a tedious and complicated procedure at the field level. By introgression of the nanoparticles, regeneration through tissue culture could be bypassed. Direct editing of shoot apical meristems advances the production of editing progenies. The application of the NPs is not complicated and can be readily sprayed by the NPs to avoid regulatory hurdles during delivery by producing non-GMO edits (Fig. 4). Farmers are reluctant to use GMO crops and thus came up to use this method in crops like rice, and soybeans to avoid GMO restrictions. But the farmers are favorably benefitted since there is less usage of pesticides, and minimization of fertilizer uses thus lowering expenses. There are fewer harmful environmental impacts due to the reduction in the usage of chemicals. However, the gap to large-scale production will require further validation, regulation approvals and interaction with the researchers for the increment of the crop production. Till now there are very few food crops involved in this procedure. Transformation and development of nanotechnology could provide further advancement in agriculture in the near future days.

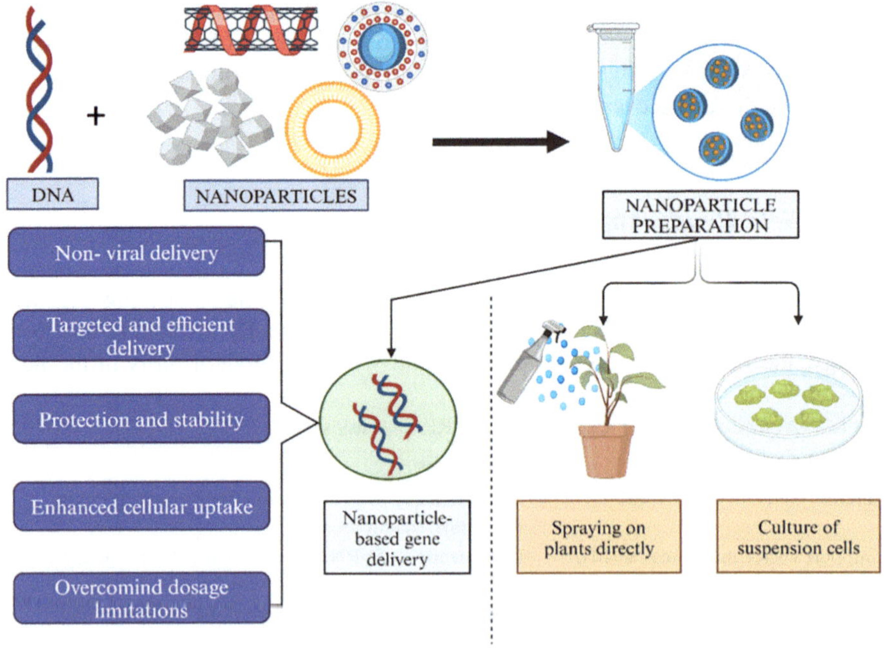

Fig. 4 The prepared nanoparticles can be applied via foliar spraying for field-grown plants or integrated into plant cell suspension cultures, paving the way for scalable and field-ready genome editing strategies in agriculture

6.3 *Bridging Conventional and Advanced Technologies*

While the advancement, scalability and environmental safety studies are ongoing in research, innovations like layer-by-layer nanoparticle assembly aim to stagnant Cas9 release for higher efficiency. Nanoparticle-mediated genome editing has immense potential but it also faces challenges like inefficient delivery, scalability issues, phytotoxic effects, etc. Integration of nanoparticles with conventional methods will pave a strategic turn to overcome the limitations faced by NPs in the field level of work and research. Conventional approaches like the use of Agrobacterium, biolistics, and cross-breeding could not be completely discarded since the potential of nanoparticles over crops is still emerging and it is an ongoing research talk even to date. Agrobacterium subsequently introduces the homology-directed repair in the targeted genes. The NPs along with Agrobacterium will edit accurately while maintaining high transformation rates in the crops. The release of NPs should be optimally dosed due to their high reactivity and less stability. The transformation procedure might be easy but the optimal dosage application is a major concern even now. A low concentration of NPs will provide weak interactions with the cargo and will cause pre-destined delivery. Also, high concentrations can cause oxidative stress within the cell. Thus, NPs if applied by the use of biolistics might reduce this crucial limitation. Virus-mediated genome editing played a pivotal role in triggering the crops to accelerate their growth and development, which was not found in NP-mediated delivery. Thus, the combination of nanoparticles with the virus vectors could be a smarter way to promote early flowering, early growth, or faster development.

Nanoparticles still hold a novel tool to probe the fundamentals of plants by introducing perturbations in an efficient manner. With the advent of next-generation omics technologies, the effect of NPs on plant transcriptome and metabolic pathways could be captured by observing the molecular alterations. However, there is a comprehensive lack of data in this arena. Navigating the regulatory authorities' environmental and safety reviews may be difficult due to the knowledge gaps regarding ecotoxicity, exposure pathways, lack of high throughput delivery, lifecycle impact assessments of nanoparticles, etc. Additional political, legal, consumer acceptance, economic, business, and ethical issues surrounding the usage of nanoparticles accompany this [76]. In order to harness more on this topic, not only we need more research exposure but also social and environmental acceptance all over the world. Although this field is still in its infancy, the accelerating change in the landscape of the usage of nanoparticles may indicate the fateful adoption in agriculture on an abundant scale.

7 Conclusion

The integration of nanotechnology with CRISPR/Cas9 genome editing represents a transformative breakthrough in plant biotechnology. While traditional delivery methods have laid the groundwork for genetic engineering, their limitations—including tissue damage, species-specific inefficiencies, and inconsistent delivery—highlight the need for advanced alternatives. Nanotechnology emerges as a next-generation solution, enabling precise, minimally invasive, and highly efficient delivery of genome-editing components. This synergy not only overcomes prior technical constraints but also unlocks unprecedented opportunities for crop improvement.

By integrating CRISPR/Cas9 with nanotechnology, researchers are actively developing resilient crop varieties designed to enhance both qualitative and quantitative traits. These innovations have demonstrated significant improvements in agronomic performance, including elevated nutritional profiles, optimized photosynthetic efficiency, and enhanced stress tolerance. Furthermore, the precise targeting afforded by these systems allows for functional gene characterization and the optimization of gene sequences, driving crop productivity and quality toward sustainable agricultural goals. Such advancements underpin novel agroecological strategies aimed at addressing food security challenges amid a rapidly expanding global population.

Despite this promise, critical challenges remain. Environmental stability, regulatory hurdles, and concerns over nanoparticle toxicity require resolution through rigorous interdisciplinary research and global collaboration. Additionally, biocompatibility, scalability for field applications, and public acceptance are pivotal factors in transitioning nanotechnology-driven CRISPR systems from laboratory breakthroughs to real-world agricultural solutions. Addressing these barriers will demand transparent dialogue between scientists, policymakers, and communities to ensure ethical and equitable adoption.

In conclusion, the strategic fusion of nanotechnology and CRISPR/Cas9 delivery holds immense potential to redefine modern agriculture. By advancing crop resilience, yield, and sustainability, this technology paves the way for a future where agriculture meets the demands of a growing population while safeguarding ecological balance. Realizing this vision will depend on sustained innovation, responsible governance, and a commitment to harnessing science for global good.

References

1. Duan L, Ouyang K, Xu X, Xu L, Wen C, Zhou X, Liang Y (2021) Nanoparticle delivery of CRISPR/Cas9 for genome editing. Front Genet 12:673286
2. Song G, Jia M, Chen K, Kong X, Khattak B, Xie C, Mao L (2016) CRISPR/Cas9: a powerful tool for crop genome editing. Crop J 4(2):75–82
3. Rautela I, Uniyal P, Thapliyal P, Chauhan N, Sinha VB, Sharma MD (2021) An extensive review to facilitate understanding of CRISPR technology as a gene editing possibility for enhanced therapeutic applications. Gene 785:145615
4. Shen S, Loh TJ, Shen H, Zheng X, Shen H (2017) CRISPR as a strong gene editing tool. BMB Rep 50(1):20
5. Ahmar S, Mahmood T, Fiaz S, Mora-Poblete F, Shafique MS, Chattha MS, Jung KH (2021) Advantage of nanotechnology-based genome editing system and its application in crop improvement. Front Plant Sci 12:663849

6. Cunningham FJ, Goh NS, Demirer GS, Matos JL, Landry MP (2018) Nanoparticle-mediated delivery towards advancing plant genetic engineering. Trends Biotechnol 36(9):882–897
7. Givens BE, Naguib YW, Geary SM, Devor EJ, Salem AK (2018) Nanoparticle-based delivery of CRISPR/Cas9 genome-editing therapeutics. AAPS J 20:1–22
8. Cho HY, Yoo M, Pongkulapa T, Rabie H, Muotri AR, Yin PT, Lee KB (2024) Magnetic nanoparticle-assisted non-viral CRISPR-Cas9 for enhanced genome editing to treat Rett syndrome. Adv Sci 11(24):2306432
9. Khalid MF, Iqbal Khan R, Jawaid MZ, Shafqat W, Hussain S, Ahmed T, Alina Marc R (2022) Nanoparticles: the plant saviour under abiotic stresses. Nano 12(21):3915
10. Pramanik B, Sar P, Bharti R, Gupta RK, Purkayastha S, Sinha S, Mitra D (2023) Multifactorial role of nanoparticles in alleviating environmental stresses for sustainable crop production and protection. Plant Physiol Biochem 201:107831
11. Mazurov D, Ramadan L, Kruglova N (2023) Packaging and uncoating of CRISPR/Cas ribonucleoproteins for efficient gene editing with viral and non-viral extracellular nanoparticles. Viruses 15(3):690
12. Rai GK, Khanday DM, Kumar P, Magotra I, Choudhary SM, Kosser R, Pandey S (2023) Enhancing crop resilience to drought stress through CRISPR-Cas9 genome editing. Plants 12(12):2306
13. Wang C, Zhou Y, Wang Y, Jiao P, Liu S, Guan S, Ma Y (2025) CRISPR-Cas9-mediated editing of ZmPL1 gene improves tolerance to drought stress in maize. GM Crops Food 16(1):1–16
14. Borisjuk N, Kishchenko O, Eliby S, Schramm C, Anderson P, Jatayev S, Shavrukov Y (2019) Genetic modification for wheat improvement: from transgenesis to genome editing. Biomed Res Int 2019(1):6216304
15. Komarova T, Ilina I, Taliansky M, Ershova N (2023) Nanoplatforms for the delivery of nucleic acids into plant cells. Int J Mol Sci 24(23):16665
16. Martínez-Carmona M, Lozano D, Colilla M, Vallet-Regí M (2018) Lectin-conjugated pH-responsive mesoporous silica nanoparticles for targeted bone cancer treatment. Acta Biomater 65:393–404
17. Liu J, Chang J, Jiang Y, Meng X, Sun T, Mao L, Wang M (2019) Fast and efficient CRISPR/Cas9 genome editing in vivo enabled by bioreducible lipid and messenger RNA nanoparticles. Adv Mater 31(33):1902575
18. Wu H, Li Z (2022) Nano-enabled agriculture: how do nanoparticles cross barriers in plants? Plant Commun 3(6):100346
19. Iqbal Z, Rehman K, Xia J, Shabbir M, Zaman M, Liang Y, Duan L (2023) Biomaterial-assisted targeted and controlled delivery of CRISPR/Cas9 for precise gene editing. Biomater Sci 11(11):3762–3783
20. Ryu N, Kim MA, Park D, Lee B, Kim YR, Kim KH, Kim UK (2018) Effective PEI-mediated delivery of CRISPR-Cas9 complex for targeted gene therapy. Nanomedicine 14(7):2095–2102
21. Chen CK, Huang PK, Law WC, Chu CH, Chen NT, Lo LW (2020) Biodegradable polymers for gene-delivery applications. Int J Nanomedicine 15:2131–2150
22. Rui Y, Wilson DR, Choi J, Varanasi M, Sanders K, Karlsson J, Green JJ (2019) Carboxylated branched poly (β-amino ester) nanoparticles enable robust cytosolic protein delivery and CRISPR-Cas9 gene editing. Sci Adv 5(12):eaay3255
23. Ma P, Wang Q, Luo X, Mao L, Wang Z, Ye E, Wu YL (2023) Recent advances in stimuli-responsive polymeric carriers for controllable CRISPR/Cas9 gene editing system delivery. Biomater Sci 11(15):5078–5094
24. Li L, Yang Z, Zhu S, He L, Fan W, Tang W, Chen X (2019) A RAtionally designed semiconducting polymer brush for NIR-II imaging-guided light-triggered remote control of CRISPR/Cas9 genome editing. Adv Mater 31(21):1901187
25. Lee CS, Kim TW, Kang Y, Ju Y, Ryu J, Kong H, Kim TH (2022) Targeted drug delivery nanocarriers based on hyaluronic acid-decorated dendrimer encapsulating gold nanoparticles for ovarian cancer therapy. Mater Today Chem 26:101083

26. Farsani AM, Mokhtari N, Nooraei S, Bahrulolum H, Akbari A, Farsani ZM, Ahmadian G (2024) Lipid nanoparticles: the game-changer in CRISPR-Cas9 genome editing. Heliyon 10(2):e24606
27. Hurtado Y, Franco CA, Riazi M, Cortés FB (2020) Improving the stability of nitrogen foams using silica nanoparticles coated with polyethylene glycol. J Mol Liq 300:112256
28. Senger K, Haley B (2022) Non-viral alternatives to delivery of CRISPR-Cas gene editing components. In: CRISPR-/Cas9 based genome editing for treating genetic disorders and diseases. Routledge, London, pp 183–223
29. Tao Y, Yi K, Hu H, Shao D, Li M (2021) Coassembly of nucleus-targeting gold nanoclusters with CRISPR/Cas9 for simultaneous bioimaging and therapeutic genome editing. J Mater Chem B 9(1):94–100
30. Tao W, Cheng X, Sun D, Guo Y, Wang N, Ruan J, Zhao M (2022) Synthesis of multi-branched au nanocomposites with distinct plasmon resonance in NIR-II window and controlled CRISPR-Cas9 delivery for synergistic gene-photothermal therapy. Biomaterials 287:121621
31. Zhou S, Li Y, Wu Q, Gong C (2024) Nanotechnology-based CRISPR/Cas9 delivery system for genome editing in cancer treatment. MedComm 3(1):e70
32. Zhu H, Zhang L, Tong S, Lee CM, Deshmukh H, Bao G (2019) Spatial control of in vivo CRISPR–Cas9 genome editing via nanomagnets. Nat Biomed Eng 3(2):126–136
33. Watson A, Ghosh S, Williams MJ, Cuddy WS, Simmonds J, Rey MD, Hickey LT (2018) Speed breeding is a powerful tool to accelerate crop research and breeding. Nat Plants 4(1):23–29
34. LaBauve AE, Saada EA, Jones IK, Mosesso R, Noureddine A, Techel J, Negrete OA (2023) Lipid-coated mesoporous silica nanoparticles for anti-viral applications via delivery of CRISPR-Cas9 ribonucleoproteins. Sci Rep 13(1):6873
35. Mahmoud LM, Kaur P, Stanton D, Grosser JW, Dutt M (2022) A cationic lipid mediated CRISPR/Cas9 technique for the production of stable genome edited citrus plants. Plant Methods 18(1):33
36. Zafar K, Sedeek KE, Rao GS, Khan MZ, Amin I, Kamel R, Mahfouz MM (2020) Genome editing technologies for rice improvement: progress, prospects, and safety concerns. Front Genome Ed 2:5
37. Qi LS, Larson MH, Gilbert LA, Doudna JA, Weissman JS, Arkin AP, Lim WA (2013) Repurposing CRISPR as an RNA-guided platform for sequence-specific control of gene expression. Cell 152(5):1173–1183
38. Mout R, Ray M, Yesilbag Tonga G, Lee YW, Tay T, Sasaki K, Rotello VM (2017) Direct cytosolic delivery of CRISPR/Cas9-ribonucleoprotein for efficient gene editing. ACS Nano 11(3):2452–2458
39. Wang HX, Li M, Lee CM, Chakraborty S, Kim HW, Bao G, Leong KW (2017) CRISPR/Cas9-based genome editing for disease modeling and therapy: challenges and opportunities for nonviral delivery. Chem Rev 117(15):9874–9906
40. Hendel A, Bak RO, Clark JT, Kennedy AB, Ryan DE, Roy S, Porteus MH (2015) Chemically modified guide RNAs enhance CRISPR-Cas genome editing in human primary cells. Nat Biotechnol 33(9):985–989
41. Zaidi SSEA, Mahas A, Vanderschuren H, Mahfouz MM (2020) Engineering crops of the future: CRISPR approaches to develop climate-resilient and disease-resistant plants. Genome Biol 21(1):289
42. Jinek M, Jiang F, Taylor DW, Sternberg SH, Kaya E, Ma E, Doudna JA (2014) Structures of Cas9 endonucleases reveal RNA-mediated conformational activation. Science 343(6176):1247997
43. Deng H, Huang W, Zhang Z (2019) Nanotechnology based CRISPR/Cas9 system delivery for genome editing: Progress and prospect. Nano Res 12:2437–2450
44. Demirer GS, Silva TN, Jackson CT, Thomas JB, Ehrhardt W, Rhee D, Landry MP (2021) Nanotechnology to advance CRISPR–Cas genetic engineering of plants. Nat Nanotechnol 16(3):243–250
45. Hu P, An J, Faulkner MM, Wu H, Li Z, Tian X, Giraldo JP (2020) Nanoparticle charge and size control foliar delivery efficiency to plant cells and organelles. ACS Nano 14(7):7970–7986

46. Santana I, Wu H, Hu P, Giraldo JP (2020) Targeted delivery of nanomaterials with chemical cargoes in plants enabled by a biorecognition motif. Nat Commun 11(1):2045
47. Barrangou R, Doudna JA (2016) Applications of CRISPR technologies in research and beyond. Nat Biotechnol 34(9):933–941
48. Zhang H, Cao Y, Xu D, Goh NS, Demirer GS, Cestellos-Blanco S, Yang P (2021) Gold-nanocluster-mediated delivery of siRNA to intact plant cells for efficient gene knockdown. Nano Lett 21(13):5859–5866
49. Kwak SY, Lew TTS, Sweeney CJ, Koman VB, Wong MH, Bohmert-Tatarev K, Strano MS (2019) Chloroplast-selective gene delivery and expression in planta using chitosan-complexed single-walled carbon nanotube carriers. Nat Nanotechnol 14(5):447–455
50. Doyle C, Higginbottom K, Swift TA, Winfield M, Bellas C, Benito-Alifonso D, Whitney HM (2019) A simple method for spray-on gene editing in planta. BioRxiv, 805036
51. Law SSY, Liou G, Nagai Y, Giménez-Dejoz J, Tateishi A, Tsuchiya K, Numata K (2022) Polymer-coated carbon nanotube hybrids with functional peptides for gene delivery into plant mitochondria. Nat Commun 13(1):2417
52. Park CW, Choi JY, Son YJ, Kim DH, Li H, Liang W, Kim YJ (2024) Magnetofected pollen gene delivery system could generate genetically modified Cucumis sativus. Hortic Res 11(8):uhae179
53. Zhang H, Demirer GS, Zhang H, Ye T, Goh NS, Aditham AJ, Landry MP (2019) DNA nanostructures coordinate gene silencing in mature plants. Proc Natl Acad Sci 116(15):7543–7548
54. Jiang L, Ding L, He B, Shen J, Xu Z, Yin M, Zhang X (2014) Systemic gene silencing in plants triggered by fluorescent nanoparticle-delivered double-stranded RNA. Nanoscale 6(17):9965–9969
55. Imai R, Hamada H, Liu Y, Linghu Q, Kumagai Y, Nagira Y, Taoka N (2020) In planta particle bombardment (iPB): a new method for plant transformation and genome editing. Plant Biotechnol 37(2):171–176
56. Acharya P, Jayaprakasha GK, Crosby KM, Jifon JL, Patil BS (2020) Nanoparticle-mediated seed priming improves germination, growth, yield, and quality of watermelons (Citrullus lanatus) at multi-locations in Texas. Sci Rep 10(1):5037
57. Wang F, Li K, Shi Z (2021) Phosphorus fertilization and mycorrhizal colonization change silver nanoparticle impacts on maize. Ecotoxicology 30(1):118–129
58. Lew TTS, Koman VB, Gordiichuk P, Park M, Strano MS (2020) The emergence of plant nanobionics and living plants as technology. Adv Mater Technol 5:1900657. https://doi.org/10.1002/admt.201900657
59. Nerkar G, Devarumath S, Purankar M, Kumar A, Valarmathi R, Devarumath R, Appunu C (2022) Advances in crop breeding through precision genome editing. Front Genet 13:880195
60. Chen F, Chen L, Yan Z, Xu J, Feng L, He N, Liu C (2024) Recent advances of CRISPR-based genome editing for enhancing staple crops. Front Plant Sci 15:1478398
61. Iwamoto M (2022) In-frame editing of transcription factor gene RDD1 to suppress miR166 recognition influences nutrient uptake, photosynthesis, and grain quality in rice. Sci Rep 12(1):10795
62. Zhang J, Zhang H, Li S, Li J, Yan L, Xia L (2021) Increasing yield potential through manipulating of an ARE1 ortholog related to nitrogen use efficiency in wheat by CRISPR/Cas9. J Integr Plant Biol 63(9):1649–1663
63. Rahim AA, Uzair M, Rehman N, Fiaz S, Attia KA, Abushady AM, Khan MR (2024) CRISPR/Cas9 mediated TaRPK1 root architecture gene mutagenesis confers enhanced wheat yield. J King Saud Univ-Sci 36(2):103063
64. Kis A, Polgári D, Dalmadi Á, Ahmad I, Rakszegi M, Sági L, Havelda Z (2024) Targeted mutations in the GW2. 1 gene modulate grain traits and induce yield loss in barley. Plant Sci 340:111968
65. Toinga-Villafuerte S, Vales MI, Awika JM, Rathore KS (2022) CRISPR/Cas9-mediated mutagenesis of the granule-bound starch synthase gene in the potato variety Yukon gold to obtain amylose-free starch in tubers. Int J Mol Sci 23(9):4640

66. Wang Q, Gao H, Liu K, Wang H, Zhang F, Wei L, Yuan H (2024) CRISPR/Cas9-mediated enhancement of semi-dwarf glutinous traits in elite Xiangdaowan rice (Oryza sativa L.): targeting SD1 and Wx genes for yield and quality improvement. Front Plant Sci 15:1333191
67. Wang Y, Liu X, Zheng X, Wang W, Yin X, Liu H, Wang F (2021) Creation of aromatic maize by CRISPR/Cas. J Integr Plant Biol 63(9):1664–1670
68. Ray PC, Yu H, Fu PP (2009) Toxicity and environmental risks of nanomaterials: challenges and future needs. J Environ Sci Health C 27(1):1–35
69. Vats S, Kumawat S, Brar J, Kaur S, Yadav K, Magar SG, Deshmukh R (2022) Opportunity and challenges for nanotechnology application for genome editing in plants. Plant Nano Biol 1:100001
70. Balusamy SR, Joshi AS, Perumalsamy H, Mijakovic I, Singh P (2023) Advancing sustainable agriculture: a critical review of smart and eco-friendly nanomaterial applications. J Nanobiotechnol 21(1):372
71. Cao Y, Turk K, Bibi N, Ghafoor A, Ahmed N, Azmat M, Ahanger MA (2025) Nanoparticles as catalysts of agricultural revolution: enhancing crop tolerance to abiotic stress: a review. Front Plant Sci 15:1510482
72. Gan WC, Ling AP (2022) CRISPR/Cas9 in plant biotechnology: applications and challenges. Biotechnologia 103(1):81
73. Shinwari ZK, Tanveer F, Khalil AT (2018) Ethical issues regarding CRISPR mediated genome editing. Curr Issues Mol Biol 26(1):103–110
74. Räsch A, Hunsche M, Mail M, Burkhardt J, Noga G, Pariyar S (2018) Agricultural adjuvants may impair leaf transpiration and photosynthetic activity. Plant Physiol Biochem 132:229–237
75. Chen K, Wang Y, Zhang R, Zhang H, Gao C (2019) CRISPR/Cas genome editing and precision plant breeding in agriculture. Annu Rev Plant Biol 70(1):667–697
76. Yang T, Duncan TV (2021) Challenges and potential solutions for nanosensors intended for use with foods. Nat Nanotechnol 16(3):251–265

Carbon Nanoparticles in Plant Genetic Engineering

Rooma Waqar, Javed Iqbal, Banzeer Ahsan Abbasi, Abdul Samad Mumtaz, Akhtar Munir, Muhammad Kaleem, Muhammad Anas, Muhammad Muzamil Sattar, Asifa Farooqi, Shumaila Ijaz, Zakir Ullah, Ghulam Murtaza, Rashid Iqbal, Sana Basharat, and Muhammad Waseem

Abstract Nanoparticles are aggregates that range in size from 1 to 100 nm and can be made either naturally or intentionally. They are useful in a variety of domains, including genetic engineering and medicine. Plant cells can be altered by carbon-based nanoparticles, specifically carbon nanodots (CNDs), which provide a biocompatible and economical option for conventional genetic modification (GM) approaches. *Arabidopsis thaliana* and *Triticum aestivum* were employed as model species in this study to assess the effectiveness of CNDs in gene delivery. Many transformation techniques, including foliar sprays, vacuum infiltration, and seed treatments, were evaluated. Multiple foliar sprays produced the highest transformation efficiency; seed treatments were ineffective. This chapter explores the increasing

Rooma Waqar and Javed Iqbal contributed equally with all other contributors.

R. Waqar
Department of Botany, Faculty of Sciences, Mohi-ud-Din Islamic University, Nerian Sharif, Azad Kashmir, Pakistan

J. Iqbal
Department of Botany, Bacha Khan University, Charsadda, Khyber Pakhtunkhwa, Pakistan

B. A. Abbasi
Department of Botany, Rawalpindi Women University, Rawalpindi, Pakistan

A. S. Mumtaz · M. Kaleem · M. Anas · M. M. Sattar · Z. Ullah
Department of Plant Sciences, Faculty of Biological Sciences, Quaid-i-Azam University Islamabad, Islamabad, Pakistan
e-mail: asmumtaz@qau.edu.pk; mkaleem@bs.qau.edu.pk; muhammadanas@bs.qau.edu.pk; zakirullah@bs.qau.edu.pk

A. Munir
Department of Chemistry, Quaid-i-Azam University, Islamabad, Islamabad, Pakistan
e-mail: akhtarmunir@qau.edu.pk

A. Farooqi
Department of Environmental Sciences, Faculty of Biological Sciences, Quaid-i-Azam University Islamabad, Islamabad, Pakistan

J.-T. Chen (ed.), *Nano-Delivering for Plant Genetic Engineering*, Smart Nanomaterials Technology, https://doi.org/10.1007/978-981-95-4165-2_3

significance of carbon-based nanomaterials as non-viral vectors for CRISPR delivery by addressing problems with viral vectors and demonstrating innovative design strategies to optimize interactions between CRISPR components and nanomaterials. The results highlight how carbon-based nanomaterials can improve genetic engineering and gene delivery, opening the door to safer and more effective gene-editing techniques.

Keywords Nanoparticles · Carbon based nanotubes · Nanodots · CRISPR · Genetic modification

1 Introduction

Nanoparticles, defined as tiny aggregates between 1 and 100 nm, have been present throughout human history as both naturally occurring volcanic eruptions and waste products from industrial processes [1]. Although nanotechnology has been of interest since the 1970s [2], it has only been used extensively in the past 10 years. The goal of nanoscale science is to improve production and society. Between 1997 and 2003, global investment in nanotechnology research and development

S. Ijaz
Department of Burn and Plastic Surgery, Shenzhen Institute of Translational Medicine, The First Affiliated Hospital of Shenzhen University, School of Biomedical Engineering, Shenzhen University Medical School, Shenzhen University, Shenzhen, China

G. Murtaza
School of Agriculture, Yunnan University, Kunming, Yunnan, China

R. Iqbal
Department of Agronomy, Faculty of Agriculture and Environment, The Islamia University of Bahawalpur, Bahawalpur, Pakistan

Department of Life Sciences, Western Caspian University, Baku, Azerbaijan
e-mail: rashid.iqbal@iub.edu.pk

S. Basharat
School of Breeding and Multiplication (Sanya Institute of Breeding and Multiplication), Sanya, China

College of Tropical Agriculture and Forestry, Hainan University, Sanya, China

M. Waseem (✉)
School of Breeding and Multiplication (Sanya Institute of Breeding and Multiplication), Sanya, China

College of Tropical Agriculture and Forestry, Hainan University, Sanya, China

Fang Zhiyuan Academician Team Innovation Center, Haikou, Hainan, China

increased sixfold. Nanoparticles are becoming more specialized and varied due to their recent engineering for usage in a variety of industries, including medicine and energy [1]. Numerous nanomaterials have found significant in vitro applications in medicine, and attempts to modify their properties have resulted in products that are extremely soluble, non-toxic, and remarkably versatile [3]. As evidenced by the way the mitotic spindle is arranged inside a cell utilizing nanotubules, these components suggest that nanomaterials may be used to augment or replace biological processes [4]. Companies involved in energy storage and construction have made investments in the creation of nanoparticles, proving their use across a range of sectors [5]. Other industries that have profited from nanotechnology's versatility include environmental services like pest control and water and air purification [5].

As a result of these developments, researchers are now considering agriculture as the next logical application of nanomaterials that increase productivity and solve important issues in the field. Carbon-based nanoparticles are the most appropriate for use in agriculture due to their low toxicity and increased biocompatibility, even though there are many various types of nanoparticles, including ones made from heavy metals [3, 6].

2 Methods for Nanomaterial Preparation

One method for producing the nanomaterial physically, chemically, or biologically is top-down or bottom-up synthesis [6]. The top-down approach usually involves a physical process that breaks down the bulk materials into small molecules, which are subsequently further converted into NPs as needed. By creating tiny nanostructures, which are often accomplished by chemical processes, nanomaterials can be produced directly using the bottom-up technique. Green synthesis methods that produce NPs from plant extracts or other microorganisms have recently become a viable technique. Nowadays, common heavy metal ions including Ag, Cu, Co, and Ni are used to create mental oxide nanoparticles [7]. By using plant extracts instead of more conventional physical and chemical processes, NPs can be produced at a lower cost and with less toxicity [8]. As highlighted by a large body of literature, the size, shape, and physical, chemical, and biological properties of nanoparticles (NPs) directly affect how effective transportation is [9]. Therefore, in order to change the properties of NPs, it is essential to understand the design and synthesis process [10]. The unique synthesis strategy may be the most effective platform for producing NPs suitable for intracellular biomolecule delivery due to its simplicity, efficiency, and lack of pollution. Table 1 presents a summary of the different methods of delivery.

Table 1 A summary of the various methods of delivery

Transformation method	Species	Targeted tissues	Delivery type	Limitations
Nanoparticles	Monocots, dicotyledons	Leaf, protoplast	DNA	Ineffectiveness
Agrobacterium	Monocots, dicotyledon	Root, apical meristem, shoot	DNA	Limitations of the host
Electroporation	Green algae,	Leaf, meristem, protoplast, pollen grain	DNA, miRNA siRNA	Non-specific tissue damage
Biolistic particle	Monocots, dicotyledons	Callus, apical meristem, shoot	DNA, miRNA siRNA	Low expression levels, poor integration efficiency, and cell damage
PEG	Monocots, dicotyledons	Protoplast	DNA, miRNA siRNA	Low regeneration and cytotoxicity

3 NPs Transportation in Plants

Different heavy metal ions which include Cu, Ag, Co, and Ni are used to form metal oxide nanoparticles [7]. NPs are generated easily by exploiting extracts of plants rather than more standard chemical and chemical processes [8]. The physical, biological, and chemical properties of nanoparticles (NPs) affect transportation effectiveness, as underlined by a large body of literature [11]. To change the properties of NPs, it is necessary to know the design and process of synthesis [10]. The innovative synthesis approach's simplicity and lack of pollution make it possibly the best platform for producing nanoparticles appropriate for intracellular biomolecule delivery. For example, it was investigated how stomatal opening or closing affected the absorption of ZnO NPs in plants (wheat leaves). The plant species, particle size, particle concentrations, surface charge, and duration of exposure are some of the variables that restrict the transport and uptake of nanoparticles in plants [12]. Furthermore, plants absorb NPs by their roots, and the differing traits (morphological) of these organelles cause NPs can be transported through different barriers [13]. Leaves of plants have a supportive cuticle covering their epidermis, which is made up of an impregnated, soluble wax-covered insoluble epidermal membrane, NPs that are larger than 10 nm enter into the cell by stomata. The activity of stomata varies by species and is affected by different environmental factors, so the absorption of NPs must take these factors into account [14]. For instance, it was studied how stomatal closing or opening influences the absorption of ZnO NP in leaves [15]. The results showed that stomatal diameter decreases in tandem with detectable zinc concentrations in the cytoplasm and chloroplasts falling by 8.3% and 33.2%, respectively [16]. In addition to the root's epidermal cells, the Casparian strip in the endodermis also limits the flow of NPs when it is absorbed by the roots. NPs can pass through the endodermis thanks to sites and passage cells; in other research, the suspension of NPs is purposefully delivered into the plant cells through the injection feeding through the leaf cuticle [13]. Although techniques work well for directing foreign materials into cells, their poor throughput and tedious nature have prevented their widespread use [17]. Current research has shown that certain NPs can drill holes in the cuticle itself, thus injection-free penetration is a more widely used technique [7]. The concentration (detectable) of zinc in the cytoplasm and chloroplasts are decreased by 8.3%, and 33.2% respectively, accompanied by a decrease in stomatal diameter [16]. Large NPs are limited by the endodermis' Casparian strip and the root's epidermal cells when they are absorbed through the root; they can only pass through affected (damaged) roots, and passage cells. In some studies, the NP's suspension is purposefully transferred into plant cells by injection [13]. While these methods are effective means of channeling foreign matter into cells, they are laborious and have low throughput, which is why they are not used widely [17].

Although injection-free penetration is used, new research has shown that some NPs can really puncture the cuticle [7]. NPs first enter through the cuticle and then the mesophyll cells in order to enter the circulatory system and travel great distances. In order for NPs to reach the vascular system in plants, they must first go through the cuticle and then the mesophyll cells [18]. The cell membrane and cell wall are the main and secondary obstacles to NP internalization in mesophyll cells of plants, respectively [19]. Diffusion, the plasmodesmata, endocytosis, or chemical and physical degradation are the ways that nanoparticles pass through cell membranes and cell walls [20]. The barriers to NPs in mesophyll cells of plants are the internalization of the cell wall and cell membrane, respectively [19]. Diffusion, endocytosis, plasmodesmata, and chemical degradation are among the ways that nanoparticles may pass through cell walls and membranes [20]. The Change of NP osmotic buffer's composition changes the strength of the cell wall and tension, which in turn modify the cell wall's local water potential and facilitates the internalization of NPs [17]. However, in order for some DNA-coated nanoparticles, for example, gold microparticles, which penetrate cells through tissue, biomolecule delivery methods like gene guns are essential [21].

Additionally, for NPs that are delivered without the assistance of an external source, a basic knowledge of the interactions between different membranes and NPs within the plant is also essential [22]. Various research suggested a variety of techniques to move nanoparticles that cross the lipid bilayer. These mechanisms are still not well understood, though, especially in view of recent assertions that efficient cargo delivery is possible without nanoparticle cells internalizing [23]. Furthermore, the charge state and shape of nanoparticles can also affect how they enter the mesophyll tissues of plants. Rod-shaped gold nanoparticles are more easily absorbed by plant (*Arabidopsis*) leaves than spherical nanoparticles of the same size. It is also necessary to consider electrostatic attraction because when plant cell walls are anionic-charged, the adsorption of cationic-charged nanoparticles is stronger than that of anionic-charged nanoparticles [24].

NPs are transported into the mesophyll cells via the apoplast, a system of cell walls, and intercellular spaces, that surrounds the protoplast of plant cells, and the symplast, a totality of interrelated protoplasm that connects by plasmodesmata which passes through the cell wall. In general, fluids that pass through the apoplastic channels face less resistance. It has been hypothesized that the apoplast can be a translocation channel of NPs in plants, as recent research shows that the plant will restrict the flow of fluid in the apoplast once NPs are introduced [18].

In order to know the mechanism of NP cell trafficking, some studies use mathematical models [21]. Stiffness and the NP aspect ratio are essential variables in the transport of nanomaterials across cell walls [7]. Additionally, NP transport is affected by the chemical and physical properties of nanoparticles, and the morphological differences across plant species also affect the translocation of NP [25]. Therefore, it is important to clear the mechanisms of NP transport in living plant cells in order to enhance NP-based agri-technologies and plant genetic engineering applications [21, 26]. The overall procedure for preparing and delivering nanoparticles (NPs) for plant transformation is shown in Fig. 1. Applications of transformation mediated by nanoparticles in different species of plants are presented in Table 2.

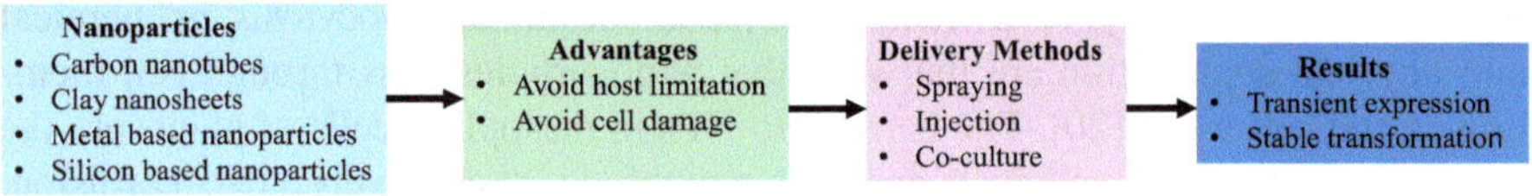

Fig. 1 Characterization and the overall procedure for preparing and delivering nanoparticles (NPs) for plant transformation

Table 2 Application of transformation mediated by nanoparticles in a variety of plant species

Target plant species and tissues	Nanoparticle type	Stable transient expression/ transformation	Delivery method	References
N. benthamiana	Carbon nanotube	Transient expression	Injection	[27]
E. sativa	Carbon nanotube	Transient expression	Injection	[27]
T. aestivum	Carbon nanotube	Transient expression	Injection	[27]
N. benthamiana	Layered dihydroxide clay nanosheets	Transient expression	Spraying	[28]
A. thaliana	Organically functionalized MSNs	Transient expression	Passive diffusion	[29]
S. lycopersicum	Mesoporous silica nanoparticles (MSNs)	Transient expression	Spraying and injection	[30]
Z. mays	Cationic fluorescence nanoparticle	Stable transformation	Biolistic	[31]
C. sativa	PEI-au@SiO2	Transient expression	Infiltration	[32]

4 Carbon-Based Nanoparticles

Carbon nanotubes and carbon nanodots are the two principal kinds of carbon-based nanoparticles that are now available. Initially, carbon nanotubes (CNTs) were invented for employment in electronics as conductors in small devices [2] and electrocatalysts [33]. Scientists are trying to make carbon nanotubes (CNTs) with one or more walls that can pass through the cell walls of a variety of plant species, showing that they can be significantly altered [34] showed how CNTs could be applied to a variety of crop species, describing how they might either promote or prevent root elongation and how this seemed to vary by species.

The fact that the CNTs lingered outside of the plants and formed a coating on the root surface caused worries regarding biocompatibility, cell penetration, and possible harm to the physiology of the roots. Water-soluble CNTs (wsCNTs) were produced and used to crop species in order to solve these difficulties [35]. This was not only very effective, but they also found that the nanotubes grew the roots and shoots of crop species by traveling through the xylem [35]. Because they appear to have a major influence on the solubility and toxicity of nanoparticles, surface modifications and coatings have been a major emphasis in the development of future materials for use in both agriculture and medicine [3]. Because carbon nanodots (CNTs) still have problems penetrating cell walls and cell membranes, researchers have begun developing highly soluble, non-penetrating nanomaterials that are highly biocompatible, have low cytotoxicity, and can penetrate cell membranes without causing significant harm to the cells [36].

Because of their photoluminescence and low cost of manufacture, CNDs have drawn the attention of researchers ever since they were discovered in 2004 [37]. With one technique that uses microwave irradiation and only takes 3 min, the creation of CNDs has become easier [38]. The topic of genetic modification (GM) is one of the many areas of interest for nanoparticles [39]. Nanoparticles are thought to be able to solve a number of the current issues with GM, and because of their high biocompatibility and low cytotoxicity, carbon-based nanodots are promising candidates for the next big technique [39].

4.1 Carbon Nanotubes

The two main types of carbon nanotubes (CNTs) are single-walled carbon nanotubes (SWCNTs) and multi-walled carbon nanotubes (MWCNTs). MWCNTs consist of multiple SWCNTs, which have a diameter of 220 nm, whereas SWCNTs have a graphene layer which is rolled into a cylindrical nanostructure that has a diameter of 0.7–3.0 nm [40]. SWCNTs are considered organelle membrane-passing gene-delivery vehicles in plant genetic engineering. Their exceptional tensile strength, high aspect ratio, high surface area-to-volume ratio, and biocompatibility are just a few advantages [40]. Due to their high surface area to volume ratio,

SWCNTs are loaded with large amounts of DNA that transfer into plant cells [41]. Additionally, SWCNTs inhibit the DNA from degradation in mammalian systems and deliver plasmid DNA and small interfering RNA (siRNA) into nonmodel and model plant cells for the silencing of genes [42]. Lastly, the cargo–nanoparticle complexes formed by SWCNTs and their loaded cargoes have strong intrinsic near-infrared (nIR) fluorescence, which allows them to be tracked in real-time [40].

Catharanthus roseus protoplasts have been used to test MWCNTs' capacity for penetration of cells because the use of carbon nanotubes in plant cells is limited due to the cell wall. Although preparation of protoplast takes time, the findings showed that MWCNTs can pass through membranes of protoplast and that short MWCNTs (less than 100 nm) target the plastids, vacuoles, and nuclei [43]. In order to enable cup-stacked carbon nanotubes (CSCNTs) that pass through the cell wall, [44] and are mixed with cellulase. SWCNTs have the intriguing ability to enter protoplasts and mesophyll cell walls without requiring energy [45], expanding the use of this technique in plant genetic engineering.

In previous studies on CNTs in plants, such as *G. hirsutum*, *Eruca sativa*, *Nicotiana benthamiana*, and *Triticum aestivum*, DNA–CNT conjugates not only permit DNA that enters the plant but also restricts the degradation of polynucleotide [42]. Since this success in delivering DNA into plant cells through CNTs, researchers have studied the verifying different payloads and activities of genes. siRNA is essential for the degradation of mRNA during the RNA interference (RNAi) process. While agrobacterium-mediated transformation is a successful method for the delivery of siRNA, it is limited by plant species therefore, a new technique is needed to solve this problem. A SWCNT-mediated siRNA delivery method has been reported to have high efficiency, and high gene silencing efficiency at the mRNA level [42]. SiRNA may mute multiple genes at once, and autofluorescence can be utilized to track the passive delivery. Gene-gun transformation and agrobacterium-mediated transformation are examples of traditional plant genetic engineering approaches that transfer DNA into the nucleus, where it integrates with a genome that generates gene migration from transgenic plants. This limitation is overcome by genetic engineering of the chloroplasts and mitochondria since it is inherited from the mother. Certain plants, such as Nasturtium officinale (watercress), Spinacia oleracea (spinach), and Nicotiana tabacum (tobacco), require specific DNA transport to their chloroplasts. Kwak [46] created chitosan-complexed SWCNTs. Excellent effectiveness, no discernible toxicity, and no transgene integration into the nuclear genome were all demonstrated by the studies of the CNT approaches mentioned above. These investigations can be conducted without the use of machinery and are appropriate for both model and non-model plant species. However, using CNTs is not without its challenges. The potential uses of carbon nanotubes (CNTs) in plants may be limited by their inert nature, entangled bundle formation, and low water dispersibility. Furthermore, it's not entirely apparent how CNTs and plants interact. Furthermore, little is known about CNTs' possible toxicity to the environment. Investigations on how CNTs affect the growth and development of plants are shown in Table 3.

Table 3 Investigations on how CNTs affect the growth and development of plants

CNT type	Plant species	Growth conditions	Effects	References
CNT	Tomato, radish, onion, turnip	Laboratory and greenhouse	Enhanced germination of seeds and development of seedlings	[26]
CNT	Date palm	In vitro	Increased leaf count and shoot length	[47]
CNT	Tomato	MS media	Improved plant development and seed germination	[48]
SWCNT	Tomato	In vitro	Increased height, weight, and chlorophyll content of roots and shoots	[49]
SWCNT	Maize	In vitro	Enhanced development of seminal roots	[50]
SWCNT	Salvia	Laboratory and greenhouse	Increased the rate at which seeds germinated	[51]
MWCNT	Red spinach	Hydroponic culture	No impact on the growth of plants	[52]
MWCNT	Tobacco	MS media	Increased dry biomass and growth in plants	[53]
MWCNT	Mustard	In vitro	Larger roots and more rapid plant growth	[54, 55]

4.2 Graphene

A single layer of carbon with partially filled sp2-orbits makes up graphene, that is found in both above and below the plane. It contains a huge surface area, great elastic strength, optical clarity, and thermal conductivity. It is employed as a semiconductor. The interaction between electrons and graphene results in the formation of new quasi-particles. The two most crucial elements in this class are quantum dots and graphene nanoribbons. Ballistic transport is demonstrated by these particles, that allow electrons to traverse up to micrometres without scattering [56]. Graphene oxide and its derivatives are used in sol-gel chemistry and found to a suitable in the fabrication of biocompatible nanocomposites.

5 *Arabidopsis thaliana* Genetic Manipulation with Carbon Nanodots

Arabidopsis thaliana has been the focus of extensive study in the field of plant science since it is considered a model plant. Its ease of growth, short generation time, and ability to be maintained in a lab setting made it suitable for a range of investigations, including genetic studies. Since its gene function and gene regulatory networks are well established, this plant species—the first to have its entire genome sequenced—has been essential to genetic modification (GM) research [57]. The GM of Arabidopsis now uses several methods. Zinc-finger nucleases (ZFNs) and transcription activator-like effector nucleases (TALENs) are used to cleave non-specific DNA domains. They have been shown to be sequence-specific, and offer a broad GM range [58]. The transformation which is based on ZFN and TALEN has some issues. ZFNs are costly because they need specialized construction and design. This approach's success rate is not known, and ZFNs' significant cytotoxicity has limited their application in cells (living) [59]. TALENs can target any sequence in an organism because they are simpler to make. Despite having a greater success rate than ZFNs, they are not more effective, and despite their lower levels of cytotoxicity, there are still concerns regarding their misuse in living cells. Agrobacterium-mediated transformation and the CRISPR/Cas system are current methods. Agrobacterium-mediated methods, that uses a strain of Agrobacterium tumefaciens, frequently used in many GM experiments due to their reliability, and ease of application [59].

It has been studied to be effective at gene activation, repression, and gene editing in plants (Arabidopsis) and is thought to be not more complicated than ZFNs and TALENs [59]. The technique has been used to genetically change the germline in addition to successfully achieving transient transformation; T2 plants have been demonstrated to be 22% ± homogeneous for the alteration of genes in the plants (parent) [60]. However, CRISPR/Cas depends on the uptake of the cell since it needs the Cas protein to enter cells. The cell walls' existences in plants inhibit the system's ability to switch species. In medical research, nanotechnology has proved itself to be highly versatile and has been employed to tackle these issues. Early studies focused on toxicity to determine the effect of nanoparticles on plants.

According to [61], some nanoparticles, particularly those containing heavy metals, have been shown to promote plant growth and development, while others have been shown to prevent it. Heavy metals are known to be harmful, thus research was done to find out how carbon-based nanoparticles affected plants. It was demonstrated that injecting medium to high concentrations of carbon nanotubes into protoplasts could result in cell death, hydrogen peroxide accumulation, and cell aggregation. Low dosages were demonstrated to either increase growth [45] or have no impact [62]. *Arabidopsis thaliana* does not require a new transformation method because it can be transformed successfully with existing techniques, like floral dipping [63].

6 The Genetic Modification of Triticum Aestivum Using Carbon Nanodots

One of the main staple grains for animal and human consumption, wheat accounts for more than 17% of cultivatable land worldwide [64]. With the world's population expected to reach 10 billion people by 2050 and the human population growing at a rate of about 100 million per year, sustainable food production is the main focus [65]. Traditional breeding methods are not able to meet the projected targets, so the agricultural industry must look to alternative methods to increase crop yield, and production, and reduce environmental impact [66]. GM presents the ability which the boost performance of agronomics through customizing species which increases yield, vitamins, stress tolerance, and resistance against various diseases, which mention a few of the areas of research being performed [64, 66]. Despite socio-political worries concerning GM the land committed to GM crops has expanded by approximately 90 million hectares in under 10 years, and benefits have already been shown with a reduction in the use of pesticides as plants are designed to be resistant [65]. PEG-mediated gene transfer, electroporation, microbeam laser punctures, and DNA application to pollen are some of the GM techniques now used on wheat plants [67]. The two main techniques for genetically modifying wheat are agrobacterium-mediated techniques and particle bombardment, both of which can achieve up to 10% efficiency. Larger amounts of DNA can be transferred using these less expensive, simpler techniques [64, 67].

CRISPR/Cas9 has been used on wheat more recently with varying degrees of success. The technique was previously used on *Arabidopsis*, but research has now demonstrated that it can also change wheat embryos and protoplasts [68]. Although this method has demonstrated promise in that it can target a wide variety of sites and facilitates engineering and multiplexing, its low efficiency is caused by guiding RNA targeting [69]. In an effort to improve efficacy and efficiency, the CRISPR/Cas system has undergone modifications.

Due to the complexity of the wheat genome, genetically modifying wheat is very difficult, and not much progress has been made [66]. The most agronomically significant genes are not always triplicated, and the genome is a massive hexaploid, comprising about 16,000 Mb and a huge number of transposable elements and repetitive sequences [65]. Due to the intricacy of genetic work done to embryos, and protoplasts, that improves the success rates of first transformation and leads to the necessity to renew the plant fully. Figure 2 represents the genetic modification techniques for *Triticum aestivum*.

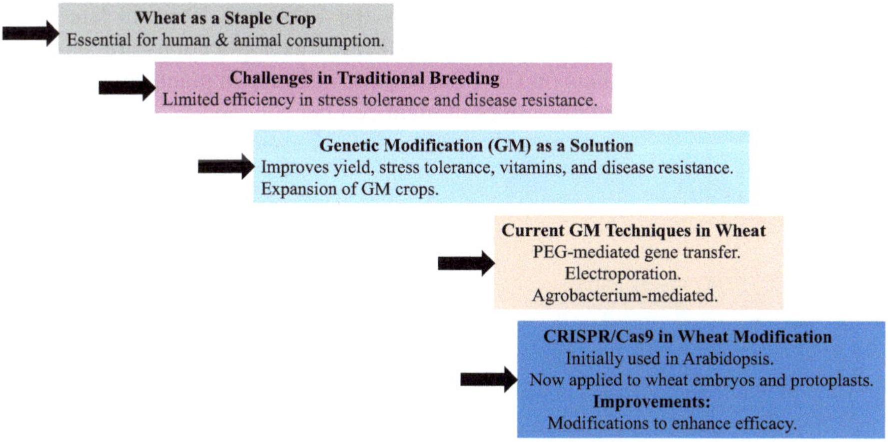

Fig. 2 Genetic modification techniques for *Triticum aestivum*: methods and applications

7 Nanomaterials Based on Carbon for CRISPR/Cas Delivery

Gene therapy is a state-of-the-art medical approach that uses therapeutic nucleic acids as potent agents to treat a range of ailments. Within the intricate realm of gene therapy, two fundamental approaches to nucleic acid delivery stand out. The ex vivo procedure involves introducing therapeutic nucleic acids into particular cells, growing them outside the body to multiply and then reinfused into the patient. However, with the in vivo method, nucleic acids are encapsulated in nanocarriers to enable direct administration into the patient's cells [70, 71]. One revolutionary approach in the pursuit of increased efficiency is the conjugation of genes onto delivery vectors. A significant consideration is that this intentional fusion increases cell transfection rates, which is important because naked DNA plasmids are prone to destruction due to their very short half-lives. An effective delivery vector must therefore meet a number of criteria, including ensuring patient safety, preventing enzymatic attacks on the sensitive nucleic acids, successfully entering the target cell's cytoplasm or nucleus in a stimuli-responsive manner, and offering a financially viable solution [72]. The landscape of gene therapy delivery methods is shaped by this dynamic interaction between state-of-the-art research and strict practical requirements [73].

A significant breakthrough in molecular tools, the creation of CRISPR/Cas systems has revolutionized gene editing and genetic engineering. The remarkable uses of these systems in synthetic biology, bioremediation, gene therapy, agriculture, pharmaceutical development, and diagnostics have garnered a lot of interest. They are essential for gene modification in a wide range of animals, from sophisticated beings like humans to microbes like bacteria, due to their superiority in accuracy, efficacy, and adaptability to different tar circumstances [74, 75]. The trajectory of the development of this technology has been nothing short of extraordinary since the first ground-breaking demonstration of programmable genome editing using CRISPR/Cas in 2012. This rapid progress has led to a significant accomplishment: the discovery of the first ex vivo cell treatment based on CRISPR, a ground-breaking discovery that will be approved by the FDA just 10 years later [76]. Advancements in plant gene therapy and CRISPR/Cas applications are shown in Fig. 3. Different experiments on CNT-based gene delivery in plants are presented in Table 4. Limitations and objectives of plant cell genetic modification are shown in Fig. 4.

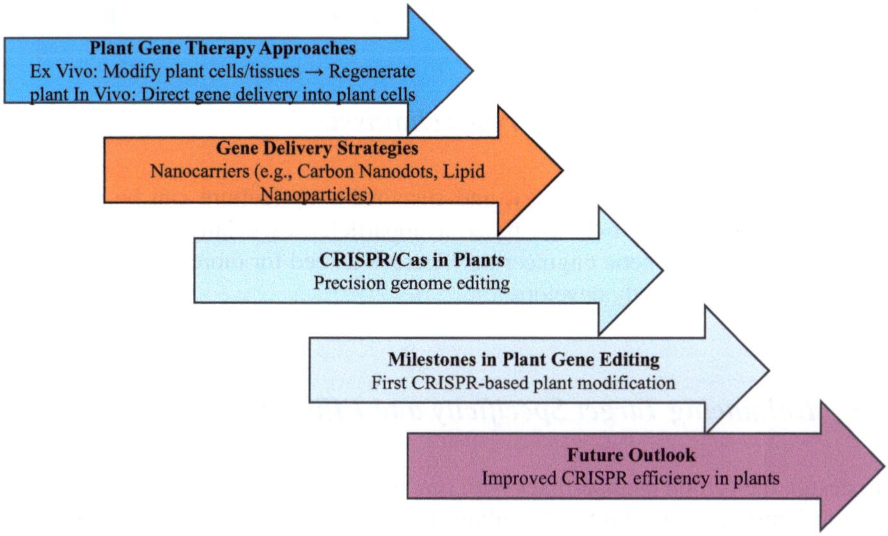

Fig. 3 Advancements in plant gene therapy and CRISPR/Cas applications

Table 4 An overview of several experiments on CNT-based gene delivery in plants

CNT type	Plant species	Target organ	Delivery method	Cargo	References
SWCNT	*N. tobacum*	Root	Passive	GFP plasmid	[55]
SWCNT	*N. tabacum*	Cell wall and membrane	Fluidic-phase endocytosis	SWCNT/ FITC	[77]
SWCNT	*N. benthamiana*	Leaf	Passive	ssRNA	[78]
SWCNT	*N. tabacum*	Cell wall and membrane	Fluidic-phase endocytosis	SWCNT/ FITC	[79]
CNTs	*N. tabacum*	Protoplast and cell wall	Passive	DNA	[80]
CNTs	*N. tabacum*	Protoplast, leaf	Passive	YFP plasmid	[81]
CNTs	Arugula, wheat, cotton	Leaves	Passive	Plasmid	[27]
CNTs	*N. tabacum*	Protoplast, leaf	Infiltration with syringe	GFP, Cy3 DNA	[82]

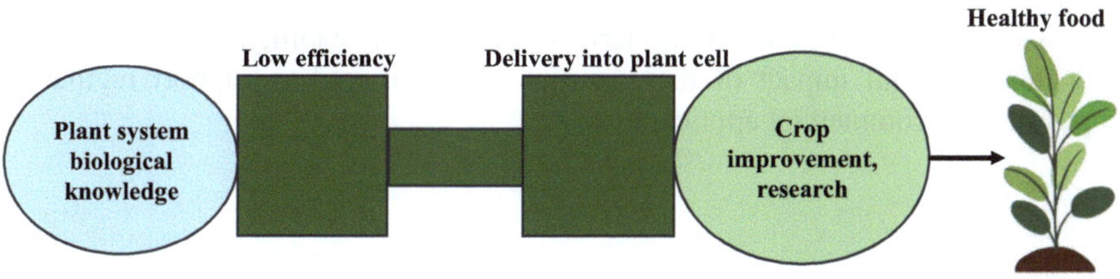

Fig. 4 Limitations and objectives of plant cell genetic modification. The capacity to create and supply a range of goods, from food to fuel, is restricted by the two main obstacles

8 Future Perspectives of Carbon Nanoparticles in Plant Genetic Engineering

8.1 Optimization of Delivery Mechanisms

Crop improvement, biotechnology, and sustainable agriculture can be beneficial from the incorporation of carbon-based nanoparticles, especially carbon nanodots (CNDs), into plant genetic engineering. There is a need for more investigation and work as this field of study develops.

8.2 Enhancing Target Specificity and Efficiency

To design the carbon nanoparticles with functionalized surfaces, it would be necessary to improve their affinity of binding to certain plant cells, or genetic material. Improving their efficiency of gene transport and reducing the likelihood of off-target effects, could make them a more precise tool for plant genetic alteration.

8.3 Integration with CRISPR and Other Gene-Editing Technologies

Carbon-based nanoparticles have shown potential as carriers of the CRISPR-Cas system in plant genome editing. In order to produce more precise and efficient genetic modifications, future research should look at how variations in the size, charge, and composition of nanoparticles can improve the stability and controlled release of CRISPR components in plant cells.

8.4 Long-Term Stability and Biosafety Assessments

Safe implementation of carbon nanoparticles in agriculture will rely on knowledge of their interactions within plant systems, possible risks of bioaccumulation, and effects on soil microbiomes. Long-term studies on the stability, biodegradability, and environmental impact of these particles are required to support regulatory approvals and commercial applications.

8.5 *Scaling Up for Agricultural Applications*

Even while carbon nanoparticles have shown promise in laboratory settings, scaling up their production and use for widespread agricultural applications is fraught with financial and technological challenges. Their widespread application in crop biotechnology will depend on the creation of cost-effective synthesis methods and sustainable nanoparticle formulations.

8.6 *Expanding Applications to Diverse Crop Species*

Most studies on gene transfer using carbon nanoparticles have focused on model plants such as Arabidopsis thaliana and staple crops such as Triticum aestivum. Future research should expand to include other economically significant crops, such as fruits, vegetables, and bioenergy plants, in order to maximize the potential benefits of this.

9 Conclusion

This study highlights the potential of carbon-based nanoparticles, namely carbon nanodots (CNDs), as effective non-viral vectors for the genetic modification of plants. CNDs offer a potential alternative to more conventional methods like particle bombardment and Agrobacterium-mediated transformation, which often have disadvantages like cytotoxicity and high cost, because of their efficacy and biocompatibility. The experiment's findings showed that many foliar sprays had the best transformation efficiency, while seed treatments were largely ineffective and other methods, such as vacuum infiltration and dipping, only somewhat efficient. Furthermore, gene editing using the Spo11-Cas9 plasmid demonstrated successful changes, demonstrating the potential of CNDs in CRISPR-based genome engineering.

These findings contribute to the increasing interest in employing nanotechnology for gene editing and genetic transformation in crops like Triticum aestivum, where improved genetic modification techniques can increase production, disease resistance, and climate adaptability. To enhance nanoparticle design, optimize delivery systems, and guarantee the stability and long-term impacts of transformations facilitated by nanoparticles, more study is necessary. To advance the use of carbon-based nanomaterials in plant biotechnology and sustainable agriculture, several problems must be resolved.

References

1. Monica RC, Cremonini R (2009) Nanoparticles and higher plants. Caryologia 62(2):161–165
2. Tans SJ, Verschueren AR, Dekker C (1998) Room-temperature transistor based on a single carbon nanotube. Nature 393(6680):49–52
3. Brunner TJ et al (2006) In vitro cytotoxicity of oxide nanoparticles: comparison to asbestos, silica, and the effect of particle solubility. Environ Sci Technol 40(14):4374–4381
4. Ball P (2002) Natural strategies for the molecular engineer. Nanotechnology 13(5):R15
5. Salamanca-Buentello F et al (2005) Nanotechnology and the developing world. PLoS Med 2(5):e97
6. Baig N, Kammakakam I, Falath W (2021) Nanomaterials: a review of synthesis methods, properties, recent progress, and challenges. Mater Adv 2(6):1821–1871
7. Banerjee A et al (2021) Nanotechnology: an emerging hope in crop improvement. Lett Appl NanoBioSci 10(4):2784–2803
8. Khan F et al (2022) Green nanotechnology: plant-mediated nanoparticle synthesis and application. Nano 12(4):673
9. Roy A et al (2019) Green synthesis of silver nanoparticles: biomolecule-nanoparticle organizations targeting antimicrobial activity. RSC Adv 9(5):2673–2702
10. Sanzari I, Leone A, Ambrosone A (2019) Nanotechnology in plant science: to make a long story short. Front Bioeng Biotechnol 7:120
11. Hu P et al (2020) Nanoparticle charge and size control foliar delivery efficiency to plant cells and organelles. ACS Nano 14(7):7970–7986
12. Wang X et al (2023) Nanoparticles in plants: uptake, transport and physiological activity in leaf and root. Materials 16(8):3097
13. Su Y et al (2019) Delivery, uptake, fate, and transport of engineered nanoparticles in plants: a critical review and data analysis. Environ Sci Nano 6(8):2311–2331
14. Ali S, Mehmood A, Khan N (2021) Uptake, translocation, and consequences of nanomaterials on plant growth and stress adaptation. J Nanomater 2021(1):6677616
15. Minkenberg B, Wheatley M, Yang Y (2017) CRISPR/Cas9-enabled multiplex genome editing and its application. Prog Mol Biol Transl Sci 149:111–132
16. Zhu J et al (2020) Mechanism of zinc oxide nanoparticle entry into wheat seedling leaves. Environ Sci Nano 7(12):3901–3913
17. Hubbard JD, Lui A, Landry MP (2020) Multiscale and multidisciplinary approach to understanding nanoparticle transport in plants. Curr Opin Chem Eng 30:135–143
18. Avellan A et al (2021) Critical review: role of inorganic nanoparticle properties on their foliar uptake and in planta translocation. Environ Sci Technol 55(20):13417–13431
19. Demirer GS et al (2021) Nanotechnology to advance CRISPR–Cas genetic engineering of plants. Nat Nanotechnol 16(3):243–250
20. Wang JW et al (2021) Nanoparticles for protein delivery in planta. Curr Opin Plant Biol 60:102052
21. Lew TTS et al (2018) Rational design principles for the transport and subcellular distribution of nanomaterials into plant protoplasts. Small 14(44):1802086
22. Khan MR et al (2019) Nanoparticle–plant interactions: two-way traffic. Small 15(37):1901794
23. Chiu YTE, Choi CHJ (2021) Enabling transgenic plant cell–derived biomedicines with nanotechnology. Adv NanoBiomed Res 1(1):2000028
24. Zhu J et al (2021) Role of charge and size in the translocation and distribution of zinc oxide particles in wheat cells. ACS Sustain Chem Eng 9(34):11556–11564
25. Zhang P et al (2020) Nanomaterial transformation in the soil–plant system: implications for food safety and application in agriculture. Small 16(21):2000705
26. Haghighi M, Teixeira da Silva JA (2014) The effect of carbon nanotubes on the seed germination and seedling growth of four vegetable species. J Crop Sci Biotechnol 17:201–208
27. Demirer GS et al (2019) Carbon nanotube–mediated DNA delivery without transgene integration in intact plants. Nat Protoc 14(10):2954–2971

28. Mitter N et al (2017) Clay nanosheets for topical delivery of RNAi for sustained protection against plant viruses. Nat Plants 3(2):1–10
29. Chang F-P et al (2013) A simple plant gene delivery system using mesoporous silica nanoparticles as carriers. J Mater Chem B 1(39):5279–5287
30. Hajiahmadi Z et al (2019) Enhancement of tomato resistance to Tuta absoluta using a new efficient mesoporous silica nanoparticle-mediated plant transient gene expression approach. Sci Hortic 243:367–375
31. Martin-Ortigosa S et al (2014) Mesoporous silica nanoparticle-mediated intracellular Cre protein delivery for maize genome editing via loxP site excision. Plant Physiol 164(2):537–547
32. Ahmed S et al (2021) Nanoparticle-based genetic transformation of Cannabis sativa. J Biotechnol 326:48–51
33. Zhao Y et al (2013) Nitrogen-doped carbon nanomaterials as non-metal electrocatalysts for water oxidation. Nat Commun 4(1):2390
34. Cañas JE et al (2008) Effects of functionalized and nonfunctionalized single-walled carbon nanotubes on root elongation of select crop species. Environ Toxicol Chem 27(9):1922–1931
35. Tripathi S, Sonkar SK, Sarkar S (2011) Growth stimulation of gram (Cicer arietinum) plant by water soluble carbon nanotubes. Nanoscale 3(3):1176–1181
36. Wang Z et al (2014) Facile preparation of oligo (ethylene glycol)-capped fluorescent carbon dots from glutamic acid for plant cell imaging. Mater Lett 129:122–125
37. Xu X et al (2004) Electrophoretic analysis and purification of fluorescent single-walled carbon nanotube fragments. J Am Chem Soc 126(40):12736–12737
38. Hill SA et al (2016) Three-minute synthesis of sp3 nanocrystalline carbon dots as non-toxic fluorescent platforms for intracellular delivery. Nanoscale 8(44):18630–18634
39. Liu G et al (2010) Functional nanoparticles for molecular imaging guided gene delivery. Nano Today 5(6):524–539
40. Hendler-Neumark A, Bisker G (2019) Fluorescent single-walled carbon nanotubes for protein detection. Sensors 19(24):5403
41. Burlaka O et al (2015) Plant genetic transformation using carbon nanotubes for DNA delivery. Cytol Genet 49(6):349–357
42. Demirer GS et al (2019) Nanotubes effectively deliver siRNA to intact plant cells and protect siRNA against nuclease degradation. BioRxiv 2019:564427
43. Serag MF et al (2011) Trafficking and subcellular localization of multiwalled carbon nanotubes in plant cells. ACS Nano 5(1):493–499
44. Serag MF et al (2012) Introducing carbon nanotubes into living walled plant cells through cellulase-induced nanoholes. RSC Adv 2(2):398–400
45. Yuan H et al (2011) Single walled carbon nanotubes exhibit dual-phase regulation to exposed Arabidopsis mesophyll cells. Nanoscale Res Lett 6:1–9
46. Kwak S-Y et al (2019) Chloroplast-selective gene delivery and expression in planta using chitosan-complexed single-walled carbon nanotube carriers. Nat Nanotechnol 14(5):447–455
47. Taha RA et al (2016) Carbon nanotubes impact on date palm in vitro cultures. Plant Cell Tissue Organ Cult 127:525–534
48. Khodakovskaya M et al (2009) Carbon nanotubes are able to penetrate plant seed coat and dramatically affect seed germination and plant growth. ACS Nano 3(10):3221–3227
49. Alimohammadi M et al (2011) Physiological responses induced in tomato plants by a two-component nanostructuralsystem composed of carbon nanotubes conjugated with quantum dots and its in vivomultimodal detection. Nanotechnology 22(29):295101
50. Yan S et al (2013) Single-walled carbon nanotubes selectively influence maize root tissue development accompanied by the change in the related gene expression. J Hazard Mater 246:110–118
51. Pourkhaloee A et al (2011) Carbon nanotubes can promote seed germination via seed coat penetration. Seed Technol 33:155–169
52. Begum P, Fugetsu B (2012) Phytotoxicity of multi-walled carbon nanotubes on red spinach (Amaranthus tricolor L) and the role of ascorbic acid as an antioxidant. J Hazard Mater 243:212–222

53. Khodakovskaya MV et al (2012) Carbon nanotubes induce growth enhancement of tobacco cells. ACS Nano 6(3):2128–2135
54. Mondal A et al (2011) Beneficial role of carbon nanotubes on mustard plant growth: an agricultural prospect. J Nanopart Res 13:4519–4528
55. Golestanipour A et al (2018) Gene delivery to tobacco root cells with single-walled carbon nanotubes and cell-penetrating fusogenic peptides. Mol Biotechnol 60:863–878
56. Crevillen AG, Escarpa A, García CD (2018) Carbon-based nanomaterials in analytical chemistry. RSC, London
57. Bevan M, Walsh S (2005) The Arabidopsis genome: a foundation for plant research. Genome Res 15(12):1632–1642
58. Gaj T, Gersbach CA, Barbas CF (2013) ZFN, TALEN, and CRISPR/Cas-based methods for genome engineering. Trends Biotechnol 31(7):397–405
59. Jiang W et al (2013) Demonstration of CRISPR/Cas9/sgRNA-mediated targeted gene modification in Arabidopsis, tobacco, sorghum and rice. Nucleic Acids Res 41(20):e188
60. Feng Z et al (2014) Multigeneration analysis reveals the inheritance, specificity, and patterns of CRISPR/Cas-induced gene modifications in Arabidopsis. Proc Natl Acad Sci 111(12):4632–4637
61. Lee CW et al (2010) Developmental phytotoxicity of metal oxide nanoparticles to Arabidopsis thaliana. Environ Toxicol Chem 29(3):669–675
62. Shen CX et al (2010) Induction of programmed cell death in Arabidopsis and rice by single-wall carbon nanotubes. Am J Bot 97(10):1602–1609
63. Koornneef M, Meinke D (2010) The development of Arabidopsis as a model plant. Plant J 61(6):909–921
64. Xia L et al (2012) GM wheat development in China: current status and challenges to commercialization. J Exp Bot 63(5):1785–1790
65. Bhalla PL (2006) Genetic engineering of wheat–current challenges and opportunities. Trends Biotechnol 24(7):305–311
66. Vasil IK (2007) Molecular genetic improvement of cereals: transgenic wheat (Triticum aestivum L.). Plant Cell Rep 26:1133–1154
67. Ding L et al (2009) Optimization of agrobacterium-mediated transformation conditions in mature embryos of elite wheat. Mol Biol Rep 36:29–36
68. Belhaj K et al (2015) Editing plant genomes with CRISPR/Cas9. Curr Opin Biotechnol 32:76–84
69. Shan Q et al (2014) Genome editing in rice and wheat using the CRISPR/Cas system. Nat Protoc 9(10):2395–2410
70. Ikawa Y et al (2019) Gene therapy of hemoglobinopathies: progress and future challenges. Hum Mol Genet 28(R1):R24–R30
71. Maestro S et al (2021) Novel vectors and approaches for gene therapy in liver diseases. JHEP Rep 3(4):100300
72. Ashrafizadeh M et al (2022) Photoactive polymers-decorated cu-Al layered double hydroxide hexagonal architectures: a potential non-viral vector for photothermal therapy and co-delivery of DOX/pCRISPR. Chem Eng J 448:137747
73. Duan L et al (2021) Exosome-mediated delivery of gene vectors for gene therapy. Nanoscale 13(3):1387–1397
74. Xu Z et al (2022) Carrier strategies boost the application of CRISPR/Cas system in gene therapy. Exploration 2:20210081
75. Song X et al (2021) Delivery of CRISPR/Cas systems for cancer gene therapy and immunotherapy. Adv Drug Deliv Rev 168:158–180
76. Huang J et al (2022) CRISPR/Cas systems: delivery and application in gene therapy. Front Bioeng Biotechnol 10:942325
77. Tiwari D et al (2013) Interaction of carbon nanotubes with mineral nutrients for the promotion of growth of tomato seedlings. Nano Stud 7:87–96
78. Demirer GS et al (2020) Carbon nanocarriers deliver siRNA to intact plant cells for efficient gene knockdown. Sci Adv 6(26):eaaz0495

79. Liu Q et al (2009) Carbon nanotubes as molecular transporters for walled plant cells. Nano Lett 9(3):1007–1010
80. Ochoa-Olmos OE et al (2016) Transformation of plant cell suspension cultures with amine-functionalized multi-walled carbon nanotubes. J Nanosci Nanotechnol 16(7):7461–7471
81. Burlaka O et al (2016) Non-covalent functionalization of carbon nanotubes for efficient gene delivery. In: Nanophysics, nanophotonics, surface studies, and applications: selected proceedings of the 3rd international conference nanotechnology and nanomaterials. Springer, Cham
82. Demirer GS et al (2019) High aspect ratio nanomaterials enable delivery of functional genetic material without DNA integration in mature plants. Nat Nanotechnol 14(5):456–464

Metallic Nanoparticles in Plant Genetic Engineering

Sanam Ashraf, Javed Iqbal, Zubaria Tul Ain, Banzeer Ahsan Abbasi, Muhammad Anas, Tariq Mahmood, Sana Basharat, and Muhammad Waseem

Abstract Plant genetic engineering achieved a breakthrough with metallic nanoparticles (MNPs) because they enable innovative methods to move genetic material into plant cells. Plants require different traditional transformation methods Agrobacterium-mediated transformation and biolistic particle delivery which limit success due to species restriction together with reduced efficacy and risks of tissue destruction. MNPs establish themselves as novel delivery tools because they possess adjustable surface toppings and large surface-to-volume dimensions and cell wall penetration abilities without mechanical pressure. The DNA delivery properties of nucleic acids using nanoparticles have been investigated through studies of various types of MNPs including gold (Au), copper oxide (CuO), titanium dioxide (TiO_2), and iron oxide (Fe_3O_4). The different processes through which MNPs enable gene delivery occur concurrently. Genetic material enters the plant cell cytoplasm and nucleus through two methods which include wall penetration via endocytosis

S. Ashraf · Z. T. Ain · T. Mahmood
Department of Plant Sciences, Faculty of Biological Sciences, Quaid-i-Azam University Islamabad, Islamabad, Pakistan
e-mail: Zubaria@bs.qau.edu.pk; tmahmood@qau.edu.pk

J. Iqbal
Department of Botany, Bacha Khan University, Charsadda, Khyber Pakhtunkhwa, Pakistan

B. A. Abbasi (✉)
Department of Botany, Rawalpindi Women University, Rawalpindi, Pakistan

M. Anas
Department of Plant Sciences, Faculty of Biological Sciences, Quaid-i-Azam University Islamabad, Islamabad, Pakistan

Department of Soil Science and Plant Nutrition, University of Selcuk Campus, Konya, Turkey
e-mail: muhammed.anas@selcuk.edu.tr

S. Basharat · M. Waseem (✉)
School of Breeding and Multiplication (Sanya Institute of Breeding and Multiplication), Sanya, China

College of Tropical Agriculture and Forestry, Hainan University, Sanya, China

Fang Zhiyuan Academician Team Innovation Center, Haikou, Hainan, China

J.-T. Chen (ed.), *Nano-Delivering for Plant Genetic Engineering*, Smart Nanomaterials Technology, https://doi.org/10.1007/978-981-95-4165-2_4

and diffusion. This mechanism allows the laboratory to perform either temporary or permanent genetic alterations which helps advance plant biotechnological possibilities. Moreover, the transformative process of plant tissue using MNP-based methods decreases toxicity levels and improves genetic stability helping scientists attain higher control of delivery processes leading to predictable gene expression levels. Current research focuses on resolving important issues regarding nanoparticle phytotoxicity, dosage optimization, and environmental safety despite their advantages. This chapter demonstrates how metallic nanoparticles used with specific properties can bring major changes to plant genetic engineering platforms.

Keywords Gene expression · Gene delivery · Genetic materials · Genetic transformation · Nanocarriers

1 Introduction

The combination of climate changes with diminished crop yield and expanding diseases and pests threatens to make thousands of people vulnerable, especially among those who earn their living from agricultural activities in low-income areas [1]. Our conventional agricultural system faces increased pressure from two main factors: the growing number of people as well as our rising dependence on plants for energy production alongside food needs and medical and chemical requirements. Cultivating food security and agricultural sustainability demands the immediate development of crop varieties that can withstand biotic and abiotic stress factors and increase their valuable outputs [2, 3]. Traditional breeding methods that develop new cultivars take too much time together with resources to support sustainable agriculture into the future while only acquiring direct control of traits important to breeding [4]. Genetic engineering of crops improves upon traditional breeding methods because traditional breeding is both time-consuming and demands significant human labor together with imprecise genotype and phenotype control [5].

Manipulation of cells through genetic engineering aims at changing their genome along with affecting the way genes express themselves. Techniques related to genetic engineering lead to recombinant protein expression together with changes in gene expression levels and permanent gene knockout actions and targeted mutations in host genes and insertion of substantial foreign DNA segments into the host genome [6, 7]. Genome modifications often result in temporary or lasting effects as well as genetic inheritance while using biomolecules mainly consisting of RNA, DNA and proteins, that cells absorb directly but need various delivery systems such as gene guns, microinjection, electroporation, sonoporation and nanoparticle-based solutions and engineered bacterial and viral vectors to penetrate cells effectively [8, 9]. The cell wall of plant cells poses obstacles to genetic engineering that force scientists to use delivery systems that require high specificity to plant hosts and easy plant regeneration capabilities as presented in Fig. 1 [10]. Scientists use unwalled plant cells known as protoplasts for genetic engineering but the process of obtaining

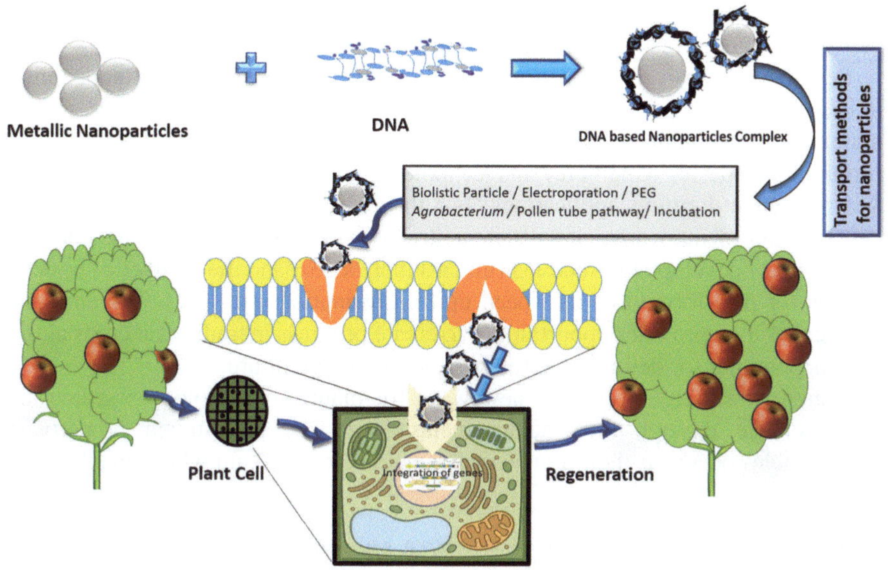

Fig. 1 Demonstrate how to deliver DNA complex nanoparticles to plant cells via transport techniques

mature transgenic plants from treated protoplasts requires time-intensive multiple stages [11].

Plant genetic engineering represents a modern plant science technological advancement that functions as an important instrument for crop quality development and yield production and secondary metabolite level increase in medicinal plants and the creation of sustainable crops. Nanoparticle-mediated gene transformation solves plant cell membrane limitations through precise DNA or RNA delivery thereby generating either temporary or permanent transformations [12]. The application of genetic engineering allows for the development of crops that demonstrate higher yields while containing nutritional improvements and resistance against both chemical herbicides and insects as well as plant diseases and physiological stresses such as drought and heat. The pharmaceutical industry uses genetically modified plants to synthesize both important small-molecule drugs and recombinant proteins as reported by Wang et al. [13]. The selection of nuclear genome or plastid genome modification depends on genetic modification requirements because these genome types show fundamental functional differences. DNA editing in plants is possible through foreign gene integration for trait enhancement and specific gene silencing with RNAi and nuclease-based approaches [14]. The techniques of Agrobacterium-mediated transformation and gene-gun transformation prove effective but Agrobacterium displays host-specific characteristics [15] and Biolistic DNA delivery employs high-pressure gene-gun technology to deliver DNA-loaded gold microparticles that penetrate the chromosomes through both cell membrane and cell wall. The damage to tissues and random multiple additions throughout the genetic sequence occur when tissues sustain injuries. The various delivery methods are shown in Table 1 [16].

Scientists extensively use nanoparticles to achieve cell transformation in animal biology but their application remains limited to plant biology [15]. The present manipulation techniques demand extensive work and they function specifically for certain plant species which prevents technological progress in agricultural sciences and molecular farming applications. For the field to advance we need ongoing development of innovative and efficient strategies to bypass this barrier. The field requires new developments and instruments that can expand the number of possible species for transformation and speed up transformation rates [17]. The emerging nanotechnology provides opportunities to create progress in traditional crops alongside its potential for defining the future of modern agriculture.

Nanoparticles create entirely new ways for plant biotechnology research and plant science through their achievements [18]. The exceptional property of nanoparticles occurs as their dimensions remain below 100 nm which enables them

Table 1 The various delivery methods are shown with targeted species, delivery types, and limitations

Transformation method	Targeted tissues	Species	Delivery type	Limitations
Biolistic particle	Depends on tissue type+/ callus, embryo, leaf	Monocots, dicotyledons, Depends on tissue type+/calli, embryos, leaves	DNA, siRNA, miRNA, ribonucleo proteins	The procedure of targeting leaves involves plant detachment yet presents limitations for effect observation time and embryo targeting demands cumbersome regeneration techniques with species-specific dependency
Electroporation	Leaf, protoplast, meristem, pollen grain	Monocots, dicotyledons,	DNA, siRNA, miRNA, proteins	The nanoparticles generate damage to tissue through non-specific processes at the same time they expose limited space for biological materials
PEG	Protoplast	Monocots, dicotyledons	DNA, siRNA, miRNA	Cytotoxicity, low regeneration
Agrobacterium	Root, shoot apical meristem, leaf, flower, hypocotyl, cotyledon	Narrow range of plant species	DNA	Host limitation, the delivery to leaves produces temporary effects because genetic changes do not cross over to subsequent generations
Pollen tube pathway	pollen tube	Monocots, dicotyledons	DNA	Low efficiency
Nanoparticles	Leaf, protoplast	Monocots, dicotyledons	DNA, siRNA, miRNA	Low efficiency

to permeate through cell walls (~5–20 nm-sized) and cellular membranes (300–500 nm-sized). They enter cells and organelles efficiently without requiring biolistic, ultrasound, or electroporation forces [16]. Nanoparticles demonstrate promising potential as biomolecular transport agents because they can cross plant cell walls autonomously and feature adjustable properties for binding different cargos and accepting diverse host cell types [19]. DNA/RNA protection against nucleases and machine vibrations occurs through nanoparticles while they maintain a superior gene delivery rate. The large surface area and biologically safe properties of nanoparticles allow them to carry significant genetic material while directing delivery towards different plant domains and cell structures including both mature and immature cell types including leaves, roots, protoplasts as well as organelles such as the nucleus and chloroplast [20].

A partial solution to regenerate transformed plants exists through the introduction of developmental regulators Wuschel2 (Wus2) and baby boom (Bbm) genes. The genes activate somatic embryogenesis by promoting embryo development thus enabling better transformation regeneration. The plant meristem stem cell maintaining gene Wus2 enables extended plant development and Wbm facilitates embryogenic change in cells during early embryogenesis. These genes can reach protoplasts and modified tissues through various nanoparticle types including DNA nanoparticles, magnetic nanoparticles, carbon nanotubes, clay nanosheets and peptide nanoparticles which enhances their efficiency of regeneration [21, 22]. Various factors affect the delivery success of NPs including DNA/NP ratio and exposure periods as well as cargo loading systems and biological compatibility of the NPs. Desirable results for sufficient genetic material support and stable NP carriers require precise coordination of all parameters. Academic research into triggered biomolecule release from NPs shows great promise in monitoring gene expression levels based on surrounding environmental factors [23]. Generally, these metal nanoparticles including gold silver and iron possess characteristics that facilitate their successful entrance into plant tissues beyond their protective membranes. The engineered nanoparticles possess the ability to transport DNA RNA or proteins between cells for efficient plant genome modification. The introduction of gene delivery cargo through nanoparticles occurs because of endocytosis and plasmodesmata diffusion and temporary cell wall displacement mechanisms [18, 24].

Gold nanoparticles can be used for delivering CRISPR-Cas9 components into plant cells through functionalization while achieving precise genome editing without unintended mutation effects. The implementation of MNPs eliminates some traditional method challenges including species protocol requirements as well as inherited genetic mutations [25, 26]. MNPs show great potential for improving both efficiency and specificity during plant genetic transformation procedures according to recent scientific developments. The use of MNPs in plant genetic engineering faces multiple barriers during actual implementation. The cytotoxicity and environmental effects of nanoparticles require complete evaluation and design improvement of nanoparticle formulations. The long-term fate and integration of genetic material carried through nanoparticles need extensive research to become properly understood. The full realization of metallic nanoparticles to develop improved genetically modified plants depends on resolving their identified limitations [27, 28]. Plant genetic engineering utilizes metallic nanoparticles as an advanced solution against the problems encountered by established gene delivery systems. These nanoparticles present unique properties that enable researchers to deliver genetic material efficiently and specifically because of which scientists can develop improved crops for sustainable agriculture.

2 Types of Metallic Nanoparticles Used in Plant Genetic Engineering

The delivery system in plant genetic engineering uses metallic nanoparticles (MNPs) because of their unique properties states [29]. The commonly used metal nanoparticles for plant genetic operations are gold nanoparticles (AuNPs) that exist in nanospheres and nanorods and nanocluster formats because they provide stability alongside easy functionalization capabilities [30]. The delivery of iron oxide (Fe_2O_3) and zinc oxide (ZnO) and copper oxide (CuO) metal oxide nanoparticles (MONPs) represents another method in plant genetic engineering because iron oxide nanoparticles specifically benefit from magnetofection through controlled plant cell entry by utilizing magnetic fields [31, 32]. Various types of nanomaterials, specific metallic nanoparticles, their delivery methods, and applications in plant genetic engineering are shown in Table 2. The scientists used a magnetic NP for delivering DNA into pollen so they could create stable transgenic seeds which avoid the conventional regeneration process [33].

According to Martin-Ortigosa et al. [34] used Au-MSN composite gold-modified mesoporous silica nanoparticles to successfully transport two proteins or two different genes or a protein-DNA combination between onion, tobacco and teosinte plants. The plasma membrane of plant cells received fluorescently tagged bovine serum albumin (BSA) and enhanced green fluorescent protein (eGFP) placed within composite nanoparticles that used Au-MSN amounts of 625 mg/g Au-MSN and 150 mg/g Au-MSN. The 30-min bombardment period of plant cells revealed the presence of BSA fluorescent signals yet detection of eGFP fluorescence occurred after 1 day. Scientists verified the simultaneous DNA introduction into plant cells through experiments conducted with Au-MSN coated by two plasmids encapsulating the red fluorescent m-cherry protein gene and eGFP gene (pLMNC95). Laboratory observations showed that matched fluorescent signals proved the successful combination and functioning of both genetic sequences. The onion tissue exposed to eGFP and mCherry plasmid DNA/Au-MSN showed simultaneous fluorescent green and red signals which indicated these particles delivered both molecules together. This research established Au-MSN delivery as a novel tool for plant genetic transformation by delivering plasmid DNA and proteins simultaneously through biolistic procedures. It achieved the distinction of becoming the first method for co-introducing DNA plasmids and proteins in plant cells by using Au-MSN materials. Scientific research indicated successful evidence of nanomagnetic beads

Table 2 The type of nanomaterials, specific metallic nanoparticles, their delivery methods, and applications in plant genetic engineering

Type of nanomaterial	Metallic nanoparticle example	Delivery method	Application in plant genetic engineering
Metallic nanoparticles (MNPs)	Gold (AuNPs)	Biolistic, foliar spray	Genetic delivery systems, transient or stable transformation technologies
	Silver (AgNPs)	Syringe infiltration	Antimicrobial supports, enhanced gene transfer efficiency techniques
	Iron oxide (Fe_2O_3, Fe_3O_4)	Magnetofection (magnetic field-assisted delivery)	Targeted transformation methods, efficient gene transport mechanisms
Metal oxide nanoparticles (MONPs)	Zinc oxide (ZnO)	Biolistic, ultrasound-mediated	DNA/RNA nanocarrier systems and improved genetic transformation
	Copper Oxide (CuO)	Foliar spray, seed priming	Antimicrobial support in gene delivery, co-delivery of biomolecules
Silicon-based nanoparticles (SiNPs)	Silicon oxide (SiO_2)	Vacuum infiltration, passive diffusion	Biocompatible nano-gene carriers, DNA delivery
Carbon-based nanomaterials	Carbon nanotubes (CNTs)	Needle-free penetration, spray method	Efficient gene editing, high stability in plant cells

delivering foreign genes into maize pollen by utilizing the natural pathway of pollen germination pores. The biocompatible properties of carbon nanotubes including single-walled carbon nanotubes and multi-walled carbon nanotubes have been documented in the research field [24]. Zhao et al. [33] conducted a study that used magnetofection to produce transgenic cotton seeds (Gossypium hirsutum). Research investigators bound plasmid DNA measuring 15 kb with magnetic nanoparticles from 0.1:1 to 1:1 mass ratios before applying these DNA-delivery agents through magnetic field assistance. The intensive wall barrier protecting pollen grains did not discourage DNA-loaded particles from passing through the apertures which represent weak and soft cell wall areas.

3 Synthesis and Characterization of Metallic Nanoparticles

The next natural step in the development of technology is said to be nanotechnology. This has resulted in a significant increase in research activity related to the advancement and basic comprehension of nanoscale assembly and processes [35]. The top-down or bottom-up synthesis method can be used to create the nanomaterial physically, chemically, or biologically as shown and summarized in Fig. 2 and

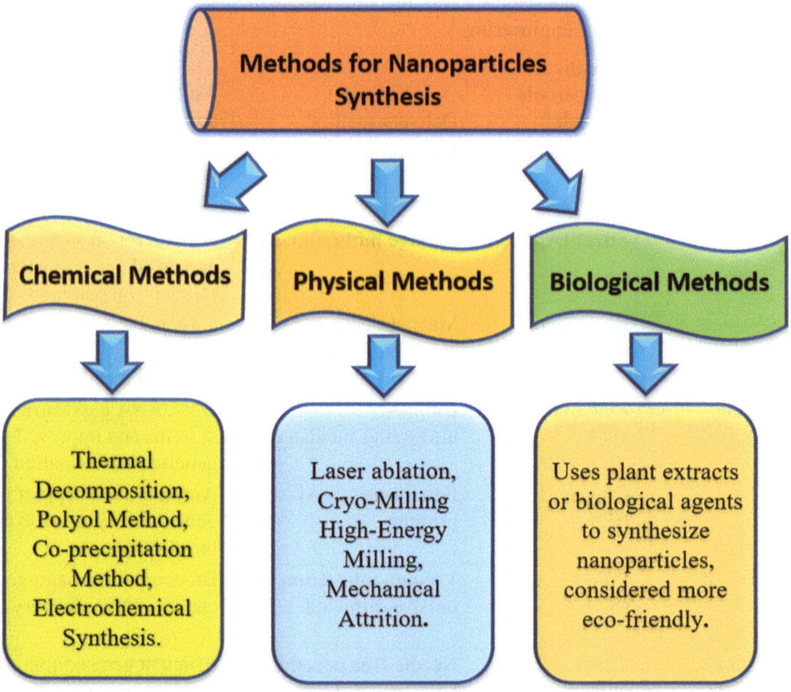

Fig. 2 Various synthesis methods of metallic nanoparticles

Table 3. The bulk materials are often broken down into small molecules via a physical process for the top-down approach, and these molecules are then further transformed into NPs as required. By creating tiny nanostructures, which are often accomplished by chemical means, nanomaterials can be directly manufactured via the bottom-up technique [36]. Scientists are particularly interested in the green synthesis of nanoparticles utilizing plants and other microbes since it is easy, affordable, and environmentally benign. Biologically produced nanoparticles hold great promise for the development of nanopesticides, nanoinsecticides, and nanofertilizers. Additionally, it reduces the amount of harmful environmental pollutants produced by agricultural fields and is biodegradable [37, 38]. Nanoparticles are thought to be the next generation of gene transfer vehicles. In order to transfer foreign DNA or gene sequences and successfully alter the target genes, nanocapsules, nanofibres, and nanoparticles are employed [39].

The effectiveness of NP transport depends heavily on size characteristics in addition to physical and chemical and biological attributes of NPs according to existing published findings. Learning how to manufacture NPs that possess specific attributes becomes essential because of this requirement. Novel synthesis demonstrates the best ability to create biomolecules transporting NPs inside cells through its high efficiency and ease of operation without generating contaminants [40]. Scientists use morphology and size along with surface charge and optical properties as the

Table 3 Provides a valuable illustration of synthesis methods and characterization of metallic nanoparticles for plant genetic engineering

Aspect	Details	Examples of nanoparticles	Characterization techniques
Synthesis methods	Scientists utilize different techniques for producing metallic nanoparticles that work in genetic engineering applications		
Chemical synthesis	Nanoparticles result from reducing metal salts found in solution	Gold (Au), silver (Ag), platinum (Pt)	UV-Vis spectroscopy, transmission electron microscopy (TEM), dynamic light scattering (DLS)
Physical synthesis	The generation of nanoparticles through laser ablation and sputtering as well as other physical forces constitutes one synthesis method	Gold, silver	Atomic force microscopy (AFM), scanning electron microscopy (SEM)
Green synthesis	The production method relies on plant extracts and biological agents which results in environmentally beneficial nanoparticles synthesis	Silver, gold, copper (Cu)	X-ray diffraction (XRD), fourier-transform infrared spectroscopy (FTIR)
Biological synthesis	Scientists produce nanoparticles using bacteria together with fungi or algae as part of the synthetic process	Zinc oxide (ZnO), silver, gold	TEM, XRD, zeta potential analysis
Characterization techniques	The assessment procedures for measuring properties of manufactured nanoparticles		
Particle size and distribution	Measurement of both nanoparticle dimensions and the way they distribute within the solution occurs through this method	All nanoparticles	Dynamic light scattering (DLS), nanoparticle tracking analysis (NTA)
Morphology and surface features	Researchers inspect the dimensional properties as well as exterior attributes of nanoparticles	Gold, silver, copper	Scanning electron microscopy (SEM), atomic force microscopy (AFM)
Crystallinity and structure	The method reveals both crystal structures and phase conditions of produced nanoparticles	Gold, silver	X-ray diffraction (XRD), selected area electron diffraction (SAED)
Surface charge and stability	Measures the surface charge (zeta potential) and stability of nanoparticles in solution	All nanoparticles	Zeta potential analysis, TEM
Elemental composition	This method identifies both the basic elements and purity composition of nanoparticles	All nanoparticles	Energy dispersive X-ray spectroscopy (EDX), X-ray photoelectron spectroscopy (XPS)

main criteria to characterize nanoparticles. Various nanotechnology devices like SEM and ESEM and the combination of TERS alongside STM and TEM unite to examine nanoparticles through surface features analysis. The testing of optical properties uses spectral analysis but researchers utilize XRD and EDX spectroscopy alongside FTIR and FCS for uncovering basic properties of nanoparticles (NPs). The discussion examines how these techniques evaluate nanoparticles [41].

4 Entry, Internalization, and Physiological Effects of Nanoparticles in Plants

Modern agriculture benefits from nanotechnology as an emerging innovative technology that enables improvements in crop management according to Kim et al. [42]. The possible benefits of nanoparticles (NPs) for enhancing agrochemical performance and crop production and soil health exceed expectations although scientists lack fundamental knowledge about how these NPs enter plant systems and interact with their environment [43]. Traditional spraying methods with agrochemicals through soil broadcasting and foliar treatments experience low target-site success rates since leaching occurs together with photodegradation and hydrolysis and microbial decomposition. Multiple rounds of repeated applications become necessary to maintain effective pest control since it leads to soil and water contamination. Nano-encapsulated agrochemicals have been designed to fix these problems by creating agents which enhance solubility and stability along with using controlled release systems that react to environmental cues. Nanoformulations contain mechanisms to deliver specific treatment patterns while minimizing environmental threats and contribute straightforward administration procedures that minimize application quantities [44–46]. Using nanoencapsulated herbicides represents a significant nanotechnology application in crop protection by enabling both parasitic weed control and decreased harm toward non-target plants [47].

The way nanoparticles interact with plant tissue depends on the cell wall physical structure since this wall serves as an organic protective barrier. Nanoparticles smaller than 20 nm as well as loosely bound nanoparticles can effectively pass through the 5–20 nm pores that exist in plant tissue cell walls to access plasma membranes. Engineered nanoparticles can cause the expansion of pores or generate new pores through which the nanoparticles enter cellular structures more effectively. Nanoparticles pass through cell membranes because endocytosis allows them to form vesicles that remain sealed inside the cell. Nanoparticles have two alternative pathways to cross membranes through transport proteins or ion channels. Inside the cytoplasm nanoparticles come into contact with different organelles which leads to changes in metabolic activities in those specific locations according to [48, 49]. Leaf surface applications of nanoparticles lead to their penetration through stomatal openings and trichome bases until they reach internal tissue locations as shown in Fig. 3 [50]. Large amounts of accumulated material on photosynthetic surfaces both heat leaves and hinder stomatal gas exchange through stomatal site blockage. The

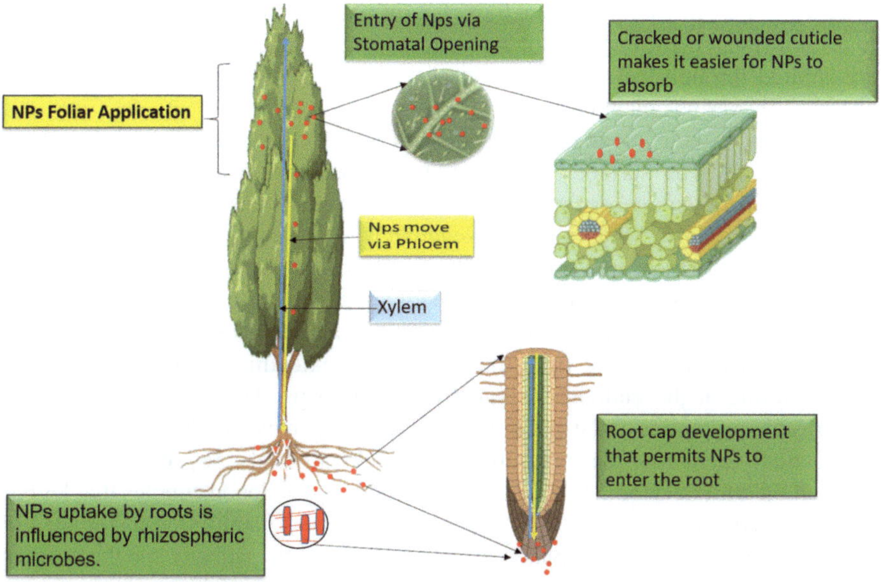

Fig. 3 Nanoparticles enter through the leaf surface's stomatal apertures

physiological changes lead to cellular dysfunction that decreases photosynthetic performance. Accurate dosing techniques combined with proper nanoparticle construction must be implemented because they stop undesirable side effects from occurring. Scientists have initiated new research programs to examine plants as facilities for nanoparticle synthesis. Plants naturally activate their reducing abilities and stabilizing functions to create environmentally friendly metal nanomaterials including silver, gold and copper. Through this green production method, scientists can produce non-harmful nanoparticles which can be applied in agricultural systems according to Da Silva et al. [51].

The cuticular pathway allowed Khan et al. [52] to prove that plants absorb nanoparticles with sizes below 4.8 nm. Researchers studied leaf surface uptake while explaining that NP penetration through the wax barrier depends on their dimensions. The protective cuticle functions as a waterproof layer which usually blocks material transportation through its surface.

Smaller NP-sized particles measuring less than 4.8 nm present ease for passing through the barrier to enter leaf tissues. Plant application of nanoparticles through leaves proves effective as an input method when researchers select targeted particle dimensions that enable cuticular penetration. It is unclear which entry pathways enable larger NPs (above 5 nm) to accumulate in plant tissues because scientists have yet to identify them. Therefore, research into this area remains necessary. Multiple external conditions apart from natural events determine how nanoparticles move through plant leaves. Necrosis and chlorosis along with leaf blight and leaf mold produce leaf surface damages (cuticle, epidermis) and occasionally affect deep mesophyll tissues [53]. The development of NPs uptake depends on plant leaf diseases or pest infestations since they disrupt entry barriers thereby enabling NPs to enter through the foliage. Plant protection using nanopesticides works when NPs are applied through leaf spray to fight pests while enhancing the amount of NPs absorbed by various plant tissues.

NPs initiate their interaction process with the root surface before they become adsorbed to it. The surface of roots maintains roughness because of root hairs together with mucilage or organic acid secretion that generates negative surface charges according to Ma et al. [54]. Negative NPs find optimal conditions to stick and absorb due to these root surface features. The root surface adsorbs nanoparticles before they progress through various physiological barriers to access the xylem for plant absorption and distribution. Nanoparticles must penetrate through five structural plant barriers starting from the root cuticle extending to the epidermis and cortex ending at the endodermis and Casparian strips. The root cuticle displays comparable properties to the leaf cuticle but scientists have not determined whether NPs actually penetrate the root cuticle [55]. Schwab et al. [56] reported that root hairs along with developing root tips possess poorly developed cuticles that let NPs access the epidermis directly. The nanoparticles access plant tissue inside the epidermis by moving across the plant through the symplastic pathway and the apoplastic pathway. The complex relationship between nanoparticles and plant roots becomes understood through this pathway as it demonstrates how nanoparticles enter and transport through the roots. The specified surface modification of nanocapsules supports their better penetration ability across plant cuticles while maintaining a gradual release pace of active components. The use of environmentally friendly encapsulants enables the development of easy-to-handle and safer concentrated agrochemicals for growers [57]. Research demonstrates that surface-modified hydrophobic nanosilica manages successfully throughout agricultural insect pest control. The interaction of lipophilic nanosilica with insect cuticular lipids occurs through physiosorption until the protective wax layer that contains fatty acids and lipids is disrupted. The insects perish due to desiccation because of the inflicted damage [58] [59]. The nanobiopesticide methodology presents environmental advantages over normal pesticides because it features decreased toxic effects and decreased pollution. The uptake behavior of NPs among plant species differed because of the specific plant nature together with their physiological characteristics and metabolic capabilities. Research demonstrated that dicotyledons displayed superior CeO NP translocation in comparison to monocotyledons [60].

5 Applications of Metallic Nanoparticles in Plant Genetic Engineering

Due to their distinctive properties such as their small size and large surface area and their molecules-interacting ability metallic nanoparticles (MNPs) represent effective genetic engineering equipment for plants. These applications represent the main functions of metallic nanoparticles for plant genetic engineering:

5.1 Gene Editing (CRISPR-Cas9 Delivery)

Accurate plant DNA sequence modifications are now possible because scientists developed CRISPR/Cas9 technology along with other plant research tools. Scientists utilize CRISPR/Cas9 to modify plant genome sequences for creating transgene-free crops because of its targeting capabilities [61]. The targeted nature of CRISPR/Cas9 eliminates random gene integration because it destroys unwanted foreign genetic material and multiple gene copy insertion during conventional transformation methods. The Cas endonuclease controls CRISPR/Cas9 by processing protospacer adjacent motifs (PAMs) near target sequences to start its operation. A pre-made guide RNA (gRNA) acts in coordination with the Cas9 protein to guide the protein to particular genomic locations where a double-strand break gets created. After a double-strand break occurs the plant cellular machinery can use either non-homologous end joining (NHEJ) which results in insertions or deletions (indels) or homology-directed repair (HDR) when it utilizes a homologous DNA template [62]. The delivery vehicle used to insert CRISPR components affects the overall efficiency of CRISPR/Cas9 technology to modify plant cells. Researchers have traditionally employed plasmid DNA to deliver CRISPR/Cas9 consisting of Cas9 gene and single-guide RNA (sgRNA) sequences. Plasmid-based delivery of CRISPR components remains inefficient because it takes too much time to complete and causes multiple undesired effects. Taking advantage of less time-consuming mRNA-based approaches is gaining popularity in the field but scientists face the difficulty of brief mRNA survival driven by ribonucleases. The delivery of Cas9 protein with gRNA in complex form as ribonucleoproteins represents a recent promising approach to boost genome editing efficiency at the same time reducing accidental effects [16]. The advancement of CRISPR/Cas9 delivery depends on nanocarriers such as lipid nanoparticles, cationic polymers, and gold nanoparticles. Nanoparticles manufactured for this purpose grant sprotection to CRISPR components during biological transport while enabling quick entry into plant cells. The combination of cationic polymers with lipid nanoparticles demonstrates exceptional properties in developing Cas9/gRNA complexes because of their electrostatic bonding capacity [63, 64]. Gold nanoparticles serve as effective delivery vehicles for CRISPR/Cas9 systems because their biocompatible nature and high stability along with adjustable sizes make them ideal for transporting larger gene constructs successfully [65]. Nanoparticles enable scientists to regulate when and where CRISPR components are released in order to precisely edit plant genes. This method enhances HDR efficiency because it provides complete sets of donor DNA sequences required at the target location. The use of nanoparticles for plant CRISPR/Cas9 applications stands at an early stage of development despite their numerous advantages. The application of nanoparticles for CRISPR editing in plants remains insufficiently investigated because documented successful instances are scarce to date. Researchers actively optimize nanoparticles while determining their size alongside their charge properties and encapsulation efficiency because these features affect delivery outcomes and reduce adverse side effects [66].

5.2 DNA and RNA Delivery for Genetic Modification

Plant genetic modification requires DNA and RNA molecule delivery by utilizing nanoparticle technologies according to [19]. The application of magnetic nanoparticles (MNPs) as well as other nanoparticle types has become a widely studied method for genetic material transfer of plasmid DNA and RNA to plants [67]. Nanoparticles function as carriers that enable desirable genetic traits to transfer DNA material into plant cells. Nanoparticle-based DNA/RNA delivery systems that were recently developed have transformed into multiple therapeutic applications including gene delivery and drug delivery and other therapeutic agents [68, 69]. DNA nanostructures surpass traditional carriers because of their distinct set of benefits. The unique aspect of DNA genetic material makes it a suitable choice as a natural component that the body can metabolize because it ensures safe performance and nonadaptation. DNA nanostructures demonstrate high stability when faced with biodegradation attacks from nucleases because of their inherent resistance to enzyme damage which positions them perfectly for gene delivery operations. The block quote content also demonstrates superior biocompatibility because DNA nanostructures connect effectively with target tissues along with avoiding immune protection activation [70, 71]. DNA nanostructures provide an easy-to-use modular platform for gene delivery operations that performs well when delivering genetic material to plants. Plant cell genetic material delivery utilizes various DNA nanostructure patterns consisting of nanostrings and nanocubes and nanostars and three-dimensional shapes like icosahedrons and tetrahedrons. The DNA nanoparticles use their architectural design to hold DNA/RNA content inside or bind them to their external layer which then allows them to pass efficiently through cell membrane lipid bilayers and access the nucleus for performing their cellular functions [72]. The primary challenge for plant gene delivery methods exists in the plant cell wall because it blocks the entry of large molecules consisting of DNA and RNA. Zhang et al. [18] created DNA nanoparticles as a delivery solution for siRNA delivery into plants. Proof obtained in this research revealed that DNA nanostructure properties including size and shape together with stiffness and compactness and attachment site can determine siRNA efficiency in plant gene silencing. The successful tracking of nanoparticle internalization in E. sativa, Nasturtium officinale, Nicotiana benthamiana, and Nicotiana tabacum plant species became possible through DNA strand attachment using fluorophore conjugation at their targeted locations. The author found that how effectively nanoparticles penetrate plant cells combined with siRNA binding conditions determined the overall silencing performance. This DNA nanoparticle delivery system demonstrates great potential because it works with many plant species for enhanced gene transfer at an optimized level. The integration of nanoparticle-based systems creates novel prospects to engineer plants without transgenes while silencing genes and improving crops which establishes them as essential tools for plant biotechnology.

The detailed study conducted by Rustgi et al. [73] examined single-walled carbon nanotubes (SWCNTs) as one of the prominent classes of nanomaterials for nucleic acid delivery in plants. The highly compatible nanoparticles demonstrate extended shapes and adjustable chemical surfaces which qualifies them as promising delivery agents for non-genome-integrative transfer processes. Scientists discovered that DNA plasmids bound efficiently to plant nuclei and chloroplasts when SWCNTs received cationic polymer functionality through covalent bonding with polyethyleneimine (PEI) and chitosan (CHI). Researchers constructed pioneering approaches that allowed reporter gene delivery across various plant species including eudicots (tobacco, arugula, spinach) and monocots (wheat) through a delivery method that did not depend on the plant species. Gene expression stability of 7–10 days indicated this may become a suitable brief gene study approach as well as an alternative to using transgenic organisms. Second-generation studies optimize SWCNT properties including size features along with their structural ratios while studying surface chemistry to enhance cargo delivery outcomes and biological results at the cell level. Researchers have used these developments to create better scalable RNA and DNA delivery systems which bring opportunities to functional genomics and genome editing and synthetic biology applications in plant biotechnology.

5.3 Virus-Induced Gene Silencing (VIGS) Using MNPs

Post-transcriptional gene silencing in plants receives essential support from VIGS technology which allows researchers to investigate genetic functions as well as search for new plant components and improve crop characteristics. Traditional VIGS methods encounter technical limitations for perennial and non-model crop species in combination with their dependency on specific plant species and reduced efficiency [74, 75]. (Gu^+-siRNA) nanoparticles containing guanidinium disulfide molecules (GDM) effectively deliver siRNA to model plants and multiple crop species according to Lin et al. [76] and Zhou et al. [77], Guanidinium units create electrostatic bonds with siRNA phosphate groups which stabilizes the siRNA while improving cell uptake of siRNA particles. Nearly all cellular uptake occurs through an endocytosis-independent mechanism while the nanoparticles continue to maintain high levels of siRNA availability. The key advantage of Gu^+-siRNA nanoparticles allow effective transportation of RNA across the plant vascular system which leads to widespread gene silencing throughout the system. Researchers successfully demonstrated this delivery method when testing it in Arabidopsis thaliana plants and rice crops where it showed successful siRNA distribution throughout the whole plant system [77]. The new innovation overcomes traditional biolistics limitations by providing plants with an environmentally friendly species-unspecified system for broad-scale RNAi control. The protective capacity of MNPs enhances siRNA stability and enables effective transport system-wide that offers promising possibilities for plant trait improvement. The technology opens new possibilities for agriculture by boosting crop output and boosting functioning genomics exploration capabilities together with better stress response in more plant species. The advancement of this research includes the application of metallic nanoparticles (MNPs) to transport RNA interference (RNAi) molecules such as small interfering RNA (siRNA) directly into plant cells which improves the delivery effectiveness and spatial and systemic spread of siRNA through plant tissues.

6 Advantages of Metallic Nanoparticles in Genetic Engineering

MNPs or metallic nanoparticles are revolutionizing the field of plant genetic engineering because they offer several benefits over the traditional techniques of transformation—like Agrobacterium mediated transfer, or biolistic methods [78]. The unique physicochemical properties of MNPs-such as size and shape, surface charge, and functionalization capabilities—leads to improved delivery efficiency and enhanced stability on the genetic level. Out of countless advantages offered by MNPs, one of the most notable is their ability to efficiently penetrate the rigid plant cell wall and membrane without any sort of external force due to their ultra-small size and alterable surfaces. This effortless form of penetration improves the transformation procedures while also reducing harm to plant tissues [12, 79]. In addition, nanoparticles like gold (AuNPs), iron oxide (Fe_3O_4 NPs), and mesoporous silica coated metal particles have a high binding affinity for DNA and RNA molecules, protecting them from enzymatic digestion and providing stability for long periods of time [80]. For MNPs, gold nanoparticles (AuNPs), for instance, have shown high efficiency in DNA delivery of around 40%, while also being less phytotoxic to plants than larger particles used in conventional methods. Due to the matter's nanoscale dimensions, these particles can overcome plant cell walls due to no force being applied from outside the cell. Also important for stability and efficiency of delivery is the zeta potential of the nanoparticles. Nano-delivery systems have high transfer efficiency when net zeta potential values exceed 30 mV, as less favorable values lead to aggregation while more favorable values increase stability [81, 82]. Surface modification through ligand and peptide and polymer conjugation enables better performance because nanoparticles target specific cell organelles that include chloroplasts and nuclei thus decreasing unintended effects while improving transformation efficiency [83]. MNP-mediated gene delivery approaches produce reduced chromosomal disruption together with genetic anomalies because they operate with enhanced precision when compared to conventional delivery methods [84]. Research exists which demonstrates how gold nanoparticles produce reduced tissue damage when used instead of traditional macro-sized gold particles employed in gene guns and provide enhanced transformation rates. Permanent gene expression is not necessary for applications like gene function testing and pathogen protection since MNP gene delivery approaches enable valuable non-integrative gene expression outcomes that permit bypassing regulatory restrictions [85].

Recent research has proven that MNPs have the ability to precisely deliver CRISPR/Cas9 components to plant cells. The efficient delivery of Cas9 protein and guide RNAs occurs via iron oxide nanoparticles together with mesoporous silica particles without needing bacterial vectors or complex instrumentation. Gene-edited crops can be developed at higher speed and greater safety through this discovery [86, 87]. MNPs facilitate the movement of genetic material through plant vascular systems delivering heritage materials to distant tissues from solitary application sites which results in amplified delivery performance as well as less work [88]. The economic value of MNP-based systems increases with their scaling potential and affordable production when using green chemistry synthesis from plant-based or microbial synthesis which preserves environment resources while maintaining cost-effectiveness. The environmental risk from green-synthesized MNPs remains minimal because they breakdown through natural processes while chemical transfection agents like polyethylene glycol (PEG) create toxicity and environmental persistence issues [89, 90]. Moreover, the costs associated with developing infrastructure in agricultural technologies are further minimized by MNP-mediated transformation not needing advanced particle bombardment devices or electroporation systems. Taken together, the increasing importance of metallic nanoparticles in genetic engineering of plants is due to their efficiency, cost effectiveness and reduced environmental impact because of the high precision with which biological, structural, and physical alterations are made, and the differential targeting that is possible in functionalized nanoparticles [91, 92]. These advantages highlight their immense potential to revolutionize crop improvement programs and sustainable agriculture by enhancing the efficiency and effectiveness of developing resilient, high-yielding, climate-adaptive plant varieties with precision and far fewer constraints than traditional methods would allow.

7 Challenges and Limitations

Plant genetic engineering achieved its breakthrough through the integration of metallic nanoparticles into agricultural biology practices. The implementation of Metallic Nanoparticles requires resolution of numerous challenges together with limitations to achieve both security and operational effectiveness. The main obstacles in the use of metallic nanoparticles in plant genetic engineering include cytotoxicity and phytotoxicity alongside stability and biodegradability challenges and regulatory and ethical obstacles.

7.1 Cytotoxicity and Phytotoxicity Concerns

Plant genetic engineering faces substantial risks because MNPs show potential toxicity toward plant cells and the environments surrounding them. Nanoparticle exposure produces cytotoxic effects against plant cellular elements and produces phytotoxic outcomes which damage overall plant growth and development. The toxicities affect plants through three mechanisms that involve reactive oxygen species (ROS) generation and cellular organelle damage along with nutrient uptake disruptions [17, 93]. Oxidative stress creates damage to plant cellular structures along with reduced photosynthetic function when plants interact with silver nanoparticles (AgNPs). High concentrations of AgNPs at 100 mg/L adversely affected zucchini plants by diminishing their biomass production by 41% as well as their transpiration activity. Plant studies indicate that DNA damage together with chromosomal aberrations occurs when plants are exposed to copper oxide (CuO) and zinc oxide (ZnO) nanoparticles which demonstrate their genotoxic capabilities. Toxic effects of nanoparticles depend on various influencing characteristics including size, concentration, surface charge and species selection between plants [94, 95].

7.2 Stability and Biodegradability Issues

The duration in which metal nanoparticles remain stable in the environment along with their breakdown rates determines how long they survive nature and affect biological systems [96]. Nanoparticle aggregation susceptibility reduces functional properties which results in application performance decline during gene delivery processes. Biodegradable nanoparticles vanish from the environment while non-degradable nanoparticles tend to stay and present long-term risks to ecological systems [97]. Research efforts have dominated the development of biodegradable coatings to improve MNP stability through green synthesis methods for environmental integrity. Plant-derived biological polymers function as sustainable nanocarriers for DNA delivery to plant cells because they provide both improved degradation behavior and reduced toxicities [98, 99]. Bio-based nanoparticles derived from natural substances have become more significant because they show the potential to boost plant development with minimal environmental harm [100]. The production methods focus on designing nanomaterials with active capabilities throughout their use phase before their components transform into harmless substances [101].

7.3 Regulatory and Ethical Concerns

The implementation of MNPs in genetic engineering projects for plants generates major issues for regulators along with ethical concerns. The lack of custom-made regulations addressing nanomaterial usage in agriculture creates difficulties in assessing safety measures and functional reliability. Characterizing the ethics involves analyzing both ecological impacts and unforeseen effects on other organisms and the permanent future use of nanoparticles [102]. The practices of public reception determine how nanotechnology finds its place in agricultural development [103]. The application of MNP technology in plant biotechnology demands open risk evaluations together with definitive guidelines alongside complete stakeholder collaboration for maintaining accountable research efforts [104]. Various obstacles prevent the widespread use of metallic nanoparticles (MNPs) in plant genetic engineering because they present cytotoxicity alongside phytotoxicity and require regulatory oversight. The current problem in nanoparticle science stems from the absence of standardized methods to manufacture nanoparticles and implement application procedures. Result inconsistencies become common due to variations in particle size alongside surface charge and shape aspects and functionalization variables [105]. Scientists have limited knowledge about the long-term impacts that nanoparticle exposure has on plants and how exposure affects future plant generations. The academic community has identified short-term enhancement of transformation efficiency but research into long-term epigenetic modifications of future plant generations or genomic instability is minimal [106]. Soil microbiota presents another major issue regarding their reaction to MNPs. Nanoparticles at elevated doses negatively affect microbial populations by both diminishing diversity and blocking nitrogen-fixing bacteria together with distorting rhizosphere functions [107, 108]. From a practical perspective, scalability and cost-effectiveness also pose substantial limitations. Jointly, synthesizing uniform functional nanoparticles in large quantities while maintaining high expenses proves difficult to scale agriculture applications, especially within limited-resource agricultural environments [109]. The technique of horizontal gene transfer (HGT) increases biosafety risks because unintentional genetic material transfers from nanoparticle-based vectors to non-target organisms and soil microbes might generate unanticipated effects on the ecosystem. The combination of the mentioned issues proves why we need extended ecological research along with standardized regulatory framework development while creating environmentally friendly cost-effective nanomaterial designs so plants can receive safe genetic engineering processes. Plant genetic engineering will benefit significantly from magnetic nanoparticle technology yet requires full resolution of cytotoxicity together with phytotoxicity and stability and degradability and regulatory and ethical obstacles to use them safely and sustainably.

8 Conclusion and Future Prospect

The inclusion of metallic nanoparticles in plant genetic engineering drives a revolutionary change that handles the problems of normal gene delivery systems. MNPs deliver genetic material to plant cells using their nanoscale symmetries along with distinctive physicochemical properties that make them an efficient and non-biological and adjustable delivery vector. Various studies demonstrate that metallic nanoparticles including gold and iron oxide together with copper oxide and titanium dioxide and others exhibit effective foreign gene expression along with low toxicity and reduced species limitations. Plant biotechnology obtains next-generation tools through the DNA and RNA and genome-editing CRISPR/Cas9 systems compatibility of MNPs. Many of the transformation cycles become shorter through MNP use which improves targeted gene modifications to create superior crop varieties having higher production capacity alongside increased stress and nutritional content resistance. Multiple challenges continue to exist despite the significant achievements made. All problems linked to nanoparticle durability and distribution consistency together with toxicity from high doses and probable environmental problems demand detailed analysis for resolution. To reach reproducible results and maintain biosafety standardization should be implemented in every step of nanoparticle fabrication and surface modification and both delivery techniques. New regulations must be created to track the agricultural applications of nanomaterials specifically for genetically modified organisms within the system. Scientific groups need to establish research priorities on designing smart sustainable nano-particles that enable accurate drug delivery systems with decay and plant-specific delivery approaches. The combination of precision agriculture and plant genome reprogramming will become possible on broad scopes through cohort work between nanotechnology and synthetic biology and systems biology. Plant breeding techniques along with functional genomics applications are set to change direction through novel transient expression systems combined with multiplex gene editing methods delivered using nanocarrier delivery systems. The examination of hybrid research using metallic nanoparticles will enable essential tools for plant genetic engineering to guarantee food security while sustaining sustainable agricultural methods under conditions of climate change and population expansion.

References

1. Getahun D (2020) Predictions of climate change on agricultural insect pests Vis-à-Vis food crop productivity: a critical review. Ethiop J Sci Sustain Dev 7(1):18–26
2. Dixon JA, Gibbon DP, Gulliver A (2001) Farming systems and poverty: improving farmers' livelihoods in a changing world. FAO, Rome
3. Singh R, Singh GS (2017) Traditional agriculture: a climate-smart approach for sustainable food production. Energy Ecol Environ 2:296–316
4. Ahmar S, Gill RA, Jung KH, Faheem A, Qasim MU, Mubeen M, Zhou W (2020) Conventional and molecular techniques from simple breeding to speed breeding in crop plants: recent advances and future outlook. Int J Mol Sci 21(7):2590
5. Acquaah G (2024) Conventional plant breeding to modern plant breeding: evolution, achievements, and limitations. In: Plant molecular breeding in genomics era: concepts and tools. Springer, Cham, pp 1–42
6. Lanigan TM, Kopera HC, Saunders TL (2020) Principles of genetic engineering. Genes 11(3):291
7. Heshmatifar S (2023) Towards extracellular release of recombinant protein from Escherichia coli during fermentation: utilisation of antisense technology in bioprocesses (Doctoral dissertation, UCL (University College London))
8. Liyanage VR, Jarmasz JS, Murugeshan N, Del Bigio MR, Rastegar M, Davie JR (2014) DNA modifications: function and applications in normal and disease states. Biology 3(4):670–723
9. Demirer GS (2020) Developing nanomaterial platforms for gene delivery and transgene-free plant genetic engineering. University of California, Berkeley
10. Mahmood MA, Naqvi RZ, Rahman SU, Amin I, Mansoor S (2023) Plant virus-derived vectors for plant genome engineering. Viruses 15(2):531

11. Reed KM, Bargmann BO (2021) Protoplast regeneration and its use in new plant breeding technologies. Front Genome Edit 3:734951
12. Lv Z, Jiang R, Chen J, Chen W (2020) Nanoparticle-mediated gene transformation strategies for plant genetic engineering. Plant J 104(4):880–891
13. Wang JW, Grandio EG, Newkirk GM, Demirer GS, Butrus S, Giraldo JP, Landry MP (2019) Nanoparticle-mediated genetic engineering of plants. Mol Plant 12(8):1037–1040
14. Rajpu M, Choudhary K, Kumar M, Vivekanand V, Chawade A, Ortiz R, Pareek N (2021) RNA interference and CRISPR/Cas gene editing for crop improvement: paradigm shift towards sustainable agriculture. Plants 10(9):1914
15. Billingsley MM, Singh N, Ravikumar P, Zhang R, June CH, Mitchell MJ (2020) Ionizable lipid nanoparticle-mediated mRNA delivery for human CAR T cell engineering. Nano Lett 20(3):1578–1589
16. Rani N, Kumari K, Hooda V (2025) The role of nanoparticles in transforming plant genetic engineering: advancements, challenges and future prospects. Funct Integr Genomics 25(1):1–22
17. Squire HJ, Tomatz S, Voke E, González-Grandío E, Landry M (2023) The emerging role of nanotechnology in plant genetic engineering. Nat Rev Bioeng 1(5):314–328
18. Zhang H, Demirer GS, Zhang H, Ye T, Goh NS, Aditham AJ, Landry MP (2019) DNA nanostructures coordinate gene silencing in mature plants. Proc Natl Acad Sci 116(15):7543–7548
19. Cunningham FJ, Goh NS, Demirer GS, Matos JL, Landry MP (2018) Nanoparticle-mediated delivery towards advancing plant genetic engineering. Trends Biotechnol 36(9):882–897
20. Yadav K, Sahu KK, Gnanakani SPE, Sure P, Vijayalakshmi R, Sundar VD, Pradhan M (2023) Biomedical applications of nanomaterials in the advancement of nucleic acid therapy: mechanistic challenges, delivery strategies, and therapeutic applications. Int J Biol Macromol 241:124582
21. Lowe K, Wu E, Wang N, Hoerster G, Hastings C, Cho MJ, Gordon-Kamm W (2016) Morphogenic regulators baby boom and Wuschel improve monocot transformation. Plant Cell 28(9):1998–2015
22. Xu P, Zhong Y, Xu A, Liu B, Zhang Y, Zhao A, Fu F (2024) Application of developmental regulators for enhancing plant regeneration and genetic transformation. Plants 13(9):1272
23. Osmani Z, Kulka M (2025) Form and function: the factors that influence the efficacy of nanomaterials for gene transfer to plants. Molecules 30(3):446
24. Wu K, Xu C, Li T, Ma H, Gong J, Li X, Hu X (2023) Application of nanotechnology in plant genetic engineering. Int J Mol Sci 24(19):14836
25. Padayachee J, Singh M (2020) Therapeutic applications of CRISPR/Cas9 in breast cancer and delivery potential of gold nanomaterials. Nano 7:1849543520983196
26. Duan L, Ouyang K, Xu X, Xu L, Wen C, Zhou X, Liang Y (2021) Nanoparticle delivery of CRISPR/Cas9 for genome editing. Front Genet 12:673286
27. KhokharVoytas A, Shahbaz M, Maqsood MF, Zulfiqar U, Naz N, Iqbal UZ, AlShaqhaa MA (2023) Genetic modification strategies for enhancing plant resilience to abiotic stresses in the context of climate change. Funct Integr Genomics 23(3):283
28. Mmbando GS (2024) The use of nanotechnology in genetic modification: a recent promising technology for enhancing crop productivity. J Crop Improv 38(6):618–638
29. Manna I, Bandyopadhyay M (2019) A review on the biotechnological aspects of utilizing engineered nanoparticles as delivery systems in plants. Plant Gene 17:100167
30. Alex S, Tiwari A (2015) Functionalized gold nanoparticles: synthesis, properties and applications—a review. J Nanosci Nanotechnol 15(3):1869–1894
31. Singh RP (2019) Potential of biogenic plant-mediated iron and iron oxide nanoparticles and their utility. In: Plant nanobionics: approaches in nanoparticles, biosynthesis, and toxicity. Springer, Cham, pp 77–113
32. Albukhaty S, Sulaiman GM, Al-Karagoly H, Mohammed HA, Hassan AS, Alshammari AAA, Khan RA (2024) Iron oxide nanoparticles: the versatility of the magnetic and functionalized nanomaterials in targeting drugs, and gene deliveries with effectual magnetofection. J Drug Deliv Sci Technol 99:105838

33. Zhao X, Meng Z, Wang Y, Chen W, Sun C, Cui B, Cui H (2017) Pollen magnetofection for genetic modification with magnetic nanoparticles as gene carriers. Nat Plants 3(12):956–964
34. Martin-Ortigosa S, Valenstein JS, Lin VSY, Trewyn BG, Wang K (2012) Gold functionalized mesoporous silica nanoparticle mediated protein and DNA codelivery to plant cells via the biolistic method. Adv Funct Mater 22(17):3576–3582
35. Thiruvengadathan R, Korampally V, Ghosh A, Chanda N, Gangopadhyay K, Gangopadhyay S (2013) Nanomaterial processing using self-assembly-bottom-up chemical and biological approaches. Rep Prog Phys 76(6):066501
36. Tripathy S, Rodrigues J, Shimpi NG (2023) Top-down and bottom-up approaches for synthesis of nanoparticles. Nanobiomater Perspect Med Appl Diagn Treat Dis 145:92–130
37. Banerjee A, Sarkar A, Acharya K, Chakraborty N (2021) Nanotechnology: an emerging hope in crop improvement. Lett Appl NanoBioSci 10(4):2784–2803
38. Sharma B, Lakra U, Sharma R, Sharma SR (2022) A comprehensive review on nanopesticides and nanofertilizers—A boon for agriculture. In: Nano-enabled agrochemicals in agriculture. CRC Press, Boca Raton, pp 273–290
39. Ghidan AY, Al-Antary TM, Awwad AM (2017) Aphidicidal potential of green synthesized magnesium hydroxide nanoparticles using Olea europaea leaves extract. ARPN J Agric Biol Sci 12:293–301
40. Sanzari I, Leone A, Ambrosone A (2019) Nanotechnology in plant science: to make a long story short. Front Bioeng Biotechnol 7:120
41. Wadhwa P, Sharma S, Sahu S, Sharma A, Kumar D (2022) A review of nanoparticles characterization techniques. Curr Nanomater 7(3):202–214
42. Kim DY, Kadam A, Shinde S, Saratale RG, Patra J, Ghodake G (2018) Recent developments in nanotechnology transforming the agricultural sector: a transition replete with opportunities. J Sci Food Agric 98(3):849–864
43. Rani S, Kumari N, Sharma V (2023) Uptake, translocation, transformation and physiological effects of nanoparticles in plants. Arch Agron Soil Sci 69(9):1579–1599
44. Gikas GD, Parlakidis P, Mavropoulos T, Vryzas Z (2022) Particularities of fungicides and factors affecting their fate and removal efficacy: a review. Sustainability 14(7):4056
45. Ayyaril SS, Shanableh A, Bhattacharjee S, Rawas-Qalaji M, Cagliani R, Shabib AG (2023) Recent progress in micro and nano-encapsulation techniques for environmental applications: a review. Results Eng 18:101094
46. Tsuji K (2001) Microencapsulationofpesticidesandtheirimprovedhandlingsafety. J Microencapsul 18:137–147
47. Balah MA, Pudake RN (2019) Use nanotools for weed control and exploration of weed plants in nanotechnology. In: Nanoscience for sustainable agriculture. Springer, Cham, pp 207–231
48. Fleischer MA, O'Neill R (1999) Ehwald, the pore size of non-graminaceous plant cell wall is rapidly decreased by borate ester cross-linking of the pectic polysaccharide rhamnogalacturon II. Plant Physiol 121:829–838
49. de Almeida MS, Susnik E, Drasler B, Taladriz-Blanco P, Petri-Fink A, Rothen-Rutishauser B (2021) Understanding nanoparticle endocytosis to improve targeting strategies in nanomedicine. Chem Soc Rev 50(9):5397–5434
50. Avellan A, Yun J, Morais BP, Clement ET, Rodrigues SM, Lowry GV (2021) Critical review: role of inorganic nanoparticle properties on their foliar uptake and in planta translocation. Environ Sci Technol 55(20):13417–13431
51. Da Silva LC, Oliva MA, Azevedo AA, De Araujo JM (2006) Responses of Restinga plant species to pollution from an iron pelletization factory. Water Air Soil Pollut 175:241–256
52. Khan I, Awan SA, Rizwan M, Hassan ZU, Akram MA, Tariq R, Xie W (2022) Nanoparticle's uptake and translocation mechanisms in plants via seed priming, foliar treatment, and root exposure: A review. Environ Sci Pollut Res 29(60):89823–89833
53. Sankaran S, Mishra A, Ehsani R, Davis C (2010) A review of advanced techniques for detecting plant diseases. Comput Electron Agric 72(1):1–13

54. Ma X, Geiser-Lee J, Deng Y, Kolmakov A (2010) Interactions between engineered nanoparticles (ENPs) and plants: phytotoxicity, uptake and accumulation. Sci Total Environ 408(16):3053–3061
55. Zhou D, Jin S, Li L, Wang Y, Weng N (2011) Quantifying the adsorption and uptake of CuO nanoparticles by wheat root based on chemical extractions. J Environ Sci 23(11):1852–1857
56. Schwab F, Zhai G, Kern M, Turner A, Schnoor JL, Wiesner MR (2016) Barriers, pathways and processes for uptake, translocation and accumulation of nanomaterials in plants–critical review. Nanotoxicology 10(3):257–278
57. Mahra S, Tripathi S, Tiwari K, Sharma S, Mathew S, Kumar V, Sharma S (2024) Harnessing nanotechnology for sustainable agriculture: from seed priming to encapsulation. Plant Nano Biol. 11:100124
58. Barik TK, Sahu B, Swain V (2008) Nanosilica—from medicine to pest control. Parasitol Res 103:253–258
59. Chandrashekharaiah M, Kandakoor SB, Basana Gowda G, Kammar V, Chakravarthy AK (2015) Nanomaterials: a review of their action and application in pest management and evaluation of DNA-tagged particles. In: New horizons in insect science: towards sustainable pest management. Springer, Cham, pp 113–126
60. Layet C, Auffan M, Santaella C, Chevassus-Rosset C, Montes M, Ortet P, Doelsch E (2017) Evidence that soil properties and organic coating drive the phytoavailability of cerium oxide nanoparticles. Environ Sci Technol 51(17):9756–9764
61. Grohmann L, Keilwagen J, Duensing N, Dagand E, Hartung F, Wilhelm R, Sprink T (2019) Detection and identification of genome editing in plants: challenges and opportunities. Front Plant Sci 10:236
62. Khanna K, Ohri P, Bhardwaj R (2023) Nanotechnology and CRISPR/Cas9 system for sustainable agriculture. Environ Sci Pollut Res 30(56):118049–118064
63. Chen F, Alphonse M, Liu Q (2020) Strategies for nonviral nanoparticle-based delivery of CRISPR/Cas9 therapeutics. Wiley Interdiscip Rev Nanomed Nanobiotechnol 12(3):e1609
64. Nimesh S, Gupta N, Chandra R (2011) Cationic polymer based nanocarriers for delivery of therapeutic nucleic acids. J Biomed Nanotechnol 7(4):504–520
65. Sharma B, Chauhan I, Kumar G, Bhardwaj K, Tiwari RK (2024) Nanoparticle carriers: A new era of precise CRISPR/Cas9 gene editing. Microrna 14(2):101–111
66. Demirer GS, Silva TN, Jackson CT, Thomas JB, Ehrhardt D, Rhee SY, Landry MP (2021) Nanotechnology to advance CRISPR–Cas genetic engineering of plants. Nat Nanotechnol 16(3):243–250
67. Mohamed MA, Abd-Elsalam KA (2019) Magnetic nanoparticles: a unique gene delivery system in plant science. In: Magnetic nanostructures: environmental and agricultural applications. Springer, Cham, pp 95–108
68. Yan Y, Zhu X, Yu Y, Li C, Zhang Z, Wang F (2022) Nanotechnology strategies for plant genetic engineering. Adv Mater 34(7):2106945
69. Shen H, Huang X, Min J, Le S, Wang Q, Wang X, Xiao J (2019) Nanoparticle delivery systems for DNA/RNA and their potential applications in nanomedicine. Curr Top Med Chem 19(27):2507–2523
70. Lu B, Lim JM, Yu B, Song S, Neeli P, Sobhani N, Chai D (2024) The next-generation DNA vaccine platforms and delivery systems: advances, challenges and prospects. Front Immunol 15:1332939
71. Kim T, Nam K, Kim YM, Yang K, Roh YH (2021) DNA-assisted smart nanocarriers: progress, challenges, and opportunities. ACS Nano 15(2):1942–1951
72. Lu J, Wang Y, Jin M, Yuan Z, Bond P, Guo J (2020) Both silver ions and silver nanoparticles facilitate the horizontal transfer of plasmid-mediated antibiotic resistance genes. Water Res 169:115229
73. Rustgi S, Naveed S, Windham J, Zhang H, Demirer GS (2022) Plant biomacromolecule delivery methods in the 21st century. Front Genome Edit 4:1011934

74. Wang Z, Cao S, Xu X, He Y, Shou W, Munaiz ED, Shen J (2023) Application and expansion of virus-induced gene silencing for functional studies in vegetables. Horticulturae 9(8):934
75. Manchanda P, Sharma D, Vanshika, Kaur A (2024) Genetic engineering in crop plants using tissue culture and tissue culture-free environment. In: Genetic engineering of crop plants for food and health security, vol 2. Springer Nature, Singapore, pp 67–96
76. Lin S, Zhang Q, Bai S, Yang L, Qin G, Wang L, Han X (2025) Beyond species and spatial boundaries: enabling long-distance gene silencing in plants via guanidinium-siRNA nanoparticles. Plant Biotechnol J. 23:1165–1177
77. Zhou J, Zhang J, Chen S, Lin Q, Zhu R, Wang L, Yang H (2023) Direct cytoplasmic delivery of RNAi therapeutics through a non-lysosomal pathway for enhanced gene therapy. Acta Biomater 170:401–414
78. Prasad A, Sidhic J, Sarbadhikary P, Narayanankutty A, George S, George BP, Abrahamse H (2024) Role of metal nanoparticles in organogenesis, secondary metabolite production and genetic transformation of plants under in vitro condition: a comprehensive review. Plant Cell Tissue Organ Culture 158(2):33
79. Ahmad F, Ashraf N, Ashraf T, Zhou RB, Yin DC (2019) Biological synthesis of metallic nanoparticles (MNPs) by plants and microbes: their cellular uptake, biocompatibility, and biomedical applications. Appl Microbiol Biotechnol 103:2913–2935
80. Friedrich RP, Cicha I, Alexiou C (2021) Iron oxide nanoparticles in regenerative medicine and tissue engineering. Nano 11(9):2337
81. Jat SK, Bhattacharya J, Sharma MK (2020) Nanomaterial based gene delivery: a promising method for plant genome engineering. J Mater Chem B 8(19):4165–4175
82. Wang D, Zhong M, Sun Y, Fang L, Sun Y, Qi B, Li Y (2021) Effects of pH on ultrasonic-modified soybean lipophilic protein nanoemulsions with encapsulated vitamin E. LWT 144:111240
83. Zhang Y, Huang C, Xiong R (2024) Advanced materials for intracellular delivery of plant cells: strategies, mechanisms and applications. Mater Sci Eng R Rep 160:100821
84. Huang RY, Liu ZH, Weng WH, Chang CW (2021) Magnetic nanocomplexes for gene delivery applications. J Mater Chem B 9(21):4267–4286
85. Shivashakarappa K, Marriboina S, Dumenyo K, Taheri A, Yadegari Z (2025) Nanoparticle-mediated gene delivery techniques in plant systems. Front Nanotechnol 7:1516180
86. Sandhya D, Jogam P, Allini VR, Abbagani S, Alok A (2020) The present and potential future methods for delivering CRISPR/Cas9 components in plants. J Genet Eng Biotechnol 18(1):25
87. Huq TB, Anil Kumar Jeeja P, Dam SK, Das SS, Vivero-Escoto JL (2025) Recent applications of mesoporous silica nanoparticles in gene therapy. Adv Healthc Mater 14:2404781
88. Nair R, Varghese SH, Nair BG, Maekawa T, Yoshida Y, Kumar DS (2010) Nanoparticulate material delivery to plants. Plant Sci 179(3):154–163
89. Tang GP, Zeng JM, Gao SJ, Ma YX, Shi L, Li Y, Wang S (2003) Polyethylene glycol modified polyethylenimine for improved CNS gene transfer: effects of PEGylation extent. Biomaterials 24(13):2351–2362
90. Verma SK, Kumar P, Mishra A, Khare R, Singh D (2024a) Green nanotechnology: illuminating the effects of bio-based nanoparticles on plant physiology. Biotechnol Sustain Mater 1(1):1
91. Patel J, Kumar GS, Roy H, Maddiboyina B, Leporatti S, Bohara RA (2024) From nature to nanomedicine: bioengineered metallic nanoparticles bridge the gap for medical applications. Discover Nano 19(1):85
92. Kirubakaran D, Wahid JBA, Karmegam N, Jeevika R, Sellapillai L, Rajkumar M, SenthilKumar KJ (2025) A comprehensive review on the green synthesis of nanoparticles: advancements in biomedical and environmental applications. Biomed Mater Dev 4:388–413
93. Jogaiah S, Paidi MK, Venugopal K, Geetha N, Mujtaba M, Udikeri SS, Govarthanan M (2021) Phytotoxicological effects of engineered nanoparticles: an emerging nanotoxicology. Sci Total Environ 801:149809
94. Stampoulis D, Sinha SK, White JC (2009) Assay-dependent phytotoxicity of nanoparticles to plants. Environ Sci Technol 43(24):9473–9479

95. Zannat M, Jahan I, Rahaman MA, Yeasmin NJ, & Angon PB (2024) Potentiality of metal nanoparticles in precision and sustainable agriculture
96. Tang SC, Lo IM (2013) Magnetic nanoparticles: essential factors for sustainable environmental applications. Water Res 47(8):2613–2632
97. Keles E, Song Y, Du D, Dong WJ, Lin Y (2016) Recent progress in nanomaterials for gene delivery applications. Biomater Sci 4(9):1291–1309
98. Wahhab A, Abideen ZU, Afzal W, Samad A, Farooqi HU, Ahmad R, Umair M (2025) Next-generation biodegradable polymers: innovations for A greener planet. Spectr Eng Sci 3(3):473–498
99. Vinzant K, Rashid M, Clouse DE, Ghosh P, Quadir M, Davis VA, Khodakovskaya MV (2025) From plants to plants: plant-derived biological polymers as sustainable and safe Nanocarriers for direct delivery of DNA to plant cells. Nano Lett 25:5572–5581
100. Verma C, Verma DK, Berdimurodov E, Barsoum I, Alfantazi A, Hussain CM (2024b) Green magnetic nanoparticles: a comprehensive review of recent progress in biomedical and environmental applications. J Mater Sci 59(2):325–358
101. Tolisano C, Del Buono D (2023) Biobased: biostimulants and biogenic nanoparticles enter the scene. Sci Total Environ 885:163912
102. Coles D, Frewer LJ (2013) Nanotechnology applied to European food production–A review of ethical and regulatory issues. Trends Food Sci Technol 34(1):32–43
103. Mukherjee A, Maity A, Pramanik P, Shubha K, Joshi DC, Wani SH (2019) Public perception about use of nanotechnology in agriculture. In: Advances in phytonanotechnology. Academic, pp 405–418
104. Choudharÿ V (2022) Strategies for biotech companies to scale up commercially through effective project and risk management. BULLET 1(5):952–965
105. Roubert F, Beuzelin-Ollivier MG, Hofmann-Amtenbrink M, Hofmann H, Hool A (2016) "Nanostandardization" in action: implementing standardization processes in a multidisciplinary nanoparticle-based research and development project. NanoEthics 10:41–62
106. Ma X, Wang X (2022) Multigenerational exposure of plants to engineered nanoparticles (ENPs). In: Plant exposure to engineered nanoparticles. Academic, pp 195–207
107. da Silva Júnior AH, Mulinari J, de Oliveira PV, de Oliveira CRS, Júnior FWR (2022) Impacts of metallic nanoparticles application on the agricultural soils microbiota. J Hazard Mater Adv 7:100103
108. Goswami S, Singh SK, Mukherjee S, Pandey A, Priyadarshini A, Patra A, Reddy GP (2024) Nanofertilizers: a novel technology for enhancing nutrient use efficiency of crops and a relevance to agroforestry. In: Agroforestry to combat global challenges: current prospects and future challenges. Springer Nature, Singapore, pp 293–322
109. Sundararajan N, Habeebsheriff HS, Dhanabalan K, Cong VH, Wong LS, Rajamani R, Dhar BK (2024) Mitigating global challenges: harnessing green synthesized nanomaterials for sustainable crop production systems. Global Chall 8(1):2300187

Silicon Nanoparticles in Plant Genetic Engineering

Tabasum Mohi Ud Din, Ishfaq Majid Hurrah, Loulaq Banday, and Sayanti Mandal

Abstract Silicon nanoparticles (SiNPs) have emerged as a promising tool in plant genetic engineering, offering innovative solutions for the delivery of genetic material, enhancement of transformation efficiency, and improvement of crop resilience. Their unique physicochemical properties—including high surface area, tunable porosity, biocompatibility, and ease of surface functionalization—make SiNPs highly effective nano-carriers for DNA, RNA, and genome-editing tools such as CRISPR/Cas systems. Recent advances demonstrate their potential in facilitating precise gene delivery, improving stress tolerance, and even enabling DNA-free genome editing in important crops. Despite these exciting developments, challenges remain, including optimization of delivery methods across diverse plant species, assessment of potential phytotoxicity, and evaluation of environmental and biosafety risks. This chapter provides a comprehensive overview of the properties, synthesis, and applications of SiNPs in plant biotechnology, highlighting key case studies, addressing potential risks, and offering perspectives on future research directions. Ultimately, SiNP-based technologies have the potential to revolutionize modern agriculture by contributing to sustainable crop improvement and food security.

Keywords Silicon nanoparticles · Plant genetic engineering · Nano-carriers · CRISPR/Cas delivery · Gene editing · siRNA · Abiotic stress tolerance · Biosafety · Nanotechnology · Sustainable agriculture

---

T. M. U. Din
Department of Botany, Amar Singh College, Srinagar, J&K, India

I. M. Hurrah
Plant Biotechnology Division, CSIR-Indian Institute of Integrative Medicine, Sanat Nagar, Srinagar, J&K, India

L. Banday
Department of Biotechnology, University of Kashmir, Srinagar, J&K, India

S. Mandal (✉)
Department of Chemistry and Biochemistry, Sharda School of Engineering and Sciences, Sharda University, Greater Noida, Uttar Pradesh, India

J.-T. Chen (ed.), *Nano-Delivering for Plant Genetic Engineering*, Smart Nanomaterials Technology, https://doi.org/10.1007/978-981-95-4165-2_5

1 Introduction

Agriculture today is confronted with a range of complex challenges, including climate change, declining soil health, increasing biotic and abiotic stresses, and the urgent need to ensure food security for a rapidly growing global population. To address these challenges, plant genetic engineering has emerged as a transformative approach for developing crops with improved traits such as enhanced stress tolerance, resistance to pests and diseases, improved nutrient use efficiency, and superior yield and quality [1, 2]. Traditional methods of plant transformation, including Agrobacterium-mediated transformation and biolistic (gene gun) delivery, have significantly contributed to crop improvement. However, these methods are often associated with several limitations, such as low transformation efficiency, tissue damage, species or genotype specificity, random insertion of transgenes, and regulatory and biosafety concerns. These limitations have prompted researchers to explore alternative approaches that are more efficient, precise, and broadly applicable across different plant species [3, 4]. In parallel with advances in genetic engineering, nanotechnology has emerged as a promising frontier in agriculture, offering novel solutions for crop production, protection, and improvement [5]. Among the wide array of nanomaterials investigated for agricultural applications, silicon nanoparticles (SiNPs) have garnered considerable attention due to their unique physicochemical properties. Silicon, a naturally abundant and beneficial element in many plant species, has long been known to enhance plant growth, strengthen cell walls, improve tolerance to abiotic stresses such as drought and salinity, and enhance resistance to pathogens. When silicon is engineered into the nanoscale, it exhibits properties such as a high surface-to-volume ratio, tunable particle size, controlled porosity, biocompatibility, and the ability to be surface-functionalized with biomolecules. These characteristics make SiNPs particularly attractive candidates for applications in plant biotechnology, especially as vehicles for gene delivery [6, 7].

The integration of SiNPs into plant genetic engineering offers several key advantages. First, SiNPs have demonstrated the ability to efficiently carry and deliver various types of genetic material including plasmid DNA, RNA molecules, and gene-editing components such as CRISPR/Cas systems across the formidable barriers posed by plant cell walls and membranes. Unlike traditional methods, nanoparticle-mediated delivery minimizes tissue injury and does not rely on species-specific interactions, potentially enabling the transformation of a wider range of crops, including recalcitrant or non-model species [8, 9]. Second, SiNPs offer excellent biocompatibility and relatively low toxicity compared to metallic nanoparticles, making them a safer option for applications in living systems [8]. Third, the surface of SiNPs can be easily modified with ligands, peptides, or other molecules to enable targeted delivery to specific plant tissues or subcellular compartments, thereby improving the precision and efficiency of genetic engineering interventions. Additionally, beyond their role as gene delivery vectors, SiNPs themselves may confer agronomic benefits by enhancing nutrient uptake, improving photosynthetic performance, or inducing plant defense responses thus providing a dual benefit. This chapter aims to provide a comprehensive overview of the emerging role of silicon nanoparticles in plant genetic engineering. It will begin by describing the fundamental properties of SiNPs, including their synthesis, characterization, and unique features relevant to plant systems [10, 11]. The chapter will then explore the mechanisms by which SiNPs interact with plant cells, including uptake pathways, transport mechanisms, subcellular localization, and their impact on plant physiology and gene expression. A central focus will be placed on the use of SiNPs as nanocarriers for the delivery of genetic material, highlighting recent advances, key experimental evidence, and representative case studies. The chapter will also discuss the current challenges and limitations in the use of SiNPs, including issues related to particle stability, cytotoxicity, environmental impact, and regulatory considerations. Finally, the chapter will present future perspectives on the integration of SiNPs with advanced genome-editing technologies, the development of smart delivery systems, and the broader implications for sustainable agriculture and crop improvement [12, 13].

The key objectives of this chapter are to provide an in-depth understanding of silicon nanoparticles and their properties relevant to plant systems; to review the mechanisms of SiNP uptake and interaction with plants; to evaluate their potential and demonstrated applications in plant genetic engineering; to identify current research gaps and challenges; and to outline promising directions for future research and practical application. By the end of this chapter, readers will gain a comprehensive understanding of how SiNPs can revolutionize plant genetic transformation strategies and contribute to the development of next-generation crops capable of meeting the pressing agricultural challenges of the twenty-first century (Fig. 1).

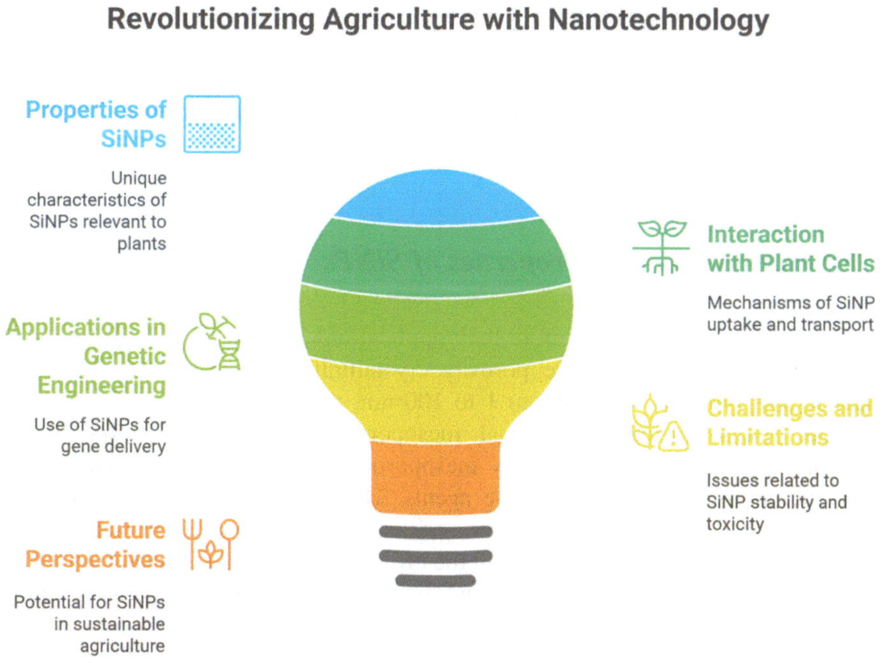

Fig. 1 Application of nanotechnology in agriculture

Table 1 Properties and synthesis methods of silicon nanoparticles (SiNPs)

Property	Description	References
Particle size	Typically 1–100 nm; smaller sizes improve cellular uptake	[14, 15]
Surface area	High surface-to-volume ratio; enhances loading capacity for nucleic acids and proteins	[16, 17]
Porosity	Mesoporous SiNPs allow high cargo loading and controlled release	[18, 19]
Biocompatibility	Generally low toxicity; biodegrades into silicic acid, a natural plant nutrient	[20, 21]
Synthesis methods	Sol–gel, microemulsion, Stöber synthesis, chemical vapor deposition; each method affects size, shape, and porosity	[22, 23]

2 Silicon Nanoparticles: Properties and Synthesis

Silicon nanoparticles (SiNPs) have emerged as one of the most promising nanomaterials for applications in plant biotechnology and genetic engineering due to their unique physicochemical properties, biocompatibility, and environmental friendliness. Silicon (Si) itself is the second most abundant element in the Earth's crust and plays an important role in plant biology, contributing to cell wall strengthening, stress alleviation, and pathogen resistance in many species. When scaled down to the nanoscale (typically <100 nm), silicon exhibits distinct characteristics that make SiNPs particularly attractive for biological applications. SiNPs possess a high surface area-to-volume ratio, tunable pore sizes, and excellent mechanical, optical, and thermal properties. Their surfaces can be readily modified or functionalized with biomolecules such as DNA, RNA, peptides, or antibodies, enhancing their ability to act as targeted delivery systems (Table 1). Furthermore, SiNPs are biodegradable and tend to degrade into orthosilicic acid, a bioavailable form of silicon that is generally regarded as non-toxic to plants and animals, making them an environmentally safe choice compared to many metallic nanoparticles [24, 25].

2.1 *Physicochemical Properties of SiNPs*

Properties that are most relevant to their applications in plant systems include particle size, shape, surface charge, porosity, crystallinity, and surface chemistry. The particle size typically ranges from 1 to 100 nm, and depending on the synthesis method, SiNPs can exhibit spherical, mesoporous, or amorphous morphologies. The large surface area provided by mesoporous SiNPs allows for high loading capacities of genetic cargo or active agents. The surface charge (often modified using cationic or anionic groups) plays a critical role in mediating interactions with plant cell membranes and determining the efficiency of cellular uptake. Crystalline SiNPs may exhibit different optical and electronic properties compared to amorphous forms, which can influence their behavior in biological systems. Importantly, the ability to modify the SiNP surface with functional groups or ligands provides opportunities to engineer nanoparticles for specific targeting or controlled release applications in plants [26, 27].

2.2 *Common Methods of SiNP Synthesis*

The synthesis of silicon nanoparticles has been extensively studied, and several methods are employed depending on the desired properties, scalability, and application. Broadly, SiNP synthesis approaches can be divided into physical, chemical, and biological (or "green") methods.

2.2.1 Physical Methods

Among the various approaches for synthesizing silicon nanoparticles (SiNPs), physical methods hold a prominent place due to their simplicity and ability to produce high-purity nanomaterials without the need for chemical reagents. Two of the most widely used physical methods are laser ablation and ball milling, each offering distinct advantages and limitations. Laser ablation involves the use of high-energy laser pulses directed onto a silicon target, typically placed in a liquid or gaseous environment. When the laser beam strikes the silicon surface, it causes rapid heating, melting, and vaporization of the material. As the vaporized silicon cools in the surrounding medium, it condenses into nanoparticles, forming a stable colloidal suspension if performed in liquid, or a nanoparticle powder if performed in gas. This technique is particularly valued for its ability to generate ultra-pure SiNPs with controlled surface chemistry and minimal contamination, as it avoids the use of surfactants or chemical precursors. Additionally, parameters such as laser wavelength, pulse duration, energy density, and the type of surrounding medium can be finely tuned to influence particle size, morphology, and crystalline structure. However, despite these advantages, laser ablation faces challenges related to low production yields, high equipment and energy costs, and limited scalability, which restrict its application mainly to laboratory-scale synthesis and niche applications [8, 28].

Ball milling, in contrast, is a top-down mechanical approach where bulk silicon materials are ground into nanoscale particles using high-energy milling media, such as steel or ceramic balls, in a rotating cylindrical chamber. The repeated collision and friction between the balls and the silicon chunks gradually reduce the particle size, eventually producing SiNPs. Ball milling is considered a relatively straightforward and cost-effective method, and it can be scaled up for industrial-level production. Additionally, it allows the incorporation of process control agents or dopants during milling to modify the physicochemical properties of the resulting nanoparticles. However, one of the main drawbacks of ball milling is the broad particle size distribution often produced, necessitating subsequent post-processing steps such as sieving, centrifugation, or sonication to achieve uniform size and dispersion. Moreover, the mechanical nature of the process can introduce defects, contamination from the milling media, and changes in crystallinity, which may require additional purification or annealing steps [28, 29].

2.2.2 Chemical Methods

Chemical methods are among the most widely employed and versatile approaches for the synthesis of silicon nanoparticles (SiNPs), offering excellent control over particle size, morphology, surface characteristics, and porosity. Among these, the sol-gel process is one of the most established and extensively studied techniques. In the sol-gel method, silicon alkoxides—most commonly tetraethyl orthosilicate (TEOS)—serve as the silicon precursor. The process begins with the hydrolysis of TEOS in the presence of water and a catalyst, typically under acidic or basic conditions, leading to the formation of silanol groups (Si–OH). These groups then undergo a condensation reaction to form siloxane bonds (Si–O–Si), gradually creating a three-dimensional silica network or gel. This wet gel is then subjected to drying and calcination at elevated temperatures, which remove the residual organic groups and water, resulting in the formation of dry SiNPs [8, 30].

One of the key advantages of the sol-gel process is its fine tenability, by adjusting reaction parameters such as pH, temperature, precursor concentration, type of solvent, catalyst, and aging time, researchers can achieve precise control over the size, porosity, surface area, and surface functional groups of the resulting nanoparticles. Furthermore, the sol-gel method is compatible with a variety of dopants, surfactants, or templating agents, enabling the synthesis of mesoporous or functionalized SiNPs tailored for specific applications. However, the method also has limitations, including long processing times, the need for post-synthesis heat treatments, and potential shrinkage or cracking during drying [24, 31]. Another prominent chemical approach is chemical vapor deposition (CVD), a gas-phase method that involves the thermal decomposition of volatile silicon-containing precursors, such as silane (SiH_4), at high temperatures in the presence of a substrate. The precursor gases react and decompose, leading to the nucleation and growth of SiNPs on the substrate surface or within the gas phase, depending on the reactor configuration. CVD is renowned for producing high-purity, crystalline SiNPs with excellent uniformity and reproducibility, making it particularly valuable for electronic, optical, and energy-related applications. Furthermore, the process can be scaled for continuous production and allows the incorporation of dopants during synthesis. Nevertheless, the CVD process has notable challenges, including the need for specialized reactors, high vacuum or controlled gas environments, elevated temperatures (typically above 600 °C), and significant energy input. These factors make CVD more suitable for industrial or high-tech manufacturing applications rather than routine laboratory synthesis [32, 33].

2.2.3 Green Synthesis

In recent years, green synthesis approaches have gained substantial attention as eco-friendly and cost-effective alternatives to conventional physical and chemical methods for synthesizing silicon nanoparticles (SiNPs). Unlike traditional methods, which often rely on harsh chemicals, high temperatures, and energy-intensive processes, green synthesis leverages biological agents such as plant extracts, microorganisms, or biopolymers to mediate the formation of SiNPs under mild and environmentally benign conditions [26, 34]. One of the most widely explored green synthesis strategies involves the use of plant extracts, which are rich in naturally occurring biomolecules such as polyphenols, flavonoids, alkaloids, terpenoids, sugars, and organic acids. These compounds serve dual functions: they act as reducing agents, facilitating the reduction of silicon precursors (such as sodium silicate or tetraethyl orthosilicate) to elemental silicon or silicon dioxide, and as stabilizing or capping agents, preventing nanoparticle aggregation and controlling particle size and shape. For example, leaf, stem, or flower extracts from plants such as *Azadirachta indica* (neem), Aloe vera, *Camellia sinensis* (green tea), and *Ocimum sanctum* (holy basil) have been successfully used to synthesize SiNPs with controlled morphology and desirable surface characteristics [35–37].

In addition to plant-based methods, microbial-mediated synthesis using bacteria, fungi, or algae has emerged as a promising green approach. Certain microorganisms possess the ability to uptake silicon precursors from their environment and biomineralize them into nanoscale silica structures, often as part of their natural defense or metabolic processes. For instance, species of Fusarium, Aspergillus, and Bacillus have been reported to synthesize SiNPs extracellularly or intracellularly, producing nanoparticles with unique surface functionalities. Similarly, biopolymers such as starch, chitosan, and alginate can act as green templates or scaffolds, enabling the controlled nucleation and growth of SiNPs while imparting additional biocompatibility. A major advantage of green-synthesized SiNPs is their enhanced biocompatibility and reduced toxicity compared to particles produced by conventional methods. The natural surface coatings derived from biological agents can improve the interaction of SiNPs with plant cells, reduce oxidative stress, and lower environmental risks, making them particularly attractive for agricultural and biomedical applications. Moreover, green synthesis typically operates under mild reaction conditions such as ambient temperature, aqueous solvents, and atmospheric pressure resulting in lower energy consumption and minimal hazardous waste generation [38, 39].

2.3 Characterization Techniques

The characterization of SiNPs is a critical step to confirm their size, shape, surface properties, and crystallinity before application in biological systems. Common techniques include scanning electron microscopy (SEM) and transmission electron microscopy (TEM) for morphological analysis, X-ray diffraction (XRD) for determining crystalline structure, Fourier transform infrared spectroscopy (FTIR) for surface functional groups, and dynamic light scattering (DLS) for size distribution and zeta potential measurements. These analyses help ensure that the synthesized SiNPs possess the desired characteristics for efficient and safe use in plant genetic engineering. Silicon nanoparticles represent a versatile and highly tunable class of nanomaterials with enormous potential in plant biotechnology. Their unique physicochemical properties, combined with a range of synthesis methods, offer researchers a powerful platform to design innovative delivery systems for genetic materials, paving the way for advances in plant genetic engineering and sustainable agriculture [32, 34].

3 Mechanisms of Interaction Between Silicon Nanoparticles and Plants

Understanding the interaction between silicon nanoparticles (SiNPs) and plants is essential for harnessing their potential in plant genetic engineering. Once SiNPs are introduced to the plant system either through foliar application, root uptake, seed priming, or direct tissue exposure they engage in a series of complex physical, chemical, and biological interactions that determine their fate and function inside plant tissues.

3.1 Uptake and Transport of SiNPs in Plant Tissues

The uptake and transport of silicon nanoparticles (SiNPs) in plants are critical processes that determine their effectiveness in delivering genetic material, enhancing nutrient use efficiency, or providing stress tolerance. These processes generally begin at two main entry points: the root-soil interface and the leaf surface, depending on the method of application. When SiNPs are applied to the roots, their uptake involves two major pathways: the apoplastic pathway and the symplastic pathway. The apoplastic route allows SiNPs to move through the extracellular spaces and the cell walls without crossing the plasma membrane, while the symplastic route involves the transport of SiNPs from cell to cell through plasmodesmata after crossing the plasma membrane. The plant cell wall, with pore sizes typically ranging between 5 to 20 nm, acts as the first physical barrier to SiNP entry. Nanoparticles smaller than this threshold can diffuse through the cell wall pores and reach the plasma membrane. In contrast, larger SiNPs may require alternative mechanisms, such as endocytosis, where the plasma membrane engulfs the particles to form vesicles that internalize the nanoparticles (Fig. 2). Additionally, environmental factors such as pH fluctuations, salinity, and the presence of cell wall-loosening enzymes

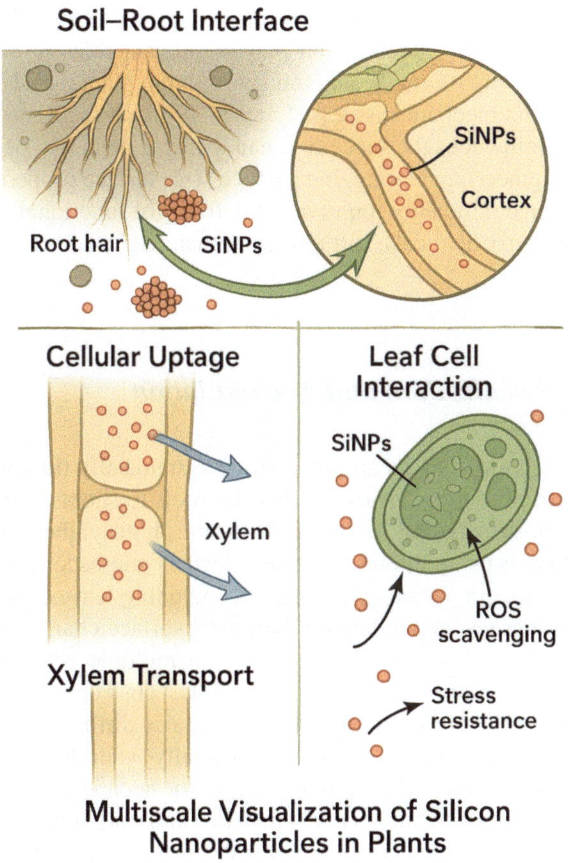

Fig. 2 Interface of silicon nanoparticles in plants

can transiently modify the architecture of the cell wall, enabling the entry of slightly larger particles [40, 41].

When SiNPs are applied to the leaf surface, they encounter a different set of entry pathways. SiNPs can penetrate through stomatal openings, which serve as natural pores for gas exchange, or through cuticular cracks and imperfections in the leaf surface. In some cases, SiNPs may also enter via direct uptake by epidermal cells, especially if the nanoparticles have been surface-functionalized with positively charged groups, peptides, or ligands that enhance adhesion to the negatively charged plant cell surfaces. This surface modification can improve the interaction between SiNPs and plant tissues, promoting more efficient penetration and translocation. Once inside the plant, the transport of SiNPs depends on the plant's vascular system [26, 31]. SiNPs can be translocated from roots to shoots via the xylem, the tissue responsible for water and mineral transport, often following the transpiration stream. When absorbed through the leaves, SiNPs may move through the phloem, facilitating redistribution to sink tissues such as roots, developing leaves, or reproductive organs. Factors such as nanoparticle size, surface charge, hydrophobicity, and the presence of surface coatings all play significant roles in determining the efficiency and specificity of uptake and transport within plant tissues [30, 42].

Understanding the mechanisms underlying SiNP uptake and movement is crucial, as it directly influences the success of nanoparticle-mediated delivery systems in plant genetic engineering. Optimizing particle size, surface properties, and application methods can significantly improve the delivery of functional cargos, such as DNA, RNA, or agrochemicals, ultimately enhancing plant performance and resilience.

3.2 Cellular Localization and Accumulation

Once inside plant tissues, SiNPs are transported systemically through the vascular system. The xylem primarily carries SiNPs from roots to shoots, while the phloem may play a role in distributing nanoparticles from photosynthetic tissues to sink organs. Recent studies using fluorescence and electron microscopy have shown that SiNPs can accumulate in various plant organs, including leaves, stems, roots, and even reproductive tissues, depending on their size, surface charge, and functionalization. At the cellular level, SiNPs interact with multiple cellular components, including the plasma membrane, cytoplasm, vacuoles, chloroplasts, and even nuclei [43, 44]. Cellular uptake of SiNPs can occur via passive diffusion, carrier-mediated transport, or endocytosis-like mechanisms, especially when the nanoparticles are functionalized with biomolecules such as peptides or nucleic acids. Upon entering the cytoplasm, SiNPs can localize near organelles and may trigger a range of cellular responses, including changes in reactive oxygen species (ROS) levels, membrane stability, enzymatic activity, and gene expression patterns. Importantly, the impact of SiNPs on cellular physiology largely depends on their dose, size, and surface characteristics. Low or optimized concentrations of SiNPs often stimulate beneficial effects, such as enhanced antioxidant activity, improved photosynthesis, and upregulation of stress-responsive genes, whereas excessive concentrations may cause cytotoxicity, membrane damage, or oxidative stress [45, 46].

3.3 Biocompatibility and Phytotoxicity Considerations

A critical aspect of using silicon nanoparticles (SiNPs) in plant systems is the careful assessment of their biocompatibility and potential phytotoxicity. While SiNPs hold great promise for enhancing plant performance and delivering genetic material, their interaction with plant tissues must be carefully managed to ensure safe and effective outcomes. Numerous studies have demonstrated that SiNPs, when applied at appropriate concentrations, generally exhibit minimal or no toxicity to plants. In fact, at optimal doses, SiNPs have been shown to enhance plant growth, improve photosynthetic efficiency, promote nutrient uptake, and increase tolerance to various abiotic stresses such as drought, salinity, and heavy metal toxicity [30, 47]. These beneficial effects are thought to result from the unique ability of SiNPs to interact with plant cell walls and membranes, modulate antioxidant enzyme activities, and improve water retention and nutrient transport within tissues. However, the positive impact of SiNPs is strongly dose-dependent, and exceeding certain concentration thresholds can lead to detrimental effects on plant health. At high doses or under unfavourable environmental conditions, SiNPs may disrupt cellular homeostasis, leading to the overproduction of reactive oxygen species (ROS), which causes oxidative stress and cellular damage. Common symptoms of SiNP-induced phytotoxicity include chlorosis (yellowing of leaves), reduced chlorophyll content, impaired root and shoot development, membrane damage, and inhibition of overall plant growth. Additionally, some studies have reported that certain surface modifications or small particle sizes can increase the likelihood of cellular penetration and accumulation, further raising the risk of toxicity [48, 49].

Beyond concentration, the physicochemical properties of SiNPs such as particle size, shape, surface charge, aggregation behaviour, and the presence of surface coatings play pivotal roles in determining their compatibility with plant systems. For example, smaller nanoparticles have a larger surface area-to-volume ratio, making them more reactive and potentially more toxic if not properly stabilized or functionalized [31, 46]. Similarly, positively charged or hydrophobic particles may interact more strongly with negatively charged plant cell membranes, increasing uptake but also potentially causing membrane disruption. To ensure the safe and responsible use of SiNPs in agricultural applications, it is essential to systematically evaluate their dose-response behaviour across different plant species and developmental stages. This includes identifying safe application ranges, understanding species-specific sensitivities, and conducting long-term assessments of their impact on plant physiology, soil health, and the surrounding environment. Furthermore, combining SiNPs with biocompatible coatings, such as polysaccharides or plant-based capping agents, can improve their stability and reduce unintended toxicity [50, 51].

3.4 *Influence on Cellular Processes and Gene Expression*

In the context of genetic engineering, the interaction of SiNPs with the plant cell wall and membrane is particularly significant because these barriers present major challenges for the delivery of genetic materials. SiNPs can facilitate the transport of nucleic acids, proteins, or CRISPR-Cas components across these barriers through several mechanisms. Surface-functionalized SiNPs can bind to genetic cargos and protect them from enzymatic degradation during transit. Upon contact with the plant cell surface, SiNPs may promote transient membrane permeabilization or trigger endocytic uptake, allowing the cargo to reach intracellular compartments such as the nucleus or plastids, where genetic modification takes place [31, 52].

Furthermore, the surface charge of SiNPs, often tailored to be positive, can enhance their affinity for the negatively charged nucleic acids and cell membranes, promoting efficient cellular internalization. The interaction between SiNPs and plants encompasses a series of sophisticated processes, including uptake through roots or leaves, systemic transport via the vascular system, cellular internalization, and modulation of physiological and molecular responses. These interactions not only determine the effectiveness of SiNPs as delivery vehicles in genetic engineering but also open avenues for broader applications in plant health and productivity. A deeper understanding of these mechanisms will allow researchers to optimize nanoparticle design and application strategies for more precise and efficient plant transformation technologies [53, 54].

4 Role of Silicon Nanoparticles in Plant Genetic Engineering

The use of silicon nanoparticles (SiNPs) in plant genetic engineering has emerged as a promising innovation to address several limitations associated with conventional transformation methods. Traditional approaches, such as Agrobacterium tumefaciens-mediated transformation and biolistic particle delivery (gene gun), have been widely used to introduce foreign genes into plants. While effective, these methods face challenges including species-specific limitations, low transformation efficiencies in some crops, random insertion of transgenes, potential for somaclonal variation, and tissue damage during delivery. SiNPs offer an alternative platform with the potential to overcome many of these hurdles by acting as nano-carriers for efficient, targeted, and minimally invasive delivery of genetic materials into plant cells.

4.1 SiNPs as Delivery Vehicles for Genetic Material

One of the major advantages of SiNPs in plant genetic engineering is their capacity to deliver a broad range of biomolecules into plant cells, including plasmid DNA, messenger RNA (mRNA), small interfering RNA (siRNA), microRNAs, proteins, and CRISPR/Cas components. This versatility allows SiNPs to support not only stable genetic transformation but also transient expression and gene-editing applications. For instance, plasmid DNA bound to the surface of functionalized SiNPs can be delivered into plant cells, where it can integrate into the genome or express transiently to produce the desired protein. More recently, the delivery of CRISPR/Cas9 ribonucleoprotein (RNP) complexes using SiNPs has attracted considerable attention because it enables precise genome editing without introducing foreign DNA, reducing regulatory concerns and improving public acceptance of genetically edited crops [55, 56]. The mechanism of delivery using SiNPs involves several coordinated steps. First, the SiNPs are typically functionalized with cationic polymers, peptides, or other biomolecules to enable strong binding to negatively charged nucleic acids or proteins. This functionalization also improves nanoparticle stability and enhances cellular uptake. Once applied to plant tissues whether by infiltration, foliar spray, seed coating, or incubation with protoplasts the SiNPs interact with the plant cell wall and plasma membrane. Their small size and surface modifications allow them to pass through cell wall pores or be internalized by endocytosis. After entering the cytoplasm, the genetic cargo can be released either passively or through controlled-release mechanisms (such as pH- or enzyme-responsive coatings) to reach its target site, including the nucleus or plastids, where gene expression or editing can occur [13, 57].

4.2 Advantages over Conventional Delivery Methods

Importantly, the use of SiNPs offers several distinct advantages over conventional delivery systems. First, SiNP-mediated delivery is often less damaging to plant tissues compared to physical methods like the gene gun, which rely on high-velocity particles to penetrate cells. Second, SiNPs provide enhanced protection to the genetic cargo, shielding it from degradation by nucleases and proteases in the extracellular environment. Third, the surface of SiNPs can be engineered for targeted delivery to specific cell types, tissues, or organelles, which improves the efficiency and precision of genetic modifications. Fourth, SiNPs enable species-independent delivery, expanding the range of transformable plant species, including recalcitrant crops that are difficult to transform using *Agrobacterium* [26, 57].

4.3 Examples of Successful Gene Delivery Using SiNPs

Recent experimental studies and proof-of-concept applications have demonstrated the feasibility and versatility of silica nanoparticles (SiNPs) in plant genetic engineering. One prominent example involves the use of mesoporous silica nanoparticles (MSNs), which possess a highly ordered pore structure and large surface area, allowing them to effectively load and transport biomolecules such as DNA, small interfering RNA (siRNA), and plasmid constructs. MSNs loaded with genetic cargo have been successfully employed to deliver DNA or siRNA into a variety of plant species, including *Arabidopsis thaliana*, tobacco (*Nicotiana tabacum*), cotton (*Gossypium hirsutum*), and maize (*Zea mays*), leading to transient gene expression or targeted gene silencing [17, 58]. These studies highlight the potential of SiNPs for facilitating both functional genomics research and trait development in important crops (Fig. 3).

In addition to conventional gene delivery, SiNPs have emerged as promising vehicles for the delivery of CRISPR/Cas systems, enabling precise genome editing in several plant models. By facilitating the targeted introduction of double-strand

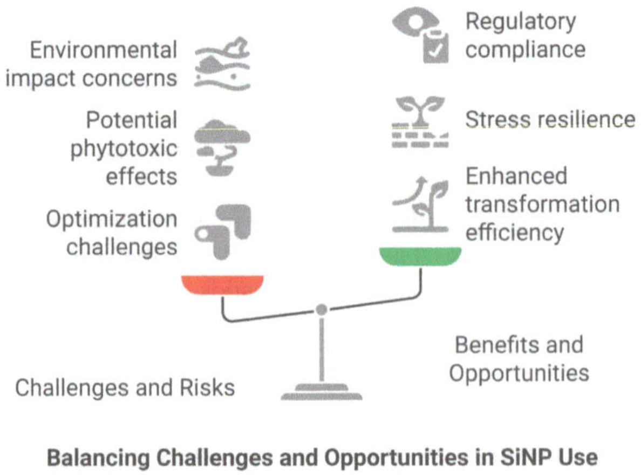

Fig. 3 Challenges and opportunities of silicon nanoparticle use in plants

breaks or base edits at specific genomic loci, SiNP-mediated CRISPR delivery has opened new avenues for developing crops with desirable traits, such as enhanced drought tolerance, improved resistance to pests and diseases, and superior nutritional profiles. This approach offers an alternative to traditional transformation methods, such as Agrobacterium-mediated transformation or biolistic bombardment, which are often limited by host specificity, tissue damage, or regulatory concerns [55, 56]. Researchers have also been exploring the integration of SiNPs with stimuli-responsive systems to further enhance delivery efficiency and achieve spatiotemporal control over gene expression. For example, the combination of SiNPs with magnetic nanoparticles enables magnetically guided delivery, improving targeting accuracy. Similarly, the incorporation of light-responsive or heat-sensitive coatings allows controlled release of genetic material in response to external cues such as laser irradiation or temperature shifts, adding an additional layer of precision to plant genetic engineering [59, 60].

Overall, SiNPs represent a highly adaptable and powerful platform for plant biotechnology, offering efficient, targeted, and minimally invasive delivery of genetic material. Their unique physicochemical properties, including tunable surface functionality, biocompatibility, and capacity for cargo loading, position them as a promising tool for advancing next-generation crop improvement strategies.

5 Applications and Case Studies

The application of silicon nanoparticles (SiNPs) in plant genetic engineering has gained significant momentum in recent years, with multiple research groups demonstrating their potential in laboratory and greenhouse settings. These applications span from enhancing transformation efficiency and delivering gene-editing tools to

Table 2 Applications of SiNPs in plant genetic engineering

Application	Plant model/crop	Outcome	References
DNA delivery (GFP, GUS reporter genes)	Arabidopsis, tobacco	Efficient transient expression	[61, 62]
siRNA delivery for gene silencing	Cotton (*Gossypium hirsutum*)	Pest resistance improvement	[57, 63]
CRISPR/Cas delivery for genome editing	Rice, wheat protoplasts	Targeted mutagenesis without DNA integration	[64, 65]
Drought tolerance enhancement	Maize, tomato	Improved stress-responsive gene expression	[66, 67]
Virus resistance (hairpin RNA delivery)	Tobacco	Protection against TMV infection	[68, 69]

improving stress resilience and crop performance (Table 2). This section highlights some of the key applications and case studies that illustrate the transformative role of SiNPs in plant biotechnology [70, 71].

5.1 Nano-Carriers for Nucleic Acid Delivery

One of the most prominent applications of SiNPs is as nano-carriers for nucleic acid delivery, particularly for DNA and RNA molecules involved in genetic transformation. For instance, mesoporous silica nanoparticles (MSNs) have been used to deliver plasmid DNA encoding reporter genes, such as green fluorescent protein (GFP) or β-glucuronidase (GUS), into *Arabidopsis thaliana* and tobacco (*Nicotiana tabacum*) leaves [61, 62]. These studies have shown efficient transient expression of the introduced genes, confirming the ability of SiNPs to protect genetic material and facilitate its uptake into plant cells. Similarly, in cotton (*Gossypium hirsutum*), SiNPs have been employed to deliver small interfering RNAs (siRNAs) targeting genes involved in pest resistance, leading to successful gene silencing and enhanced pest tolerance [57, 72].

5.2 Genome Editing Using CRISPR/Cas Systems

Another major area of application is genome editing using CRISPR/Cas systems, where SiNPs act as delivery vehicles for Cas9 protein and guide RNAs (gRNAs). For example, researchers have successfully used SiNP-based systems to deliver CRISPR-Cas9 ribonucleoproteins into rice (*Oryza sativa*) and wheat (*Triticum aestivum*) protoplasts, resulting in targeted mutagenesis of key agronomic genes. This DNA-free approach is particularly attractive because it avoids the integration of foreign DNA into the plant genome, which can simplify regulatory approval and improve public acceptance of edited crops [64, 65].

5.3 Improving Plant Tolerance to Abiotic and Biotic Stresses

SiNPs have also been investigated for improving plant tolerance to abiotic and biotic stresses, either through genetic manipulation or by directly enhancing physiological responses. For example, SiNPs loaded with drought-responsive transcription factors have been delivered into maize seedlings to induce the expression of stress tolerance genes. Separately, foliar sprays of SiNPs have been reported to enhance antioxidant enzyme activity and reduce reactive oxygen species accumulation in tomato (*Solanum lycopersicum*) under salinity stress, highlighting their dual role as both carriers and biostimulants [9, 66].

5.4 DNA Vaccines or Gene Constructs for Plant Virus Resistance

A noteworthy case study is the use of SiNPs to deliver DNA vaccines or gene constructs for plant virus resistance. In tobacco, SiNP-mediated delivery of hairpin RNA constructs targeting viral coat proteins has resulted in successful gene silencing and protection against tobacco mosaic virus (TMV) infection. This approach represents a promising alternative to traditional breeding or transgenic strategies for managing viral diseases in crops [68, 69, 73].

5.5 Seed Priming with SiNPs

In addition to experimental studies, seed priming with SiNPs has emerged as an innovative application. For example, priming wheat and rice seeds with SiNP solutions loaded with growth-promoting genes or signaling molecules has been shown to improve germination rates, seedling vigor, and stress tolerance. This strategy offers a scalable and cost-effective route for delivering genetic materials or bioactive agents directly to the next plant generation [31, 74].

5.6 Integration of SiNPs with Precision Agriculture Technologies

Researchers are exploring the integration of SiNPs with precision agriculture technologies, such as smart delivery systems that respond to environmental cues like light, temperature, or pH. These stimuli-responsive systems could enable controlled release of genetic materials at specific developmental stages or under particular stress conditions, thereby enhancing transformation precision and minimizing unintended effects. The diverse applications and case studies on SiNPs in plant genetic engineering demonstrate their vast potential to advance crop improvement, enhance stress resilience, and support sustainable agriculture. Continued research and interdisciplinary collaboration will be critical to refining these approaches, scaling them for commercial use, and ensuring their safe and effective integration into modern agricultural practices [49, 75].

6 Challenges, Risks, and Biosafety Considerations

While silicon nanoparticles (SiNPs) hold immense promise in advancing plant genetic engineering, their widespread adoption is accompanied by several challenges, risks, and biosafety concerns that must be thoroughly addressed. A clear understanding of these issues is essential to ensure the safe, efficient, and responsible use of SiNPs in agricultural biotechnology. One of the foremost challenges lies in the optimization of SiNP design and delivery systems for different plant species and tissues [76, 77]. The physical and chemical properties of SiNPs such as size, shape, surface charge, porosity, and functionalization strongly influence their uptake, transport, and interaction with plant cells. However, plants vary widely in their anatomical and physiological characteristics, including differences in cell wall composition, cuticle thickness, stomatal density, and vascular structure. What works effectively in model species like *Arabidopsis thaliana* or tobacco may not translate directly to important food crops such as wheat, maize, or rice (Fig. 4). Therefore,

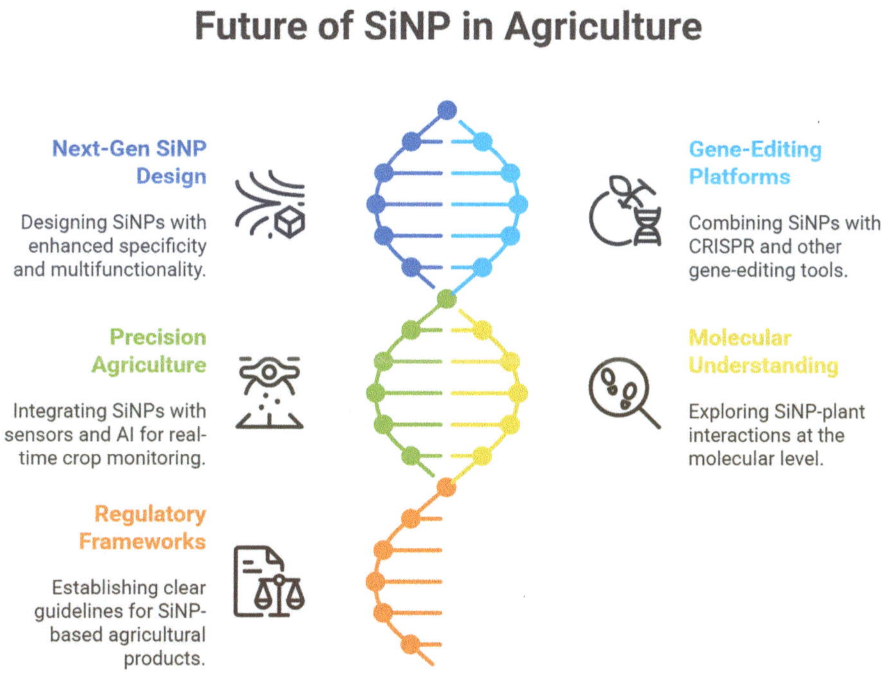

Fig. 4 Prospects of silicon nanoparticles in agriculture

tailoring SiNP formulations and delivery methods to specific crops and tissue types is a critical area of ongoing research [78, 79].

Another challenge involves controlling the dose, distribution, and fate of SiNPs inside plants. While low or optimized concentrations of SiNPs often promote beneficial outcomes such as enhanced transformation efficiency or stress resilience excessive levels may induce phytotoxic effects. These can include oxidative stress, membrane damage, nutrient imbalance, growth inhibition, or even cell death. Additionally, the long-term accumulation of SiNPs in plant tissues, including edible parts, raises concerns about potential impacts on food safety, human health, and animal consumption [72, 80]. A closely related concern is the potential environmental impact of SiNPs. Once applied in agricultural settings, nanoparticles may leach into the soil, water, and surrounding ecosystems, where they can interact with soil microbes, insects, and non-target plants. While silicon is a naturally abundant and generally considered low-toxicity element, the behaviour of engineered SiNPs with their unique nanoscale properties may differ significantly from bulk silicon or naturally occurring silica. Assessing the persistence, mobility, bioaccumulation, and ecotoxicity of SiNPs in the environment is therefore critical to minimizing unintended consequences [47, 81]. In addition, biosafety considerations related to genetic engineering itself remain central to the use of SiNPs. For example, the delivery of transgenes, gene-editing components, or RNA interference (RNAi) constructs into plants raises important questions about off-target effects, unintended gene modifications, horizontal gene transfer, and the potential for ecological disruption. While SiNPs can improve the precision and efficiency of delivery, they do not inherently eliminate these broader genetic risks (Table 3). Consequently, comprehensive risk assessment, containment measures, and molecular analyses are required to ensure that SiNP-based plant transformations meet national and international biosafety standards [82, 83].

The regulatory landscape surrounding nanotechnology and genetic engineering adds another layer of complexity. In many countries, the release of genetically modified organisms (GMOs) into the environment is subject to rigorous regulatory oversight, including environmental impact assessments, food safety evaluations, and public consultations. However, there is currently no universal framework for

Table 3 Challenges and biosafety considerations for SiNPs in agriculture

Challenge/concern	Description	References
Dose optimization	Excess SiNPs can induce oxidative stress, inhibit growth	[63, 80]
Environmental impact	Potential leaching into soil and water, interaction with soil microbiota	[47, 81]
Off-target genetic effects	Unintended gene edits or silencing during delivery of CRISPR/siRNA	[82, 83]
Regulatory uncertainty	Lack of global consensus on nanomaterial use and genome-edited crop regulation	[84, 85]
Public perception and acceptance	Concerns about GMOs and nanotechnology in food systems	[86, 87]

assessing the risks of nanomaterials in agriculture, and regulatory definitions of what constitutes a "genetically modified crop" vary across jurisdictions. For instance, crops modified through DNA-free genome editing using CRISPR-Cas systems delivered by SiNPs may fall into regulatory grey zones, with some countries classifying them as non-GMO and others treating them as regulated products. Clarifying these regulatory pathways and developing robust risk assessment protocols will be essential for the responsible commercialization of SiNP-based technologies [84, 85].

Finally, public perception and acceptance represent significant social challenges. Nanotechnology and genetic engineering are both subjects of public debate, and the combination of the two may amplify concerns about environmental safety, food security, ethical considerations, and corporate control over agricultural technologies. Transparent communication, stakeholder engagement, and inclusive governance will be critical to building public trust and ensuring that SiNP-based innovations align with societal values and priorities [86, 87]. Furthermore, while SiNPs offer transformative opportunities for plant genetic engineering, their responsible development requires careful attention to technical optimization, environmental safety, biosafety, regulatory compliance, and public engagement. Addressing these challenges proactively will pave the way for the safe and sustainable use of SiNPs in modern agriculture, unlocking their full potential to enhance crop productivity, resilience, and sustainability.

7 Future Perspectives and Conclusions

The integration of silicon nanoparticles (SiNPs) into plant genetic engineering is rapidly evolving, offering transformative potential for advancing crop improvement and sustainable agriculture. As research progresses, several future perspectives are emerging that could shape the development and application of SiNP-based technologies over the coming decades. One key future direction is the design of next-generation SiNPs with enhanced specificity, efficiency, and multi-functionality [88, 89]. Advances in materials science are expected to enable the synthesis of SiNPs with highly controlled size, shape, surface chemistry, and porosity, allowing precise tuning of their interactions with plant cells. Functionalization with targeting ligands such as peptides, antibodies, or aptamers may enable SiNPs to deliver genetic cargo specifically to particular cell types, tissues, or organelles, reducing off-target effects and improving transformation precision. Moreover, the development of stimuli-responsive SiNPs that release their cargo in response to environmental cues, such as pH, temperature, or light, could provide spatiotemporal control over gene delivery and expression [90, 91]. Another promising area is the integration of SiNPs with emerging gene-editing platforms, including CRISPR-Cas variants, base editors, and prime editors. SiNPs offer an ideal platform for delivering these advanced molecular tools into plant cells without introducing foreign DNA, helping to create precisely edited, transgene-free crops. This approach holds particular promise for addressing global agricultural challenges such as climate change, evolving pathogens, soil degradation, and the need for enhanced nutritional quality in staple crops [70, 92]. The combination of SiNPs with precision agriculture technologies and digital farming tools also offers exciting opportunities. By integrating SiNP-based delivery systems with sensors, drones, and artificial intelligence, it may become possible to monitor crop health in real time and apply genetic or protective treatments only where and when they are needed. This precision approach could significantly reduce inputs, minimize environmental impact, and increase the overall efficiency and sustainability of agricultural production systems [93, 94].

From a research standpoint, further exploration of SiNP–plant interactions at the molecular, cellular, and whole-plant levels will be crucial. Understanding the mechanisms of SiNP uptake, transport, degradation, and clearance in plants will help optimize their design and application. Multi-omics approaches, including transcriptomics, proteomics, and metabolomics, can shed light on plant responses to SiNP exposure, revealing both desired and unintended effects. On the regulatory and societal front, establishing clear guidelines for the evaluation and oversight of SiNP-based agricultural products will be essential [95, 96]. Regulatory agencies, scientists, industry stakeholders, and policymakers will need to work collaboratively to develop science-based frameworks for assessing the safety, efficacy, and environmental impact of SiNP-enabled genetic engineering. Equally important is transparent communication with the public to build trust, address ethical concerns, and promote informed decision-making about the adoption of nanotechnology in food and agriculture. In conclusion, silicon nanoparticles represent a versatile and powerful tool in the field of plant genetic engineering, with the potential to revolutionize crop breeding, stress tolerance, and productivity [85, 86]. While significant progress has been made in demonstrating their applications, challenges related to optimization, biosafety, environmental impact, and public acceptance remain. By fostering interdisciplinary research, responsible innovation, and proactive engagement with society, the full potential of SiNP-based technologies can be harnessed to contribute to global food security, environmental sustainability, and agricultural resilience in the twenty-first century.

References

1. Dayıoğlu MA, Turker U (2021) Digital transformation for sustainable future-agriculture 4.0: a review. J Agric Sci 27(4):373–399
2. Liu Y, Ma X, Shu L, Hancke GP, Abu-Mahfouz AM (2020) From industry 4.0 to agriculture 4.0: current status, enabling technologies, and research challenges. IEEE Trans Industr Inform 17(6):4322–4334
3. Bisht IS, Rana JC, Yadav R, Ahlawat SP (2020) Mainstreaming agricultural biodiversity in traditional production landscapes for sustainable development: the Indian scenario. Sustainability 12(24):10690
4. Sekhar M, Rastogi M, Rajesh CM, Saikanth DRK, Rout S, Kumar S, Patel AK (2024) Exploring traditional agricultural techniques integrated with modern farming for a sustainable future: A review. Journal of Scientific Research and Reports 30(3):185–198

5. Narang C, Kant R, Mandal S (2024) Nanosensors for soil and nutrients analytics. In: Shahzad R, Fiaz S, Qayyum A, Ul Islam M, Lee IJ (eds) Revolutionizing agriculture: A comprehensive exploration of Agri-nanotechnology. Nanotechnology in the life sciences. Springer, Cham. https://doi.org/10.1007/978-3-031-76000-6_8
6. Lv Z, Jiang R, Chen J, Chen W (2020) Nanoparticle-mediated gene transformation strategies for plant genetic engineering. Plant J 104(4):880–891
7. Yan Y, Zhu X, Yu Y, Li C, Zhang Z, Wang F (2022) Nanotechnology strategies for plant genetic engineering. Adv Mater 34(7):2106945
8. Roychoudhury A (2020) Silicon-nanoparticles in crop improvement and agriculture. International Journal on Recent Advancement in Biotechnology & Nanotechnology 3(1):2582–1571
9. Wang L, Ning C, Pan T, Cai K (2022a) Role of silica nanoparticles in abiotic and biotic stress tolerance in plants: a review. Int J Mol Sci 23(4):1947
10. Siddiqui H, Ahmed KBM, Sami F, Hayat S (2020) Silicon nanoparticles and plants: current knowledge and future perspectives. In: Sustainable agriculture reviews 41: nanotechnology for plant growth and development. Springer, Cham, pp 129–142
11. Squire HJ, Tomatz S, Voke E, González-Grandío E, Landry M (2023) The emerging role of nanotechnology in plant genetic engineering. Nature Reviews Bioengineering 1(5):314–328
12. El-Shetehy M, Moradi A, Maceroni M, Reinhardt D, Petri-Fink A, Rothen-Rutishauser B, Schwab F (2021) Silica nanoparticles enhance disease resistance in Arabidopsis plants. Nat Nanotechnol 16(3):344–353
13. Mukarram M, Petrik P, Mushtaq Z, Khan MMA, Gulfishan M, Lux A (2022) Silicon nanoparticles in higher plants: uptake, action, stress tolerance, and crosstalk with phytohormones, antioxidants, and other signalling molecules. Environ Pollut 310:119855
14. Sarkar MM, Mathur P, Mitsui T, Roy S (2022) A review on functionalized silica nanoparticle amendment on plant growth and development under stress. Plant Growth Regul 98(3):421–437
15. Wang L, Pan T, Gao X, An J, Ning C, Li S, Cai K (2022b) Silica nanoparticles activate defense responses by reducing reactive oxygen species under Ralstonia solanacearum infection in tomato plants. NanoImpact 28:100418
16. Matlou GG, Abrahamse H (2021) Hybrid inorganic-organic core-shell nanodrug systems in targeted photodynamic therapy of cancer. Pharmaceutics 13(11):1773
17. Hussein HA, Nazir MS, Azra N, Qamar Z, Seeni A, Tengku Din TADAA, Abdullah MA (2022) Novel drug and gene delivery system and imaging agent based on marine diatom biosilica nanoparticles. Mar Drugs 20(8):480
18. Punz B, Johnson L, Geppert M, Dang HH, Horejs-Hoeck J, Duschl A, Himly M (2022) Surface functionalization of silica nanoparticles: strategies to optimize the immune-activating profile of carrier platforms. Pharmaceutics 14(5):1103
19. Zhang Y, Yu Y, Yang Y, Wang Y, Yu C (2024) Engineered silica nanoparticles for nucleic acid delivery. Small Methods 8(3):2300812
20. Mohanty S, Chakraborty S, Das M, Paul S (2022) Role of nanomaterials in phytoremediation of tainted soil. In: Phytoremediation Technology for the Removal of heavy metals and other contaminants from soil and water. Elsevier, Amsterdam, pp 329–353
21. Kumari A, Rana V, Yadav SK, Kumar V (2023) Nanotechnology as a powerful tool in plant sciences: recent developments, challenges and perspectives. Plant Nano Biology 5:100046
22. Emam-Djomeh Z, Ekrami M, Mobahi N, Esmaeily R, Tousi SS (2024) Nanoparticles and methods involved in their synthesis. In: Nanotechnology in the food industry. CRC Press, Singapre, pp 1–19
23. Saha A, Mishra P, Biswas G, Bhakta S (2024) Greening the pathways: a comprehensive review of sustainable synthesis strategies for silica nanoparticles and their diverse applications. RSC Adv 14(16):11197–11216
24. Naaz H, Rawat K, Saffeullah P, Umar S (2023) Silica nanoparticles synthesis and applications in agriculture for plant fertilization and protection: A review. Environ Chem Lett 21(1):539–559
25. Rastogi A, Tripathi DK, Yadav S, Chauhan DK, Živčák M, Ghorbanpour M, Brestic M (2019) Application of silicon nanoparticles in agriculture. 3 Biotech 9:1–11

26. Bhat JA, Rajora N, Raturi G, Sharma S, Dhiman P, Sanand S, Deshmukh R (2021) Silicon nanoparticles (SiNPs) in sustainable agriculture: major emphasis on the practicality, efficacy and concerns. Nanoscale Adv 3(14):4019–4028
27. Ijaz U, Ahmed T, Rizwan M, Noman M, Shah AA, Azeem F, Ali S (2023) Rice straw based silicon nanoparticles improve morphological and nutrient profile of rice plants under salinity stress by triggering physiological and genetic repair mechanisms. Plant Physiol Biochem 201:107788
28. Yan G, Huang Q, Zhao S, Xu Y, He Y, Nikolic M, Zhu Z (2024) Silicon nanoparticles in sustainable agriculture: synthesis, absorption, and plant stress alleviation. Front Plant Sci 15:1393458
29. Ostapchuk GO (2024) Eco-friendly synthesis of silica nanoparticles and their applications. Materials Research Foundations, Millersville, p 169
30. Goswami P, Mathur J, Srivastava N (2022) Silica nanoparticles as novel sustainable approach for plant growth and crop protection. Heliyon 8(7):e09908
31. Naidu S, Pandey J, Mishra LC, Chakraborty A, Roy A, Singh IK, Singh A (2023) Silicon nanoparticles: synthesis, uptake and their role in mitigation of biotic stress. Ecotoxicol Environ Saf 255:114783
32. Shahbaz A, Ayaz M, Bin Khalid U, Liaqat L (2023) Sustainable synthesis of silica nanoparticles from agricultural waste and its utilization in modern technology: A review. Energy Sources, Part A: Recovery, Utilization, and Environmental Effects 45(1):1464–1484
33. Zeng F, Zheng K (2024) Role of silicon nanoparticles in bone regeneration. In: Silicon advances for sustainable agriculture and human health: increased nutrition and disease prevention. Cham, Springer Nature, pp 445–460
34. Karande SD, Jadhav SA, Garud HB, Kalantre VA, Burungale SH, Patil PS (2021) Green and sustainable synthesis of silica nanoparticles. Nanotechnol Environ Eng 6(2):29
35. Dheyab MA, Oladzadabbasabadi N, Aziz AA, Khaniabadi PM, Al-ouqaili MT, Jameel MS, Ghasemlou M (2024) Recent advances of plant-mediated metal nanoparticles: synthesis, properties, and emerging applications for wastewater treatment. J Environ Chem Eng 12:112345
36. Karmous I, Pandey A, Haj KB, Chaoui A (2020) Efficiency of the green synthesized nanoparticles as new tools in cancer therapy: insights on plant-based bioengineered nanoparticles, biophysical properties, and anticancer roles. Biol Trace Elem Res 196:330–342
37. Khoshnazar SM, Asadi A, Holghoomi R, Abdolmaleki A (2023) Green synthesis of silica nanoparticles/nanocomposites for biomedical applications: a Narraitive review. Biochem Suppl Ser B Biomed Chem 17(2):41–49
38. Abd-Elsalam KA, Periakaruppan R, Rajeshkumar S (2021) Agri-waste and microbes for production of sustainable nanomaterials. Elsevier, Amsterdam
39. Mbatha LS, Akinyelu J, Chukwuma CI, Mokoena MP, Kudanga T (2023) Current trends and prospects for application of green synthesized metal nanoparticles in cancer and COVID-19 therapies. Viruses 15(3):741
40. Ali S (2024) Exploring silicon nanoparticle uptake, transport, and biological functions in plants. In: Silicon advances for sustainable agriculture and human health: increased nutrition and disease prevention. Springer Nature, Cham, pp 237–263
41. Manokari M, Dey A, Faisal M, Alatar AA, Joshee N, Shekhawat MS (2023) Structural alterations of Cymbopogon citratus (DC.) Stapf leaves and roots caused by silicon nanoparticles during in vitro propagation. Ind Crop Prod 197:116648
42. Sharma S, Muddassir M, Muthusamy S, Vaishnav PK, Singh M, Sharma D, Shanmugam V (2020) A non-classical route of efficient plant uptake verified with fluorescent nanoparticles and root adhesion forces investigated using AFM. Sci Rep 10(1):19233
43. Choudhary S, Konopková AS (2024) Silicon nanoparticles vs trace elements toxicity: modus operandi and its omics bases. In: Mineral nutrition and plant stress tolerance. Frontiers Media, London
44. Younis AA, Khattab H, Emam MM (2020) Impacts of silicon and silicon nanoparticles on leaf ultrastructure and TaPIP1 and TaNIP2 gene expressions in heat stressed wheat seedlings. Biol Plant 64(1):343–352

45. Aqeel U, Aftab T, Naeem M, Gill SS (2023) Silicon nanoparticle–mediated metal stress tolerance in crop plants. In: Biostimulants in alleviation of metal toxicity in plants. Academic, NewYork, pp 321–351
46. Yadav M, George N, Dwibedi V (2023) Emergence of toxic trace elements in plant environment: insights into potential of silica nanoparticles for mitigation of metal toxicity in plants. Environ Pollut 333:122112
47. Okeke ES, Nweze EJ, Ezike TC, Nwuche CO, Ezeorba TPC, Nwankwo CEI (2023) Silicon-based nanoparticles for mitigating the effect of potentially toxic elements and plant stress in agroecosystems: a sustainable pathway towards food security. Sci Total Environ 898:165446
48. Ahmed T, Masood HA, Noman M, Al-Huqail AA, Alghanem SM, Khan MM, Li B (2023) Biogenic silicon nanoparticles mitigate cadmium (cd) toxicity in rapeseed (Brassica napus L.) by modulating the cellular oxidative stress metabolism and reducing cd translocation. J Hazard Mater 459:132070
49. Mahawar L, Ramasamy KP, Suhel M, Prasad SM, Živčák M, Brestic M, Skalický M (2023) Silicon nanoparticles: comprehensive review on biogenic synthesis and applications in agriculture. Environ Res 232:116292
50. Raza A, Khandelwal K, Pandit S, Singh M, Kumar S, Rustagi S, Prasad R (2024) Exploring the potential of metallic and metal oxide nanoparticles for reinforced disease management in agricultural systems: A comprehensive review. Environ Nanotechnol Monit Manage 22:100998
51. Tritean N, Trică B, Dima ŞO, Capră L, Gabor RA, Cimpean A, Constantinescu-Aruxandei D (2024) Mechanistic insights into the plant biostimulant activity of a novel formulation based on rice husk nanobiosilica embedded in a seed coating alginate film. Front Plant Sci 15:1349573
52. Shabbaj II, Madany MM, Tammar A, Balkhyour MA, AbdElgawad H (2021) Silicon dioxide nanoparticles orchestrate carbon and nitrogen metabolism in pea seedlings to cope with broomrape infection. Environ Sci Nano 8(7):1960–1977
53. Li Y, Xi K, Liu X, Han S, Han X, Li G, Zhu Y (2023) Silica nanoparticles promote wheat growth by mediating hormones and sugar metabolism. J Nanobiotechnol 21(1):2
54. Rashad YM, El-Sharkawy HH, Belal BE, Abdel Razik ES, Galilah DA (2021) Silica nanoparticles as a probable anti-oomycete compound against downy mildew, and yield and quality enhancer in grapevines: field evaluation, molecular, physiological, ultrastructural, and toxicity investigations. Front Plant Sci 12:763365
55. Mostafa M, Ahmed FK, Alghuthaymi M, Abd-Elsalam KA (2021) Inorganic smart nanoparticles: a new tool to deliver CRISPR systems into plant cells. In: CRISPR and RNAi systems. Elsevier, Amsterdam, pp 661–686
56. Vats S, Kumawat S, Brar J, Kaur S, Yadav K, Magar SG, Deshmukh R (2022) Opportunity and challenges for nanotechnology application for genome editing in plants. Plant Nano Biol 1:100001
57. Gad MA, Li MJ, Ahmed FK, Almoammar H (2020) Nanomaterials for gene delivery and editing in plants: challenges and future perspective. In: Multifunctional hybrid nanomaterials for sustainable agri-food and ecosystems. Elsevier, Amsterdam, pp 135–153
58. Niculescu VC (2020) Mesoporous silica nanoparticles for bio-applications. Front Mater 7:36
59. Song X, Qian H, Yu Y (2023) Nanoparticles mediated the diagnosis and therapy of glioblastoma: bypass or cross the blood–brain barrier. Small 19(45):2302613
60. Zheng X, Wu Y, Zuo H, Chen W, Wang K (2023) Metal nanoparticles as novel agents for lung cancer diagnosis and therapy. Small 19(18):2206624
61. Mujtaba M, Wang D, Carvalho LB, Oliveira JL, Espirito Santo Pereira AD, Sharif R, Fraceto LF (2021) Nanocarrier-mediated delivery of miRNA, RNAi, and CRISPR-Cas for plant protection: current trends and future directions. ACS Agric Sci Technol 1(5):417–435
62. Yong J, Wu M, Carroll BJ, Xu ZP, Zhang R (2024) Enhancing plant biotechnology by nanoparticle delivery of nucleic acids. Trends Genet 40(4):352–363
63. Wei Y, Wu Y, He Y (2024a) Soil application of SiNPs suppress pathogen population and improved plant resistance and antioxidant activity for disease management. S Afr J Bot 172:31–41

64. Liu L, Song Z, Tang J, Li Q, Sarkar B, Ellam RM, Wang H (2023) New insight into the mechanisms of preferential encapsulation of metal (loid) s by wheat phytoliths under silicon nanoparticle amendment. Sci Total Environ 875:162680
65. Gill S, Ramzan M, Naz G, Ali L, Danish S, Ansari MJ, Salmen SH (2024) Effect of silicon nanoparticle-based biochar on wheat growth, antioxidants and nutrients concentration under salinity stress. Sci Rep 14(1):6380
66. Wang T, Long H, Mao S, Jiang Z, Liu Y, He Y, Yan G (2024) Silicon nanoparticles improve tomato seed germination more effectively than conventional silicon under salt stress via regulating antioxidant system and hormone metabolism. Horticulturae 10(8):785
67. Sharf-Eldin AA, Alwutayd KM, El-Yazied AA, El-Beltagi HS, Alharbi BM, Eisa MA, Ibrahim MF (2023) Response of maize seedlings to silicon dioxide nanoparticles (SiO2NPs) under drought stress. Plants 12(14):2592
68. Madany MM, Saleh AM, Habeeb TH, Hozzein WN, AbdElgawad H (2020) Silicon dioxide nanoparticles alleviate the threats of broomrape infection in tomato by inducing cell wall fortification and modulating ROS homeostasis. Environ Sci Nano 7(5):1415–1430
69. Bhatt D, Bhatt MD, Nath M, Dudhat R, Sharma M, Bisht DS (2020) Application of nanoparticles in overcoming different environmental stresses. In: Protective chemical agents in the amelioration of plant abiotic stress: biochemical and molecular perspectives. Wiley, Hoboken, pp 635–654
70. Kumar P, Rajput VD, Singh AK, Agrawal S, Das R, Minkina T, Mishra SK (2024) Nano-assisted delivery tools for plant genetic engineering: a review on recent developments. Environ Sci Pollut Res 32:469–484
71. Sharma B, Chauhan I, Kumar G, Bhardwaj K, Tiwari RK (2024) Nanoparticle carriers: A new era of precise CRISPR/Cas9 gene editing. MicroRNA 14:101–101
72. Wei ZH, Zhao P, Ning XY, Xie YQ, Li Z, Liu XX (2024b) Nanomaterial-encapsulated dsRNA-targeting chitin pathway–A potential efficient and eco-friendly strategy against cotton aphid, Aphis gossypii (Hemiptera: Aphididae). J Agric Food Chem 72(38):20905–20917
73. Kant R, Rayaprolu V, McDonald K et al (2018) Curating viscoelastic properties of icosahedral viruses, virus-based nanomaterials, and protein cages. J Biol Phys 44:211–224. https://doi.org/10.1007/s10867-018-9491-x
74. El-Serafy RS, El-Sheshtawy ANA, Atteya AK, Al-Hashimi A, Abbasi AM, Al-Ashkar I (2021) Seed priming with silicon as a potential to increase salt stress tolerance in Lathyrus odoratus. Plants 10(10):2140
75. Silviana S, Dalanta F (2024) Advances in modified silica nanoparticles utilization for various applications: now and future. E3S Web Conf 503:09004
76. Omran BA, Baek KH (2022) Control of phytopathogens using sustainable biogenic nanomaterials: recent perspectives, ecological safety, and challenging gaps. J Clean Prod 372:133729
77. Weisany W, Razmi J, Eshaghabadi AH, Pashang D (2024) Silicon nanoparticles (SiNP): A novel and sustainable strategy for mitigating environmental stresses in plants. J Soil Sci Plant Nutr 24(2):2167–2191
78. Sidhu AK, Sharma M, Bhickchand Agrawal S, Pradip Bhavsar P, Samota MK (2024) Nanomaterial strategies for enhancing plant resilience in the face of temperature stress. CABI Agric Biosci 5(1):60
79. Ullah Z, Iqbal J, Abbasi BA, Ijaz S, Munir M, Yaseen T, Mahmood T (2025) Applications of nanoparticles in biofortification of crops: amplifying nutritional quality. In: Crop biofortification: biotechnological approaches for achieving nutritional security under changing climate. Wiley, Hoboken, pp 321–350
80. Masood HA, Qi Y, Zahid MK, Li Z, Ahmad S, Lv JM, Qi X (2024) Recent advances in nano-enabled immunomodulation for enhancing plant resilience against phytopathogens. Front Plant Sci 15:1445786
81. Jia L, Song Y, You F, Wang S, Rabiya U, Liu X, Wang L (2024) Ameliorating the detrimental effects of chromium in wheat by silicon nanoparticles and its enriched biochar. Sci Total Environ 950:175270

82. Manghwar H, Li B, Ding X, Hussain A, Lindsey K, Zhang X, Jin S (2020) CRISPR/Cas systems in genome editing: methodologies and tools for sgRNA design, off-target evaluation, and strategies to mitigate off-target effects. Adv Sci 7(6):1902312
83. Saakre M, Jaiswal S, Rathinam M, Raman KV, Tilgam J, Paul K, Pattanayak D (2024) Host-delivered RNA interference for durable pest resistance in plants: advanced methods, challenges, and applications. Mol Biotechnol 66(8):1786–1805
84. Marchant GE, Bazzi R, Bowman D, Connor J, Davis RA III, Kang E, Marchant M (2024) Learning from emerging technology governance for guiding quantum technology. UNSW Law Res. https://doi.org/10.2139/ssrn.4923230
85. Munawar N, Faheem M, Niamat A, Munir A, Khan SH, Zahoor MK, Ahmad A (2024) Regulatory, ethical, social, and biosafety concerns in genome-edited horticultural crops. In: CRISPRized horticulture crops. Academic, New York, pp 421–438
86. Dessie A, Zegeye ZB (2024) Review on: public perception of biotechnology on genetically modified crops, bio policy and intellectual property rights. Am J Polym Sci Technol 10(2):26–35
87. Krewski D, Saunders-Hastings P, Larkin P, Westphal M, Tyshenko MG, Leiss W, Coyle D (2022) Principles of risk decision-making. J Toxic Environ Health, Part B 25(5):250–278
88. Chu B, Chen Z, Shi H, Wu X, Wang H, Dong F, He Y (2023) Fluorescence, ultrasonic and photoacoustic imaging for analysis and diagnosis of diseases. Chem Commun 59(17):2399–2412
89. Nasreldin M, de Mulatier S, Delattre R, Ramuz M, Djenizian T (2020) Flexible and stretchable microbatteries for wearable technologies. Adv Mater Technol 5(12):2000412
90. Anwar S, Mir F, Yokota T (2023) Enhancing the effectiveness of oligonucleotide therapeutics using cell-penetrating peptide conjugation, chemical modification, and carrier-based delivery strategies. Pharmaceutics 15(4):1130
91. Deshmukh R, Sethi P, Singh B, Shiekmydeen J, Salave S, Patel RJ, Kumar A (2024) Recent review on biological barriers and host–material interfaces in precision drug delivery: advancement in biomaterial engineering for better treatment therapies. Pharmaceutics 16(8):1076
92. Rani N, Kumari K, Hooda V (2025) The role of nanoparticles in transforming plant genetic engineering: advancements, challenges and future prospects. Funct Integr Genomics 25(1):1–22
93. Pan Y, Gao S, Sun F, Yang H, Cao PF (2019) Polymer binders constructed through dynamic noncovalent bonds for high-capacity silicon-based anodes. Chemistry 25(47):10976–10994
94. Zhou Y, Hu L, Wang C, Ma L (2022) Evaluation and design of colored silicon nanoparticle systems using a bidirectional deep neural network. Nano 12(15):2715
95. Mardi A, Mohajjel Shoja H, Mohajel Kazemi E (2022) Comparative study of growth responses, photosynthetic pigment content, and gene expression pattern in tobacco plants treated with ZnO nano and ZnO bulk particles. J Nanopart Res 24(10):208
96. Sarraf M, Vishwakarma K, Kumar V, Arif N, Das S, Johnson R, Hasanuzzaman M (2022) Metal/metalloid-based nanomaterials for plant abiotic stress tolerance: an overview of the mechanisms. Plants 11(3):316

Bioinspired Nanoparticles in Plant Genetic Engineering

Ifrah Imran, Muhammad Waseem Sajjad, Anam Ishtiaq, Syed Ali Raza Naqvi, Naima Razzaq, Mohammad Abubakar Yasin, Mariam Akhtar, Ayesha Akram, Amna Rafique, Muhammad Arif, Muhammad Asif, Imran Amin, and Rubab Zahra Naqvi

Abstract Bioinspired Nanoparticles (BNPs) offer an innovative approach to plant genetic engineering by harnessing biological systems to create stable, biocompatible, and efficient gene delivery methods. Conventional gene delivery techniques, agrobacterium-mediated transformation and gene guns are tissue-specific, have off-target effects and have low efficacy. Whereas, BNPs allow the targeted delivery of genetic material such as RNA, DNA, and components of CRISPR-Cas, thereby improving genome editing precision. These nanoparticles (NPs) allow gene editing, particularly using CRISPR-Cas constructs and RNA methods, achieving a higher success rate with minimal cellular damage. BNPs significantly contribute to enhancing plants' disease resistance and abiotic stress tolerance, making them ideal candidates for improvements in many crop plants. Besides genetic engineering, BNPs also exhibit potential roles in biocontrol applications, especially against fungal, bacterial, and viral pathogens through targeted microbial delivery. Despite the promising potential, there are challenges such as assessing the long-term impacts of BNPs on the environment, optimizing BNPs for large-scale agricultural applications, and developing a standardized framework. This chapter explores the synthesis, applications, emerging trends, and challenges associated with BNPs to get new insights into their potential for crop improvement and sustainable agriculture.

Keywords BNPs · Biocompatible NPs · Food security · Gene delivery · Stress tolerance · Disease resistance · Green synthesis · Crop improvement · Environmental impact · Sustainable agriculture

---

I. Imran · M. W. Sajjad · A. Ishtiaq · N. Razzaq · M. A. Yasin · M. Akhtar · A. Akram · A. Rafique · M. Arif · M. Asif · I. Amin · R. Z. Naqvi (✉)
Agricultural Biotechnology Division, National Institute for Biotechnology and Genetic Engineering (NIBGE), Constituent College of Pakistan Institute of Engineering and Applied Sciences (PIEAS), Faisalabad, Pakistan

S. A. R. Naqvi
Department of Chemistry, Government College University Faisalabad, Faisalabad, Pakistan

J.-T. Chen (ed.), *Nano-Delivering for Plant Genetic Engineering*, Smart Nanomaterials Technology, https://doi.org/10.1007/978-981-95-4165-2_6

1 Background

A genetically modified (GM) crop is any plant whose genetic material has been modified specifically using genetic techniques in a manner that this alteration does not occur under natural conditions. The history of plant genetic modifications traces back approximately 10,000 years through selective breeding and artificial selection. In 1946, scientists discovered that genetic material could move across different species. In 1953, Watson and Crick presented their double-helical DNA structure and the basic concept of the central dogma of life. In 1973, Boyer and Cohen created a recombinant DNA molecule, opening a whole new era of scientific research, all this led us here, today, with advanced genome editing systems like clustered regularly interspaced short palindromic repeats (CRISPR)/CRISPR-associated protein system (Cas) as shown in Fig. 1 [4].

By 2030, the population of the world is expected to expand by 8.5 billion, increasing world food demand to 11.6 billion tons. If this trend persists, the number of people affected by hunger will surpass 670 million. Crop productivity should be increased by 60–100% to meet the need. However, various challenges like climate change and plant biotic and abiotic stresses are posing significant threats to food security [5]. Genetic engineering is a good option for coping with these and for the rapid development of improved crops, providing transgenic crops with the traits to resist various stresses and increase yield with enhanced nutritional values [6]. In the past, genome modifications relied on chemical and physical mutagenesis. The mutagens mostly used were ethyl methyl sulfonate (EMS) and fast-moving neutrons. The mutagenic populations were screened using targeted-induced local lesions in genomes (TILLING), but the problem with this method is that there is no control over the alteration of genes. Moreover, its screening was also labor-intensive [7]. To overcome this, scientists used protein-binding systems for genome editing techniques like meganucleases, zinc finger nucleases (ZFNs), transcription activator-like effector nucleases (TALENs), and RNA-binding systems like CRISPR Cas [8].

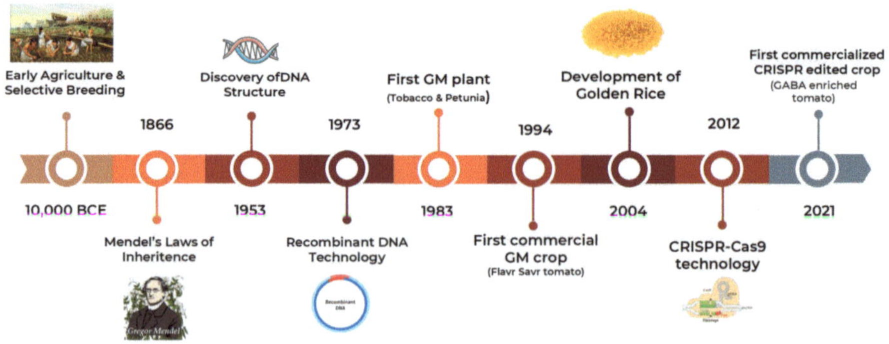

Fig. 1 This timeline depicts major events in plant biotechnology from early agriculture and Mendelian genetics to the rise of cutting-edge technologies like recombinant technology and CRISPR-Cas9

Over the past 30–40 years, the development of novel engineered materials, specifically nanomaterials, has increased significantly. These materials are applied in different fields, like food, optics, textiles, electronics, and cosmetics, influencing nearly every aspect of daily life [9]. Nanoparticles (NPs) are the fundamental units of nanotechnology, with sizes ranging from 1 to 100 nm [10]. Nanobiotechnology is a blend of nanotechnology and biotechnology and is a significantly developing field. It aims to understand how biological systems work, design advanced diagnostics and therapeutic solutions for human diseases, and create innovative materials and tools for biological use. Throughout recent history, nanobiotechnology has excelled in drug delivery systems, imaging, diagnostics, and bioinspired materials [11].

The terms "bioinspired" and "biomimetics" are often used in place of each other, with the minor difference that the former is about taking a cue from a natural system, while the latter is the direct imitation of the natural system. NPs that are synthesized biogenically have shown good size variability and stability. Additionally, the biological method facilitates the synthesis of NPs under natural pH, temperature, and pressure, making them eco-friendly and cost-effective. This makes the NP more stable, less toxic, and more efficient for use [12, 13]. BNPs, e.g., liposomes, can deliver the genetic cargo through the plant cell wall more efficiently than conventional methods like *Agrobacterium*-mediated genetic transformation due to their natural design. Moreover, these are customizable and site-specific according to specific tissue or species, since these particles mimic the natural system, so they are less likely to elicit any adverse effect [14].

2 NPs in Plant Biotechnology

Food security is challenged by climate change and population growth globally. Conventional chemical fertilizers and pesticides provide minimal benefits and substantial environmental harm [15–17]. Nanotechnology provides a solution through nano-sized agrochemicals, which bring advantages such as higher surface area, reduced nutrient or pesticide concentrations, more targeted delivery, and boosting crop productivity while curbing environmental impact [18, 19]. NPs have emerged as advanced tools, offering new ways to enhance yield, growth, and protection through precise delivery systems.

2.1 Types of Different NPs

NPs are classified into various types based on their structure, composition, and chemical properties [20]. This classification helps in understanding the varied forms and functions of NPs in industrial and scientific areas.

2.1.1 Metal NPs

Metal NPs are synthesized from metallic precursors. They are actively used in the agriculture sector for their excellent properties. Silver (Ag) NPs have bactericidal, fungicidal, antioxidant, antiviral, and anti-inflammatory properties [21, 22]. They are also involved in enhancing the rate of seed germination, chlorophyll contents, and yields [23–25]. The chitosan-coated Copper (Cu) NPs show antifungal activities against *Alternaria solani and Fusarium oxysporum* in tomato [26]. Manganese (Mn) NPs enhance the photosynthetic activity of plants [27]., ZnO-engineered, NiO-engineered and Fe2O3-engineered NPs are used as nano-fertilizers to regulate phytohormones, antioxidant action and in improving nutrient insufficiency, and [28, 29].

2.1.2 Carbon-Based NPs

Carbon-based NPs are known to have a wide range of applications in plant systems due to their unique structure, and mechanical and chemical properties. Graphene and carbon nanotubes (CNTs), penetrate plant cells and trigger reactive oxygen species (ROS) production [30–33]. They facilitate water storage in seeds and translocation to leaves in plants like rice and chickpea [34].

2.1.3 Polymeric NPs

Polymeric NPs are biodegradable and versatile nanocarriers, such as poly(lactic-co-glycolic) acid (PLGA) that improve pesticide efficiency through slow release, enhanced adhesion, and delayed degradation [35]. Alginate/chitosan NPs loaded with gibberellic acid boost tomato seed production [36], and mPEG-PLGA NPs facilitate metolachlor delivery to plants [37].

2.2 Mechanisms of Action in Plants

NPs interact with plant systems through complex mechanisms that range from molecular to organism scales. Their small size enables unique entry pathways, eluding traditional biological barriers, while surface properties permit cellular adhesion and internalization kinetics. Modern applications employ NPs as programmable platforms for genome editing cargo release, *in situ* biosensors, and abiotic stress mitigators.

NPs can penetrate plants via several modes including stomata, microscopic fissures on leaves and root hairs, and move in the plant system with the help of the plant's mechanisms of transport such as diffusion, phloem loading, and pressure-driven flow as depicted in Fig. 2. NP flow is dependent on its inclusive properties

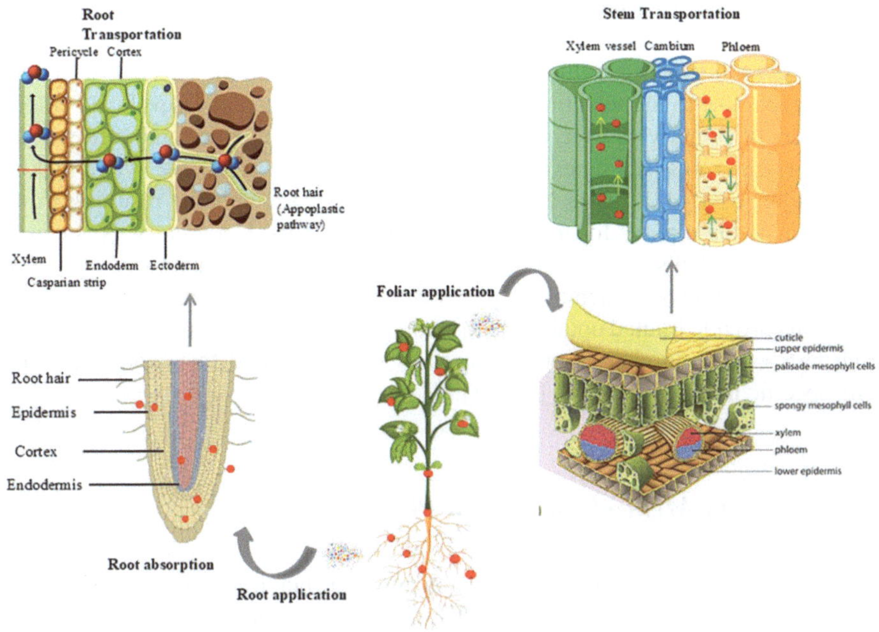

Fig. 2 A schematic diagram illustrating uptake and translocation pathways of NPs, via foliar treatment and root exposure. The foliar application enables nanoparticles to enter leaf surfaces via stomatal openings and cuticles, while root uptake initiates at root hairs followed by absorption through epidermal cells. Both pathways enable nanoparticle movement via apoplastic and symplastic routes to reach target tissues

such as form, size, and surface structure. It also depends on the environment and properties of their transport pathway [24, 25].

2.2.1 Pathways for Foliar Uptake

In agricultural applications, NPs are conventionally applied on the plants via foliar spraying and then they enter through the cuticle or stomatal openings on the leaf. NPs are sprayed on cuticles, this layer protects plants from water loss and restricts the entry of foreign substances [38, 39]. However, the surface of waxy stratum corneum has two types of channels: lipophilic and hydrophilic channels [40, 41]. The hydrophilic channel permits the movement of water-soluble NPs while the lipophilic channel allows the lipid-soluble NP to pass through via the mechanism of diffusion and infiltration [42, 43]. Despite this, there is a limit to the movement of NPs based on their size. Recently, the uptake of carbon dots having a size less than 2 nm has been proven [44]. Additionally, NPs may also enter leaves via the stomatal pathway.

2.2.2 Pathways for Root Uptake

The NP starts interacting with the plants by adsorbing onto the negatively charged root surfaces due to the secretion of exudates [45, 46]. The underdeveloped epidermal layers of primary and secondary root tips allow direct entry and access to the xylem [47]. In roots lacking exodermis, NPs can directly access the xylem making its transport easy [48]. Some NPs damage the plasma membrane resulting in the formation of large pores for internalization of large NPs. Upon entry, NPs can be absorbed by the cells via a number of mechanisms such as ion pathway, endocytosis, or binding with cell membrane proteins [45]. Hydrophilic channels of the root epidermis provide limited uptake due to small pore sizes [49]. Furthermore, NPs are transported via a number of other processes such as endocytosis and interaction with specific transport proteins [50, 51].

2.3 Delivery Methods of NPs in Plant Systems

Plant transgenic methods have enhanced many important traits like crop size, grain quality, secondary metabolites and fundamental plant biology [52–55]. Nanotechnology also advances innovative genetic engineering by using a variety of NPs like nucleic acid-based, such as DNA and RNA; carbon-based like single-walled carbon nanotubes (SWCNTs) and multiwalled carbon nanotubes (MWCNTs); metal-based, such as gold (Au), Ag and Cu; and amino acid-based, such as short peptides and silicon-based material such as silicon-carbide whiskers [56–59]. Different delivery methods of foreign DNA into the nucleus using NPs are summarized in Table 1.

Several techniques have been introduced for NP-mediated gene delivery in plants, each offering distinct advantages (Fig. 3). These include methods such as syringe infiltration, biolistic (gene gun) delivery, ultrasound-mediated delivery, vacuum infiltration, magnetofection, cellular uptake, passive diffusion, and spray-based approaches.

2.3.1 Magnetofection

Pollen magnetofection, where magnetic NPs deliver DNA into pollen under a magnetic field, thereby producing transgenic seeds through pollination without tissue culture. They achieved stable DNA integration and inheritance in cotton seeds [60].

Table 1 Summary of various methods for delivery of NPs in crops

Type of NPs	Size (nms)	Targeted species and tissues	Delivery type	Delivery method	Transient expression/ Stable transformation	Tissue culture	Transformation frequency	Reference(s)
Magnetic Fe_3O_4	~200	Cotton pollens	Plasmid DNA	Magnetic force	Stable transformation	No	~1%	[64]
PEGylated chitosan (CS)–SWNTs, carboxylated SWNTs	127–275	Chloroplast of arugula, watercress, tobacco, spinach	Plasmid DNA	Injection	Transient expression	No	~88%	[56]
SWNTs	~20	Tobacco leaves	siRNA	Injection	Transient expression	No	~95%	[65]
CNTs	~20	Tobacco, cotton, wheat, and arugula leaves	Plasmid DNA	Injection	Transient expression	No	~85%	[57–59]
		Tobacco roots	Plasmid DNA	Co-culture	Transient expression	No	–	[66]
SWCNTs	~2	Tobacco protoplasts and walled cells	Plasmid DNA	Co-culture	Stable transformation	Yes	~7%	[67]
MWCNTs	~13	Tobacco protoplasts and walled cells	Plasmid DNA	Co-culture	Stable transformation	Yes	~3%	[67]
DNA	2.4–5	Leaves of tobacco	siRNA	Injection	Transient expression	No	~80%	[68]

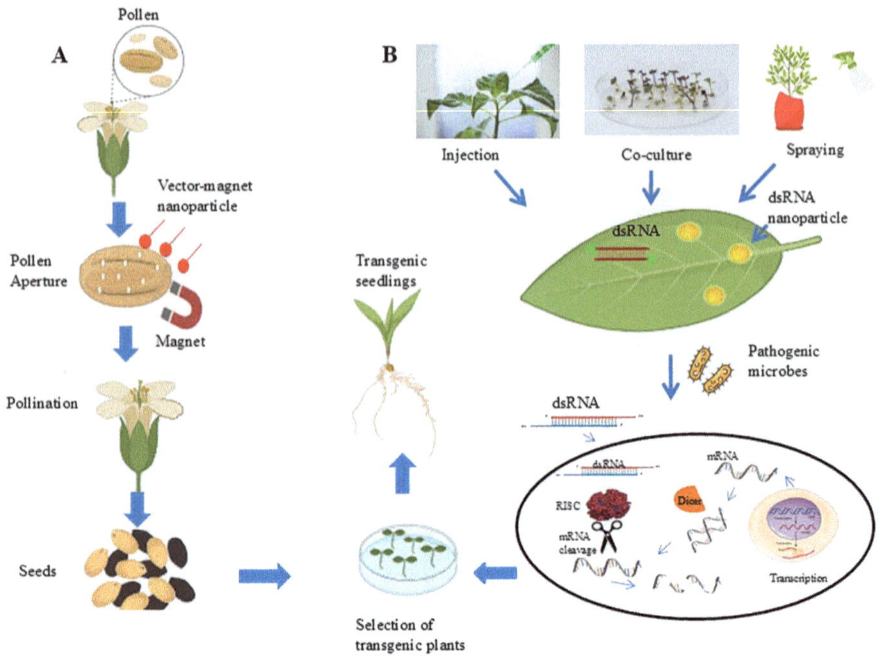

Fig. 3 Delivery methods of NPs in plant systems. (**a**) Schematic illustration of magnetofection. The NP–DNA complex is guided into the pollen aperture under a magnetic field. Transgenic seedlings are obtained under antibiotic selection. (**b**) NP-mediated RNAi delivery by injection, co-culture and spraying, where nanoparticles hauling the genetic material are introduced to plant surface for efficient absorption via incubation, infiltration and absorption through stomatal openings or cuticles followed by transgene expression. This triggers RNAi, which targets and breaks down viral pathogens or plant's endogenous mRNA

2.3.2 Clay Nanosheets

Plants can take up external RNAs like ds-RNAs, si-RNAs, and hairpin RNAs to silence pathogenic gene expression protecting against viruses, fungi, and insects [61, 62]. Leaves can be sprayed with naked dsRNA but they are more prone to degradation [63]. Layered double hydroxide (LDH) clay nanosheets offer a stable alternative, allowing sustained dsRNA release and uptake by plants to provide protection against pathogens.

2.4 NPs for Molecular Imaging in Plants

NPs provide high-precision molecular imaging in plants. Their unique optical, magnetic, and chemical properties enable researchers to visualize and track biological processes at cellular and tissue levels. These interact with plants through various

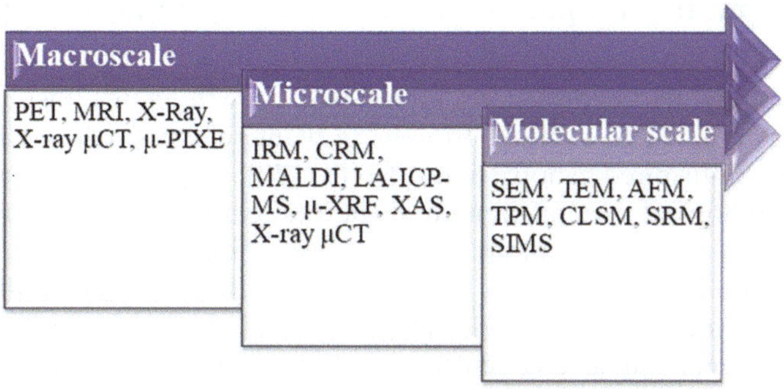

Fig. 4 Imaging tools for plant nanobiotechnology at the micro-to-macro scale. Microscopic imaging techniques including electron and optical microscopes (SEM, TEM, CLSM) along with XAS and SIMS-are used to observe nanomaterial interactions at the cellular level and subcellular level. While MALDI, μ-PIXE, LA-ICP-MS, MRI, PET and X-ray μCT track their distribution in specific plant organs, tissues and the whole body at the macroscopic scale (Abbreviations: TEM, Transmission electron microscopy; SEM, Scanning Electron Microscope; AFM, Atomic Force Microscopy; SRM, Super-resolution microscopy; TPM, Two-photon microscopy; CLSM, Confocal Laser Scanning Microscope; CRM, Confocal Raman microscopy; XAS, Synchrotron X-ray absorption spectroscopy; IRM, Infrared spectral microscope; μ-XRF, Micro-X-ray fluorescence; LA-ICP-MS, Laser ablation inductively coupled plasma mass spectrometry; SIMS, Secondary ion mass spectrometry; MALDI, Matrix-assisted laser desorption ionization; micro-Particle-Induced X-Ray Emission; μ-PIXE PET, positron emission tomography; X-ray μCT, MRI, magnetic resonance imaging; X-ray microcomputed tomography) [69]

mechanisms like passive transport, where SWCNTs penetrate plant cells, localize in chloroplasts, boost photosynthesis, and enable near-infrared monitoring of reactive oxygen species (ROS) and nitric oxide [1, 2]. Through the plasma membrane's electrochemical gradient, negatively charged Poly(acrylic acid)-functionalized nanoceria are transported to chloroplasts through non-endocytic pathways [1, 2]. They scavenge ROS and enhance photosynthesis under abiotic stress. Contrast-based imaging employs metal NPs for applications like confocal laser scanning microscopy (CLSM), transmission electron microscopy (TEM), and many more enabling precise localizations at cell level at different levels, as shown in Fig. 4 [3].
Tiny Tech, Great Impact!
NPs are more than just carriers; they are excellent imaging tools!
SWCNTs can go into plant cells and observe stress by glowing under infrared light. Gold and silver NPs bring the bling as they enhance tissue visibility under microscopes, while nanoceria not only highlights health but also reduces oxidative stress in plants. Nanowires are so cool, they deliver DNA without alteration of genome, making them ideal for safe and temporary molecular tracking [1–3].

3 Bioinspired Nanoparticles; Eco-Friendly Engineering

Nanotechnology has experienced significant growth in the last few years, emerging as a new and exciting frontier in plant sciences, particularly within the scope of bioinspired nanoparticles (BNPs) [70], they are nanostructures created using biological systems or natural sources. Unlike traditional chemical or physical techniques, bioinspired synthesis uses biological entities, including botanical extracts, algae, fungi, bacteria, and different biocompatible molecules, to generate NPs in an environmentally benign and affordable way [71]. Unique physicochemical characteristics of these NPs are controlled dimensions, form, biocompatibility, and surface functionality, which make them very efficient in plant genetic engineering, crop protection, and stress reduction [72, 73].

BNPs have been made from several sources, including plants and bacteria. During NPs creation, plants provide a rich source of secondary chemicals, like phenols, flavonoids, and terpenoids, which serve as harmless reducing and stabilizing agents [74]. For transformation and antibacterial purposes, Au and Ag NPs have been produced using *Azadirachta indica* (neem) leaf extract [75]. Using enzymes or secreted proteins, microorganisms can also intracellularly or extracellularly convert metal ions into NPs. *Bacillus subtilis* has been used to create zinc oxide (ZnO) NPs that improve salt tolerance and disease resistance [76]. Furthermore, because of their fast growth and bioaccumulation potential, marine and freshwater algae provide a sustainable platform for NP generation as well. Cu NPs created by *Sargassum muticum* have helped to improve drought resistance in wheat [77].

3.1 Green Synthesis of BNPs

Green synthesis is utilized for the creation of BNPs utilizing eco-friendly biological sources, removing hazardous chemicals and high-energy input [78]. Its mechanism mostly consists of three fundamental processes. Reduction is the first step; in this process, biomolecules, including ascorbic acid, polyphenols, and sugars in biological extracts, lower metal ions (e.g., Ag^{1+}, Au^{3+}, Zn^{2+}) to their elemental state [79]. The second is stabilization, which involves proteins and polysaccharides covering the NP surface to stop aggregation and enhance dispersion [78]. Essentially a control phase, the last one determines NP size, shape, and function using pH, temperature, extract concentration, and reaction duration. Being non-toxic, environmentally benign, scalable, affordable, and highly biocompatible in plants makes it advantageous [79].

Lately, various instances of green synthesized BNPs have appeared. One type of Ag NPs designed to manage *Fusarium wilt* in tomato utilizing *Camellia sinensis* (green tea) as the source [80]. *Trichoderma harzianum* has been used to create CuO NPs that promote salinity tolerance in maize [81]. Likewise, iron oxide NPs are being created from *Moringa oleifera* leaf extract to cause nano-priming in wheat under drought stress [81].

3.2 Comparison of Traditional and Bioinspired Production Approaches

For NPs, their preparation is a crucial step and there are two primary methods: chemical and green synthesis. Chemical synthesis is for traditional NPs, it requires solvents, and reducing, and stabilizing agents and they are often used for scalable production in large quantities. For research purposes, they have several advantages, including surface modification and functionalization, large-scale production, colloidal stability and monodispersity, diversity in material choice, reliability and flexibility for scaling. However,chemical preparation involves stringent conditions like elevated temperature, energy demand, and waste production, and causes hazardous health and environmental effects. On the other hand, biosynthesis, or green synthesis, for BNPs, is an environmentally sustainable approach. They require moderate conditions of pressure and temperature, being eco-friendly and cost-effective. They are bio-compatible and stable and have enhanced efficiency leading to increased performance in energy storage and conversion systems like solar cells, batteries, and fuel cells using biological materials [82, 83].

3.3 BNPs in Plant Systems for Crop Improvement

Nanotechnology has introduced revolutionary methods in agriculture, especially in the manufacturing of fertilizers and pesticides. Due to their greater surface area, reactivity, and regulated release capabilities nanomaterials greatly improve the delivery of agrochemicals and nutrients to plants. Their application has led to improved nutrient absorption, reduced environmental pollution, and increased agricultural productivity. To produce bioinspired nanomaterials green synthesis approaches are used by environmentally friendly or mostly natural methods. Table 2 shows a variety of nanomaterials used as pesticides or fertilizers, their production methods, and specific functions in different plant systems [84, 85].

Table 2 Applications of nanomaterials in plant systems: Fertilizers and pesticides

Application	Synthesis method	Nanomaterial	Function in plant system	Organism	Reference
Fertilizer	Traditionally made	Urea hydroxyapatite	30% increase in the amount of nitrogen	*Oryza sativa*	[86]
		Hydroxyapatite	Elongation of plant roots	*Solanum lycopersicum*	[87]
		Chitosan	Enhanced crop yield	*Triticum aestivum L.*	[88]
		Alginate/chitosan and chitosan/tripolyphosphate	90% enhanced germination, 35–107% increment in shoot length, four-fold increment in fruit production	*Solanum lycopersicum*	[36]
		Nitrogen-doped carbon-dots	High yield and promoted plant growth	*Vigna radiata*	[89]
		Carbon dots	Elevated disease resistance, 14.8% improved yield	*O. sativa*	[90]
		NaYF4:Yb, Er-Carbon dots	The photosynthetic rate increased by two-fold, and transpiration rate and stomatal conductance improved by 144 and 148%	*Vigna radiata*	[91]
		MWCNTs	Enhanced water uptake, plant growth, and biomass, faster sprouting of seeds	Wheat, maize, peanut, garlic	[92]
		(MWCNTs)	Improved translocation	*Zea mays, Glycine max*	[93]
		(MWCNTs)	Effective nutrient delivery and enhanced growth	O. *sativa*	[94]
		Graphene and (MWCNTs)	28% increase in total biomass, reduced salt stress symptoms	Sorghum and switchgrass	[95]
		(MWCNTs)	Increases in oxidative stress	*Salvia verticillata*	[96]
		Cu-Carbon nanofibers	Improved germination rate, root length, and water-holding capacity	*Cicer arietinum*	[97]
		(MWCNTs)	Enhancement in drought tolerance	*Cucumis sativus*	[98]
		(MWCNTs)-COOH	90% augmentation in seed germination	*Alnus viridis*	[99]
		Carbon dots	Promoted growth and photosynthesis	*Vigna radiata*	[100]
		Chitosan-silicon	Enhanced yield	*Z. mays*	[101]
		Chitosan-selenium	Increased salinity tolerance	*Momordica charantia*	[102]
	Green synthesis	Phosphorus-based nanomaterials (PNMs)	Enhanced physiological growth, metabolism, nutrient absorption	–	[103]
		Chitosan (produced by seafood bio-waste)	Increased nutrients payloads	–	[104]
		Carbon Dots (CDs) (by waste tea residues)	Improved length of buds, and length of roots	*Morus alba*	[105]
		CDs (from dragonfruit peels)	Enhancement of plant growth	*Ipomoea aquatic*	[106]

Pesticides	Traditionally made	MWCNTs	Enhanced inhibition in spore elongation and germination	*Fusarium graminearum*	[107]
		Clay NPs	Controlled pesticide release	-	[108]
		Norcantharidin	Microemulsion serves as biopesticide	Diamondback moth (DBM), *Plutella xylostella* (Linnaeus)	[109]
		polyethylene glycol- camptothecin polymer	Insecticidal potential	*Brevicoryne brassicae*, *Tetranychus cinnabarinus*, and *Bursaphelenchus xylophilus*	[110]
		Carlina oxide	Effective larvicidal	*Culex quinquefasciatus*	[111]
	Green synthesis	Fe^0 and Fe_2O_3 (synthesized by *Ficus natalensis* aqueous extract)	Negligible off-target impacts on larvivorous fishes	*C. quinquefasciatus*	[112]
		Fe_3O_4 NPs (synthesized by *Syzygium cumini* seed extract)	Antiviral potential against TMV, increased expression of salicylic acid genes	*Nicotiana benthamiana*	[113]
		Fe_2O_3 (biogenic synthesis via *Skimmia laureola*)	Bacterial wilt severity was markedly reduced with 6% w/v of NPs	*Ralstonia solanacearum*	[114]
		Silver NPs (synthesized using *Aegle marmelos* extracts)	Antifungal, and antimicrobial effects	*F. solani, S. aureus*	[115]
		Silver chitosan nanocomposite (synthesized by *sumac aqueous* extract)	Anti-fungal activity	*Pseudomonas syringae* pv.	[116]
		Ag NPs (bio fabricated from *Pseudomonas fluorescens*)	89% reduced disease	*Tobacco mosaic virus* (TMV)	[117]
		Ag NPs (by *Moringa oleifera* and *Ficus exasperate* extracts)	23–0% mortality rate	*Moringa oleifera, Ficus exasperate*	[118]
		Ag NPs (synthesized by aqueous extract of *Origanum Marjorana)*	37–44% reduction in green boll	*E. insulana*	[119]
		Ag NPs (by endophytic bacteria isolated from garlic)	Protection of wheat from fungal infection	*F. graminearum*	[120]
		Ag NPs (synthesized by using bacteria, fungi, and sugarcane husk)	Antifungal activity recorder against sugarcane fungal pathogens in sugarcane	*Colletotricum falcatum, F. moniliforme*	[121]
		Ag NPs (biosynthesized using *Malva parviflora*)	65.3%–88.6% reduction in fungal growth	*Helminthosporium rostratum, F. solani*	[122]
		Zn NPs (synthesized by using Eucalyptus globules)	55.2–76.7% disease inhibition	*Alternaria mali, Botryosphaeria dothidea, Diplodia seriata*	[123]
		Mt NPs (*Ananas comosus* peel–mediated green synthesized)	Inhibition of fungal growth and reproduction	*F. verticilliodes*, *Aspergillus flavus*, and *A. alternata*	[124]

4 BNPs in Plant Genetic Engineering

BNPs are considered ideal for biomedical applications due to their biocompatibility and stability. They have brought about an immense level of advancement in various vital aspects of plant genetic engineering, including plant transformation, genome editing, biocontrol applications, and stress resistance [125]. They can also greatly facilitate the high and efficient delivery of genetic materials (DNA, RNA) and proteins, which is far more difficult to achieve with traditional methods [126].

4.1 Gene Delivery and Genome Editing

Delivering of the desired gene/construct through genome editing techniques is seen as critical for improving various traits, such as yield improvement, disease resistance and stress tolerance. BNPs have been demonstrated to be effective in improving the efficiency of CRISPR-Cas9 constructs delivery in plants. Historically, the gene delivery methods most commercially used are transformation with *Agrobacterium* and using gene guns [127]. These methods have limitations with respect to efficiency and tissue specificity, as well as possible injury to the cells. In contrast, the BNPs demonstrate high efficiency, selective distributions and low cell cytotoxicity [128].

4.1.1 Delivery of CRISPR/Cas9 Constructs in Plants

For delivery of genome editing via CRISPR/Cas9 technology to be successful, delivery of Cas9 protein and custom-designed short guide RNA (sgRNA) into the nucleus of plant cells is necessary. BNPs offer this while avoiding processing, which can lead to degradation of either the Cas9 or sgRNA components, as well as improved efficiency of delivery. Various nanoparticle types have been described and used to deliver CRISPR/Cas9 constructs to plant cells [129]. One of the types is lipid-based NPs (LNPs), which have gained popularity for use in medicine and are now being used for plant genetic engineering as well. One of its types, cationic LPNs, has been used successfully in *O. sativa* with increased editing efficiency up to 70% as compared to conventional methods [129]. Similarly, LNPs have been used to deliver ribonucleoproteins (RNPs) in *Zea mays* that resulting in a 60% higher success rate compared to *Agrobacterium*-based methods [70]. Mesoporous silica NPs (MSNs) are another type of nanoparticle used for plant genetic engineering due to their ability to release biomolecules (Cas9 and sgRNA) at a controlled rate due to their high surface area and optimal pore size. They have also been used successfully for delivering CRISPR-based constructs in *Triticum aestivum* and *Arabidopsis thaliana* with a success rate of more than 50% [130, 131].

4.1.2 Targeted Delivery of DNA, RNA, and Proteins into Plant Cells

Delivering nucleic acids and proteins is challenging in plants due to issues like the rigid plant cell wall, the complex intracellular cytoplasm, and the nucleoplasm of plant cells. However, different BNPs along with peptides, ligands, and surface coatings have helped in overcoming these challenges to achieve targeted delivery [132]. One of the prominent examples is the use of AuNPs because of their high stability and biocompatible nature. A recent study reported the use of AuNPs along with peptides to deliver plasmid into rice and maize protoplasts. The overall efficiency increase was 55% more compared to non-functionalized Au NPs [133]. Similarly, plant-specific ligands and Au NPs were combined for the delivery of RNA in *Glycine max* roots, which enhanced gene expression by almost 40% [134]. Another important nanotechnology approach used especially for RNA and protein delivery is CNTs because of their high mechanical strength. A recent study used polyethylene glycol (PEG) coated on CNTs for delivering GFP-tagged proteins into *Spinacia oleracea,* which resulted in increased protein expression. CNTs have also been used for RNA delivery into *Nicotiana tabacum,* resulting in high stability and transformation success [135].

4.2 Plant Transformation and Agrobacterium-Free Transformation

Plant transformation through nanotechnology approaches, especially through BNPs, has been used in recent years as a more efficient, targeted and more precise approach. They offer a more biocompatible and stable approach than *Agrobacterium*-mediated or other conventional transformation methods and are used against a variety of traits for crop improvement as mentioned in Table 3 [136].

4.2.1 Direct NP-Mediated Transformation

Directly modifying plants by a NP-mediated mechanism enables the delivery of foreign DNA directly into the plant without a biological vector. These NPs are designed to effectively traverse the rigid cell wall of plants and deliver their contents in plant nuclei or chloroplasts. Gold and silica NPs are the most commonly used for this purpose due to their biocompatible nature [137]. Gold NPs have been exploited for the delivery of CRISPR/Cas9 constructs and proteins, and recently, gold NPs have also been exploited in *Solanum lycopersicum* and *Zea mays* for high transformation and efficient genome editing [138]. Silica NPs have also been used successfully for short interfering RNA (siRNA) delivery into *Nicotiana tabacum* and plasmid DNA delivery into *Triticum aestivum* [126, 139].

Table 3 Successful examples of the use of different types of BNPs against specific traits for crop improvement

Nanoparticle Type	Biological Source	Targeted Gene(s)	Plant Species	Cargo Delivered	Outcome	Reference
Gold NPs (Au^0)	Neem leaf extract	*TaDREB2*	*Triticum aestivum*	CRISPR/Cas9	Enhanced drought tolerance	[156]
CNTs	Chitosan (fungal-derived)	*PDS*	*Nicotiana benthmiana*	CRISPR/Cas9	Induction of albino phenotype	[157]
Lipid-based NPs	Phospholipids from plant oil	*SlMLO1*	*Solanum lycopersicum*	CRISPR/Cas9	Improved resistance against powdery mildew	[158]
Chitosan NPs	Crustacean exoskeleton	*OsHKT1;5*	*Oryza sativa*	CRISPR/Cas9	Improved salt tolerance	[159]
LDH nanosheets	*Ocimum sanctum* (medicinal plant)	*TMV* genome	*Nicotiana benthamiana*	dsRNA	Increased virus resistance through RNA interference	[160]
Polyethyleneimine-coated gold NPs	Dextran	*FAD2*	*Camelina sativa*	Plasmid DNA	Modified fatty acid profile	[161]
SWCNTs	Chitosan (fungal derived)	*CBF1*	*Spinacia oleracea*	Plasmid DNA	Enhanced cold tolerance	[162]
Silicon carbide NPs	Rice husk	*ZmPIF3*	*Zea mays*	CRISPR/Cas9	Modified plant morphology and physiology under shade	[163]
Polyamidoamine-based dendrimers	Plant oil	*GmNAC20*	*Glycine max*	Plasmid DNA	Improved drought tolerance	[164]
Silver and copper NPs	*Moringa oleifera*	*BnFAD3*	*Brassica napus*	Plasmid DNA	Increased oleic acid content	[165]
Silicon dioxide NPs	Rice husk	*DXS*	*Dioscorea zingiberenin*	Plasmid DNA	Enhanced diosgenin content	[166]
Silver NPs	Onion extract	*NtACO1*	*Nicotiana tabacum*	Plasmid DNA	Extended shelf life of fruits	[167]

Nanoparticle Type	Biological Source	Targeted Gene(s)	Plant Species	Cargo Delivered	Outcome	Reference
Chitosan NPs	Shrimp shells	*AtPAP1*	*Arabidopsis thaliana*	Plasmid DNA	Increased anthocyanin content	[168]
Silicon dioxide NPs	Rice husk	*NAC1*	*Brassica napus*	CRISPR/Cas9	Enhanced heat tolerance	[169]
Magnetic NPs	Bacterial metabolites	*GUS*	*Brassica juncea*	Plasmid DNA	Stable genetic transformation	[170]
Poly-L-lysine-coated starch NPs	Starch from a tuber	*GFP*	*Dioscorea zingiberensis*	Plasmid DNA	Successful gene delivery and expression	[164]
Mesoporous silica NPs	*Quillaja saponaria*	*mCherry*	*Arabidopsis thaliana*	Plasmid DNA	Successful gene delivery and transient expression in root cells	[171]
Silver NPs	*Eucalyptus globulus* leaf extract	*hrpA*	*Solanum lycopersicum*	Antimicrobial peptides	Reduced bacterial infection	[172]
Copper NPs	*Citrus sinensis* peel extract	*rpfF*	*Oryza sativa*	siRNA	Reduced bacterial biofilm formation	[173]
Selenium NPs	Onion extract	*Tri5*	*Triticum aestivum*	CRISPR/Cas9	Reduced mycotoxin production	[173]
Chitosan NPs	Shrimp shell	*PiAvr3a*	*Solanum tuberosum*	RNAi construct	Reduced infection of late blight	[174]
Silicon dioxide NPs	Rice husk	*ABC1*	*Brassica juncea*	Salicylic acid	Increased resistance to fungal attack	[175]
Iron oxide NPs	*Moringa oleifera* seed extract	*Bcpg1*	*Arabidopsis thaliana*	CRISPR-RNPs	Enhanced resistance to fungal pathogenesis	[176]
Graphene oxide NPs	*Punica granatum* (plant extract)	*Coat Protein*	*Capsicum annuum*	RNAi construct	Reduced viral load and spread	[177]

4.2.2 Chitosan and Carbon-Based NPs

Chitosan is a plant polysaccharide derived from chitin and holds significance as a nanoparticle because of its biodegradable and biocompatible nature. Chitosan NPs are positively charged and can bind negatively charged DNA or RNA for uptake via endocytosis. In addition, chitosan NPs are also involved in enhancing plant immunity and stress tolerance response. One of its successful examples is the delivery of the CRISPR/Cas9 ribonucleoprotein complex in *O. sativa* with highly targeted gene knockout [140]. Another example is the delivery of siRNA via chitosan NPs conjugated with folic acid, ultimately leading to the downregulation of targeted genes in different crops [129]. Along with chitosan-based NPs, another widely used NP for *Agrobacterium*-free transformation is carbon-based NPs. They come in different forms like carbon dots, CNTs, and graphene oxide which all have strong interactions with biomolecules as well as a large surface area. Recently, carbon dots have been used for the delivery of plasmid in protoplasts of *Nicotiana tabacum* with high transformation efficiency [131]. Similarly, they have also been successfully used in *Triticum aestivum* to deliver siRNA for downregulation of target genes [141]. On the other hand, CNTs have been recently successfully used in the delivery of plasmid DNA in *Zea mays* and CRISPR/Cas9-based ribonucleoproteins in *Glycine max* for precise gene editing [142].

4.3 *Stress Resistance and Crop Improvement*

Stress tolerance is considered very critical in plants to tackle different stresses. Different abiotic stresses like heat, salinity, drought, UV, and many others have been tackled using BNPs [143].

4.3.1 Drought and Salt Stress Tolerance

To achieve higher yield and productivity in plants, drought and salinity are some major abiotic stress factors. BNPs have been proven and demonstrated to strengthen water absorption, ion stability and antioxidant-rich functioning in plants through the increase in the expression of genes that respond to stress. For this purpose, different NPs have been used [142]. One of the major examples is the use of Cu or selenium NPs for increased drought tolerance in *Solanum lycopersicum* [144]. Similarly, silicon-based NPs have been used to improve drought tolerance in *Triticum aestivum* [145]. CNTs have also been used in *O. sativa* to improve water retention and promote root hair growth [146].

4.3.2 UV and Heat Stress Tolerance

Prolonged exposure to UV and high-temperature damages plant tissues, particularly in their guard cells. Several BNPs have recently been applied for plant protection against thermal and UV-induced damage. For example, AuNPs have been shown to increase the production of flavonoids and anthocyanins in *Solanum lycopersicum* plants, which protect them from damage due to oxidation from UV [147]. Additionally, Ag NPs stimulate the development of protective enzymes and heat shock proteins (HSPs) for thermal tolerance in *Triticum aestivum* [148]. Likewise, ZnO NPs have been shown to have protective qualities in *Cucumis sativus* plants against exposure to UV due to decreasing lipid permeability and increasing the proportion of chlorophyll [149].

4.4 Biocontrol and Antimicrobial Applications

BNPs also show great potential for the management of bacterial, viral, and fungal plant diseases. Their potential as biocontrol agents is attributed to their ability to disrupt the cell walls of microbes and prevent subsequent propagation [150].

4.4.1 Antibacterial and Antifungal Activity

Different lines of bacterial and fungal strains are major biotic threats or destroyers for plants, at which point several BNPs have recently been used to manage them. Copper NPs have been used for bacterial leaf blight control in *Oryza sativa* by preventing biofilm formation [151]. Silver NPs have also been used to target pathogenic fungi, *Fusarium* spp., in *Triticum aestivum*. Recently, graphene oxide-based NPs have been used for the inhibition of fungal growth in *Solanum lycopersicum* [152]. Copper oxide NPs have also been used to inhibit fungal growth in *Oryza sativa*. ZnONPs have also been reported to control bacterial blight in *Gossypium hirsutum* [153].

4.4.2 Antiviral Activity

Because of the high rates of transmission and quick mutation, viruses are difficult to manage. It has been demonstrated that BNPs can prevent viral entrance and reproduction. In recent years, gold NPs coupled with antiviral peptides have been used in *Cucumis sativus* to tackle cucumber mosaic virus by preventing viral entry [154]. Similarly, in *Nicotiana tabacum*, ZnO NPs have been used to tackle tobacco mosaic virus by inhibiting viral RNA replication [155].

5 Challenges and Limitations

5.1 *Toxicity and Environmental Concerns*

NPs have emerged as a promising tool in plant biotechnology. BNPs derived from biological sources hold significant potential in advancing agriculture, environmental protection, and healthcare technology. However, their interactions with plant systems raise significant toxicological concerns. Therefore, biosafety assessment is necessary to avoid any unexpected toxicity to the environment [178].

5.1.1 Toxicological Aspects of NPs in Plants

Studies have shown that the uptake, translocation, and accumulation of NPs in plants can lead to oxidative stress, genotoxicity, and disruptions in physiological processes [179]. Auxin is a crucial plant growth hormone that significantly regulates growth and development, including geotropism in plants. AgNPs are well-known for their antimicrobial properties, but their phytotoxic effects have raised concerns. Studies have shown that the toxicity of AgNPs in *Arabidopsis thaliana* revealed significant physiological and molecular alterations [180]. Exposure to AgNPs resulted in stunted root growth, disrupted hormone signaling, and increased ROS accumulation. Additionally, AgNP-treated plants exhibited differential gene expression linked to stress response pathways. These findings underscore the need for the controlled use of AgNPs to mitigate potential environmental hazards [181, 182].

5.1.2 Environmental Impact of BNPs

BNPs offer a greener alternative to synthetic NPs. Green synthesis of BNPs eliminates the need for hazardous chemicals and minimizes energy consumption, thereby reducing their ecological footprint [183]. For instance, rice husk-derived silica NPs efficiently purify water without releasing toxic residues [184]. However, BNPs also raise environmental concerns as they do not degrade in acidic or saline environments. This poses a serious threat to soil and aquatic life, leading to potential ecological disruption and bioaccumulation hazards [185].

5.2 *Regulatory Issues and Public Perception*

BNPs exhibit distinct behavior due to their nanoscale size and biological interfaces, making it challenging to assess their safety through standard protocols. Regulatory bodies in various countries often lack specific guidelines tailored for BNPs, which results in gaps in risk assessment, labeling, and approval processes. Regulations primarily focus on preventing nanoparticle bioaccumulation and potential toxicity to plants and animals.

5.2.1 Regulatory Framework for Nanoparticle-Based GM Plants

The regulatory landscape for NP-based GM plants is complex and still evolving, as existing frameworks remain fragmented across different regions. They often struggle to keep pace with continuous development and product innovation [186]. International organizations such as Codex Alimentarius and the Organisation for Economic Co-operation and Development (OECD) provide overarching guidelines for the regulation of nanotechnology and GMOs, but national approaches differ significantly; countries like the United States, the European Union, and India implement these within the framework of their respective biotechnology and nanomaterial regulations to varying degrees (Codex Alimentarius, 2011; OECD, 2010; USDA, 2020; EFSA, 2021; DBT, 2017). For instance, in the European Union, stringent regulations under the GMO Directive 2001/18/EC require extensive risk assessments. However, the U.S. follows a product-based regulatory approach under the Coordinated Framework for Biotechnology.

Overall, current frameworks lack harmonization and often rely on case-by-case assessments. Therefore, it highlights the need for updated, integrated regulatory approaches to address these emerging technologies regarding biosafety, environmental impact, and human health risks.

5.2.2 Public Perception and Acceptance of Bioinspired Nanoparticle-Based Approaches

Public acceptance of bioinspired nanoparticle applications in agriculture varies widely. While some consumers recognize their potential to enhance crop resilience and reduce chemical inputs, others express concerns over unknown health and environmental risks. Factors influencing public perception include media representation, regulatory transparency, and prior attitudes toward GM technology. Effective science communication and stakeholder engagement are crucial for fostering public trust in NP-based innovations [187].

5.3 Technical Limitations

5.3.1 Challenges in Scaling up Nanoparticle Applications in Agriculture

Studies have shown promising results for NP-based agricultural solutions. However, large-scale field applications of NP technologies face several challenges. These include high production costs, limited reproducibility in field conditions, and difficulties ensuring targeted NP delivery to plants. Furthermore, the complex interactions between NPs and soil components can alter their bioavailability and efficacy, requiring further research to optimize field applications [188].

5.3.2 Standardization Issues for BNPs in Plant Genetic Engineering

The lack of standardized protocols for BNPs in plant genetic engineering hinders their widespread adoption. Variability in synthesis methods, NP characterization, and application techniques lead to inconsistent study results. Establishing regulatory guidelines and standardized methodologies for NP formulation, stability assessment, and toxicity evaluation is essential for advancing their use in agricultural biotechnology [189].

6 Advancing BNPs; Integration with Genetic Tools and the Road to Commercialization

Traditional methods for the delivery of CRISPR components often face limitations due to species specificity, low transformation efficiency and off-target effects. Integrating nanotechnology, especially BNPs, with CRISPR offers innovative solutions to the above-mentioned problems. Recently, several types of nanomaterials such as liposomes, biogenically synthesized modified metal NPs, magnetic NPs, mesoporous silica NPs (MSNs), LDHs, and CNTs have been used to deliver genome engineering tools (DNA, RNA, Proteins, RNPs) into plants devoid of their genotype with high efficiency [126]. Nanomaterial-DNA/RNA complexes were delivered to plant cells via injection, spraying, and co-culturing methods. For instance, bioinspired magnetic nanomaterials were successfully used to transfer *Bt* resistance genes to cotton pollens and the offspring were highly pest-resistant and a similar method has been reported in maize [64, 190]. Despite their promising potential, bioinspired nano-vectors still face many limitations, they often lack integration genes, expression efficiency may not be enough for genetic engineering applications, delivery efficiency is limited, and no screening methods for genetically engineered plants [126].

BNPs have demonstrated efficacy in multiple agricultural aspects but their widespread commercialization faces critical challenges like, scalability of production, regulatory approvals and critical understanding of their long-term environmental and ecological impacts. BNPs synthesized through the green synthesis method may not easily be optimized for large-scale production [191]. Secondly, as this field is still evolving, it needs regulatory approvals and deployment. Thirdly, the persistence of bNPs in the soil and plant system and their toxicological impact on the environment microorganisms, plants, and animals need further investigation.

7 Conclusion

The integration of nanotechnology and plant genetic engineering has altered agricultural biotechnology. Biocompatible, non-toxic, and sustainable BNPs are more compatible with plant cells than chemical or physical NPs. The biocompatibility enhances genetic material diffusion over plant tissue barriers and reduces synthetic nanomaterial-induced physiological problems. BNPs' flexibility from chitosan and liposome-based gene delivery to plant-extract-mediated metallic NPs for disease control and abiotic stress tolerance is incredible. These NPs are ideal for *Agrobacterium*-independent plant transformation and CRISPR/Cas9-mediated gene editing. BNPs may improve global food security by improving crops' resilience to all the stresses, especially due to climate change. Despite their potential, there are still challenges in the way, like regulatory issues, field toxicity, and scalable and reproducible synthesis. Clear communication, stakeholder participation, and extensive safety evaluations are needed to overcome these challenges, especially due to the varied public opinion about nanotechnology in food systems. Nanotechnology, precision agriculture, and synthetic biology will provide effective and smart delivery systems in plants. BNPs will transform plant genome modification and sustainable agriculture in the next phase of precision engineering.

References

1. Kim J, Lee N, Hyeon T (2017) Recent development of nanoparticles for molecular imaging. Philos Trans R Soc A Math Phys Eng Sci 375(2107):20170022
2. Buriak JM, Liz-Marzán LM, Parak WJ, Chen X (2022) Nano and Plants. ACS Nano 16(2):1681–1684. https://doi.org/10.1021/acsnano.2c01131
3. Sun XD, Ma JY, Feng LJ, Duan JL, Yuan XZ (2025) Precise tracking of nanoparticles in plant roots. Nat Protoc 20(1):248–271. https://doi.org/10.1038/s41596-024-01044-5
4. Abdul Aziz M, Brini F, Rouached H, Masmoudi K (2022) Genetically engineered crops for sustainably enhanced food production systems. Front Plant Sci 13:1027828
5. Hamdan MF, Tan BC (2024) Genetic modification techniques in plant breeding: a comparative review of CRISPR/Cas and GM technologies. Hortic Plant J
6. Kamthan A, Chaudhuri A, Kamthan M, Datta A (2016) Genetically modified (GM) crops: milestones and new advances in crop improvement. Theor Appl Genet 129:1639–1655
7. Amritha P, Shah JM (2021) Can genetic engineering-based methods for gene function identification be eclipsed by genome editing in plants? A comparison of methodologies. Mol Genet Genomics 296(3):485–500
8. Abdallah NA, Hamwieh A, Radwan K, Fouad N, Prakash C (2021) Genome editing techniques in plants: a comprehensive review and future prospects toward zero hunger. GM Crop Food 12(2):601–615
9. Contera S, Bernardino de la Serna J, Tetley TD (2020) Biotechnology, nanotechnology and medicine, vol 4. Portland Press Ltd, pp 551–554
10. Gebre SH (2023) Bio-inspired synthesis of metal and metal oxide nanoparticles: the key role of phytochemicals. J Clust Sci 34(2):665–704
11. Kaymaz SV, Nobar HM, Sarıgül H, Soylukan C, Akyüz L, Yüce M (2023) Nanomaterial surface modification toolkit: principles, components, recipes, and applications. Adv Colloid Interf Sci 322:103035

12. Wu G et al (2022) Recent advancement of bioinspired nanomaterials and their applications: a review. Front Bioeng Biotechnol 10:952523
13. Chaudhry N et al (2018) Bio-inspired nanomaterials in agriculture and food: current status, foreseen applications and challenges. Microb Pathog 123:196–200
14. Cunningham FJ, Goh NS, Demirer GS, Matos JL, Landry MP (2018) Nanoparticle-mediated delivery towards advancing plant genetic engineering. Trends Biotechnol 36(9):882–897
15. Raliya R, Saharan V, Dimkpa C, Biswas P (2017) Nanofertilizer for precision and sustainable agriculture: current state and future perspectives. J Agric Food Chem 66(26):6487–6503
16. Tudi M et al (2021) Agriculture development, pesticide application and its impact on the environment. Int J Environ Res Public Health 18(3):1112
17. Dincă LC, Grenni P, Onet C, Onet A (2022) Fertilization and soil microbial community: a review. Appl Sci 12(3):1198
18. Kalwani M, Chakdar H, Srivastava A, Pabbi S, Shukla P (2022) Effects of nanofertilizers on soil and plant-associated microbial communities: emerging trends and perspectives. Chemosphere 287:132107
19. Hussain M et al (2023) Integrating nanotechnology with plant microbiome for next-generation crop health. Plant Physiol Biochem 196:703–711
20. Altammar KA (2023) A review on nanoparticles: characteristics, synthesis, applications, and challenges. Front Microbiol 14:1155622
21. Cho K-H, Park J-E, Osaka T, Park S-G (2005) The study of antimicrobial activity and preservative effects of nanosilver ingredient. Electrochim Acta 51(5):956–960
22. Kale SK, Parishwad GV, Patil ASHAS (2021) Emerging agriculture applications of silver nanoparticles. ES Food & Agroforestry 3:17–22
23. Shelar GB, Chavan AM (2015) Myco-synthesis of silver nanoparticles from Trichoderma harzianum and its impact on germination status of oil seed. Biolife 3(1):109–113
24. Hatami M, Ghorbanpour M (2013) Effect of nanosilver on physiological performance of pelargonium plants exposed to dark storage. J Horticult Res 21(1):15–20
25. Sadak MS (2019) Impact of silver nanoparticles on plant growth, some biochemical aspects, and yield of fenugreek plant (Trigonella foenum-graecum). Bull Natl Res Cent 43(1):1–6
26. Saharan V et al (2015) Synthesis and in vitro antifungal efficacy of cu–chitosan nanoparticles against pathogenic fungi of tomato. Int J Biol Macromol 75:346–353
27. Pradhan S et al (2014) Manganese nanoparticles: impact on non-nodulated plant as a potent enhancer in nitrogen metabolism and toxicity study both in vivo and in vitro (in eng). J Agric Food Chem 62(35):8777–8785. https://doi.org/10.1021/jf502716c
28. Rui M et al (2016) Iron oxide nanoparticles as a potential iron fertilizer for peanut (Arachis hypogaea). Front Plant Sci 7:815
29. Rameshraddy G, Pavithra J, Mahesh S, Geetha K, Shankar A (2017) Seed priming and foliar spray with nano zinc improves stress adaptability and seed zinc content without compromising seed yield in ragi (finger millet). Int J Pure Appl Biosci 5(3):251–258
30. Paramo LA, Feregrino-Pérez AA, Guevara R, Mendoza S, Esquivel K (2020) Nanoparticles in agroindustry: applications, toxicity, challenges, and trends. Nanomaterials 10(9):1654
31. Saleh TA (2020) Nanomaterials: classification, properties, and environmental toxicities. Environ Technol Innov 20:101067
32. Mathur RB, Singh BP, Pande S (2016) Carbon nanomaterials: synthesis, structure, properties and applications. Taylor & Francis
33. Zaytseva O, Neumann G (2016) Carbon nanomaterials: production, impact on plant development, agricultural and environmental applications. Chem Biol Technol Agric 3:1–26
34. Nair R et al (2012) Effect of carbon nanomaterials on the germination and growth of rice plants. J Nanosci Nanotechnol 12(3):2212–2220
35. Shakiba S, Astete CE, Paudel S, Sabliov CM, Rodrigues DF, Louie SM (2020) Emerging investigator series: polymeric nanocarriers for agricultural applications: synthesis, characterization, and environmental and biological interactions. Environ Sci Nano 7(1):37–67
36. Pereira A d ES, Oliveira HC, Fraceto LF (2019) Polymeric nanoparticles as an alternative for application of gibberellic acid in sustainable agriculture: a field study. Sci Rep 9(1):7135

37. Tong Y et al (2017) Polymeric nanoparticles as a metolachlor carrier: water-based formulation for hydrophobic pesticides and absorption by plants. J Agric Food Chem 65(34):7371–7378
38. Yang C, Powell CA, Duan Y, Shatters R, Zhang M (2015) Antimicrobial nanoemulsion formulation with improved penetration of foliar spray through citrus leaf cuticles to control citrus huanglongbing. PLoS One 10(7):e0133826
39. Pérez-de-Luque A (2017) Interaction of nanomaterials with plants: what do we need for real applications in agriculture? Front Environ Sci 5:12
40. Avellan A et al (2019) Nanoparticle size and coating chemistry control foliar uptake pathways, translocation, and leaf-to-rhizosphere transport in wheat. ACS Nano 13(5):5291–5305
41. Eichert T, Kurtz A, Steiner U, Goldbach HE (2008) Size exclusion limits and lateral heterogeneity of the stomatal foliar uptake pathway for aqueous solutes and water-suspended nanoparticles. Physiol Plant 134(1):151–160
42. Banerjee K, Pramanik P, Maity A, Joshi D, Wani SH, Krishnan P (2019) Methods of using nanomaterials to plant systems and their delivery to plants (mode of entry, uptake, translocation, accumulation, biotransformation and barriers). In: Advances in phytonanotechnology. Elsevier, pp 123–152
43. Bussieres P (2014) Estimating the number and size of phloem sieve plate pores using longitudinal views and geometric reconstruction. Sci Rep 4(1):4929
44. Hu P et al (2020) Nanoparticle charge and size control foliar delivery efficiency to plant cells and organelles. ACS Nano 14(7):7970–7986
45. Lv J, Christie P, Zhang S (2019) Uptake, translocation, and transformation of metal-based nanoparticles in plants: recent advances and methodological challenges. Environ Sci Nano 6(1):41–59
46. Zhao L et al (2012) Transport of Zn in a sandy loam soil treated with ZnO NPs and uptake by corn plants: electron microprobe and confocal microscopy studies. Chem Eng J 184:1–8
47. Rajput V et al (2020) Accumulation of nanoparticles in the soil-plant systems and their effects on human health. Ann Agri Sci 65(2):137–143
48. Péret B et al (2009) Arabidopsis lateral root development: an emerging story. Trends Plant Sci 14(7):399–408
49. Li L et al (2020) Effective uptake of submicrometre plastics by crop plants via a crack-entry mode. Nat Sustain 3(11):929–937
50. Liu Q et al (2009) Carbon nanotubes as molecular transporters for walled plant cells. Nano Lett 9(3):1007–1010
51. Grillo R, Mattos BD, Antunes DR, Forini MM, Monikh FA, Rojas OJ (2021) Foliage adhesion and interactions with particulate delivery systems for plant nanobionics and intelligent agriculture. Nano Today 37:101078
52. Si SL, LiZhen et al (2016) OsSPL13 controls grain size in cultivated rice. Nat Genet 48:447
53. Lv Z et al (2019) Interaction of bZIP transcription factor TGA6 with salicylic acid signaling modulates artemisinin biosynthesis in Artemisia annua. J Exp Bot 70(15):3969–3979
54. Wang S et al (2015) The OsSPL16-GW7 regulatory module determines grain shape and simultaneously improves rice yield and grain quality. Nat Genet 47(8):949–954
55. Wang Y et al (2017) NIP1; 2 is a plasma membrane-localized transporter mediating aluminum uptake, translocation, and tolerance in Arabidopsis. Proc Natl Acad Sci 114(19):5047–5052
56. Kwak S-Y et al (2019) Chloroplast-selective gene delivery and expression in planta using chitosan-complexed single-walled carbon nanotube carriers. Nat Nanotechnol 14(5):447–455
57. Demirer GS et al (2019) High aspect ratio nanomaterials enable delivery of functional genetic material without DNA integration in mature plants. Nat Nanotechnol 14(5):456–464
58. Demirer GS, Zhang H, Goh NS, Chang R, Landry MP (2019) Nanotubes effectively deliver siRNA to intact plant cells and protect siRNA against nuclease degradation. bioRxiv:564427
59. Demirer GS, Zhang H, Goh NS, González-Grandío E, Landry MP (2019) Carbon nanotube–mediated DNA delivery without transgene integration in intact plants. Nat Protoc 14(10):2954–2971

60. Lv Z, Jiang R, Chen J, Chen W (2020) Nanoparticle-mediated gene transformation strategies for plant genetic engineering. Plant J 104(4):880–891
61. Dubrovina AS, Kiselev KV (2019) Exogenous RNAs for gene regulation and plant resistance. Int J Mol Sci 20(9):2282
62. Kaldis A et al (2018) Exogenously applied dsRNA molecules deriving from the zucchini yellow mosaic virus (ZYMV) genome move systemically and protect cucurbits against ZYMV. Mol Plant Pathol 19(4):883–895
63. Gogoi A, Sarmah N, Kaldis A, Perdikis D, Voloudakis A (2017) Plant insects and mites uptake double-stranded RNA upon its exogenous application on tomato leaves. Planta 246:1233–1241
64. Zhao X et al (2017) Pollen magnetofection for genetic modification with magnetic nanoparticles as gene carriers. Nat Plants 3(12):956–964
65. Demirer GS, Zhang H, Goh NS, Pinals RL, Chang R, Landry MP (2020) Carbon nanocarriers deliver siRNA to intact plant cells for efficient gene knockdown. Sci Adv 6(26):eaaz0495
66. Golestanipour A, Nikkhah M, Aalami A, Hosseinkhani S (2018) Gene delivery to tobacco root cells with single-walled carbon nanotubes and cell-penetrating fusogenic peptides. Mol Biotechnol 60:863–878
67. Burlaka O, Pirko YV, Yemets A, Blume YB (2015) Plant genetic transformation using carbon nanotubes for DNA delivery. Cytol Genet 49(6):349–357
68. Zhang H et al (2019) DNA nanostructures coordinate gene silencing in mature plants. Proc Natl Acad Sci 116(15):7543–7548
69. Zhao B, Luo Z, Zhang H, Zhang H (2022) Imaging tools for plant nanobiotechnology. Front Genome Ed 4:1029944
70. Yan Y, Zhu X, Yu Y, Li C, Zhang Z, Wang F (2022) Nanotechnology strategies for plant genetic engineering. Adv Mater 34(7):2106945
71. Velusamy P, Kumar GV, Jeyanthi V, Das J, Pachaiappan R (2016) Bio-inspired green nanoparticles: synthesis, mechanism, and antibacterial application. Toxicol Res 32(2):95–102
72. Barhoum A et al (2022) Review on natural, incidental, bioinspired, and engineered nanomaterials: history, definitions, classifications, synthesis, properties, market, toxicities, risks, and regulations. Nano 12(2):177
73. Barbinta-Patrascu M-E, Bita B, Negut I (2024) From nature to technology: exploring the potential of plant-based materials and modified plants in biomimetics, bionics, and green innovations. Biomimetics 9(7):390
74. Godeto YG, Ayele A, Ahmed IN, Husen A, Bachheti RK (2023) Medicinal plant-based metabolites in nanoparticles synthesis and their cutting-edge applications: an overview. In: Secondary metabolites from medicinal plants, pp 1–34
75. Chinnasamy G, Chandrasekharan S, Koh TW, Bhatnagar S (2021) Synthesis, characterization, antibacterial and wound healing efficacy of silver nanoparticles from Azadirachta indica. Front Microbiol 12:611560
76. Wang Z, Wang S, Ma T, Liang Y, Huo Z, Yang F (2023) Synthesis of zinc oxide nanoparticles and their applications in enhancing plant stress resistance: a review. Agronomy 13(12):3060
77. Abideen Z et al (2022) Algal-mediated nanoparticles, phycochar, and biofertilizers for mitigating abiotic stresses in plants: a review. Agronomy 12(8):1788
78. Madani M et al (2022) Green synthesis of nanoparticles for varied applications: Green renewable resources and energy-efficient synthetic routes. Nanotechnol Rev 11(1):731–759
79. Roy M, Mukherjee P (2021) Bio-inspired synthesis of nanomaterials. In: Handbook on Synthesis Strategies for Advanced Materials: Volume-I: Techniques and Fundamentals, pp 589–622
80. Gomaa NA, Mahdy AM, Fawzy RN, Ahmed GA (2021) Green synthesis of silver nanoparticle by plant extracts to control tomato wilt disease caused by fusarium oxysporum f. sp. lycopersici. Int J Sci Res Sustain Dev 4(3):1–14
81. Shah IH et al (2023) Co-application of copper oxide nanoparticles and Trichoderma harzianum with physiological, enzymatic and ultrastructural responses for the mitigation of salt stress. Chemosphere 336:139230

82. Jain K, Takuli A, Gupta TK, Gupta D (2024) Rethinking nanoparticle synthesis: a sustainable approach vs. traditional methods. Chem Asian J 19(21):e202400701
83. Banjara RA, Kumar A, Aneshwari R, Satnami ML, Sinha S (2024) A comparative analysis of chemical vs green synthesis of nanoparticles and their various applications. Environ Nanotechnol Monit Manag 22:100988
84. Tang Y et al (2023) Nano-pesticides and fertilizers: solutions for global food security (in eng). Nanomaterials (Basel) 14(1). https://doi.org/10.3390/nano14010090
85. Arora PK et al (2024) Next-generation fertilizers: the impact of bionanofertilizers on sustainable agriculture. Microb Cell Fact 23(1):254. https://doi.org/10.1186/s12934-024-02528-5
86. Kottegoda N, Munaweera I, Madusanka N, Karunaratne V (2011) A green slow-release fertilizer composition based on urea-modified hydroxyapatite nanoparticles encapsulated wood. Curr Sci:73–78
87. Marchiol L et al (2019) Influence of hydroxyapatite nanoparticles on germination and plant metabolism of tomato (Solanum lycopersicum L.): preliminary evidence. Agronomy 9(4):161
88. Corradini E, De Moura M, Mattoso L (2010) A preliminary study of the incorparation of NPK fertilizer into chitosan nanoparticles. Express Polym Lett 4(8):509–515
89. Wang A et al (2019) Facile synthesis of nitrogen-rich carbon dots as fertilizers for mung bean sprouts. Adv Sustain Syst 3(3):1800132
90. Li H et al (2018) Impacts of carbon dots on rice plants: boosting the growth and improving the disease resistance. ACS Appl Bio Mater 1(3):663–672
91. Xu X et al (2020) Promoting the growth of mung bean plants through uptake and light conversion of NaYF4: Yb, Er@ CDs nanocomposites. ACS Sustain Chem Eng 8(26):9751–9762
92. Srivastava A, Rao D (2014) Enhancement of seed germination and plant growth of wheat, maize, peanut and garlic using multiwalled carbon nanotubes. Eur Chem Bull 3(5):502–504
93. Zhai G, Gutowski SM, Walters KS, Yan B, Schnoor JL (2015) Charge, size, and cellular selectivity for multiwall carbon nanotubes by maize and soybean. Environ Sci Technol 49(12):7380–7390
94. Yatim NM, Shaaban A, Dimin MF, Mohamad N, Yusof F (2019) Urea functionalized multi-walled carbon nanotubes as efficient nitrogen delivery system for rice. Adv Nat Sci Nanosci Nanotechnol 10(1):015011
95. Pandey K, Lahiani MH, Hicks VK, Hudson MK, Green MJ, Khodakovskaya M (2018) Effects of carbon-based nanomaterials on seed germination, biomass accumulation and salt stress response of bioenergy crops. PLoS One 13(8):e0202274
96. Rahmani N, Radjabian T, Soltani BM (2020) Impacts of foliar exposure to multi-walled carbon nanotubes on physiological and molecular traits of Salvia verticillata L., as a medicinal plant. Plant Physiol Biochem 150:27–38
97. Ashfaq M, Verma N, Khan S (2017) Carbon nanofibers as a micronutrient carrier in plants: efficient translocation and controlled release of cu nanoparticles. Environ Sci Nano 4(1):138–148
98. Kim TY, Lee S-H, Ku H, Lee S-Y (2019) Enhancement of drought tolerance in cucumber plants by natural carbon materials. Plants 8(11):446
99. Ali MH et al (2020) Carbon nanoparticles functionalized with carboxylic acid improved the germination and seedling vigor in upland boreal forest species. Nano 10(1):176
100. Wang H et al (2018) Carbon dots promote the growth and photosynthesis of mung bean sprouts. Carbon 136:94–102
101. Kumaraswamy R et al (2021) Chitosan-silicon nanofertilizer to enhance plant growth and yield in maize (Zea mays L.). Plant Physiol Biochem 159:53–66
102. Sheikhalipour M et al (2021) Chitosan–selenium nanoparticle (Cs–se NP) foliar spray alleviates salt stress in bitter melon. Nano 11(3):684
103. Zhu G et al (2023) Phosphorus-based nanomaterials as a potential phosphate fertilizer for sustainable agricultural development. Plant Physiol Biochem 205:108172
104. Prajapati D et al (2022) Chitosan nanomaterials: a prelim of next-generation fertilizers; existing and future prospects. Carbohydr Polym 288:119356

105. Waghmare RD, Gore AH, Anbhule PV, Sohn D, Kolekar GB (2020) Dataset on the shooting and rooting ability of Morus alba using waste tea residue derived carbon dots as an alternative of growth plant stimulator. Data Brief 29:105345
106. Aji MP et al (2020) Carbon dots from dragonfruit peels as growth-enhancer on ipomoea aquatica vegetable cultivation. Adv Nat Sci Nanosci Nanotechnol 11(3):035005
107. Wang X, Zhou Z, Chen F (2017) Surface modification of carbon nanotubes with an enhanced antifungal activity for the control of plant fungal pathogen. Materials 10(12):1375
108. Ianchis R et al (2017) Novel hydrogel-advanced modified clay nanocomposites as possible vehicles for drug delivery and controlled release. Nano 7(12):443
109. Shao H, Xi N, Zhang Y (2018) Microemulsion formulation of a new biopesticide to control the diamondback moth (Lepidoptera: Plutellidae). Sci Rep 8(1):10565
110. Zhang Z, Shang X, Yang L, Shi Y, Liu Y (2020) Engineering of peglayted camptothecin into nanomicelles and supramolecular hydrogels for pesticide combination control. Front Chem 7:922
111. Pavela R et al (2021) Encapsulation of Carlina acaulis essential oil and carlina oxide to develop long-lasting mosquito larvicides: microemulsions versus nanoemulsions. J Pest Sci 94:899–915
112. Murugan K et al (2018) Iron and iron oxide nanoparticles are highly toxic to Culex quinquefasciatus with little non-target effects on larvivorous fishes. Environ Sci Pollut Res 25:10504–10514
113. Cai L, Cai L, Jia H, Liu C, Wang D, Sun X (2020) Foliar exposure of Fe3O4 nanoparticles on Nicotiana benthamiana: evidence for nanoparticles uptake, plant growth promoter and defense response elicitor against plant virus. J Hazard Mater 393:122415
114. Alam T, Khan RAA, Ali A, Sher H, Ullah Z, Ali M (2019) Biogenic synthesis of iron oxide nanoparticles via Skimmia laureola and their antibacterial efficacy against bacterial wilt pathogen Ralstonia solanacearum. Mater Sci Eng C 98:101–108
115. Sriramulu M, Balaji, Sumathi S (2021) Photo catalytic, antimicrobial and antifungal activity of biogenic iron oxide nanoparticles synthesised using Aegle marmelos extracts. J Inorg Organomet Polym Mater 31:1738–1744
116. Shahryari F, Rabiei Z, Sadighian S (2020) Antibacterial activity of synthesized silver nanoparticles by sumac aqueous extract and silver-chitosan nanocomposite against pseudomonas syringae pv. Syringae. J Plant Pathol 102:469–475
117. Ahsan T (2020) Biofabrication of silver nanoparticles from Pseudomonas fluorescens to control tobacco mosaic virus. Egypt J Biol Pest Contr 30(1):66
118. Idowu ET et al (2021) Larvicidal potential of two silver nano-particles (Moringa oleifera and Ficus exasperata) against laboratory and field strains of Anopheles gambiae (Diptera: Culicidae) in Lagos, Nigeria. J Basic Appl Zool 82:1–9
119. Al Shater H, Moustafa HZ, Yousef H (2020) Synthesis, phytochemical screening and toxicity measuring against Earias insulana (Boisd.)(Lepidoptera: Noctuidae) of silver nano particles from Origanum marjorana extract in the field. Egypt Acad J Biol Sci F Toxicol Pest Contr 12(1):175–184
120. Ibrahim E et al (2020) Green-synthesization of silver nanoparticles using endophytic bacteria isolated from garlic and its antifungal activity against wheat fusarium head blight pathogen fusarium graminearum. Nano 10(2):219
121. Amna et al (2021) Characterization of bio fabricated silver nanoparticles for distinct antifungal activity against sugarcane phytopathogens. Microsc Res Tech 84(7):1522–1530
122. Al-Otibi F et al (2021) Biosynthesis of silver nanoparticles using Malva parviflora and their antifungal activity. Saudi J Biol Sci 28(4):2229–2235
123. Ahmad H et al (2020) Green synthesis and characterization of zinc oxide nanoparticles using eucalyptus globules and their fungicidal ability against pathogenic fungi of apple orchards. Biomolecules 10(3):425
124. Akpomie KG, Ghosh S, Gryzenhout M, Conradie J (2023) Ananas comosus peel–mediated green synthesized magnetite nanoparticles and their antifungal activity against four filamentous fungal strains. Biomass Convers Biorefinery 13(7):5649–5660

125. Giri VP et al (2023) A review of sustainable use of biogenic nanoscale agro-materials to enhance stress tolerance and nutritional value of plants. Plants 12(4):815
126. Zhi H, Zhou S, Pan W, Shang Y, Zeng Z, Zhang H (2022) The promising nanovectors for gene delivery in plant genome engineering. Int J Mol Sci 23(15):8501
127. Alghuthaymi MA et al (2021) Exosome/liposome-like nanoparticles: new carriers for CRISPR genome editing in plants. Int J Mol Sci 22(14):7456
128. Prasad S et al (2021) A review on bio-inspired synthesis of silver nanoparticles: their antimicrobial efficacy and toxicity. Eng Sci 16(16):90–128
129. Mujtaba M et al (2021) Nanocarrier-mediated delivery of miRNA, RNAi, and CRISPR-Cas for plant protection: current trends and future directions. ACS Agri Sci Technol 1(5):417–435
130. Lu X et al (2020) Stimulation of photosynthesis and enhancement of growth and yield in Arabidopsis thaliana treated with amine-functionalized mesoporous silica nanoparticles. Plant Physiol Biochem 156:566–577
131. Liu S et al (2025) Nanodelivery of nucleic acids for plant genetic engineering. Disc Nano 20(1):31
132. Chen L, Hong W, Ren W, Xu T, Qian Z, He Z (2021) Recent progress in targeted delivery vectors based on biomimetic nanoparticles. Signal Transduct Target Ther 6(1):225
133. Yong J, Wu M, Carroll BJ, Xu ZP, Zhang R (2024) Enhancing plant biotechnology by nanoparticle delivery of nucleic acids. Trends Genet 40(4):352–363
134. Ijaz M et al (2023) Nanobiotechnology to advance stress resilience in plants: current opportunities and challenges. Mater Today Bio 22:100759
135. Zare H et al (2021) Carbon nanotubes: Smart drug/gene delivery carriers. Int J Nanomedicine Volume 16:1681–1706
136. Squire HJ, Tomatz S, Voke E, González-Grandío E, Landry M (2023) The emerging role of nanotechnology in plant genetic engineering. Nat Rev Bioeng 1(5):314–328
137. Kumar P et al (2024) Nano-assisted delivery tools for plant genetic engineering: a review on recent developments. Environ Sci Pollut Res:1–16
138. Demirer GS et al (2021) Nanotechnology to advance CRISPR–Cas genetic engineering of plants. Nat Nanotechnol 16(3):243–250
139. Niazian M, Molaahmad Nalousi A, Azadi P, Ma'mani L, Chandler SF (2021) Perspectives on new opportunities for nano-enabled strategies for gene delivery to plants using nanoporous materials. Planta 254(4):83
140. Caprifico AE, Foot PJ, Polycarpou E, Calabrese G (2022) Advances in chitosan-based CRISPR/Cas9 delivery systems. Pharmaceutics 14(9):1840
141. Rice CL (2023) Optimisation of carbon dot-facilitated wheat transformation. University of Bristol
142. Rehman A et al (2024) Exploring the nano-wonders: unveiling the role of nanoparticles in enhancing salinity and drought tolerance in plants. Front Plant Sci 14:1324176
143. Hayat F et al (2023) Nanoparticles and their potential role in plant adaptation to abiotic stress in horticultural crops: a review. Sci Hortic 321:112285
144. Mubashir A et al (2023) Effect of foliar application of nano-nutrients solution on growth and biochemical attributes of tomato (Solanum lycopersicum) under drought stress. Front Plant Sci 13:1066790
145. Raza MAS et al (2023) Morpho-physiological and biochemical response of wheat to various treatments of silicon nano-particles under drought stress conditions. Sci Rep 13(1):2700
146. Kumar A et al (2024) Carbon nanoparticle exposure strengthens water-relation parameters by stimulating abscisic acid pathway and aquaporins genes in rice. Plant Stress 13:100540
147. Selvakesavan RK, Kruszka D, Shakya P, Mondal D, Franklin G (2023) Impact of nanomaterials on plant secondary metabolism. In: Nanomaterial interactions with plant cellular mechanisms and macromolecules and agricultural implications, pp 133–170
148. Iqbal M et al (2024) Insight into carbohydrate metabolism, protein quantification and mineral regulation in wheat (Triticum aestivum L.) by the action of green synthesized silver nanoparticles (AgNPs) against heat stress. J Biomol Struct Dyn:1–15

149. Ghani MI et al (2022) Foliar application of zinc oxide nanoparticles: an effective strategy to mitigate drought stress in cucumber seedling by modulating antioxidant defense system and osmolytes accumulation. Chemosphere 289:133202
150. Dong B-r, Jiang R, Chen J-f, Xiao Y, Lv Z-y, Chen W-s (2023) Strategic nanoparticle-mediated plant disease resistance. Crit Rev Biotechnol 43(1):22–37
151. Chauhan H, Patel M, Patel P, Tiwari S, Jinal HN, Amaresan N (2023) Assessment of copper (cu) nanoparticle for their biocontrol activity against Xanthomonas oryzae pv. Oryzae, growth promotion, and physiology of rice (Oryza sativa L.) plants. Lett Appl Microbiol 76(1):ovac066
152. Jian Y et al (2022) Toxicity and action mechanisms of silver nanoparticles against the mycotoxin-producing fungus fusarium graminearum. J Adv Res 38:1–12
153. Verma N, Kaushal P, Sidhu AK (2024) Harnessing biological synthesis: zinc oxide nanoparticles for plant biotic stress management. Front Chem 12:1432469
154. Abdelkhalek A et al (2022) Ocimum basilicum-mediated synthesis of silver nanoparticles induces innate immune responses against cucumber mosaic virus in squash. Plants 11(20):2707
155. Cai L, Liu C, Fan G, Liu C, Sun X (2019) Preventing viral disease by ZnONPs through directly deactivating TMV and activating plant immunity in Nicotiana benthamiana. Environ Sci Nano 6(12):3653–3669
156. Mondal S et al (2022) Nanotechnology in developing abiotic stress resilience in crops: a physiological implication. In: Advancements in developing abiotic stress-resilient Plants. CRC Press, pp 351–370
157. Cunningham FJ (2022) Conjugating CRISPR-cas9 machinery to single-walled carbon nanotubes for plant cellular delivery. University of California, Berkeley
158. Larriba E, Yaroshko O, Pérez-Pérez JM (2024) Recent advances in tomato gene editing. Int J Mol Sci 25(5):2606
159. Fu C et al (2023) Mechanisms of nanomaterials for improving plant salt tolerance. Crop Environ 2(2):92–99
160. Venu E et al (2024) Exogenous dsRNA-mediated RNAi: mechanisms, applications, delivery methods and challenges in the induction of viral disease resistance in plants. Viruses 17(1):49
161. Gogoi N, Susila H, Leach J, Müllner M, Jones B, Pogson BJ (2024) Developing frameworks for nanotechnology-driven DNA-free plant genome-editing. Trends Plant Sci
162. Yemets AI, Karpets YV, Kolupaev YE, Blume YB (2019) Emerging technologies for enhancing ROS/RNS homeostasis. In: Reactive oxygen, nitrogen and sulfur species in plants: Production, metabolism, signaling and defense mechanisms, pp 873–922
163. Li G, Ma Z, Zhang N, Li M, Li W, Mo Z (2024) Foliar application of silica nanoparticles positively influences the early growth stage and antioxidant defense of maize under low light stress. J Soil Sci Plant Nutr 24(2):2276–2294
164. Sheikh Mohamed M, Sakthi Kumar D (2019) Application of nanotechnology in genetic improvement in crops. In: Nanoscience for sustainable agriculture, pp 3–24
165. Rameen S et al (2024) Foliar application of silver (ag-NPs) and copper (cu-NPs) nanoparticles enhances phenotypic traits and oil quality in Brassica napus L. Sci Rep 14(1):27618
166. Estiasih T, Kuliahsari DE, Martati E, Ahmadi K (2022) Cyanogenic compounds removal and characteristics of non-and pregelatinized traditional detoxified wild yam (Dioscorea hispida) tuber flour. Food Sci Technol 42:e119121
167. Sharma MMM, Kapoor D, Loyal A, Kumar R, Sharma P, Husen A (2024) Plant response to silver nanoparticles in terms of growth, development, production, and protection: an overview. In: Plant Response to Silver Nanoparticles: Plant Growth, Development, Production, and Protection, pp 1–22
168. Mattioli R, Francioso A, Mosca L, Silva P (2020) Anthocyanins: a comprehensive review of their chemical properties and health effects on cardiovascular and neurodegenerative diseases. Molecules 25(17):3809
169. Lohani N, Jain D, Singh MB, Bhalla PL (2020) Engineering multiple abiotic stress tolerance in canola, Brassica napus. Front Plant Sci 11:3

170. Elhefnawy SM, Elsheery NI (2023) Use of nanoparticles in improving photosynthesis in crop plants under stress. In: Photosynthesis. Elsevier, pp 105–135
171. Zhao Y et al (2024) Rational nanoparticle design for efficient biomolecule delivery in plant genetic engineering. Nanoscale 16(46):21264–21278
172. Farvardin A et al (2023) Solanum lycopersicum heme-binding protein 2 as a potent antimicrobial weapon against plant pathogens. Sci Rep 13(1):20336
173. Pattanayak SP, Bose P, Sunita P (2023) Interaction between nanoparticles and phytopathogens. In: Agricultural and environmental nanotechnology: novel technologies and their ecological impact. Springer, pp 169–220
174. Khalid S, Siddique R, Bakhsh A (2023) Smart plant breeding for potato in the post-genomics era. In: Smart plant breeding for vegetable crops in post-genomics Era. Springer, pp 337–356
175. Ali S, Tyagi A, Rajarammohan S, Mir ZA, Bae H (2023) Revisiting Alternaria-host interactions: new insights on its pathogenesis, defense mechanisms and control strategies. Sci Hortic 322:112424
176. Rahman FU et al (2024) Current insights into the biocontrol and biotechnological approaches for postharvest disease management of Botrytis cinerea. Postharvest Biol Technol 216:113055
177. Kumari N, Sharma V, Patel P, Sharma P (2023) Pepper mild mottle virus: a formidable foe of capsicum production—a review. Front Virol 3:1208853
178. Ying S et al (2022) Green synthesis of nanoparticles: current developments and limitations. Environ Technol Innov 26:102336
179. Gao M, Chang J, Wang Z, Zhang H, Wang T (2023) Advances in transport and toxicity of nanoparticles in plants. J Nanobiotechnol 21(1):75. https://doi.org/10.1186/s12951-023-01830-5
180. Chauhan D et al (2023) Silver nanoparticles trigger physiological and molecular perturbations in Arabidopsis thaliana. Sci Total Environ 857:159720. https://doi.org/10.1016/j.scitotenv.2022.159720
181. Wang L et al (2020) Silver nanoparticles regulate Arabidopsis root growth by concentration-dependent modification of reactive oxygen species accumulation and cell division. Ecotoxicol Environ Saf 190:110072
182. Sun J, Wang L, Li S, Yin L, Huang J, Chen C (2017) Toxicity of silver nanoparticles to Arabidopsis: Inhibition of root gravitropism by interfering with auxin pathway," (in eng). Environ Toxicol Chem 36(10):2773–2780. https://doi.org/10.1002/etc.3833
183. Mishra M, Patil H (2021) Bioinspired nanoparticles: synthesis, applications, and environmental impact. In: Green synthesis in nanomedicine and human health. Springer, pp 245–266
184. Malafatti JOD, Tavares FA, Neves TR, Mascarenhas BC, Quaranta S, Paris EC (2023) Modified silica Nanoparticles from Rice husk supported on Polylactic acid as adsorptive membranes for dye removal. Materials 16(6):2429. [Online]. Available: https://www.mdpi.com/1996-1944/16/6/2429
185. Fadeel B et al (2018) Advanced tools for the safety assessment of nanomaterials (in eng). Nat Nanotechnol 13(7):537–543. https://doi.org/10.1038/s41565-018-0185-0
186. J. Allan et al., "Regulatory landscape of nanotechnology and nanoplastics from a global perspective (in eng),", Regul Toxicol Pharmacol, vol. 122, p. 104885, 2021, https://doi.org/10.1016/j.yrtph.2021.104885
187. Suhag D, Kaushik S, Taxak VB (2024) Challenges and future directions. In: Handbook of biomaterials for medical applications, volume 1: fundamentals. Springer Nature Singapore, Singapore, pp 329–355
188. Rana L et al (2024) Nexus between nanotechnology and agricultural production systems: challenges and future prospects. Discover Appl Sci 6(11):1–19
189. Kumari R et al (2023) Regulation and safety measures for nanotechnology-based agri-products (in eng). Front Genome Ed 5:1200987. https://doi.org/10.3389/fgeed.2023.1200987
190. Wang ZP et al (2022) Efficient and genotype independent maize transformation using pollen transfected by DNA-coated magnetic nanoparticles. J Integr Plant Biol 64(6):1145–1156
191. Xu L, Zhu Z, Sun D-WJA n (2021) Bioinspired nanomodification strategies: moving from chemical-based agrosystems to sustainable agriculture. ACS Nano 15(8):12655–12686

Introduction to Plant Nano-Biotechnology: Advances and Prospects

Rabia Kousar, Sana Fatima, Majid Anwar, and Khawaja Shafique Ahmad

Abstract Developing countries strongly depend on agriculture as their essential sector while facing crucial problems from growing food requirements together with deteriorating soil conditions and climate-induced pressures. Traditional farming falls behind present requirements therefore society must adopt progressive alternatives. Plant nano-biotechnology serves as an innovative technology that combines nanotechnology and biotechnology systems for developing improved farming methods and pest control strategies alongside sustainable agricultural practices. This chapter investigates the past development of nano-biotechnology by defining interaction patterns between nanoparticles and plants combined with a description of multiple nanomaterials and their applications through techniques such as nano-fertilizers, nano-pesticides, precision agriculture, and gene delivery systems. New developments in responsive nanocarrier systems are described with emphasis on their multi-functional properties. The present research field maintains its attention on environmentally smart systems, CRISPR applications, and biodegradable nanomaterials.

Keywords Plant nano-biotechnology · Nanomaterials · Nano-fertilizers · Nano-pesticides · Nanosensors · Smart nanocarriers · Gene delivery · CRISPR-Cas systems

R. Kousar · S. Fatima
Department of Botany, The Govt. Sadiq College Women University Bahawalpur, Bahawalpur, Pakistan

M. Anwar
Department of Pharmacology, Multan University of Science and Technology, Multan, Pakistan

K. S. Ahmad (✉)
Department of Botany, University of Poonch Rawalakot, Rawalakot, Pakistan
e-mail: ahmadks@upr.edu.pk

J.-T. Chen (ed.), *Nano-Delivering for Plant Genetic Engineering*, Smart Nanomaterials Technology, https://doi.org/10.1007/978-981-95-4165-2_7

1 Introduction

Agriculture is the backbone of most of the developing countries in the world providing food directly and indirectly to mankind, but present-day agriculture is facing challenges [44]. Traditional farming strategies do not meet today's population needs because the growing food demand along with declining soil quality and climate-induced pest outbreaks and disease occurrences lead to insufficient agricultural production systems [54].

The urgent agricultural needs required revolutionary scientific and technological solutions which plant nano-biotechnology stands as a leading candidate. The combination of nanotechnology advancements with biotechnology principles creates innovative solutions for crop advancement along with pest protection and eco-friendly agricultural practices [19]. Agricultural transformation is possible through plant nano-biotechnology because of nanomaterial-specific properties which lead to enhanced efficiency and sustainability.

Plant nano-biotechnology combines nanotechnology alongside plant biology and biotechnology to produce innovative agricultural tools that confront agricultural challenges [47]. Through the manipulation of nanomaterials, this innovative science system targets three critical agricultural needs which include food security alongside environmental sustainability as well as climate resilience [7]. Nanomaterials show unique physicochemical properties which include elevated surface area and reactivity and adjustable optical attributes because they measure at least one dimension below 100 nm [24].

Applications of nanobiotechnology in agriculture have spread to a number of fields, such as (1) the nanotechnology of encapsulated agro-chemicals, (2) the use of nanobiosensors to monitor various environmental stresses and crop conditions, (3) the enhancement of crop productivity and the protection of plants from diseases, and (4) the resolution of various environmental issues.

Some novel agrochemicals, such as nanopesticides and nanofertilizers, could also increase agricultural yield. In addition to suppressing plant pests and diseases and preventing pollution of the environment, these agrochemicals are highly successful at delivering encapsulating nanoparticles and increasing agricultural production. However, many nano-agrochemicals or foods that have undergone nanoprocessing may allow nanoparticles to infiltrate the food chain. Therefore, before deploying various bionanotechnological instruments in the agro-production sector, a number of ways are required, such as the uptake of nanoparticles by plants and their entry and bio-distribution into the food chain [59].

The nanofabrication, biosystem, and applications of bionanotechnology for plants are all included in the field of plant nanobiotechnology. Many studies have been carried out in vitro using nanoparticles of metals/metalloids or metal oxides such as selenium, titanium dioxide, silver oxide, copper oxide, zinc oxide, etc., making plant tissue culture with nanoparticles or plant bionanotechnology potentially useful tools. Additionally, the use of nanoparticles in in vitro studies has yielded numerous advantages, such as (1) increasing the rate of multiplication, (2) removing contamination, (3) inducing hard-rooting plants, (4) reducing the effects of stressors like salinity and drought, and (5) promoting plant growth at low nanoparticle concentrations [1].

Numerous scientific disciplines have combined with nanotechnology to provide novel goods and technologies that enhance crop plant characteristics [16]. Using effectively managed inputs, such as fertilizers, herbicides, and agrochemicals, facilitated by real-time monitoring systems, agri-nanotechnology seeks to increase crop productivity. At every stage of agriculture, from seed germination to post-harvesting, a range of nanotechnology applications will improve crop quality and yield [59].

The main participants in the agri-tech revolution are engineered nanomaterials, or ENMs. These materials' nanoscale size (less than 100 nm) allows them to get past biological barriers and guarantee effective pesticide and nutrition delivery through foliar or root treatments [15]. Moreover, NM-based optical and electrochemical sensors can be used to identify environmental contaminants and heavy metals in soil media. Large-scale crop health monitoring at high geographical and temporal precision is made possible by ENMs, which are perfectly suited to bringing about this transformation since they can function as sensors in plants [22].

Abiotic stressors have a major impact on plant growth and result in a 50% reduction in agricultural output. Plant growth, development, and yield production are significantly reduced as a result of these stressors, which disrupt the metabolism of the plant. Among other consequences, one of the main causes of crop losses brought on by abiotic stressors is the excessive generation of reactive oxygen species (ROS). Scavenging ROS and enhancing tolerance to abiotic stressors have been accomplished by the use of nutrients, osmoprotectants, and stress-signaling chemicals [31].

On the other hand, these materials may work better when applied at nanoscales. For instance, there is mounting evidence that the aforementioned compounds, when delivered as nanomaterials, enhance plant resistance to stressors such as heat, drought, salt, and trace metals. When given to plants, nanomaterials' high surface energy and high surface/volume ratio enhance their metabolic activity, reactivity, and efficacy. Nanomaterials are used in plant defense because they both protect against ROS and cause oxidative stress. The antioxidant defense mechanism of plants is stimulated by the latter. Novel characteristics of nanoparticles (NPs) show promise for managing abiotic stressors as well as crops [5]. NPs have been produced using a variety of metals and their oxides, such as silver (Ag), copper (Cu), gold (Au), iron (Fe), titanium (Ti), and zinc (Zn). These NPs have recently been employed in the green synthesis of NPs using the DNA of viruses, including diatoms, microorganisms, and their membranes, and plants and their extracts. AgNPs are one type of green-synthesized NP that is frequently employed in agriculture [25].

Nanobiotechnology provides farmers with technological tools for precise nutrient and agrochemical administration systems while enhancing stress tolerance in plants and enabling live monitoring of biological indicators. Nanobiotechnology acts as a transformative tool for agricultural development since it supports applications starting from soil management through post-harvest processing [10].

2 Historical Development and Evolution

In 1959 physicist Richard Feynman established the fundamental concepts of nanotechnology through his introduction of atomic and molecular matter manipulation [63]. Research integration between nanotechnology and both biological sciences and agriculture started to pick up speed during the early 2000s. Scientists first studied nanoparticle production methods and evaluated their harmful effects on organisms. Practical approaches in agricultural applications started gaining priority because of inefficiencies regarding chemical fertilizers and pesticides [6].

3 Theoretical Foundations of Nanoparticle-Plant Interactions

3.1 Physicochemical Interactions

Plant nano-biotechnology requires thorough comprehension of nanoparticle-plant system interactions which function at both cell and subcellular levels. Nanoparticle size together with shape as well as surface charge, solubility, and composition serves as crucial elements that affect these interactions according to Zhao et al. [75]. The characteristics influence how nanoparticles enter plants through roots or leaves and their movements and modifications inside plant tissue after entry.

3.2 Mechanisms of Uptake and Transport

Plant cells permit nanoparticle penetration through endocytosis as well as carrier proteins and aquaporins but also via natural stomatal openings. Plants allow nanoparticles to move through their tissues using two pathways which include the apoplastic movement through cell walls and intercellular spaces and symplastic movement through cytoplasm and plasmodesmata. Nanoparticles use these methods for their transportation to particular plant tissues so agrochemicals or nutrients can reach their targeted destination [69, 70]. Knowledge of these uptake and transportation pathways plays a crucial role in designing nanocarriers because it enables controlled distribution and enhanced accessibility of agro-inputs targeting specific areas which enhances crop yields while decreasing environmental effects (Fig. 1).

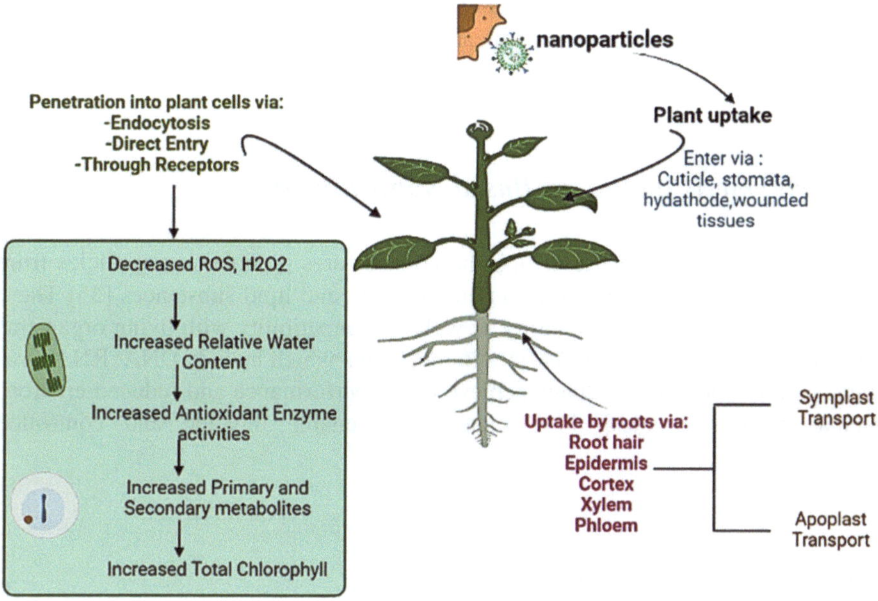

Fig. 1 Schematic representation of nanoparticle uptake and transport in plants

4 Types of Nanomaterials in Plant Sciences

Different nanomaterial varieties have been developed for plant science applications which bring unique physical characteristics to create new solutions for agricultural use.

4.1 Metallic and Metal Oxide Nanoparticles

The category of metallic nanoparticles includes pure metals but metal oxide nanoparticles contain metals linked to oxygen atoms. Silver (Ag), zinc oxide (ZnO) and titanium dioxide (TiO_2) and iron oxide (Fe_2O_3) nanoparticles demonstrate antimicrobial and optical and magnetic properties as part of their unique features. Scientists extensively study these nanoparticles because they improve plant development but serve as nutrient transportation systems and combat pathogens [61].

4.2 Carbon-Based Nanomaterials

The body of carbon-based nanomaterials contains only carbon atoms which create different shapes such as nanotubes and graphene sheets as well as spherical fullerenes [12]. The combination of electrical conductivity and mechanical strength alongside a large surface area makes carbon nanotubes graphene oxide and fullerenes suitable for cell delivery applications and biosensing and stress resistance improvement in plants [9].

4.3 *Polymeric and Lipid-Based Nanocarriers*

Nanocarriers with polymeric or lipid-based structures produce nanoparticles from organic polymers including chitosan and alginate and lipid substances [35] These nanocarriers maintain degradability as well as compatibility with living organisms while offering encapsulation services to molecules which include DNA, RNA, proteins, and agrochemicals. Plants achieve better performance and reduced environmental hazards because these nanocarriers enable directed and controlled delivery [37].

4.4 *Biosynthesized Nanoparticles*

Biosynthesized nanoparticles emerge from biological systems which include bacteria, fungi, algae, or plant extracts to provide environmentally friendly production methods. Biosynthesized nanoparticles demonstrate natural biocompatibility together with reduced toxicity levels which makes them suitable for environmentally friendly agricultural usage [17].

Studying the wide assortment of nanomaterials provides the necessary knowledge for developing sustainable agricultural strategies incorporating nano-enabled approaches.

5 Advances of Nanobiotechnology in Agriculture

5.1 *Nano-Fertilizers*

The transformative potential of nano-fertilizers serves to boost nutrient use efficiency together with maintaining sustainable agricultural methods. These nanofertilizers provide regulated and extended delivery of nutrients that prevent ecological losses from leaching and volatilization. Science is utilizing zeolite- and clay-based nanocarriers to deliver macronutrients (N, P, K) and micronutrients (Fe, Zn, Mn) according to Mahesha et al. [38]. The versatility of nano-fertilizers grows through formulations made of organic compounds as well as hybrids and carbon-based substances. The use of these materials strengthens crop output while minimizing fertilizer-related environmental damage according to Yadav et al. [72, 73].

Nanotechnology promotes plant growth as it facilitates direct nanoscale nutrient uptake, has a facility for targeted nutrient delivery, and provides systems of controlled release for increased efficacy [50]. Common nanoparticles such as AuNPs and AgNPs are highly prized primarily for bioconjugation and experimental reliability, though their nutrition supply capabilities are not their main focus. To satisfy plant-specific needs, nanoparticles from metals like zinc, silver, copper, iron, gold, and nickel, and metal oxides like aluminum oxide, silicon dioxide, titanium dioxide, and iron oxide have been developed [26]. Nanoscale iron enhances germination and physiological measures in maize and rice, TiNPs and ZnNPs also increase photosynthesis, and seed germination in diverse species of plants [49].

Copper carbon nanofertilizers (Cu-CNFs) enable efficient transport of nutrients and prevent excessive ion release, which in turn enhance water uptake and biological pigment content and plant growth. Polymeric mixtures with Cu -Zn- CNFs have also been found to promote plant growth. The application of carbon-based nanomaterials including graphite oxide and graphene quantum dots alters the root morphology that enhances plant growth [33].

Quantum scale fertilizers help minimize waste because they thus release nutrients more accurately. The use of synthetic apatite nanoparticles and surfactant-modified zeolites increases the bioavailability of phosphorus and improves crop productivity, and small-sized copper and iron increase plant nutrition, too [32]. In addition, micronutrients provided in the nanoscale promote not only an increased uptake of macronutrients but also increase plant resistance to abiotic stresses. Low doses of nanoscale fertilizers passively applied via aerosol sprays and aqueous suspensions show that they have been found to improve crop yield, as evidenced by field trials [5].

5.2 Nano-Pesticides

The application of nano-pesticides advances disease and pest control through improved properties of active ingredients that enhance stability and increased solubility and targeted concentration levels. Nanotechnology creates conditions for pesticides to work more efficiently with reduced doses which simultaneously decreases environmental impact and pesticide resistance risk. The designed controlled-release nano-formulations extend their pest-control effect which decreases the number of applications required in agricultural fields [65].

Sustainable crop production must be achieved on the basis of decreasing crop losses because of pathogens, pests, and environmental threats, and nano-enabled pesticides describe a promising solution. Standard pesticide application delivers active ingredients at an extremely low efficacy rate of less than 0.1% but nano-formulations offer a huge accuracy or efficacy benefit for applying pesticides [51]. Metallic nanoparticles that contain zinc, silver, iron, manganese and copper have been found to be able to stimulate defense mechanisms in crops such as wheat (*Triticum aestivum*) from pathogens such as Pseudocercosporella herpotrichoides.

Silver nanoparticles (Ag NPs) have shown antifungal activity against Bipolaris sorokiniana and Magnaporthe grisea but the antifungal power of the nanoparticles differs in concentration and formulation. Biofunctionalized manganese nanoparticles (MnNPs) have also shown efficacy during application in the control of Fusarium wilt on watermelon by interrupting infection, stimulating defense mechanisms, and modulation of soil microbiota [4]. By using enzyme-responsive nanocarriers, CHT@MSNs- β-glucans improved chlorothalonil delivery and bioactivity to protect against M. grisea.

The application of chitosan, which is valued for its biocompatibility and sustainability, has given rise to the antifungal activity of the pathogens such as Aspergillus niger, Alternaria alternata, Rhizopus oryzae, Phomopsis asparagi, and Rhizopus stolonifer [30]. Furthermore, Cu-chitosan nanoparticles have improved vegetation and resistance to diseases in *Zea mays* (maize) where Curvularia leaf spot is present. Although the use of nano-pesticides has practicality today, more studies have to be conducted to get over barriers like formulation stability and the interaction with the environment. Providing agriculture, and protecting the environment, with safety and effectiveness requires proper consideration of risks associated with nanoparticles in terms of plant health and food safety [13].

5.3 Nanosensors for Precision Agriculture

Precision agriculture relies heavily on nanosensors to achieve its operational practices. The small size scale of these devices allows them to detect various stresses including droughts and limited nutrients as well as disease infections with precise detection capabilities. The addition of nanosensors into agricultural monitoring systems enables time-saving decisions that optimize water use besides optimizing fertilizer and pesticide applications leading to enhanced resource utilization efficiency [66].

Modern agriculture has come a long way by automating several tasks to automate the crop management process, increase the sustainability of crops and reduce the reliance on manual labor [52]. These nano-sensors provide an ingenious method for growers to track crops by identifying nutritional shortages and indications of stress with subsequent presentation of these results either as electrical or optical output. Nano-biosensors, GPS, and GNSS systems make it possible to provide continuous instant evaluation for the health, stress, and supply of nutrients for the plants reducing the usage of traditional inspection of fields [8].

Change of focus to precision farming enhances the efficiency of resource use, improves the quality of crops, and promotes judicious usage of agrochemicals in contributing to a more sustainable future of agriculture. Nano-sensors are able to detect the presence of pathogens and soil conditions, measure pesticide levels, and sense the environment's various factors in the rhizosphere that underpins disease prevention and maximized yield through intelligent nutrient application.

Current studies demonstrate that agar technology of nano-sensors centers largely on laboratory-based research [28]. Wu et al. [71] monitored the changes in H_2O_2 concentration level in the leaves of Arabidopsis thaliana in an effort to detect stress markers, using near-infrared fluorescent single-walled carbon nanotubes (SWCNTs) conjugated with DNA aptamers (HeAptDNA-SWCNTs). Through MnNP use on Capsicum annuum, Shi et al. [58] remediated salinity stress at germination in a 100 mM sodium chloride (NaCl) setting.

Spectroscopy and imaging indicated that MnNPs entered seed coat cells to bind to proteins. In addition, the incorporation of plasmonic nanomaterials into smartphone technologies has enabled non-invasive diagnosis of the disease both within laboratories and in greenhouses. Although intelligent nano-sensors are believed to be valuable assets for smart farming, the effectiveness and responsiveness of nanosensors in the real world, which is constantly changing, have not been shown yet [23]. Further probing is essential to facilitate these technologies for practical application in real farming setups.

5.4 Nanotechnology for Gene Delivery and Genetic Engineering

Nanocarrier systems introduced new pathways that enhance genetic substance penetration into plant cells effectively. The delivery of DNA, RNA, and genome-editing components to plant cells occurs through nanoparticulate carriers that include mesoporous silica and carbon nanotubes and lipid-based carriers as non-transgenic genetic transformation solutions [62]. Nanocarrier technologies enable the development of genetically modified crops through their ability to deliver new desirable characteristics such as drought and disease resistance and production enhancement [40].

Nanotechnology is gaining increasing attention as a valuable player in gene delivery in plants and enabling advanced methodologies in plant genetic engineering. An important challenge to plant gene editing is the efficient delivery of foreign genetic material to the robust plant cell wall. Nanomaterials offer potential antagonistic solutions for direct and specific delivery of exogenous molecules, such as DNA, RNA, and proteins, into plant cells avoiding the use of conventional biolistic or pathogenic techniques. In terms of CRISPR-Cas9-based genome editing, nanocarriers are critical for the delivery of DNA plasmids or RNA molecules coding CRISPR components and thus allow reaching a targeted and defined-mediated activity of Cas9 [68].

Moreover, nanotechnology plays a large role in supporting RNAi approaches through the protection of dsRNA molecules from degradation. RNA interference (RNAi) is a useful technology for pest and pathogen suppression by silencing the target genes, but topical dsRNA approaches often experience diminished effectiveness because of the degradation of dsRNA under environmental conditions.

Nanocarriers such as layered double hydroxides (LDH) or chitosan-based polymers protect dsRNA from degradation and ensure its entering plant cells, enhancing its efficacy and delaying its protective potential [11].

Further, the ability for dsRNA to penetrate cellular barriers and access the cytoplasm is enhanced by these carriers, which allow dsRNA to be converted into small interfering RNA (siRNA) for silencing genes. Moreover, nanomaterials, which are biodegradable and biocompatible at the same time, are preferred as they have a minor impact on the environment and are not very expensive. Chitosan serves as both a genetic material protector and an attractive and environmentally friendly option for large-scale agricultural applications. While some progress has been made there are continuing challenges; namely, establishing and addressing long-term risks and accumulative effects of the non-biodegradable nanocarriers used in delivering non-genetic agrochemicals [14]. Additional examination is essential to understand how engineered nanoparticles perform at the plant's biological and ecological levels and to establish better vectors of gene delivery that do not cause harm and work efficiently.

5.5 *Smart Nanocarrier Systems in Agriculture*

5.5.1 Stimuli-Responsive Delivery Systems

The engineered smart nanocarriers contain release mechanisms that activate due to changes in environmental factors like pH value, light intensity, temperature conditions, and enzymatic activities. The on-demand delivery system happens when nanocarriers detect specific stimuli for efficient controlled releases that reduce accidental environmental chemical leakage. Precise agricultural systems require quick adjustments to field changes because of their critical nature during implementation [57].

The use of photoresponsive nanocarriers activated by UV or near-infrared light has improved targeting in agrochemical delivery on the farm, providing refined localized and timely control of the application. Responsive to both pH and redox potential materials are being used in the root zone in order to maximize nutrient use given the fluctuating rhizospheric conditions. By supplying nutrients and pesticides at the same time on key stages of plant growth, these innovations allow to reduce input use and enhance effectiveness [36].

5.5.2 Multifunctional Nanoparticles

Nanoparticles with multiple functions combine several tasks when designed for concurrent actions which either fuse pesticide function with fertilizer abilities or combine antimicrobial properties with delivery capabilities. Silver and copper nanoparticles possess dual qualities that work as micronutrient supplementations and antimicrobial agents at the same time. Field management benefits from simplified operations along with reduced production expenses because these multiple functions can be achieved with single inputs [74].

Nanocomposites such as ZnO/chitosan and CuO-MgO NPs are used as micronutrient suppliers, antifungal agents and plant growth stimulants simultaneously. Moreover, nanoparticles, with hormone functionalization features, for instance, silica particles possessing auxin have been effective in promoting root development and reducing stress for cereal plants, including rice and wheat [29]. These multitasking systems reduce the need for chemicals, increase the efficiency of nutrients, and increase the competency of a plant under stress tolerance.

5.5.3 Integration of Nanotechnology with CRISPR and Genetic Tools

Nanotechnology combined with CRISPR-Cas systems generates an unprecedented scientific breakthrough. Nanocarriers provide a vectorless delivery system for CRISPR components into plant cells so genetic modifications can be achieved precisely with few unintended alterations. The technique demonstrates potential as an approach to create non-transgenic genetically resistant crops that could help resolve regulatory as well as public acceptance issues [43].

Single-walled carbon nanotubes and lipid-based nanoparticles have facilitated the successful delivery of CRISPR-Cas9 ribonucleoproteins (RNPs) in *Arabidopsis thaliana* and *Zea mays* leading to targeted genome editing without DNA integration [67]. The study of plant-derived exosome-like nanocarriers is still in progress due to their good biocompatibility and elevated intracellular delivery of genes that allow for achieving precise and transient genetic modifications with a lower level of toxicity.

5.6 Enhancing Plant Physiology and Stress Tolerance

5.6.1 Photosynthesis Enhancement and Water Use Efficiency

Nanoparticles demonstrate their potential to boost photosynthetic capabilities because they optimize the absorption of light, chlorophyll efficiency and electron transport operations. Some nanomaterials help plants optimize water intake along with better stomatal regulation which improves drought survival and boosts water conservation efficiency. The physiological benefits of nanomaterials lead to elevated crop yields no matter the growth environment conditions [34].

5.6.2 Plant Nanobionics

Plant nanobionics implements functional nanomaterials inside plant tissues to strengthen their biological functions naturally. The field of plant nanobionics serves three main functions including photosynthetic efficiency improvement and real-time environmental monitoring and building new communication systems between plant structures [18]. The developing field demonstrates promise to develop future agricultural systems that sense and adapt to their environment through intelligent behavior.

5.7 Sustainability and Post-Harvest Applications

5.7.1 Biogenic and Biodegradable Nanoparticles

Plant-derived, bacterial-derived, fungal-derived, and algal-derived nanoparticles function as environmentally safe nanomaterial replacements for chemical nanomaterials. Nanoparticles synthesized from biological sources feature both natural compatibility and decay properties which resolve environmental safety obstacles of nanoparticles in agricultural ecosystems. Biogenic nanoparticles serve as an agricultural solution that supports sustainable intensification goals at a global level [41].

6 Challenges and Risk Assessment

6.1 Nanotoxicity and Environmental Risks

Plant systems that use nanoparticles create hazardous effects on biological systems. Biological toxicity toward plants along with chemical accumulation in soil organisms and microbial community disruptions represent essential problems [42]. Plant growth together with the entire ecosystem receives negative effects from these occurrences. Our current understanding of nanomaterial persistence and their dual relations with organic and nonorganic environmental factors in the environment remains incomplete. The utilization of nanomaterials in pig farming depends heavily on complete risk assessments and life cycle evaluations for maintaining secure practices [69, 70].

6.2 Economic and Technical Barriers

Production costs and scalability problems act as substantial obstacles to nanotechnology adoption in agricultural applications. The complex manufacturing process combined with large-scale application complexity puts hurdles in the way of commercial nanoparticle adoption. Research results show inconsistencies because there are no established standardized protocols for nanoparticle characterization. The adoption of nanotechnology in agriculture needs improved cost-effective manufacturing practices and standardized testing procedures to succeed [60].

6.3 *Regulatory and Ethical Issues*

Nanomaterials in agriculture face an absent regulatory agreement between nations worldwide. The absence of global regulations creates barriers to the deployment of safe nanotechnology solutions. Public awareness alongside transparency issues along with socioeconomic inequalities must be dealt with as ethical matters [39]. The successful employment of nano-biotechnology requires defined protocols together with enhanced participation from interested parties to maintain responsible utilization practices.

7 Current Research Trends and Innovations

Recent studies in plant nanotechnology focus on several cutting-edge innovations. Nanocarriers with environmental-responsive systems (pH and light-sensitive) improve the delivery management of agrochemicals and nutrients [21]. Studies investigate two types of plant-friendly nanoparticles that break down in nature and react to plant requirements to lower overall environmental harm. Nanocarriers unite with CRISPR-Cas9 methods to provide accurate trait enhancements, specifically drought defense mechanisms and disease immunity. The research of nano-based plant growth regulators provides better precision and efficiency to enhance crop productivity. Modern agricultural sustainability has received new transformative power through recent advances in nanotechnology usage [3].

8 Future Prospects

Plenty of potential exists for plant nanobiotechnology to revolutionize global agriculture because it creates sustainable farming systems that are efficient yet climate-resistant. Research about nanoplant interactions along with improved nanoparticle development creates new chances to implement precise farming practices and enhance real-time diagnosis while providing environment-friendly input delivery methods [72, 73].

Progress in nanomaterials is driving advances in techniques for real-time observation and analysis of plant cell activities. With the use of quantum dots and fluorescent nanoprobes, scientists are beginning to watch and quantify cellular physiological events such as calcium movement, oxidative stress, and hormonal signals. Such approaches provide for anticipatory stress treatment because the plant distress is identified prior to the sign's appearance, which minimizes the chances of losses in crops. Additionally, the addition of particular ligands to biocompatible nanosensors enables the direct observation of hormone levels, such as abscisic acid and auxins, and offers a profound insight into plant growth and stress tolerance [20].

The future development of digital agriculture becomes stronger through nanotechnology implementation. Using nanosensors integrated with IoT platforms along with machine learning algorithms enables real-time crop health monitoring and precise nutrient tracking along with predictive yield optimization [55]. Such advancements will allow worldwide farmers to practice improved and efficient agricultural techniques in their fields.

Furthermore, we are conducting studies on combining AI diagnostics with nanosensor data, so that the nutrient imbalances and the plant diseases can be automatically identified. Through plants' health data, the systems can predict yield loss and make specific agrochemical interventions before misuse and reduce environmental damage. The synergies of nanosensor networks and cloud-based platforms allow farmers to take advantage of decision-making instruments that can vary the tendencies of the farmers according to constant changes in the environment [8].

The advancement of nanocarriers that can transport nutrients together with pesticides and growth regulators at the same time will transform agricultural management practices [6]. Plant-derived and biodegradable nanomaterials currently enter widespread use because they offer improved biocompatibility together with reduced environmental impact thus obtaining broader public acceptance from regulatory bodies [27].

There is also remarkable development in engineering nanocarriers that are trigger-sensitive to many environmental inputs for payload release in a tuned manner as a function of pH, temperature, and enzyme activity. With such an approach, root-targeted delivery is facilitated ensuring fertilizers or pesticides are applied exactly when and where, thus enhancing plant nutrition and reducing runoff pollution. The application of polymer nanocapsules for the encapsulation of the fertilizers is a clear case of precise nutrient delivery that harmonizes with the plant uptake hence maximizing crop yield even in plants such as maize and rice [2].

The adoption of climate-smart agriculture represents a promising path as nanotechnology shows potential to improve plant stress resistance against environmental threats especially drought and salt exposure together with high temperatures. Tailored nanomaterials are essential to help plants adapt to climate change challenges because they guarantee food security during unpredictable environmental circumstances.

Research efforts have focused on materials such as graphene oxide and carbon dots, evidencing their ability to increase plant tolerance to abiotic stress by the regulation of antioxidant enzymes and improved water conservation in tissues. Experiments found that these materials trade stimulating root expansion and chlorophyll preservation in plants in drought and salinity conditions, which make it essential to continue plant health and productivity in deteriorated soil [76].

Nanobiotechnology together with gene editing and epigenetic regulation holds promising potential for non-transgenic modifications of plants [53]. Nano-enabled gene delivery systems establish a safe plant genetic modification approach that multiple public audiences can embrace because they support the improvement of crop properties like disease resistance alongside nutrient content [46].

Researchers are also exploring the possibility of using nanomaterials as epigenetic modulators, which will allow temporary delivery of small RNA remodeling agents, which will not be permanently integrated into the genome. This practice enables temporary improvement of desirable attributes, meaning plants can present advantages such as pest resistance or drought tolerance only within sensitive periods and therefore have reduced associated long-term risks to the environment due to natural disasters which lead to genetic change [64].

Nano-based food packaging technologies demonstrate a transformative impact on the food industry which will completely transform how they operate. Nanotechnology provides solutions for preserving food through extended shelf life together with microbial resistance capabilities while delivering atmosphere control technology which results in sustainable food conservation systems [45].

Other research examines edible nanocoatings in the form of natural polymers and nanoclays that can be used for post-harvest fruits and vegetables to increase their quality and lengthen them. These coatings add value by offering barriers against moisture, oxygen, and pathogens, and the incorporation of natural antimicrobial agents such as essential oils. The application of nanochitosan films has created an almost 50% increase in the shelf lives of perishable crops; a strong commercial viability for incorporating such technology into the fresh food distribution chain [56].

References

1. Abdalla N, Ragab MI, Fári M, El-Ramady H, Alshaal T, Elhawat N et al (2018) Nanobiotechnology for plants: needs and risks. Environ Biodiv Soil Sec 2:155–174
2. Abd-Elsalam KA, Hashim AF, Abdelkhalek TE, Hassan RK (2024) Polysaccharide biopolymer nanoparticles for soil nutrient delivery. In: Biopolymeric nanoparticles for agricultural applications. Springer Nature, Cham, pp 79–105
3. Achari GA, Kowshik M (2018) Recent developments on nanotechnology in agriculture: plant mineral nutrition, health, and interactions with soil microflora. J Agric Food Chem 66(33):8647–8661
4. Ali J, Alam D, Noori R, Faridi S, Sardar M (2024) Metal nanoparticles: management and control of Phytopathogenic fungi. In: Advances in antifungal drug development: natural products with antifungal potential. Springer Nature, Singapore, pp 411–437
5. Al-Khayri JM, Rashmi R, Surya Ulhas R, Sudheer WN, Banadka A, Nagella P et al (2023) The role of nanoparticles in response of plants to abiotic stress at physiological, biochemical, and molecular levels. Plants 12(2):292
6. An C, Sun C, Li N, Huang B, Jiang J, Shen Y et al (2022) Nanomaterials and nanotechnology for the delivery of agrochemicals: strategies towards sustainable agriculture. J Nanobiotechnol 20(1):11
7. Atanda SA, Shaibu RO, Agunbiade FO (2025) Nanoparticles in agriculture: balancing food security and environmental sustainability. Discover Agriculture 3(1):26
8. Bharti A, Jain U, Chauhan N (2024) From lab to field: Nano-biosensors for real-time plant nutrient tracking. Plant Nano Biol 9:100079
9. Chen Z, Zhao J, Cao J, Zhao Y, Huang J, Zheng Z et al (2022) Opportunities for graphene, single-walled and multi-walled carbon nanotube applications in agriculture: A review. Crop Design 1(1):100006

10. Chugh G, Siddique KH, Solaiman ZM (2021) Nanobiotechnology for agriculture: smart technology for combating nutrient deficiencies with nanotoxicity challenges. Sustainability 13(4):1781
11. Das S, Debnath N, Cui Y, Unrine J, Palli SR (2015) Chitosan, carbon quantum dot, and silica nanoparticle mediated dsRNA delivery for gene silencing in Aedes aegypti: a comparative analysis. ACS Appl Mater Interf 7(35):19530–19535
12. Díez-Pascual AM (2021) Carbon-based nanomaterials. Int J Mol Sci 22(14):7726
13. Dorjee L, Gogoi R, Kamil D, Kumar R, Mondal TK, Pattanayak S, Gurung B (2023) Essential oil-grafted copper nanoparticles as a potential next-generation fungicide for holistic disease management in maize. Front Microbiol 14:1204512
14. El-Naggar NEA, Saber WI, Zweil AM, Bashir SI (2022) An innovative green synthesis approach of chitosan nanoparticles and their inhibitory activity against phytopathogenic Botrytis cinerea on strawberry leaves. Sci Rep 12(1):3515
15. Farooq MA, Hannan F, Islam F, Ayyaz A, Zhang N, Chen W et al (2022) The potential of nanomaterials for sustainable modern agriculture: present findings and future perspectives. Environ Sci Nano 9(6):1926–1951
16. Fraceto LF, Grillo R, de Medeiros GA, Scognamiglio V, Rea G, Bartolucci C (2016) Nanotechnology in agriculture: which innovation potential does it have? Front Environ Sci 4:20
17. García-Ovando AE, Pina JER, Naranjo EUE, Chávez JAC, Esquivel K (2022) Biosynthesized nanoparticles and implications by their use in crops: effects over physiology, action mechanisms, plant stress responses and toxicity. Plant Stress 6:100109
18. Gautam K, Singh H, Sinha AK (2025) Nanotechnology in plant Nanobionics: mechanisms, applications, and future perspectives. Adv Biol:2400589
19. Goel RK, Yadav CS, Vishnoi S, Rastogi R (2021) Smart agriculture–urgent need of the day in developing countries. Sustain Comput Inform Syst 30:100512
20. Halicka K, Meloni F, Czok M, Spychalska K, Baluta S, Malecha K et al (2022) New trends in fluorescent nanomaterials-based bio/chemical sensors for neurohormones detection— a review. ACS Omega 7(38):33749–33768
21. Hegde NG (2023) Nanotechnology: cutting-edge tool for increasing agricultural production. Asian J Agric Hortic Res 10(1):20–28
22. Ijaz M, Khan F, Ahmed T, Noman M, Zulfiqar F, Rizwan M et al (2023) Nanobiotechnology to advance stress resilience in plants: current opportunities and challenges. Mater Today Bio 22:100759
23. Johnson MS, Sajeev S, Nair RS (2021, March) Role of nanosensors in agriculture. In: 2021 international conference on computational intelligence and knowledge economy (ICCIKE). IEEE, pp 58–63
24. Joudeh N, Linke D (2022) Nanoparticle classification, physicochemical properties, characterization, and applications: a comprehensive review for biologists. J Nanobiotechnology 20(1):262
25. Kaningini AG, Nelwamondo AM, Azizi S, Maaza M, Mohale KC (2022) Metal nanoparticles in agriculture: a review of possible use. Coatings 12(10):1586
26. Kanwar MK, Sun S, Chu X, Zhou J (2019) Impacts of metal and metal oxide nanoparticles on plant growth and productivity. In: Nanomaterials and plant potential. Springer, pp 379–392
27. Karnwal A, Jassim AY, Mohammed AA, Sharma V, Al-Tawaha ARMS, Sivanesan I (2024) Nanotechnology for healthcare: plant-derived nanoparticles in disease treatment and regenerative medicine. Pharmaceuticals 17(12):1711
28. Karuvelan M, Lakshman S, Boopathy U, Chelliah R, Oh DH (2025) Role of nanosensors for sustainable agricultural practices. In: Agricultural crop improvement: plant and soil relationships, vol 23. Taylor & Francis
29. Kathuria A, Zhang S (2022) Sustainable and repulpable barrier coatings for fiber-based materials for food packaging: A review. Front Mater 9:929501
30. Kaziem AE, Yang L, Lin Y, Kazem AE, Xu H, Zhang ZX (2021) Pathogenic invasion-responsive carrier based on mesoporous silica/β-glucan nanoparticles for smart delivery of fungicides. ACS Sustain Chem Eng 9(27):9126–9138

31. Kopecká R, Kameniarová M, Černý M, Brzobohatý B, Novák J (2023) Abiotic stress in crop production. Int J Mol Sci 24(7):6603
32. Kottegoda N, Nimanka S, Fernando N, de Silva M, Munaweera I (2023) Climate smart agriculture: the role of fertilizer innovations and efficient plant nutrient management. Vidyodaya J Sci 1(s1)
33. Kumar R, Ashfaq M, Verma N (2018) Synthesis of novel PVA–starch formulation-supported cu–Zn nanoparticle carrying carbon nanofibers as a nanofertilizer: controlled release of micronutrients. J Mater Sci 53(10):7150–7164
34. Li P, Xia Y, Song K, Liu D (2024) The impact of nanomaterials on photosynthesis and antioxidant mechanisms in gramineae plants: research progress and future prospects. Plants 13(7):984
35. Lu H, Zhang S, Wang J, Chen Q (2021) A review on polymer and lipid-based nanocarriers and its application to nano-pharmaceutical and food-based systems. Front Nutr 8:783831
36. Ma C, Li G, Xu W, Qu H, Zhang H, Bahojb Noruzi E, Li H (2024) Recent advances in stimulus-responsive nanocarriers for pesticide delivery. J Agric Food Chem 72(16):8906–8927
37. Machado TO, Grabow J, Sayer C, de Araújo PH, Ehrenhard ML, Wurm FR (2022) Biopolymer-based nanocarriers for sustained release of agrochemicals: A review on materials and social science perspectives for a sustainable future of Agri-and horticulture. Adv Colloid Interf Sci 303:102645
38. Mahesha KN, Singh NK, Amarshettiwar SB, Singh G, Gulaiya S, Das H, Kumar J (2023) Entering a new agricultural era through the impact of Nano-fertilizers on crop development: A review. Int J Plant Soil Sci 35(20):94–102
39. Mishra AK, Das R, Sahoo S, Biswal B (2022) Global regulations and legislations on nanoparticles usage and application in diverse horizons. In: Comprehensive analytical chemistry, vol 99. Elsevier, pp 261–290
40. Mmbando GS (2024) The use of nanotechnology in genetic modification: a recent promising technology for enhancing crop productivity. J Crop Improv 38(6):618–638
41. Morgan RN, Aboshanab KM (2024) Green biologically synthesized metal nanoparticles: biological applications, optimizations and future prospects. Future Sci OA 10(1):FSO935
42. Mubeen B, Hasnain A, Wang J, Zheng H, Naqvi SAH, Prasad R et al (2023) Current progress and open challenges for combined toxic effects of manufactured nano-sized objects (MNO'S) on soil biota and microbial community. Coatings 13(1):212
43. Mujtaba M, Wang D, Carvalho LB, Oliveira JL, Espirito Santo Pereira AD, Sharif R et al (2021) Nanocarrier-mediated delivery of miRNA, RNAi, and CRISPR-Cas for plant protection: current trends and future directions. ACS Agric Sci Technol 1(5):417–435
44. Munjal P, Munjal R (2022) Agricultural nanobiotechnology: current possibilities and constraints. In: Nanotechnology in agriculture and environmental science. CRC Press, pp 1–14
45. Mustafa F, Andreescu S (2020) Nanotechnology-based approaches for food sensing and packaging applications. RSC Adv 10(33):19309–19336
46. Niazian M, Molaahmad Nalousi A, Azadi P, Ma'mani L, Chandler SF (2021) Perspectives on new opportunities for nano-enabled strategies for gene delivery to plants using nanoporous materials. Planta 254(4):83
47. Imran, Ortas I (2024) Nano biotechnology: exploring the potential of nano materials in enhancing crop health and combating abiotic stress. Commun Soil Sci Plant Anal 55(16):2509–2528
48. Pisano R, Durlo A (2023) Feynman's frameworks on nanotechnology in historiographical debate. In: Handbook for the historiography of science. Springer, Cham, pp 441–478
49. Prerna DI, Govindaraju K, Tamilselvan S, Kannan M, Vasantharaja R, Chaturvedi S, Shkolnik D (2021) Influence of nanoscale micro-nutrient α-Fe2O3 on seed germination, seedling growth, translocation, physiological effects and yield of rice (*Oryza sativa*) and maize (*Zea mays*). Plant Physiol Biochem 162:564–580
50. Rehman A, Qunyi T, Usman M, Manzoor MF, Zhao L, Feng J, Jafari SM (2022) Nano-enabled agrochemicals for sustainable agriculture. In: Nano-enabled agrochemicals in agriculture. Academic Press, pp 291–306

51. Sabzevari S, Hofman J (2022) A worldwide review of currently used pesticides' monitoring in agricultural soils. Sci Total Environ 812:152344
52. Saravanadevi K, Devi NR, Dorothy R, Joany RM, Rajendran S, Nguyen TA (2022) Nanotechnology for agriculture: an introduction. In: Nanosensors for smart agriculture. Elsevier, pp 3–23
53. Sashidhar P, Arya SS, Das RK, Dubey MK, Lenka SK (2021) Nanobiotechnology for plant genome engineering and crop protection. In: Genetically modified crops in asia pacific, pp 279–310
54. Sekhar M, Rastogi M, Rajesh CM, Saikanth DRK, Rout S, Kumar S, Patel AK (2024) Exploring traditional agricultural techniques integrated with modern farming for a sustainable future: A review. J Sci Res Rep 30(3):185–198
55. Sharma K, Shivandu SK (2024) Integrating artificial intelligence and internet of things (IoT) for enhanced crop monitoring and management in precision agriculture. Sensors Int 5:100292
56. Sharma A, Thakur A, Sharma A, Thakur M, Sharma S, Sharma H et al (2024) Nano-edible coatings for quality enhancement and shelf-life extension of fruits and vegetables. J Food Sci Technol:1–16
57. Shen M, Liu S, Jiang C, Zhang T, Chen W (2023) Recent advances in stimuli-response mechanisms of nano-enabled controlled-release fertilizers and pesticides. Eco-Environ Heal 2(3):161–175
58. Shi LN, Lu LX, Ye JR, Shi HM (2022) The endophytic strain ZS-3 enhances salt tolerance in Arabidopsis thaliana by regulating photosynthesis, osmotic stress, and ion homeostasis and inducing systemic tolerance. Front Plant Sci 13:820837
59. Singh H, Sharma A, Bhardwaj SK, Arya SK, Bhardwaj N, Khatri M (2021) Recent advances in the applications of nano-agrochemicals for sustainable agricultural development. Environ Sci Process Impact 23(2):213–239
60. Singh M, Upadhyay L, Badekhan A, Shil S, Gautam SK, Anand A (2024) Leveraging remote sensing and nanotechnology to overcome barriers to agroforestry adoption by small holder farmers. J Sci Res Rep 30(5):113–140
61. Soltys L, Olkhovyy O, Tatarchuk T, Naushad M (2021) Green synthesis of metal and metal oxide nanoparticles: principles of green chemistry and raw materials. Magnetochemistry 7(11):145
62. Squire HJ, Tomatz S, Voke E, González-Grandío E, Landry M (2023) The emerging role of nanotechnology in plant genetic engineering. Nat Rev Bioeng 1(5):314–328
63. Srivastava S, Bhargava A (2022) Green nanotechnology: an overview. In: Green nanoparticles: the future of nanobiotechnology. Springer, Singapore, pp 1–13
64. Sun C, Ali K, Yan K, Fiaz S, Dormatey R, Bi Z, Bai J (2021) Exploration of epigenetics for improvement of drought and other stress resistance in crops: a review. Plants 10(6):1226
65. Tang Y, Zhao W, Zhu G, Tan Z, Huang L, Zhang P et al (2023) Nano-pesticides and fertilizers: solutions for global food security. Nanomaterials 14(1):90
66. Tong SW, Goh WP, Jiang C (2023) Recent advances in nanosensors for precision agriculture. J Electrochem Soc 170(12):127507
67. Vats S, Kumawat S, Brar J, Kaur S, Yadav K, Magar SG et al (2022) Opportunity and challenges for nanotechnology application for genome editing in plants. Plant Nano Biol 1:100001
68. Wang JW, Grandio EG, Newkirk GM, Demirer GS, Butrus S, Giraldo JP, Landry MP (2019) Nanoparticle-mediated genetic engineering of plants. Mol Plant 12(8):1037–1040
69. Wang Q, Zhang P, Zhao W, Li Y, Jiang Y, Rui Y et al (2023a) Interplay of metal-based nanoparticles with plant rhizosphere microenvironment: implications for nanosafety and nano-enabled sustainable agriculture. Environ Sci Nano 10(2):372–392
70. Wang X, Xie H, Wang P, Yin H (2023b) Nanoparticles in plants: uptake, transport and physiological activity in leaf and root. Materials 16(8):3097
71. Wu H, Nißler R, Morris V, Herrmann N, Hu P, Jeon SJ et al (2020) Monitoring plant health with near-infrared fluorescent H2O2 nanosensors. Nano Lett 20(4):2432–2442

72. Yadav A, Yadav K, Abd-Elsalam KA (2023a) Nanofertilizers: types, delivery and advantages in agricultural sustainability. Agrochemicals 2(2):296–336
73. Yadav N, Garg VK, Chhillar AK, Rana JS (2023b) Recent advances in nanotechnology for the improvement of conventional agricultural systems: a review. Plant Nano Biol 4:100032
74. Yin J, Su X, Yan S, Shen J (2023) Multifunctional nanoparticles and nanopesticides in agricultural application. Nanomaterials 13(7):1255
75. Zhao B, Luo Z, Zhang H, Zhang H (2022) Imaging tools for plant nanobiotechnology. Front Genome Ed 4:1029944
76. Zia MU, Sambasivam PT, Chen D, Bhuiyan SA, Ford R, Li Q (2024) A carbon dot toolbox for managing biotic and abiotic stresses in crop production systems. EcoMat 6(5):e12451

Nanoparticles in Plant Sciences: Types, Synthesis and Applications

Hajra Abid, Khawaja Shafique Ahmed, Ansar Mehmood, Muhammad Faraz Khan, Sajjad Hussain, and Hummera Nawaz

Abstract Nanotechnology has emerged as a recent, efficient and transformative tool in the domain of plant sciences, offering novel solutions to long-lasting challenges in agriculture, biotechnology, and environmental sustainability. Nanoparticles (NPs) are structures with small size between 1 and 100 nm that possess unique physico-chemical properties due to their high surface area to volume ratio. NPs are classified into diverse range of categories including organic, inorganic and carbon-based. The synthesis of NPs ranges from conventional physical and chemical methods to eco-friendly biological (green) synthesis using plant extracts, microbes, algae and fungi emphasizing their comparative advantages and limitations. This chapter further elaborates the multifaceted applications of NPs in plant sciences, such as enhancing nutrient delivery through nanofertilizers, improving crop protection via nanopesticides, facilitating genetic transformation, and strengthening plants' resistance to biotic and abiotic stresses. Overall, this chapter underscores the growing importance of nanotechnology as a sustainable and efficient platform for advancing plant sciences in the era of climate change and food security challenges.

Keywords Nanoparticles · Green synthesis · Stress tolerance · Disease control · Genetic engineering

H. Abid · K. S. Ahmed (✉) · A. Mehmood · M. F. Khan · S. Hussain
Department of Botany, University of Poonch Rawalakot, Rawalakot, Pakistan
e-mail: ahmadks@upr.edu.pk

H. Nawaz
Department of Botany, Division of Science and Technology, University of Education, Lahore, Pakistan

J.-T. Chen (ed.), *Nano-Delivering for Plant Genetic Engineering*, Smart Nanomaterials Technology, https://doi.org/10.1007/978-981-95-4165-2_8

1 Introduction

Nanotechnology has emerged as a transformative field with wide-ranging applications in plant sciences due to their unique physicochemical characteristics [90]. NPs are defined as structures with dimensions typically less than 100 nanometers in all spatial directions (x, y, and z), exhibit unique functionalities owing to their nanoscale size and modifiable surface chemistry. Their high surface area to volume ratio and nanoscale dimensions significantly influence molecular interactions, enabling novel functional applications across various biological systems [100]. Structurally, NPs can be categorized as zero-dimensional (0D), one-dimensional (1D), two-dimensional (2D), or three-dimensional (3D) based on their geometric configuration [91]. On the basis of chemical composition, they are broadly classified into organic, inorganic, and carbon-based NPs which encompass several subtypes, including metallic, ceramic, polymeric, and lipid-based etc., each exhibiting unique physicochemical attributes related to size, morphology, and functionality [62]. Their exceptional mechanical strength, thermal stability, and optical responsiveness make them particularly valuable in agricultural and food-related technologies, especially in response to growing global demands and environmental challenges [69, 107].

Conventional synthesis methods of NPs such as physical and chemical often enable precise control over particle size and morphology and are capable of producing large quantities but these approaches can be expensive and potentially hazardous. Commonly employed techniques include sol-gel processing [84, 118], chemical vapor deposition [16], electrospinning [95], laser ablation [149], thermal decomposition [39], sputtering [133], nanolithography [60], pyrolysis [23] and mechanical milling [3]. Conversely, green synthesis a biologically inspired approach is gaining widespread attention due to its simplicity, cost-effectiveness, lower toxicity, and minimal environmental impact [29, 55]. This method typically follows a bottom-up strategy, where individual metal ions are reduced and assembled into NPs through the action of natural reducing, stabilizing, and capping agents derived from plant extracts, microorganisms, or other biological systems. The efficiency and characteristics of green synthesized NPs are influenced by several parameters, including pH, temperature, reaction time, and precursor concentration [93]. Given the increasing concerns over the toxicity and cost of chemical reagents, researchers now favor biosynthetic routes, particularly those employing phytochemicals, for their dual role as reducing and stabilizing agents [12].

NPs exhibit broad applicability across various sectors such as agriculture, food technology, environmental remediation, and biotechnology. They are employed in wastewater treatment [146], environmental sensing [112], as antimicrobial agents [63] and as functional food ingredients [27]. The integration of NPs into agriculture and the food industry offers promising solutions to address emerging challenges related to sustainability and productivity [94]. They offer distinct advantages over traditional inputs, including a high surface area/volume ratio, improved mass transfer, and the potential for precise, continuous delivery of nutrients and agrochemicals

[61]. Their unique physicochemical properties influence plant growth by regulating physiological and biochemical pathways, facilitating controlled fertilizer release [88], targeted nutrient transport [72, 73], enhanced water absorption under saline conditions [14], and stimulation of plant defense mechanisms against biotic stressors [137]. Furthermore, they improve phytoremediation efficiency by accelerating the removal of contaminants from soil and water [44]. This chapter aims to provide a comprehensive overview of the various types of NPs, their synthesis approaches, and their multifaceted applications in plant sciences thereby offering insights into their future prospects. It also highlights their potential role in promoting sustainable and resilient agricultural systems in the face of global environmental and food security challenges.

2 Types of Nanoparticles

2.1 Organic NPs

Organic NPs are defined as nanoscale solid structures composed of either naturally derived or synthetically produced organic compounds, typically within the size range of 10–100 nm, though they may extend up to 1000 nm under broader classifications [71]. These nanostructures are primarily made from biomolecules such as proteins, lipids, polymers, and polysaccharides, which confer essential features like biocompatibility, non-toxicity, and biodegradability. These attributes make them suitable for a wide array of biomedical applications, including targeted drug delivery, molecular imaging, oncological treatments, and biosensing platforms [92]. Furthermore, physicochemical parameters such as particle dimension, geometry, and surface architecture play a pivotal role in determining their therapeutic potential and functional performance [39].

2.1.1 Polymeric NPs

Polymers composed of repeating monomeric units, exhibit substantial structural and functional diversity, with variations in size, structure, and properties [36]. This compositional versatility, along with scalable synthesis routes, has made polymeric NPs (PNPs) increasingly attractive for a wide range of applications. Their production from natural polymers enhances their biocompatibility, biodegradability, environmental safety, renewability, and cost-effectiveness particularly advantageous for the encapsulation of agrochemicals such as insecticides [43, 147]. PNPs can be fabricated using either natural or synthetic polymers. Chitosan and alginate are most commonly used natural polymers. Among synthetic polymers, polylactic acid (PLA), polyglycolic acid (PGA), and their copolymer poly lactic-co-glycolic acid (PLGA) are widely employed due to their tunable degradation profiles and established safety in biomedical and agricultural contexts.

2.1.2 Lipid Based NPs

Lipid-based NPs (LBNPs) represent a flexible and biocompatible class of nanocarriers extensively explored in biomedical research, particularly for drug delivery applications. Their inherent non-toxicity, biodegradability, and structural adaptability make them superior to many other NP types [53]. LBNPs can be engineered to incorporate antimicrobial agents, enhancing their utility in wound healing, skin tissue regeneration, and inflammation control [97]. LBNPs are broadly classified into three main categories: liposomes, solid lipid (SLNPs), and nanostructured lipid carriers (NLCs). Liposomes are synthetic, globular vesicles composed of one or more phospholipid bilayers typically containing phosphatidylserine, phosphatidylcholine, phosphatidylethanolamine, phosphatidic acid and related lipids [65]. Their composition directly influences key parameters such as size, membrane fluidity, stability, and surface charge [82]. Depending on lamellarity and size, liposomes are subclassified into small unilamellar vesicles (SUVs), large unilamellar vesicles (LUVs), multilamellar vesicles (MLVs), and multivesicular vesicles (MVs) [102]. They are widely utilized across medical and industrial domains due to their high biocompatibility and drug-loading efficiency [30, 123]. SLNs, the first generation of LBNPs, consist of a solid lipid matrix stabilized by surfactants or polymers [78]. These spherical NPs, typically 50–1000 nm in size, serve as effective colloidal carriers for controlled drug release [98]. NLCs, considered the second generation LNPs, are formulated by combining solid and liquid lipids; this hybrid lipid matrix enhances their efficiency and overcomes the limitations associated with SLNs [134].

2.1.3 Protein Based NPs

Protein based NPs (PNPs) are typically synthesized using proteins of natural origin or through recombinant production. Proteins, a diverse class of biomacromolecules, offer distinctive physicochemical properties and functional roles across material and life sciences. Their amphiphilic nature allows favorable interactions with both hydrophilic and hydrophobic substances, making them ideal candidates for NP formulation. Natural protein-derived NPs are biodegradable and metabolically compatible, and their surfaces can be chemically engineered to enhance drug loading and targeted ligand attachment [138]. Due to the abundance of reactive functional groups such as hydroxyl, amine, and carboxyl moieties, these NPs exhibit considerable potential for surface functionalization. These functional groups enable both covalent and non-covalent conjugation of therapeutic agents and targeting ligands, thereby expanding their application in targeted drug delivery systems [67, 80]. Protein based NPs also include number of NPs such as collagen, casein, albumin, milk protein and plant protein based NPs [144]. Plant-based proteins, such as zein and gliadin, possess inherent hydrophobicity, contributing to a more sustained release profile for encapsulated drugs compared to animal-derived protein carriers

[38]. Commonly utilized plant proteins for NP synthesis include zein (from maize), gliadin and glutenin (from wheat), as well as proteins extracted from legumes, peanuts, rapeseed, and soy, all of which exhibit broad applicability across pharmaceutical, agricultural, and food-related sectors.

2.1.4 Carbohydrate Based NPs

Carbohydrate-based NPs, also known as glyco NPs, are nanoscale systems primarily composed of carbohydrate molecules or their derivatives. These NPs have garnered significant interest due to their biocompatibility, biodegradability, and versatile functional properties [75]. Common carbohydrate-derived NPs include those made from starch, dextran, and cyclodextrins. Starch NPs (SNPs), often termed bio-NPs, are distinguished from natural starch by their significantly smaller size typically ranging from tens to a few hundred nanometers and their globular morphology [25]. SNPs exhibit a high surface area to mass ratio, enhanced solubility, improved bioavailability, superior adsorption properties, and efficient cellular uptake, making them suitable and less toxic for environment [21, 110]. Dextran, a neutral exopolysaccharide produced by bacteria, offers excellent biocompatibility and biodegradability. It's simple yet functional polymeric nature supports applications in drug delivery, nanomedicine, biosensing, and cell imaging. Dextran-based NPs demonstrate improved aqueous solubility, high payload capacity, low cytotoxicity, and favorable rheological properties, although they typically have limited shelf life [13]. Cyclodextrins (CDs), cyclic oligosaccharides derived from starch through enzymatic degradation. They exhibit enhanced stability, greater surface area, and superior encapsulation efficiency when compared to liposomes, making them promising carriers for pharmaceuticals and bioactives.

2.2 Inorganic NPs

Inorganic NPs consist of inorganic elements bonded through covalent or metallic interactions [70]. They are typically derived from semiconducting materials, ceramics, or magnetic metals, and their core structure is formed through the crystallization of inorganic salts into well-organized, three-dimensional lattices. This high degree of structural order enhances their stability and resistance to physical or chemical degradation. The surface area/volume ratio of these NPs plays a vital role in determining their functional reactivity, which also influences their potential cytotoxicity [41]. These properties contribute to their broad applicability across multiple fields. Owing to their efficacy in biomedical applications, particularly as therapeutic agents, inorganic NPs are among the most commercially exploited nanomaterials [15]. Inorganic NPs can be further categorized into distinct subtypes based on their structural composition and functional characteristics.

2.2.1 Magnetic NPs

Magnetic NPs (MNPs) are a category of inorganic nanomaterials that exhibit sensitivity to external magnetic fields [64]. They are primarily composed of magnetic elements such as iron, cobalt, nickel, or their respective oxides [122]. Based on their magnetic domain structure, they are classified into single-domain and multi-domain particles, with that typically measuring 10–20 nm falling under the single-domain category. MNPs demonstrate distinct physicochemical properties, including magnetic anisotropy, low Curie temperature, coercivity, and efficient heating capabilities [64]. Iron oxide NPs (IONPs) notably magnetite (Fe_3O_4) and its oxidized form maghemite (γ-Fe_2O_3) have garnered extensive attention for biomedical applications because of their biodegradable, biocompatible nature, surface functionalization potential, and cost-effective synthesis [132]. Additionally, they exhibit noteworthy antimicrobial activity and are widely utilized in targeted drug delivery.

2.2.2 Semiconductor NPs

Semiconductors are crystalline solid materials characterized by electrical conductivity and band gaps that lie between those of conductors and insulators [129]. Owing to their intermediate electronic properties, semiconductors NPs (sNPs) exhibit a combination of metallic and non-metallic behaviors. These nanoscale materials are capable of absorbing and emitting light, making them highly suitable for optoelectronic applications such as high-efficiency solar cells, light-emitting diodes (LEDs), and miniaturized transistors. Additionally, their photoactive nature and quantum confinement effects have led to their utilization in bioimaging, targeted cancer therapies, and biosensing platforms [17]. Commonly used semiconducting materials for NP synthesis include zinc oxide (ZnO) and zirconium dioxide (ZrO_2) [59]. The application of sNPs spans diverse fields such as catalysis, chemical sensing, optical coatings, ceramics, and dental materials [58, 129].

2.2.3 Ceramic NPs

Ceramic NPs are nanoscale particles composed of inorganic, non-metallic materials that are typically processed through controlled heating and cooling techniques to achieve specific physicochemical properties. These particles can exhibit various morphologies, including amorphous, polycrystalline, dense, porous, or hollow structures, and are renowned for their thermal stability, mechanical strength, and chemical durability. Ceramics are synthesized by consolidating materials often a combination of metals and non-metals through the application of heat, pressure, or both, leading to solid matrices with minimal deformation under harsh conditions [140]. Although alumina (Al_2O_3) and silica (SiO_2) are among the most widely used materials in ceramic NP fabrication, a broad range of metal oxides, sulfides, and other inorganic compounds can be engineered into nanoscale architectures with diverse shapes, sizes, and porosities [11]. A key advantage of ceramic NPs lie in their chemical inertness and structural robustness, which allows them to protect sensitive biological molecules such as enzymes, proteins, or therapeutic agents from degradation due to environmental stressors like pH or temperature fluctuations [125].

2.3 Carbon Based NPs

Fullerenes and carbon nanotubes (CNTs) represent two major classes of carbon-based NPs. Fullerenes are spherical, hollow nanostructures composed of carbon atoms arranged in a closed cage-like configuration. These structures, which are allotropes of carbon, exhibit unique properties such as high electrical conductivity, exceptional mechanical strength, electron accepting capacity, and structural versatility, making them highly attractive for diverse technological applications. The carbon atoms are sp^2 hybridized and organized into a combination of pentagonal and hexagonal rings, resulting in a highly symmetrical molecular structure. On the other hand, CNTs are cylindrical nanostructures with diameters typically ranging from 1 nm to 2 nm, formed by rolling graphene sheets into seamless tubes. Depending on the number of concentric layers, CNTs are classified as single-walled (SWCNTs), double-walled (DWCNTs), or multi-walled carbon nanotubes (MWCNTs) [152].

2.3.1 Graphene

Graphene is a carbon-based nanomaterial composed of a single layer of carbon atoms arranged in a two-dimensional hexagonal lattice. It exhibits exceptional electrochemical characteristics due to its high thermal conductivity, hydrophobic surface, and high density [108]. Among its derivatives, graphene oxide (GO) and various chemically modified forms have attracted significant research interest owing to their tunable properties [105]. GO, in particular, contains a substantial number of oxygen-rich functional groups, which facilitate surface modification and enable targeted applications across multiple scientific domains [145].

2.3.2 Fullerenes NPs

Fullerenes represent a unique allotrope of carbon composed of hollow cages formed by sp^2 hybridized carbon atoms, making them prominent among nanomaterials for diverse applications [141]. These structures typically exhibit a spherical geometry, where carbon atoms occupy the vertices of polyhedral frameworks [35]. Among the various fullerene derivatives, C_{60} is the most extensively studied and widely utilized in NP research. Structurally, C_{60} consists of 60 carbon atoms arranged into a sphere comprising 20 hexagonal and 12 pentagonal rings [104]. A defining property of C_{60}

is its remarkable radical scavenging capability, so called "radical sponge", attributed to its delocalized π-electron system and cage-like molecular structure [150]. Upon exposure to visible or ultraviolet light, it undergoes photoexcitation, leading to the generation of reactive oxygen species (ROS) [86].

2.3.3 Carbon Black NPs

Carbon black NPs (CBNPs) are nanoscale carbon-based materials composed primarily of elemental carbon, with minor quantities of hydrogen and oxygen elements [77]. They are typically synthesized through the thermal decomposition of hydrocarbon compounds either gaseous or liquid under oxygen-limited conditions, which leads to incomplete combustion or pyrolysis [151]. CBNPs are characterized by their high electrical conductivity, chemical versatility, and ease of surface functionalization, as well as their cost-effectiveness and widespread availability [10]. So they are particularly employed as reinforcing fillers in rubber manufacturing, where they significantly improve the mechanical strength and durability of rubber products [51].

2.3.4 Carbon Quantum Dots

Carbon quantum dots (CQDs) are a class of eco-friendly, low-cost, and low-toxicity carbon-based NPs typically less than 10 nm in diameter [31]. These nanoscale materials exhibit unique electronic and photoluminescent properties, which are fundamental to their fluorescence behavior and underpin their application in various technological and biomedical fields [135]. Due to their distinctive structure and surface characteristics, CQDs show promise in optoelectronics, drug delivery, biosensing, catalysis, and biomedical imaging [115]. Furthermore, CQDs are rich in surface functional groups such as amino, carboxyl, hydroxyl, and carbonyl moieties, which enhance their reactivity and enable tailored functionalization for integration into biological, organic, and polymeric systems [68].

3 Synthesis of NPs

NP synthesis can generally be categorized into two main approaches: the top-down and bottom-up methods [40] which are further carried out by physical, chemical and biological methods as shown in Fig. 1.

Top-Down Method

The top-down approach is a reductive technique that involves breaking down bulk materials into nanoscale structures. Common methods under this category include

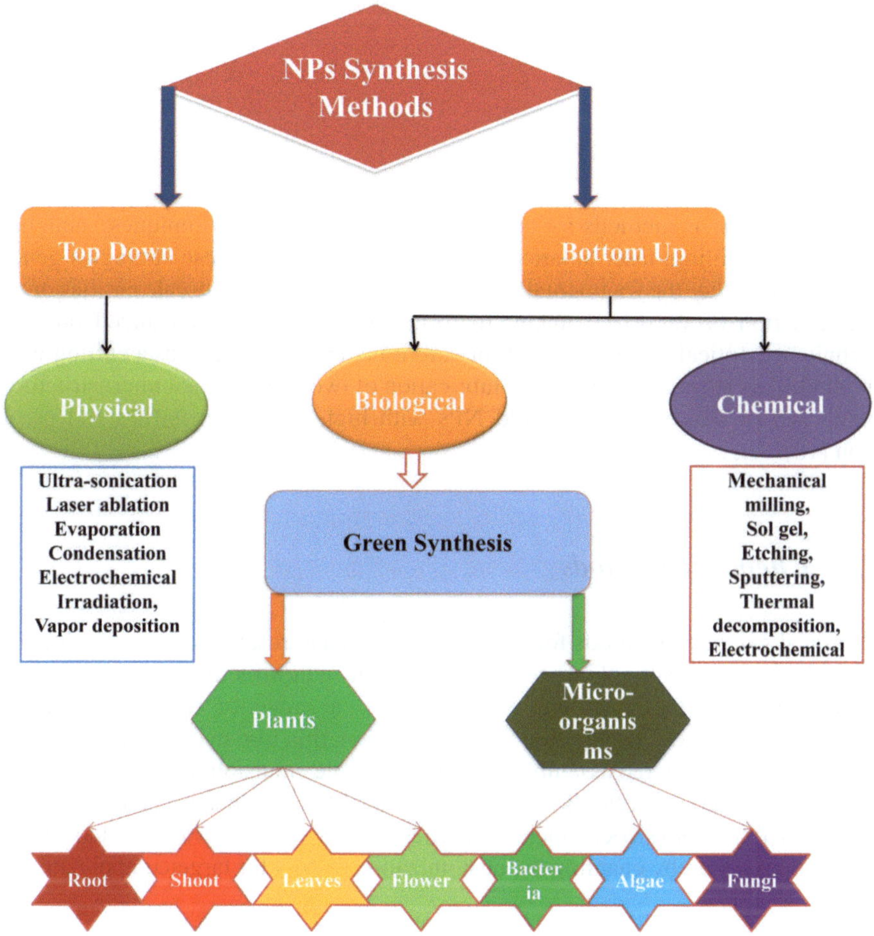

Fig. 1 Methods of NP Synthesis

lithography, laser ablation, thermal decomposition, sputtering, arc discharge, electron beam evaporation, and mechanical processes such as ball milling [89].

Bottom-Up Method

The bottom-up approach, also known as the constructive method, involves assembling materials at the atomic or molecular level to form clusters and eventually NPs. This technique enables the formation of nanostructures through the progressive organization of atoms or particles. Common bottom-up methods for NP synthesis include sol-gel processing, electrospinning, chemical vapor deposition (CVD), pyrolysis, and biosynthesis [40]. Typically, the process begins with a state of high supersaturation, leading to nucleation and subsequent particle growth [101]. Numerous researchers have explored diverse chemical and physical techniques within both top-down and bottom-up frameworks to synthesize NPs with tailored properties [22].

3.1 Physical Methods

Physical synthesis methods for NPs encompass a variety of techniques, including mechanical milling, laser ablation, and vapor deposition. These strategies generally involve either the fragmentation of bulk materials into nanoscale entities or the condensation of vaporized matter to form NPs. Among the advanced physical techniques, optical vortex-induced forward transfer has emerged as a cutting-edge method. This approach enables the fabrication of two-dimensional microstructures composed of densely packed gold NPs with high spatial resolution and precision [18].

3.2 Chemical Methods

Chemical synthesis methods for NPs rely on chemical reactions such as reduction, precipitation, and sol-gel processes to generate nanostructures. In the case of polymer-based NPs, including polymersomes, a range of techniques can be employed such as electroporation, direct hydration, thin film hydration, solvent displacement, pH-triggered assembly, single and double emulsion methods, flash nano precipitation, and microfluidic-assisted synthesis [47]. These approaches offer precise control over physicochemical properties such as size, morphology, and surface functionality, thereby enabling the customization of NPs for targeted applications.

3.3 Biological Methods/Green Synthesis

Biological synthesis, also referred to as biosynthesis or green synthesis, is a process for producing NPs utilizing living organisms or biologically derived materials. Green synthesis, the process which employs biological entities such as bacteria, fungi, algae, and plants for NPs production [9, 106], has gained significant attention due to its ecofriendly, non-toxic nature, and adaptability to biological systems [96]. This method leverages biological templates such as proteins, enzymes, and other biomolecules to facilitate the formation of nanostructures with tailored physicochemical characteristics. Biogenic synthesis of NPs has emerged as a rapidly advancing research domain due to its potential to yield nanomaterials with distinct functionalities optimized for targeted applications. Furthermore, this approach is typically more cost-effective, sustainable, and environmentally benign compared to conventional physicochemical methods [19, 24]. A wide range of natural biological entities including fungi (e.g., mushrooms), yeasts, bacteria, and plants serve as key biological agents in NP biosynthesis [49].

Fungi have been utilized to synthesize various NPs, including those of gold, silver, zinc oxide, and titanium dioxide, due to their high capacity for bioaccumulation and ease of cultivation. Fungi contribute to NP synthesis via two principal mechanisms: extracellular and intracellular pathways. In the extracellular approach, metal precursors are introduced into fungal culture filtrates, where bioactive compounds and enzymes secreted by fungal cells mediate the reduction process [56]. These secreted metabolites such as lovastatin, griseofulvin, cyclosporine, and mevastatin play significant roles in metal ion reduction and stabilization [57]. Key fungal redox enzymes involved in this reduction include acetyl-CoA carboxylase (ACCase), nicotinamide adenine dinucleotide (NADH/NADPH), and peroxidases [20], which convert metal ions into their elemental nanoparticulate form. Conversely, the intracellular method entails exposing fungal mycelial cultures to metal precursors, where metal uptake and bioreduction occur within the cellular matrix. In both methods, post-synthesis purification steps such as centrifugation, chemical washing, or membrane filtration are essential to eliminate residual fungal components and isolate the synthesized NPs [143].

Algae also play a significant role in nanomaterial synthesis. For instance, certain species like *Sargassum tenerrimum* contain a diverse range of bioactive compounds, including phenolic compounds, alkaloids, flavonoids, carbohydrates, amino acids, sterols, tannins, and saponins. After purification, these compounds are critical in modulating the size, shape, and functionality of nanomaterials. Characterizing algae is a fundamental step toward advancing future research and exploring their potential applications in NP synthesis [114].

Bacteria are widely utilized in the synthesis of NPs due to their inherent ability to reduce metal ions [5]. Proteins present in the bacterial cytoplasm and cell wall, containing amino acids like tryptophan and tyrosine, contribute to the stabilization and reduction of bacterial NPs. Moreover, certain sugars, such as aldose and ketose, exhibit stabilizing and reducing properties. These bioactive compounds, commonly found in different bacterial species, interact with metal ions to reduce their size and promote the formation of complex structures, such as spherical NPs. The ease of handling and culturing microorganisms has led to the growing popularity of bacterial NP synthesis [4]. The final size of the synthesized NPs depends on their interactions with biologically active substances and their localization, whether extracellular or intracellular [130]. Research indicates that NP biosynthesis in bacteria generally occurs through a two-step mechanism: initially, metal ions are adsorbed either onto the bacterial cell surface or internalized within the cell. Subsequently, these ions are reduced to their elemental form via enzymatic activity, primarily facilitated by bacterial reductases [142]. Among the two biosynthetic routes i.e., extracellular and intracellular; the extracellular pathway is often favored due to its relatively simpler downstream processing and ease of NP isolation. During this process, bacterial metabolites and secreted biomolecules function as natural reducing agents, reacting with metal ion solutions (e.g., silver or gold) to produce NPs [2, 7]. However, a notable limitation of microbial synthesis is its relatively slow production rate compared to phyto-mediated methods, which can pose challenges for large-scale or time-sensitive applications [116].

Plants offer a more efficient, straightforward, cost-effective, and sustainable alternative to traditional NP synthesis methods [6]. They are rich in bioactive compounds such as sugars, vitamins, and polyphenols, which facilitate the reduction of metal ions to atoms and promote the synthesis of metal and metal oxide NPs [109]. Key factors influencing the green synthesis of NPs include the concentration of the plant extract, the metal salt, the pH and the temperature [93]. The biosynthesis of NPs using plant extracts generally proceeds through three distinct phases. The initial activation phase involves the reduction of metal ions by phytochemicals present in the extract, leading to the nucleation of NPs. This is followed by the growth phase, wherein nearest NPs undergo coalescence, forming larger, more stable structures as thermodynamic equilibrium is approached. The final termination phase is crucial for defining the morphology and dimensional characteristics of the NPs, as the system stabilizes and particle shape is determined [45].

The process of green NPs synthesis can be summarized in following steps [126]:

a) Preparation of microbe or plant-derived extracts and mixing them with a metal salt solution.
b) Biochemical reduction and fabrication of the metal salt solution, where monovalent or divalent metal ions are reduced to their zero-valent state, followed by their nucleation and growth into NPs.
c) Analysis of the synthesized NPs using UV-Vis spectroscopy to confirm their formation.
d) Characterization of NPs using advanced techniques such as transmission electron microscopy (TEM), scanning electron microscopy (SEM), and X-ray diffraction (XRD), followed by their application in various fields. Synthesis of NPs by plant parts are shown in Fig. 2.

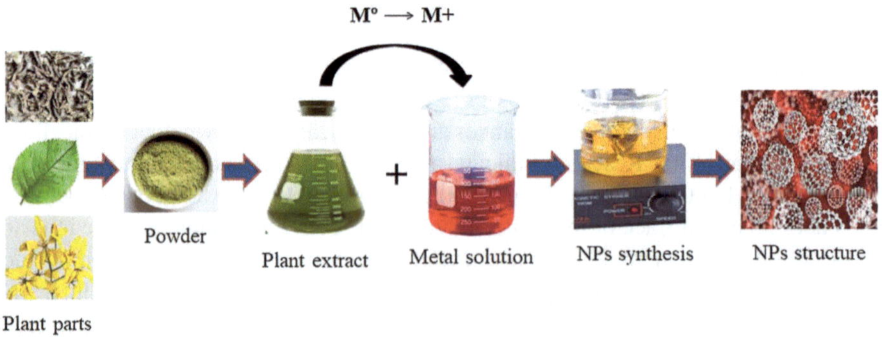

Fig. 2 Synthesis of NPs from plant parts

4 Applications of NPs in Plants

4.1 Growth and Yield Enhancement

Nanotechnology offers innovative solutions to agricultural challenges through the development of nanopesticides, nanofertilizers, and nanomaterials aimed at enhancing plant growth, genetic transformation, and transgene expression [34, 72, 73]. NPs, particularly bio-based variants like nano-biopesticides and nano-fertilizers, serve as efficient nutrient sources, outperforming conventional fertilizers in promoting seed germination, leaf morphology, root and shoot development, and biomass accumulation [8, 52]. Due to their high surface area and water retention capacity, NPs can improve soil moisture availability and storage, benefiting plant hydration [127]. Furthermore, they enhance chloroplast light absorption and utilization, leading to improved photosynthesis, biomass, and yield [119]. Their antimicrobial properties help suppress plant pathogens, supporting crop health and minimize crop losses [79]. Chitosan-based nanofertilizers have shown efficacy in increasing seedling vigor and enzymatic activity related to nutrient mobilization. Foliar applications of these nanofertilizers also elevated antioxidant enzyme levels, reduced lipid peroxidation, and boosted chlorophyll content, collectively enhancing maize productivity via sustained release of copper and salicylic acid [120]. Additionally, a green-synthesized novel phosphorus nanofertilizer, incorporating hydroxyapatite NPs with pomegranate peel and coffee ground extracts, was shown to enhance photosynthesis, metabolic activity, carbohydrates, and biocompatibility while minimizing cytotoxic and genotoxic effects in *Punica granatum* L. [1].

4.2 Nano-pesticides and Disease Control

Nanotechnology, particularly through the application of nanomaterials, offers innovative strategies for plant disease management via three primary mechanisms: [1] direct antimicrobial action, [2] stimulation of plant innate immunity, and [3] targeted delivery of active agents such as pesticides and micronutrients [48]. Among these materials, metallic NPs such as Ag, zinc oxide (ZnO), titanium dioxide (TiO_2), magnesium oxide (MgO), and copper oxide (CuO) have demonstrated potent antimicrobial effects against phytopathogenic bacteria and fungi, making them highly valuable in agricultural disease control [26]. Bio-based NPs also exhibit strong antimicrobial properties, enabling plants to resist a broad range of pathogens, including fungi, viruses and bacteria [37]. These NPs inhibit microbial growth and proliferation by disrupting pathogen membranes preventing their entry into plant tisues [99]. They also generate ROS and other cytotoxic compounds that kill or stop the growth of microbes, ultimately reducing disease incidence and improving plant health [28]. For example, tomato bacterial wilt, soil borne disease, caused by *Ralstonia solanacearum*, was effectively managed using synthesized metal oxide NPs such as CuO, ZnO, and FeO. Notably, CuO NPs significantly reduced pathogen load and improved physiological performance in infected tomato plants, indicating their effectiveness against soil-borne bacterial infections [66, 76]. Likewise, MgO NPs have demonstrated dual functionality acting as both a nematicide and a plant growth enhancer. When applied via root dip, foliar spray, or soil drenching at 100 ppm, MgO NPs reduced nematode reproduction, minimized gall formation, enhanced pigment synthesis (chlorophyll and carotenoids), and increased protein and nitrogen content in roots and shoots and enhancing growth response [128].

4.3 Abiotic Stress Tolerance

NPs play a crucial role in enhancing plant tolerance to abiotic stresses by modulating key physiological and biochemical processes. They influence gene expression related to antioxidant systems and stress responses, thereby improving plant resilience under adverse conditions [148]. By activating specific signaling pathways, NPs enhance stress insight and adaptive responses, regulating numerous genes involved in oxidative stress defense and boosting physiological and biochemical adaptation. Salinity is a major abiotic stressor that hampers plant growth, development and productivity. In rapeseed plants, selenium and ZnO NPs synthesized using phytohormones improved germination under salt stress by downregulating abscisic acid and upregulating gibberellic acid levels, promoting early seedling development and germination [42]. Similarly, foliar application of chitosan–selenium NPs on bitter gourd (*Momordica charantia*) under saline conditions increased antioxidant enzyme activity, proline accumulation, water and potassium content, and reduced malondialdehyde (MDA), hydrogen peroxide, and sodium levels, enhancing salinity tolerance [121]. Yttrium-doped and stabilized γ-Fe_2O_3 (maghemite) NPs applied via nutrient solution improved drought tolerance in *Brassica napus* by decreasing lipid peroxidation and H_2O_2 levels while boosting chlorophyll content and growth, outperforming traditional chelated iron fertilizers as compared to chelated iron [103]. In banana, the foliar application of chitosan NPs improved growth parameters, nutrient retention, photosynthetic pigment levels and increased accumulation of osmoprotectants, while lowering MDA and ROS accumulation, enhancing cold stress tolerance [136]. FeO NPs have been shown to alleviate heavy metal toxicity (e.g., Cd, Pb) by improving biomass production, chlorophyll synthesis, and triggering antioxidant defense system [85]. Silicon NPs mitigate arsenic stress in maize by maintaining photosynthetic pigments, protein content, and photosystem II efficiency, while also reducing toxic ion accumulation in rice, wheat, and pea [50, 131]. The application of multiwalled CNTs has found to promote the biosynthesis of heat shock proteins and gene expression, further increasing the capability of plants to tolerate elevated temperature stress [117].

4.4 Biotic Stress Management

Biotic stress is a major factor contributing to crop losses, thereby threatening global food security and economic stability through the detrimental effects of pathogens, pests, and weeds on plants [83]. NPs offer a precise, efficient, and eco-friendly strategy for managing biotic stress, holding significant promise for advancing sustainable agriculture, environmental protection, and public health [113]. The interaction of NPs with plants occurs at physiological, biochemical, and molecular levels, collectively enhancing plant resilience against biotic challenges [46]. AgNPs, renowned for their potent antibacterial and antifungal properties, effectively suppress various fungal pathogens such as *Botrytis cinerea* and *Alternaria alternata* and inhibit bacterial species including *Fusarium graminearum* and *Erwinia* spp., highlighting their broad-spectrum antibacterial and antifungal activities [54]. Their efficacy extends to both Gram-positive and Gram-negative bacteria, encompassing several foodborne microbes [81]. CuNPs also exhibit strong antimicrobial effects, comparable to AgNPs and silver-chitosan composites, particularly against *Pseudomonas syringae* pv. *syringae*. Moreover, magnesium oxide (MgO) and silicon dioxide (SiO_2) NPs demonstrate antimicrobial properties against diverse bacterial and fungal pathogens, further illustrating the multifaceted role of NPs in mitigating biotic stresses [124].

4.5 Genetic Engineering and Plant Transformation

Nanomaterials have been widely employed for the targeted delivery and controlled release of genes and proteins into plant cells, facilitating advanced genetic engineering techniques. Mesoporous silica NPs serve as carriers for Cre recombinase protein delivery into maize cells via biolistic methods. Upon intracellular release, Cre induces recombination at loxP sites, leading to targeted gene excision. Successful recombination events were visually selected to regenerate fertile plants, demonstrating that mesoporous silica NPs can effectively deliver functional enzymes to specific cells, offering a DNA-free alternative for plant genome editing [87]. In another approach, pristine and chemically modified nanomaterials enabled efficient biomolecule diffusion and delivery in intact plants across multiple species. These nanomaterials facilitated robust protein expression and DNA delivery without transgene integration, while also protecting nucleic acids from nuclease degradation in leaves of *Nicotiana benthamiana*, *Eruca sativa*, *Gossypium hirsutum*, and *Triticum aestivum* [32]. Chitosan-complexed SWCNTs have proven effective for plasmid DNA delivery into chloroplasts of various plant species, including *Nicotiana tabacum*, *Spinacia oleracea*, *Nasturtium officinale*, and *Eruca sativa*, as well as isolated *Arabidopsis thaliana* mesophyll protoplasts, without requiring additional biological or chemical facilitators [74, 139]. Similarly, carbon nanotube-based NPs have been successfully employed to deliver DNA into *Eruca sativa*, cotton and wheat, resulting in elevated protein expression [33]. Moreover, AgNPs (approximately 100 nm) have shown superior gene carrier efficiency compared to larger gold micro-carriers (0.6 μm) in *Nicotiana tabacum* via biolistic transformation, under varied helium pressures (450, 650, 900, 1100 psi) and target distances (6–9 cm), achieving the highest transient GUS reporter expression [111].

5 Conclusion

The integration of nanotechnology into plant sciences marks a significant advancement in our ability to address the complex challenges of modern agriculture and environmental management. This chapter has explored the diverse array of NPs highlighting their various types, synthesis methods, and wide-ranging applications across plant growth promotion, crop protection, genetic engineering, and stress management. NPs, owing to their small size, high surface area, and scalable properties, offer unique advantages over conventional agrochemicals and genetic tools. The development of eco-friendly or green synthesis methods using biological agents offers a sustainable approach to NPs production and provides promising solutions in the field of plant sciences to some of the most pressing challenges ranging from improving nutrient use efficiency and stress tolerance to enhancing disease resistance and promoting sustainable crop production. The application of NPs in plants improve crop growth and development by enhancing the uptake of nutrients and water absorption, photosynthetic efficiency, disease controlling by acting as antimicrobials, increase stress tolerance by inducing physiological and biochemical mechanisms and targeted delivery of genes and proteins for genetic engineering. However, the potential risks associated with nanoparticle accumulation in plants, soil ecosystems, and the food chain cannot be overlooked. The chapter underscores the need for continued research into NP-plant interactions, toxicity thresholds, and long-term environmental effects. Conclusively, the future of NPs in plant sciences is undeniably promising while ensuring sustainable agriculture, food security, and climate resilience.

References

1. Abdelmigid HM, Morsi MM, Hussien NA, Alyamani AA, Alhuthal NA, Albukhaty S (2022) Green synthesis of phosphorous-containing hydroxyapatite NPs (nHAP) as a novel Nano-fertilizer: preliminary assessment on pomegranate (Punica granatum L.). Nanomat 12:1527
2. Abdel-Raouf N, Al-Enazi NM, Ibraheem IBM, Alharbi RM, Alkhulaifi MM (2019) Biosynthesis of silver nanoparticles by using of the marine brown alga Padina pavonia and their characterization. Saudi J Biol Sci 26(6):1207–1215
3. Abid N, Khan AM, Shujait S, Chaudhary K, Ikram M, Imran M, Haider J, Khan M, Khan Q Maqbool M (2022) Synthesis of nanomaterials using various top-down and bottom-up approaches, influencing factors, advantages, and disadvantages: a review. Adv Colloid Interf Sci 300:102597, Adv Colloid Interface Sci

4. Adelere IA, Lateef A (2016) A novel approach to the green synthesis of metallic nanoparticles: the use of agro-wastes, enzymes, and pigments. Nanotechnol Rev 5:567–587
5. Agarwal H, Venkat Kumar S, Rajeshkumar S (2017) A review on green synthesis of zinc oxide nanoparticles—an eco-friendly approach. Resour-Eff Technol 3:406–413
6. Akhtar MS, Panwar J, Yun YS (2013) Biogenic synthesis of metallic nanoparticles by plant extracts. ACS Sustain Chem Eng 1:591–602
7. Ali J, Ali N, Wang L, Waseem H, Pan G (2019) Revisiting the mechanistic pathways for bacterial mediated synthesis of noble metal nanoparticles. J Microbiol Methods 159:18–25
8. Al-Juthery HW, Lahmod NR, Al-Taee RA (2021) Intelligent, nano-fertilizers: a new technology for improvement nutrient use efficiency (article review). In: IOP conference series: earth and environmental science, vol 735. IOP Publishing, p 012086
9. Andrade-Zavaleta K, Chacon-Laiza Y, Asmat-Campos D, Raquel-Checca N (2022) Green synthesis of superparamagnetic iron oxide nanoparticles with eucalyptus globulus extract and their application in the removal of heavy metals from agricultural soil. Molecules 27:1367
10. Arduini F, Cinti S, Mazzaracchio V et al (2020) Carbon black as an outstanding and affordable nanomaterial for electrochemical (bio)sensor design. Biosens Bioelectron 156:112033
11. Armatas GS, Kanatzidis MG (2006) Mesostructured germanium with cubic pore symmetry. Nature 441:1122–1125
12. Arumugam A, Karthikeyan C, Hameed ASH et al (2015) Synthesis of cerium oxide nanoparticles using Gloriosa superba L. leaf extract and their structural, optical and antibacterial properties. Mater Sci Eng C 49:408–415, Mater Sci Eng C Mater Biol Appl
13. Banerjee A, Bandopadhyay R (2016) Use of dextran nanoparticle: a paradigm shift in bacterial exopolysaccharide based biomedical applications. Int J Biol Macromol 87:295–301
14. Barzana G, Garcia-Gomez P, Carvajal M (2022) Nanomaterials in plant systems: smart advances related to water uptake and transport involving aquaporins. Plant Nano Biology 1:100005
15. Bhatti R, Shakeel H, Malik K et al (2022) Inorganic nanoparticles: toxic effects, mechanisms of cytotoxicity and phytochemical interactions. Adv Pharm Bull 12:757–762
16. Bhaviripudi S, Mile E, Steiner SA (2007) CVD synthesis of single-walled carbon nanotubes from gold nanoparticle catalysts. J Am Chem Soc 129(60):1516–1517
17. Biju V, Itoh T, Anas A, Sujith A, Ishikawa M (2008) Semiconductor quantum dots and metal nanoparticles: syntheses, optical properties, and biological applications. Anal Bioanal Chem 391:2469–2495
18. Blal N, Guarnieri D (2024) Preparation method and in vitro characterization of nanoparticles sensitive to tumor microenvironment. Methods Mol Biol 2748:85–98
19. Brar KK, Magdouli S, Othmani A, Ghanei J et al (2022) Green route for recycling of low-cost waste resources for the biosynthesis of nanoparticles (NPs) and nanomaterials (NMs)-a review. Environ Res 207:112202
20. Brenelli LB, Persinoti GF, Cairo JPLF et al (2019) Novel redox-active enzymes for ligninolytic applications revealed from multiomics analyses of Peniophora sp. CBMAI 1063, a laccase hyper-producer strain. Sci Rep 9(1):17564
21. Campelo PH, Sant'Ana AS, Pedrosa Silva Clerici MT (2020) Starch nanoparticles: production methods, structure, and properties for food applications. Curr Opin Food Sci 33:136–140
22. Chandrakala V, Aruna V, Angajala G (2022) Review on metal nanoparticles as nanocarriers: current challenges and perspectives in drug delivery systems. Emerg Mater 5(6):1593–1615
23. Chang BP, Gupta A, Mekonnen TH (2021) Flame synthesis of carbon nanoparticles from corn oil as a highly effective cationic dye adsorbent. Chemosphere 282:131062
24. Chaudhary R, Nawaz K, Khan AK et al (2020) An overview of the algae-mediated biosynthesis of nanoparticles and their biomedical applications. Biomolecules 10(11):1498
25. Chavan P, Sinhmar A, Nehra M et al (2021) Impact on various properties of native starch after synthesis of starch nanoparticles: a review. Food Chem 364:130416
26. Chen J, Wu L, Lu M, Lu S et al (2020) Comparative study on the fungicidal activity of metallic MgO nanoparticles and macroscale MgO against soilborne fungal phytopathogens. Front Microbiol 11:365

27. Chen J, Guo Y, Zhang X et al (2023) Emerging nanoparticles in food: sources, application, and safety. J Agric Food Chem 71:3564–3582
28. Cruz-Luna AR, Vásquez-López A, Rojas-Chávez H, Valdés-Madrigal MA, Cruz-Martínez H, Medina DI (2023) Engineered metal oxide nanoparticles as fungicides for plant disease control. Plants 12(13):13
29. Dabhane H, Ghotekar S, Tambade P et al (2021) A review on environmentally benevolent synthesis of CdS nanoparticle and their applications. Environ Chem Ecotoxicol 3:209–219
30. Daraee H, Etemadi A, Kouhi M, Alimirzalu S, Akbarzadeh A (2016) Application of liposomes in medicine and drug delivery. Artif Cells Nanomed Biotechnol 44:381–391
31. Das R, Bandyopadhyay R, Pramanik P (2018) Carbon quantum dots from natural resource: a review. Mater Today Chem 8:96–109
32. Demirer GS, Zhang H, Matos JL et al (2019a) High aspect ratio nanomaterials enable delivery of functional genetic material without DNA integration in mature plants. Nat Nanotechnol 14:456–464
33. Demirer GS, Zhang H, Goh NS et al (2019b) Carbon nanotube–mediated DNA delivery without transgene integration in intact plants. Nat Protoc 14:2954–2971
34. Demirer GS, Silva TN, Jackson CT et al (2021) Nanotechnology to advance CRISPR–Cas genetic engineering of plants. Nat Nanotechnol 16:243–250
35. Dhall S, Nathawat R, Sood K (2023) Carbon-based nanomaterials. In: Carbon nanomaterials and their nanocomposite based Chemiresistive gas sensors: applications, fabrication and commercialization. Elsevier, Amsterdam, The Netherlands, pp 3–39
36. Dilliard SA, Siegwart DJ (2023) Passive, active and endogenous organ-targeted lipid and polymer nanoparticles for delivery of genetic drugs. Nat Rev Mater 8:282–300
37. Dong B-R, Jiang R, Chen J-F, Xiao Y, Lv Z-Y, Chen W-S (2023) Strategic nanoparticle-mediated plant disease resistance. Crit Rev Biotechnol 43(1):22–37
38. Durowoju IB, Bhandal KS, Hu J, Carpick B, Kirkitadze M (2017) Differential scanning calorimetry—a method for assessing the thermal stability and conformation of protein antigen. J Vis Exp 2017:e55262
39. Ealia SAM, Saravanakumar MP (2017) A review on the classification, characterisation, synthesis of nanoparticles and their application. In: IOP conference series: materials science and engineering, vol. 263:3 IOP publishing, p. 032019
40. Ealias AM, Saravanakumar MP (2017) A review on the classification, characterisation, synthesis of nanoparticles and their application. In: Proceedings of the IOP conference series: materials science and engineering, vol 263. Institute of Physics Publishing, London
41. Egbuna C, Parmar VK, Jeevanandam J et al (2021) Toxicity of nanoparticles in biomedical application: Nanotoxicology. J Toxicol:2021:9954443
42. El-Badri AM, Batool M, Wang C et al (2021) Selenium and zinc oxide nanoparticles modulate the molecular and morpho-physiological processes during seed germination of Brassica napus under salt stress. Ecotoxicol Environ Saf 225:112695
43. Elmowafy M, Shalaby K, Elkomy MH et al (2023) Polymeric nanoparticles for delivery of natural bioactive agents: recent advances and challenges. Polymers 15:1123
44. El-Saadony MT, Almoshadak AS, Shafi ME et al (2021) Vital roles of sustainable nano-fertilizers in improving plant quality and quantity-an updated review. Saudi J Biol Sci 28(12):7349–7359
45. El-Seedi HR, El-Shabasy RM, Khalifa SA et al (2019) Metal nanoparticles fabricated by green chemistry using natural extracts: biosynthesis, mechanisms, and applications. RSC Adv 9(42):24539–24559
46. Etesami H, Fatemi H, Rizwan M (2021) Interactions of nanoparticles and salinity stress at physiological, biochemical and molecular levels in plants: a review. Ecotoxicol Environ Saf 225:112769
47. Fonseca M, Jarak I, Victor F, Domingues C, Veiga F, Figueiras A et al (2024) Polymersomes as the next attractive generation of drug delivery systems: definition, Synthesis and Applications Materials 17(2):319

48. Fu L, Wang Z, Dhankher OP, Xing B (2020) Nanotechnology as a new sustainable approach for controlling crop diseases and increasing agricultural production. J Exp Bot 71:507–519
49. Gahlawat G, Choudhury AR (2019) A review on the biosynthesis of metal and metal salt nanoparticles by microbes. RSC Adv 9(23):12944–12967
50. Gao M, Zhou J, Liu H et al (2018) Foliar spraying with silicon and selenium reduces cadmium uptake and mitigates cadmium toxicity in rice. Sci Total Environ 631:1100–1108
51. Gao M, Zheng F, Xu J, et al (2019) Surface modification of Nano-sized carbon black for reinforcement of rubber. Nanotechnol Rev 8:405–414, Nanotechnology Rev
52. García-Ovando AE, Pina JER et al (2022) Biosynthesized nanoparticles and implications by their use in crops: effects over physiology, action mechanisms, plant stress responses and toxicity. Plant Stress 6:100109
53. García-Pinel B, Porras-Alcalá C, Ortega-Rodríguez A et al (2019) Lipid-based nanoparticles: application and recent advances in cancer treatment. Nanomaterials 9:638
54. Gautam N, Salaria N, Thakur K et al (2020) Green silver nanoparticles for phytopathogen control. Proc Natl Acad Sci India Sect B Biol Sci 90:439–446
55. Ghotekar S, Pansambal S, Bilal M, Pingale SS, Oza R (2021) Environmentally friendly synthesis of Cr 2 O 3 nanoparticles: characterization, applications and future perspective— a review. Case Stud Chem Environ Eng 3:100089
56. Gudikandula K, Vadapally P, Charya MS (2017) Biogenic synthesis of silver nanoparticles from white rot fungi: their characterization and antibacterial studies. OpenNano 2:64–78
57. Hoeksma J, Misset T, Wever C, Kemmink J, Kruijtzer J et al (2019) A new perspective on fungal metabolites: identification of bioactive compounds from fungi using zebrafish embryogenesis as read-out. Sci Rep 9(1):17546
58. Hossain N, Islam MA, Chowdhury MA, Alam A (2022) Advances of nanoparticles employment in dental implant applications. Appl Surf Sci Adv 12:100341
59. Hossain N, Mobarak MH et al (2023) Advances and significances of nanoparticles in semiconductor applications—a review. Results Eng 19:101347
60. Huang Z, Chi-Pong Tsui G, Deng Y, Tang CY (2020) Two-photon polymerization nanolithography technology for fabrication of stimulus-responsive micro/nano-structures for biomedical applications. Nanotechnol Rev 9(1):1118–1136
61. Hussain M, Zahra N, Lang T et al (2023) Integrating nanotechnology with plant microbiome for next-generation crop health. Plant Physiol Biochem 196:703–711
62. Ijaz I, Gilani E et al (2020) Detail review on chemical, physical and green synthesis, classification, characterizations and applications of nanoparticles. Green Chem Lett Rev 13:59–81
63. Islam F, Shohag S, Uddin MJ et al (2022) Exploring the journey of zinc oxide nanoparticles (ZnO-NPs) toward biomedical applications. Dent Mater 15:2160
64. Issa B, Obaidat IM, Albiss BA, Haik Y (2013) Magnetic nanoparticles: surface effects and properties related to biomedicine applications. Int J Mol Sci 14:21266–21305
65. Jesork A, Orwar O (2008) Liposomes: technologies and analytical applications. Annu Rev Anal Chem 1:801–832
66. Jiang H, Lv L et al (2021) Effect of the nanoparticle exposures on the tomato bacterial wilt disease control by modulating the rhizosphere bacterial community. Int J Mol Sci 23:414
67. Jin S, Li S, Wang C et al (2014) Biosafe Nanoscale Pharmaceutical Adjuvant Materials. J Biomed Nanotechnol 10:2393–2419
68. John VL, Nair Y, Vinod TP (2021) Doping and surface modification of carbon quantum dots for enhanced functionalities and related applications. Part Part Syst Charact 38:2100170
69. Khan S, Hossain MK (2022) Classification and properties of nanoparticles. In: Nanoparticle-based polymer composites. Woodhead Publishing, pp 15–54
70. Kim T, Hyeon T (2014) Applications of inorganic nanoparticles as therapeutic agents. Nanotechnology 25:012001
71. Kumar R, Lal S (2014) Synthesis of organic nanoparticles and their applications in drug delivery and food nanotechnology: a review. J Nanomater Mol Nanotechnol 3
72. Kumar A, Gupta K et al (2019a) A review on positive and negative impacts of nanotechnology in agriculture. Int J Environ Sci Technol 16:2175–2184

73. Kumar S, Nehra M et al (2019b) Nano-based smart pesticide formulations: emerging opportunities for agriculture. J Control Release 294:131–153
74. Kwak S-Y, Lew TTS et al (2019) Chloroplast-selective gene delivery and expression in planta using chitosan-complexed single-walled carbon nanotube carriers. Nat Nanotechnol 14:447–455
75. Le Corre D, Angellier-Coussy H (2014) Preparation and application of starch nanoparticles for nanocomposites: a review. React Funct Polym 85:97–120
76. Li Y, Zhang P, Li M et al (2023) Application and mechanisms of metal-based nanoparticles in the control of bacterial and fungal crop diseases. Pest Manag Sci 79:21–36
77. Lindner K, Ströbele M et al (2017) Biological effects of carbon black nanoparticles are changed by surface coating with polycyclic aromatic hydrocarbons. Part Fibre Toxicol 14:8
78. Lingayat VJ, Zarekar NS, Shendge RS (2017) Solid lipid nanoparticles: a review. Nanosci Nanotechnol Res 4:67–72
79. Liu C, Zhou H, Zhou J (2021) The applications of nanotechnology in crop production. Molecules 26(23):7070
80. MaHam A, Tang Z, Wu H et al (2009) Protein-based nanomedicine platforms for drug delivery. Small 5:1706–1721
81. Mahanta N, Dambale A et al (2019) Nutrient use efficiency through nano fertilizers. Int J Chem Stud 7(3):2839–2842
82. Makino K, Shibata A (2006) Chapter 2: surface properties of liposomes depending on their composition. Adv Planar Lipid Bilayers Liposomes 4:49–77
83. Mangena P, Adejumo SA (2023) Biotic stress and breeding of plants for stress resistance. In: Mangena P (ed) advances in legume research: physiological responses and genetic improvement for stress resistance. Bentham books, pp 1–13
84. Mann S, Burkett SL et al (1997) Sol– gel synthesis of organized matter. Chem Mater 9(11):2300–2310
85. Manzoor N, Ahmed T et al (2021) Iron oxide nanoparticles ameliorated the cadmium and salinity stresses in wheat plants, facilitating photosynthetic pigments and restricting cadmium uptake. Sci Total Environ 769:145221
86. Markovic Z, Trajkovic V (2008) Biomedical potential of the reactive oxygen species generation and quenching by fullerenes (C60). Biomaterials 29:3561–3573
87. Martin-Ortigosa S, Peterson DJ et al (2014) Mesoporous silica nanoparticle-mediated intracellular Cre protein delivery for maize genome editing via loxP site excision. Plant Physiol 164:537–547
88. Mazumder M, Roy S, Parvin S et al (2023) Nanotechnological approaches against fungal pathogens of economically important crop plants. In: Malik JA, Sadiq MJ (eds) modern nanotechnology: volume 1: environmental sustainability and remediation, pp 559–584
89. Mekuye B, Abera B (2023) Nanomaterials: an overview of synthesis, classification, characterization, and applications. Nano Select 4:486–500
90. Mgadi K, Ndaba B, Roopnarain A, Rama H, Adeleke R (2024) Nanoparticle applications in agriculture: overview and response of plant-associated microorganisms. Front Microbiol 15:1354440
91. Mikhailova EO (2020) Silver nanoparticles: mechanism of action and probable bioapplication. J Funct Biomater 11(4):84
92. Mitragotri S, Stayton P (2014) Organic nanoparticles for drug delivery and imaging. MRS Bull 39:219–223
93. Mittal AK, Chisti Y, Banerjee UC (2013) Synthesis of metallic nanoparticles using plant extracts. Biotechnol Adv 31(2):346–356
94. Mittal D, Kaur G, Singh P, Yadav K, Ali SA (2020) Nanoparticle-based sustainable agriculture and food science: recent advances and future outlook. Front Nanotechnol 2:579954
95. Mohammadi S, Harvey A, Boodhoo KV (2014) Synthesis of TiO2 nanoparticles in a spinning disc reactor. Chem Eng J 258:171–184

96. Mohd Yusof H, Mohamad R, Zaidan UH, Abdul Rahman NA (2019) Microbial synthesis of zinc oxide nanoparticles and their potential application as an antimicrobial agent and a feed supplement in animal industry: a review. J Anim Sci Biotechnol 10:1–22
97. Motsoene F, Abrahamse H, Dhilip Kumar SS (2023) Multifunctional lipid-based nanoparticles for wound healing and antibacterial applications: a review. Adv Colloid Interface Sci 321:103002
98. Mukherjee S, Ray S, Thakur RS (2009) Solid lipid nanoparticles: a modern formulation approach in drug delivery system. Indian J Pharm Sci 71:349–358
99. Munir N, Gulzar W, Abideen Z, Hancock JT, El-Keblawy A, Radicetti E (2023) Nanotechnology improves disease resistance in plants for food security: applications and challenges. Biocatal Agric Biotechnol 51:102781
100. Najahi-Missaoui W, Arnold RD, Cummings BS (2021) Safe nanoparticles: are we there yet? Int J Mol Sci 22:385
101. Nicosia A, Vento F, Pellegrino AL et al (2020) Polymer-based graphene derivatives and microwave-assisted silver nanoparticles decoration as a potential antibacterial agent. Nanomater 10(11):2269
102. Nsairat H, Khater D, Sayed U et al (2022) Liposomes: structure, composition, types, and clinical applications. Heliyon 8:e09394
103. Palmqvist NM, Seisenbaeva GA, Svedlindh P, Kessler VG (2017) Maghemite nanoparticles acts as Nanozymes, improving growth and abiotic stress tolerance in Brassica napus. Nanoscale Res Lett 12:631
104. Pan Y, Liu X, Zhang W et al (2020) Advances in Photocatalysis based on fullerene C60 and its derivatives: properties, mechanism, synthesis, and applications. Appl Catal B 265:118579
105. Patel KD, Singh RK, Kim HW (2019) Carbon-based nanomaterials as an emerging platform for Theranostics. Mater Horiz 6:434–469
106. Patil S, Chandrasekaran R (2020) Biogenic nanoparticles: a comprehensive perspective in synthesis, characterization, application and its challenges. J Gen Eng Biotechnol 18(1): 67, J Genet Eng Biotechnol
107. Prasad R, Bhattacharyya A, Nguyen QD (2017) Nanotechnology in sustainable agriculture: recent developments, challenges, and perspectives. Front Microbiol 8:1014
108. Priyadarsini S, Mohanty S, Mukherjee S, Basu S, Mishra M (2018) Graphene and graphene oxide as nanomaterials for medicine and biology application. J Nanostructure Chem 8:123–137
109. Purnomo Sari KR, Ikawati Z, Danarti R, Hertiani T (2023) Micro-titer plate assay for measurement of total phenolic and total flavonoid contents in medicinal plant extracts. Arab J Chem 16:105003
110. Qin Y, Xue L, Hu Y et al (2020) Green fabrication and characterization of Debranched starch nanoparticles via Ultrasonication combined with recrystallization. Ultrason Sonochem 66:105074
111. Rajkumari N, Alex S, Soni KB et al (2021) Silver nanoparticles for biolistic transformation in Nicotiana tabacum L. 3 Biotech 11:497
112. Rassaei L, Marken F, Sillanpää M et al (2011) Nanoparticles in electrochemical sensors for environmental monitoring. TrAC Trends Anal Chem 30:1704–1715
113. Rodriguez-Serrano C, Guzmán-Moreno J, Ángeles-Chávez C et al (2020) Biosynthesis of silver nanoparticles by fusarium scirpi and its potential as antimicrobial agent against uropathogenic Escherichia coli biofilms. PLoS One 15(3):e0230275
114. Saleh TA (2020) Nanomaterials: classification, properties, and environmental toxicities. Environ Technol Innov 20:101067
115. Samimi S, Ardestani MS, Dorkoosh FA (2021) Preparation of carbon quantum dots- Quinic acid for drug delivery of gemcitabine to breast cancer cells. J Drug Deliv Sci Technol 61:102287
116. Sanguinedo P, Fratila RM, Estevez MB, Grazú V, Alborés S (2018) Extracellular biosynthesis of silver nanoparticles using fungi and their antibacterial activity

117. Shafqat W, Jaskani MJ, Maqbool R, Chattha WS, Ali Z, Naqvi SA, Haider MS, Khan IA Vincent CI (2021) Heat shock protein and aquaporin expression enhance water conserving behavior of citrus under water deficits and high temperature conditions. Environ Exp Bot 181:104270
118. Shahjahan M, Rahman MH, Hossain MS, Khatun MA, Islam A, Begum MA (2017) Synthesis and characterization of silver nanoparticles by sol-gel technique. Nanosci Nanometrol 3(1):34–39
119. Shang Y, Hasan MK, Ahammed GJ, Li M, Yin H, Zhou J (2019) Applications of nanotechnology in plant growth and crop protection: a review. Molecules 24(14):2558
120. Sharma G, Kumar A, Devi KA et al (2020) Chitosan nanofertilizer to foster source activity in maize. Int J Biol Macromol 145:226–234
121. Sheikhalipour M, Esmaielpour B, Behnamian M et al (2021) Chitosan–Selenium Nanoparticle (Cs–Se NP) Foliar Spray Alleviates Salt Stress in Bitter Melon. Nanomaterials 11:684
122. Shubayev VI, Pisanic TR, Jin S (2009) Magnetic nanoparticles for Theragnostics. Adv Drug Deliv Rev 61:467–477
123. Shukla S, Haldorai Y, Hwang SK et al (2017) Current demands for food-approved liposome nanoparticles in food and safety sector. Front Microbiol 8:2398
124. Singh G, Kalia A (2019) Nano-enabled technological interventions for sustainable production, protection, and storage of fruit crops. In: Nanoscience for sustainable agriculture. Springer, pp 299–322
125. Singh D, Dubey P, Pradhan M, Singh MR (2013) Ceramic Nanocarriers: versatile Nanosystem for protein and peptide delivery. Expert Opin Drug Deliv 10:241–259
126. Singh J, Dutta T, Kim KH, Rawat M, Samddar P, Kumar P (2018) Green synthesis of metals and their oxide nanoparticles: applications for environmental remediation. J Nanobiotechnol 16:1–24
127. Singh D, Pandey H, Dhiman VK et al (2022) Secondary metabolite engineering for plant immunity against various pathogens. In: Metabolic engineering in plants. Springer Nature, Singapore, pp 123–143
128. Tauseef A, Hisamuddin KA, Uddin I (2020) Role of MgO nanoparticles in the suppression of Meloidogyne incognita, infecting cowpea and improvement in plant growth and physiology. Exp Parasitol 220:108045
129. Terna AD, Elemike EE, Mbonu JI, Osafile OE, Ezeani RO (2021) The future of semiconductors nanoparticles: synthesis, properties and applications. Mater Sci Eng B 272:115363
130. Thakkar KN, Mhatre SS, Parikh RY (2010) Biological synthesis of metallic nanoparticles. Nanomed Nanotechnol Biol Med 6:257–262
131. Tripathi DK, Singh S, Singh VP, Prasad SM, Chauhan DK, Dubey NK (2016) Silicon nanoparticles more efficiently alleviate arsenate toxicity than silicon in maize cultiver and hybrid differing in arsenate tolerance. Front Environ Sci 4:46
132. Tucek J, Kemp KC, Kim KS, Zboril R (2014) Iron-oxide-supported Nanocarbon in lithium-ion batteries, medical, catalytic, and environmental applications. ACS Nano 8:7571–7612
133. Vaidulych M, Hanuš J, Kousal J et al (2019) Effect of magnetic field on the formation of cu nanoparticles during magnetron sputtering in the gas aggregation cluster source. Plasma Process Polym 16(11):1900133
134. Viegas C, Patrício AB, Prata JM et al (2023) Solid Lipid Nanoparticles vs. Nanostructured Lipid Carriers: A Comparative Review Pharmaceut 15:1593, Pharmaceutics
135. Wang Y, Hu A (2014) Carbon quantum dots: synthesis, properties and applications. J Mater Chem C Mater 2:6921–6939, J Mater Chem C Mater Opt Electron Devices
136. Wang A, Li J, Al-Huqail AA et al (2021) Mechanisms of chitosan NPs in the regulation of cold stress resistance in Banana plants. Nano 11:2670
137. Wang L, Ning C, Pan T, Cai K (2022) role of silica nanoparticles in abiotic and biotic stress tolerance in plants: a review. Int J Mol Sci 23(4):1947
138. Weber C, Coester C, Kreuter J, Langer K (2000) Desolvation process and surface characterisation of protein nanoparticles. Int J Pharm 194:91–102

139. Wong MH, Misra RP, Giraldo JP et al (2016) Lipid exchange envelope penetration (LEEP) of nanoparticles for plant engineering: a universal localization mechanism. Nano Lett 16:1161–1172
140. Wu C, Zreiqat H (2010) Porous bioactive Diopside (CaMgSi2O6) ceramic microspheres for drug delivery. Acta Biomater 6:820–829
141. Xu T, Shen W, Huang W, Lu X (2020) Fullerene micro/nanostructures: controlled synthesis and energy applications. Materials Today Nano 11:100081
142. Yadav R, Saini H, Kumar D, Pasi S, Agrawal V (2019) Bioengineering of Piper longum L. extract mediated silver nanoparticles and their potential biomedical applications. Mater Sci Eng: C 104:109984, Mater Sci Eng C Mater Biol Appl
143. Yahyaei B, Pourali P (2019) One step conjugation of some chemotherapeutic drugs to the biologically produced gold nanoparticles and assessment of their anticancer effects. Sci Rep 9(1):1–15
144. Yang J, Yan S, Xie S et al (2022) Construction and application of star polycation nanocarrierbased microRNA delivery system in Arabidopsis and maize. J Nanobiotechnol 20:219
145. Yu W, Sisi L, Haiyan Y, Jie L (2020) Progress in the functional modification of graphene/graphene oxide: a review. RSC Adv 10:15328–15345
146. Zahra Z, Habib Z, Chung S, Badshah MA (2020) Exposure route of TiO2 NPs from industrial applications to wastewater treatment and their impacts on the agro-environment. Nanomaterials 10:1469
147. Zakeri A, Kouhbanani MAJ, Beheshtkhoo N, Beigi V, Mousavi SM, Hashemi SAR et al (2018) Polyethylenimine-based nanocarriers in co-delivery of drug and gene: a developing horizon. Nano Rev Exp 9(1):1488497
148. Zeeshan M, Hu YX, Guo XH et al (2023) Physiological and transcriptomic study reveal SeNPs-mediated AsIII stress detoxification mechanisms involved modulation of antioxidants, metal transporters, and transcription factors in Glycine max L.(Merr.) roots. Environ Pollut 317:120637
149. Zhang J, Chaker M, Ma D (2017) Pulsed laser ablation based synthesis of colloidal metal nanoparticles for catalytic applications. J Colloid Interface Sci 489:138–149
150. Zhang Q, Zhang Y, Hong L et al (2024) Fullerene Nanorings as nitric oxide radical scavengers for ultraviolet-induced cellular injury. ACS Appl Nano Mater 7:5689–5697
151. Nalon GH, Ribeiro JCL, Araújo ENDD et al (2020) Effects ofDifferent kinds of carbon black nanoparticles on the Piezoresistive and mechanical properties of cement-based Composites.J Build Eng 32:101724
152. Astefanei A, Núñez O, Galceran MT (2015) Characterisation and determination of fullerenes: a critical review. Anal Chim Acta 882:1–21

Nanoparticle Transportation in Plants: Mechanistic Insights and Applications

Mohsen Mohamed Elsharkawy

Abstract Nanoparticle transportation in plants is a growing study area with many potential uses in agriculture. A summary of the difficulties and opportunities related to nanoparticle transportation in plants is presented in this chapter, with an emphasis on the distribution techniques. The difficulties in delivering nanoparticles to plants are emphasized, highlighting the necessity of creative solutions to maximize transport effectiveness. Research on plants' absorption and translocation methods has shown that they react differently to various kinds of nanoparticles. Nanoparticles exhibit the ability to improve plant development, absorption of nutrients, and resilience to stress because of their diverse characteristics. Although studies have demonstrated that nanoparticles may be taken by plants and affect their physiological processes, it is still unknown how exactly they are absorbed and transported. The chapter covers the future possibilities of nanoparticles as cutting-edge products in agriculture, spurring a paradigm change towards stress-resilient systems for agriculture, and gives comprehensive information on the mechanisms involved in their transportation in plants.

Keywords Nano-deliver · Plant growth · Plant development · Nanobiotechnology · Plant stress resilience

1 Introduction

Agriculture offers food for a better living and serves as the main pillar of the growing economy. The agricultural industry is now dealing with several issues, such as erratic climate change and soil pollution from several hazardous environmental contaminants, and primarily increasing food needs due to a growing world population

M. M. Elsharkawy (✉)
Agricultural Botany Department, Faculty of Agriculture, Kafrelsheikh University, Kafr Elsheikh, Egypt
e-mail: mohsen.abdelrahman@agr.kfs.edu.eg

J.-T. Chen (ed.), *Nano-Delivering for Plant Genetic Engineering*, Smart Nanomaterials Technology, https://doi.org/10.1007/978-981-95-4165-2_9

[1]. To accommodate the expanding population, food production has to increase by almost 50%. Industrialization leads to rapid loss of natural resources, including seas, forests, and biodiversity, that are essential for human survival. It emphasizes the need for enhanced agricultural output and improved food security. Ecological biodiversity helps preserve the delicate relationship between the ecosystem and plants by strengthening tolerance to stress, which can lead to crop loss. The biodiversity includes animals, crops, aquaculture, and forests, which support individual life. Many plant and animal species that primarily support the global population for food, fiber, fuel, and fodder may be found in forests and oceans. Therefore, any problems that can lead to ecological disequilibrium have a notable influence on foodstuffs. Old breeding techniques have undoubtedly contributed significantly to the improvement of agricultural characteristics and crop output. The problem is that various techniques (such as backcrossing, single seed descent, and pure line selection) can be time-consuming, and they can have trouble transferring new characteristics between species, which could impede the quick emergence of much better varieties [2]. Furthermore, it is becoming rapidly clear that in order to solve environmental issues and maintain long-term sustainability, present agricultural techniques must be changed to improved and environmentally friendly ones [3]. Novel practices are always being improved to address these relevant difficulties. NM (nanomaterial)-created products may alter contemporary agriculture techniques. These substances are highly responsive because of their huge surface and unique features, which may be easily adjusted to control the increasing requirements.

Initially, NPs were exploited in several areas [4]. NPs are used in many customer commodities as a result of their vast abilities [1]. Therefore, the food and agricultural industries have unquestionably taken notice of the novel and developing qualities of these modern materials. These materials are helping to raise production and changing modern farming into precise farming. Although fertilizers are essential to improve yields, they can lower fertility levels by altering the soil's mineral balances. Usually applied fertilizers and pesticides may be effectively rinsed off. In addition, the high expense of producing these fertilizers and pesticides must be kept under control. Reducing nutrient losses to boost yields and lowering product quantities for plant protection are the goals of NM use in agriculture. NMs in agriculture seek to minimize nutrient losses, boost yields, and reduce the need for plant protection actions [5]. This chapter states that NMs offer a more effective approach to improving agricultural and food chains. However, we emphasize the importance of testing the harmfulness of NM items before putting them in the marketplace. Nanotechnology alternatives, including novel agrochemical release methods that use NPs to transmit active chemicals, are being developed regularly. Agricultural waste, such as wheat straw, may be turned into modern bionanocomposites with better physical and mechanical qualities for industrial applications. The practical advantages of nanotechnology, such as better procedure strategy, hazard evaluation of nanofertilizers and nanopesticides, and laws for the marketing of nanoagroproducts, still require some improvements [1]. Farming systems that are more capable of dealing with unexpected climatic fluctuation, such as novel cultivars that can tolerate drought, heat, and other stressors, represent another difficult problem that has not yet been addressed [1].

It is widely assumed that NPs may be absorbed and transferred in plants. Their absorption and transfer are affected by parameters such as surface charge, size, contact period, concentration, and plant type. NPs are capable of entering the plant system via many paths, including root hairs, stomata, and fissures in the leaf surface. NPs can migrate throughout the plant structures via phloem loading, diffusion, and bulk flow. The size and form of the nanoparticles, the availability of the solution's chemicals and ions, the solution's pH, and their surface features can all have an impact on their transport. Research focused on applying NPs by root application, leaf spraying, seed treatment, and branch injection. Some studies verified the absorption of NPs [6]. Recognizing the absorption and transport of NPs in plants is critical for developing optimum NPs for agricultural usage [7, 8]. Understanding the method of action and NPs bioaccumulation in plants could improve their biological safety and guide safe application [6]. The movement of NPs into other living organisms and their effects on consumers are not discussed in this chapter, despite research showing that NPs may persist in edible plants and transfer to consumers, impacting the health of humans.

The main actors in the agricultural rebellion are engineered nanomaterials [9]. The materials' nanoscale size allows them to get through biological barriers and guarantee effective chemical pesticides and nutrient distribution through foliar or root applications [10]. To achieve the purpose of intelligently delivering and releasing the active components to the intended site, nanomaterials' physio-chemical and structural properties may be changed through engineering to change their mode of action and delivery route [11]. CRISPR has evolved genome editing technology's possibilities. Its low intracellular distribution efficacy, which significantly reduces the success rate, is the largest obstacle to its use in genome alteration. Together with liposomes, polymers, and inorganic NPs, a few nanomaterials function as non-viral nanocarriers and have demonstrated exceptional gene transport capabilities [2]. Using specific nutrition and additive delivery, more effective resource uses, and better disease and stress control, nanotechnology additionally works to reduce pre-harvest crop losses [12]. NMs are a viable option for sustainable agriculture considering the previously described features. Additionally, NMs can enhance the utilization of nutrients through targeted delivery, the development of cost-effective resource recovery strategies, the transformation of plants by smart sensors that can anticipate impending stress, the creation of more durable varieties through genetic modification, improved crop supervision to prevent crop damages, and the improvement of the microbiome to restore soil features and boost production [13]. A broad knowledge of nanotechnology is, in our opinion, crucial to achieving the objective of supporting global agriculture. This chapter addresses the absorption and movement processes of NPs and provides complete evidence on the effects of NPs in agriculture and discusses their prospects.

2 Nanoparticles Uptake and Pathway in Roots and Leaves

In agricultural treatments, NPs are usually sprayed on the leaf surface, and then enter plants through the leaf surface's stomata or cuticle. The main components of the epidermis' waxy cuticle are pectin, wax, and cutin. Two separate channels, hydrophilic and lipophilic, are present on the surface of waxy stratum corneum [14]. Using high-spatial and temporal-resolution confocal fluorescence microscopy, Hu et al. [15] showed that carbon dots smaller than 2 nm may penetrate cotton leaves through the cuticular route. However, plants' ability to absorb NPs through the epidermis is restricted because of the small size of the cuticle pore channels. After being applied to the leaf surface, some research findings suggested that NPs could build up in the vascular tissue and leaf epidermis. Numerous studies have mentioned that NPs have the capability to enter various tissues. The researchers suggested that the stomatal channel is a way via which NPs enter plants [16].

Plant absorption of NPs is dependent on the nanoparticles' composition, the type of plant, and its environmental circumstances. The system that NPs are absorbed by leaves could be affected by their characteristics, including size, chemical structure, surface cation, and charge. NPs size is an important factor in examining NP uptake in the blade [17].

Adsorption on the root surface is the first way that NPs contact with plant roots. Positively charged NPs could probably concentrate in plant roots and readily enter plants because the root surface possesses negative charges due to the root hairs' capacity to produce organic acids or mucus [18]. The epidermis of roots functions similarly to that of leaves in terms of composition. The difference is that the epidermis on the hair surfaces and tip of the main and secondary roots is not grown completely. NPs come into close touch with the epidermis [19]. Some NPs have the capacity to rupture the plasma membrane causing cell walls to form new holes, which allows large-diameter NPs to penetrate [20]. NPs enter cells mostly via the hydrophilic channel after being absorbed by the roots. However, there are very few nanoparticle penetrations into cells in this way due to the small size of the pores [17]. Plants may absorb NPs by connecting them with transport proteins on their outermost layer [8].

3 Physicochemical Characteristics-Dependent Uptake of NPs

The physicochemical characteristics of NPs (surface charge (NP-SC), shape, and size) and possible plant interactions are the primary factors that lead to NP absorption and transformation [1].

3.1 Size-Dependent Uptake of NPs

The size of NPs should be taken into account as a crucial factor while evaluating their absorption in plants because the different barriers that are enclosed inside the plant fluctuate in size from micrometers to nanometers. The most significant element influencing the uptake of NPs is believed to be their size. The foliar epidermis consists of cells that create a cuticle membrane. The epidermis has stomata (with two guard cells) that form an opening (measuring 3–12 μm wide) during gas exchange. Consequently, stomatal holes allow NPs to pass through the plants. The permeability characteristics of the epidermal cuticle layer and stomatal trichrome vary greatly. It was shown that NPs between 4 and 100 nm in size might penetrate the cuticle [21]. Studies have shown that metal NPs (less than 50 nm in diameter) can penetrate leaves via the stomata [16]. The capability of leaves to absorb nanoparticles decreased as particle size increased. Similarly, previous research has demonstrated that NPs (size smaller than 10 nm) may be absorbed by roots (such as Au-NPs, 3.5 nm) [15]. The foliar uptake of NPs was recorded in several studies. For example, Zhu et al. [22] treated wheat leaves with ZnO-NPs (30 nm) tagged with FITC (fluorescein isothiocyanate). They discovered using confocal microscopy that ZnO-NPs mostly traveled along the stomata channel into the wheat leaf epidermis before gathering in the chloroplasts [22]. The influence of stomatal opening and closure on ZnO-NPs absorption was also examined. It was demonstrated that when stomatal width was reduced, the content of zinc in leaf cells decreased. Polymeric NPs (43 nm in diameter) can only enter the leaves of faba beans via the stomata, however, NPs (1 μm in diameter) did not pass at all [23]. NPs enter through the stomata and then accumulate on the substomatal cavity's cell wall. According to TEM experiments, tobacco plants absorbed small NPs, such as Fe_3O_4-NPs (20 nm) [24]. Therefore, NPs (smaller than 5 nm) can effectively deliver via cell walls. The biotransformation phenomena were further explained by plant transformation of $AgNO_3$ applied with Ag-NPs [4]. Researchers also discovered that plant leaves could absorb NPs composed of silica, polymers, and natural materials in addition to those based on metals. The findings of their study showed that the crucial size at which plant leaves could absorb NP was significantly influenced by its type. TEM allowed researchers to verify that SiO_2 (54 nm) could penetrate *Arabidopsis thaliana*'s leaves via the stomatal channel [25]. Cucumber leaves could absorb FITC-MSNs (FITC-labeled mesoporous silica NPs), which had a particle size of 200–300 nm. Using TEM, researchers discovered that a 93.6 nm polystyrene nanoplastic could penetrate the lettuce leaf and reach the plant's phloem [26]. Recently, scientists discovered that rice leaves absorbed and dispersed chitosan-based silicon NPs, which had a 166 nm average size [27].

The interaction between the cell wall and NPs produces huge holes that make NP penetration easier [28]. NP absorption across the plant cell membrane has been studied using plant protoplasts as a model system. Additionally, the strategy avoids the technical challenges that come with examining the absorption and quantification of NPs inside plants, which are mostly caused by the internal fluorescence of the plant cell wall and the inhibition of staining dyes by plants [29].

Recognizing the paths of NPs after entering plants is necessary because it provides insight into accumulation sites. The apoplast and the symplast are two crucial pathways that NPs move up and down in plants. The symplastic route permits movement via neighboring cells' cytoplasm [30]. On the other hand, the apoplastic route permits movement via the extracellular spaces via cell walls and xylem vessels [31]. The plasmodesmata, which act as a cytoplasmic bridge to promote particle transport between cells, allow for cell-to-cell migration. They have a diameter of 20–50 nm and are encircled by 3 nm particles [32].

3.2 Effect of NPs Surface Charge (NP-SC)

NPs are highly adaptable and mostly depend on their uptake, and moving inside the plant. Nanoparticles' access into mesophyll tissues is influenced by their structure and charge. Certain membrane proteins, transporters, and surface receptors are altered as a consequence of the physicochemical contact of NPs with plants according to their NP-SC and energy [33]. NPs with varying NP-SC have distinct surface characteristics and aggregation capacities than their natural counterparts [34]. Similarly, the absorption of NP by roots is also dependent on its NP-SC [35]. The NP-SC characteristics that may be absorbed by plant roots are influenced by the negative charge of the root's cell walls. Although they are unable to penetrate the root tissue, positively charged NPs have a tendency to gather on the root surface due to the attraction between the positively charged NPs and the negatively charged cell wall [36]. NPs utilizing the RAFT polymerization technique were created, varying in size (20–100 nm) and NP-SC [37]. They discovered that uncharged and negatively charged NPs (22 nm) may be absorbed by *Arabidopsis* roots and thereafter penetrate the root's xylem by employing confocal microscopy. In contrast, the positively charged NPs cannot enter *Arabidopsis* roots and can only gather at the epidermis [37]. Various forms of NPs have various interfacial qualities, impacting their contact angle, surface area, plant surfaces, and NP absorption [38, 39].

A biological membrane, or cell wall, with both hydrophilic and hydrophobic constituents and an uneven distribution of fixed negative charges, is found inside the leaves [40]. Therefore, one elucidation might be that the tissues' uptake of positively-charged NPs is preferred by the negatively-charged cell wall. Plant cell walls that are negatively charged serve as the surface for ion exchange, which may facilitate the uptake of cationic NPs as compared to anionic ones. However, for negatively charged NPs, the efficiency of transport is significantly increased. The NP-SC-dependent absorption and distribution of AuNPs were elucidated by demonstrating much greater adsorption of positively charged NPs at root surfaces [41]. However, it was shown that negatively charged NPs had a greater rate of internalization and translocation. In this scenario, CeO-NPs having a positive charge were significantly adsorbed onto plant root surfaces, while negatively-charged CeO_2-NPs showed limited root deposition but increased shoot internalization, mostly due to reducing electrostatic repulsion [42]. AuNPs are uptaken and transported from roots to shoots based on their surface charge [43]. It is assumed that negatively charged NPs help both apoplastic and symplastic transfer into the vascular system via roots to shoots [43]. It is assumed that negatively charged NPs help both symplastic and apoplastic transfer into the vascular system via roots to shoots. Surface charge change greatly affects the transport act of nanoparticles via plants.

When comparing rod-shaped and spherical Ag-NPs on the condition of identical particle size, rod-shaped NPs were more readily absorbed and infused by *Arabidopsis* leaves [44]. Plant leaves have the ability to absorb negatively and positively charged NPs. A recent study has assessed how surface charge affects GQDs' (graphene quantum dots) absorption by maize leaves [45] and the adsorption of ZnO-NPs by wheat leaves [39]. Positively charged NPs adsorb more strongly in the leaves than negatively charged ones do, mostly because of the electrostatic interaction between the two types of NPs and the negatively charged cell walls.

3.3 Anatomical Alterations Effect on NP Uptake

There is a strong correlation between the NP-SC and the anatomical variations in plants. One of the key elements influencing how effectively NPs can be absorbed is the species of the plant [16]. The density, size, and location of the pores in the leaves all affect how effectively NPs are absorbed. For instance, monocotyledonous plants have stomata that are uniform in shape and grouped neatly in contrast to dicotyledonous plants. Plants absorb NPs differently depending on their life cycle and development stage. Additionally, dicotyledonous plants with a greater root surface area, like lettuce and tomatoes, have a better interaction with positively charged cerium oxide NPs. Negatively and neutral-charged NPs are more effective for root-to-shoot translocation [46]. Plants have various uptake patterns of NMs due to anatomical and physiological variations. This is supported by studies showing crops from different species accumulate and absorb different classes of unique NMs [47]. Dicot plants absorb more NPs than monocots due to increased permeability in their cuticles. Nanostructures of lipid-based liquid crystalline serve as effective delivery vehicles [48].

Monocot leaves feature parallel veins, making them stronger and more resistant to ripping [15, 49] Plants' capacity to absorb NMs varies depending on their manner of administration. Roots obtain soil nutrients, whereas leaves exchange gases, and cuticles prevent free penetration [50]. Soil components including humic acid and organic matter stabilize NMs, increasing their bioavailability [51]. Additionally, the existence of bacteria and mycorrhizal fungi around plants influences the rate at which plants absorb NPs [52]. NPs can enter plants through an endocytic pathway that is independent of clathrin. A study on foliar absorption of AuNPs on wheat found that smaller NPs with a surface coating of PVPAuNPs, instead of citrate-AuNPs, were more efficiently translocated [14]. Cucumber seedlings did not absorb

SWCNT, but nanosheets were compatible with the roots' external surface [53]. The nature of the method of transport significantly impacts the absorption of NMs. A variety of techniques was used, such as direct tree addition and petiole-feeding in Mexican lime citrus trees, to examine the mobility and fate of surface-modified AgNP delivery to trees [54]. Various groups thoroughly review additional explanations of uptake and distribution in relation to the transformation and fate of NMs [18, 55].

4 NPs Transportation in Roots and Leaves

Nanoparticles can travel great distances in plants via the extracellular or plasmid pathways after entering mesophyll tissue through the stomata and cuticle [56]. The primary means of transporting the protoplast route are intercellular channels called plasmodesmata, which have a diameter of less than 20 nm [57]. Following their passage through the plasmodesmata, NPs will gather in the cytoplasm and migrate to the plant endothelium and Casparian strip [56].

The impact of NPs on the roots was the topic of more research than the application of various NPs on the leaves. NPs are absorbed by the root hair cells when they contact with the root [58]. The xylem vessels carry them to aboveground organs after they penetrate the cell wall and reach the epidermis endothelial layer via the exoplast or symplast. The vascular column consists of vascular and pericycle tissue at the center of the root [55]. The cortex comprises the exodermis and endothelium, with the endothelium connecting to the pericycle.

Prior research revealed that in contrast to NPs composed of nonmetal elements, metal-based NPs that could be transferred to other plant organs were smaller. The metal-based NPs were present in the phloem and other plant leaf components. However, it was frequently seen that a bigger non-metal NP was relocated in the leaves [59]. The PHSNs (porous hollow silica NPs) were discovered to be able to travel to roots, and immature leaves in addition to absorption by leaves, despite the fact that they were far bigger than the tissues exclusion limit [60]. Similarly, research indicates that NPs with tiny particle sizes may be easily absorbed by roots and transmitted to the plant's surface. The root apical-meristem of wheat allowed NPs to enter the vascular system because the root meristem Casparian strip was immature [17]. Following 2 h of application, they were discovered in the xylem's vascular system, demonstrating that the extracellular route was used to carry the nanoparticles.

The extracellular pathway's ability to transport NPs was influenced by their surface charge. The positively charged NPs are connected to the negatively charged cell wall due to the pectin carboxyl groups. This facilitates their passage by the extracellular paths [61]. Two types of PS NPs with varying charges were discovered to be able to efficiently aggregate on maize leaves after spraying the leaves with amino-modified polystyrene and carboxyl NPs. Among NPs accumulated in the stomata, PS-NH2 NPs with positive charge at a higher rate than PS-COOH NPs with negative charge. PS-COOH NPs could efficiently penetrate the plant's circulatory system and travel via vascular bundles to the roots once the NPs have entered the leaves [38]. The way surface charge affects NP transport via plant roots is identical to how it affects NPs transport via leaves. The first line of resistance for NPs entering roots is the negatively charged secretions and mucus released by the roots, which also prevent positively charged nanoparticles from entering the cell wall. Positively charged NPs are positioned to gather at the cell wall after absorption because the cell wall is negatively charged. The transport ability of the electrically neutral nanoparticles in the root is more evident than that of the negatively charged NPs, which will move through the root.

5 NPs Distribution Via Roots and Leaves

Numerous investigations have demonstrated that NPs sprayed through leaves may travel to other plant components, including the roots and rhizosphere soil [62]. Azo@PHSNs (PHSNPs prepared with azoline oxalate) were created and applied by spraying on tomato [63]. The second and fourth days following the action were used to identify and measure the amount of Si present in various plant sections. The findings demonstrated that the greatest Si content among different tomato tissues was found in the stem followed by the root, immature leaf and mature leaf, indicating the PHSNs' capacity for transport across various tomato tissues. Transporting NPs from the stem to the base was simpler than moving between leaves [63].

6 Methods to Study Uptake, Quantification, and Translocation of NPs

Potential methods for monitoring the interaction between plants and NMs must be established. Additionally, more crucial data is required to measure the uptake and translocation of NPs within plants as well as their release into the environment. Fe_3O_4-NPs can be monitored in plant organs using their magnetic characteristics [64]. Tracking and translocating Fe_3O_4-NPs is challenging due to the difficulty distinguishing between intact and leached ions. MPS (Magnetic particle spectrometry) can be used in conjunction with AAS (atomic absorption spectroscopy) to address this issue [65]. Critical imaging methods like computed tomography (CT) and X-ray are useful for locating and detecting different NMs in plants. Using 3-D (nano-CT) and 2-D (DF-HSI) approaches together offers improved ways to describe and assess the NP–plant interaction [49]. In a groundbreaking investigation, laser-induced, time-resolved fluorescence, and ultraviolet-visible spectra were used to observe the impact and interaction of TiO_2-NPs with garlic plants. Garlic leaves showed higher chlorophyll content than the control group. The authors observed a decrease in the intensity ratio between far-red and red chlorophyll fluorescence bands, indicating increased photosynthetic activity [66]. ICP-MS (inductively-coupled plasma-mass spectroscopy), in particular, SP-ICP-MS single-particle (ICPMS) analysis, is a promising method for the characterization, quantification, and detection of NMs [67]. SP-ICP-MS measures CuO-NP uptake in edible plants such as collard greens, lettuce, and kale [68]. Analytical techniques using mass spectrometry can differentiate between various chemical forms of NMs, which are easily transformed inside plants. One such method confirmed the status of ZnO NPs in lettuce using ESI tandem MS and SP-ICP-MS [69]. EDS, SEM and SP-ICP-MS were used to explore the concurrent uptake, distribution, and retention of Cu, ZnO, and Ag-NPs in *A. thaliana* [70].

7 NPs' Effect on Physiological Activity

NPs have become reliable tools in a variety of fields, creating new avenues for advancement in healthcare and other areas (Fig. 1). Given the benefits of NPs as nano-pesticides, and nano-carriers including their tiny size, convenience of use, prolonged storage capacity, and ability to increase the effectiveness of agricultural chemicals, their usage in agriculture is now much more widespread [1, 63]. The physiological morphology, growth, and crop production are all impacted by the interaction between NPs and plants, both positively and negatively. Plant species, application techniques, NPs dose, and concentration are all important for plant reaction.

7.1 NPs Effect on the Development of Plants

Numerous researchers have demonstrated that the application of NPs has enhanced the quality of plant products more than conventional chemicals (Table 1). Cotton's antioxidant capacities and chlorophyll level may be increased by using ZnO-NPs. Additionally, the weight and quantity of cotton bolls for each plant can be increased, and the quality criteria of cotton fiber including fiber strength and uniformity can be improved [71]. ZnO-NPs increase the uptake of nutrients (zinc and phosphorus) by tomato plants, hence improving tomato output [72]. Plant productivity can be enhanced by Fe_3O_4-NPs by raising the amount of minerals, carbohydrates, and protein in plants [72, 73]. Fe_3O_4-NPs were utilized to operate on wheat and maize plants [74, 75]. All of their findings supported the aforementioned conclusions [74, 75]. Furthermore, applying Fe_3O_4-NPs to wheat at the tillering stage resulted in a significant increase in the number of wheat seeds. TiO_2-NPs enhanced spinach biomass production [76]. CeO_2-NPs encouraged stem elongation and fruit weight [77]. TiO_2-NPs were demonstrated as an effective, non-toxic photocatalyst that can enhance light absorption by increasing the light system's energy conversion. TiO_2-NPs were able to boost wheat production. However, this enhancement is caused by

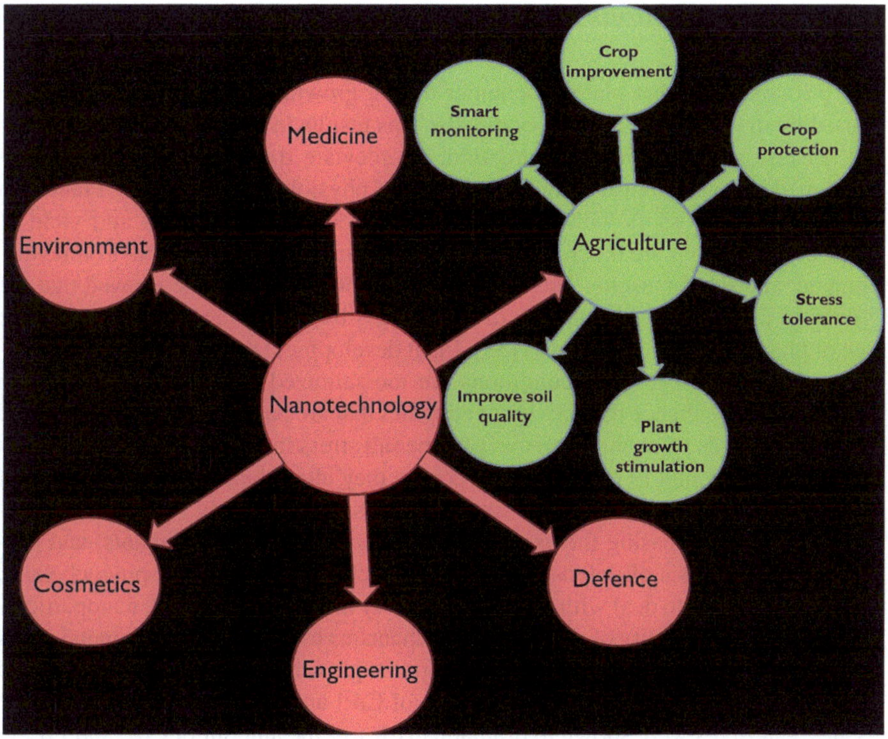

Fig. 1 A summary of the different approaches that nanotechnology is used in agriculture and other fields

Table 1 Positive effects of nanoparticles on plant development

Plant	Nanoparticle	Size	Effect	Citation
Mangifera indica	ZnO	Less than 100 nm	Increase fruit production, TSS and sugar levels	[79]
Triticum asetivum	TiO_2	32 nm	Increase growth	[112]
Solanum lycopersicum	Cu	50 nm	Increase fruit number and quality	[113]
Arabidopsis thaliana	CeO_2	15–30 nm	Increase plant growth	[114]
Capsicum chinense	ZnO	12–24 nm	Increase fruit number and quality	[115]

enhanced photosynthetic action by promoting linear and cycling phosphorylation. This increases the availability of photosynthates, which in turn boosts plant production. TiO_2-NPs increase the photoreduction action of the leaves, which increases the dry matter production of barley [78]. Additionally, the production, sugar content, and weight of mango fruit rose when ZnO-NPs and Si-NPs worked together [79].

7.2 NPs' Utilizations Against Abiotic Stresses

The primary issue that plants encounter during growth is abiotic stress, which is caused by dryness, salt, and heavy metals. This results in a loss of agricultural output because of plant growth retardation. To alleviate these stressors, plants have developed various defense systems by means of physiological pathways. Plants can reduce abiotic stressors with the use of nanoparticles. The biomass and yield of crops in arid regions are much lower than in semi-arid regions due to environmental, climatic, and other reasons. The application of ZnO-NPs to maize showed that the NPs have a moderating impact on drought stress by increasing the photosynthetic rate of plants under drought stress [38]. Plant development will be slowed and nutrients imbalanced if growing conditions are too salinized [80]. The use of NPs in agriculture can enhance the activity of enzymes that are part of plants' salt tolerance system. ZnO NPs were shown to mitigate the salt stress in wheat [81]. Heavy metals are more detrimental to plant growth because they alter plant morphology, prevent plant growth, and cause oxidative stress in plants. By managing the level of heavy metals in soil, decreasing the expression of heavy metal genes in plants, and promoting the synthesis of defense compounds (such as root exudates, phytochelatin, and organic acids) to deal with the stress of heavy metal ions in plants, nanoparticles can enhance the capacity of antioxidant mechanisms to withstand heavy metal stress [82]. Lettuce was cultivated using a hydroponic growth system with ZnO-NPs and heavy metals (Cd^{2+} and Pb^{2+}). The amount of Cd^{2+} and Pb^{2+} were measured in the tissues. ZnO-NPs were shown to dramatically lower the amount of Cd^{2+} and Pb^{2+} in lettuce roots and shoots [83]. ZnO-NPs were used to treat rice demonstrating that these NPs may enhance rice development and lessen arsenic buildup in rice. When the ZnO-NPs dose was 100 mg/L, the greatest impact was seen. Si and Se-NPs lessened the fatal effect of heavy metals on plants by lowering the absorption and buildup of Cd and Pb in rice. Simultaneously, the application of SiO_2-NPs on leaves contributes to a higher chlorophyll and a lower Cd accumulation in rice [81].

7.3 NPs' Utilizations Against Biotic Stresses

Nanotechnology has changed agricultural pest management by creating nanopesticides, which use nanoparticles to deliver pesticides in a controlled manner that allows for exact pest targeting [84–86]. This creative method optimizes resource utilization, reduces the negative ecological effects, and increases pest control effectiveness. Nanoparticles' tiny size, target specificity, and improved interaction with crops in regulated conditions allow them to effectively target certain pests [87]. MDNPs (Microbial-derived NPs) have demonstrated encouraging outcomes in controlling pests and fostering plant development [88]. Their insecticidal qualities and capacity to defend plants from phytophagous arthropods were demonstrated [89]. The creation of intelligent and ecologically conscious pesticide nanoformulations enhances pesticide targeting, lowers toxicity, limits seepage and runoff into water bodies, and lowers total pesticide use [90, 91]. Additionally, after surface chemical alteration, they exhibit antimicrobial activity, which can shorten the pesticide's half-life and encourage seed germination and growth of seedlings.

Items such as nanocapsules and nanoemulsions have become effective formulations of nanopesticides. These formulations expand the period of pest exposure and improve treatment efficacy by encapsulating active chemicals in nanocarriers, which enable progressive and prolonged release [92]. The health of non-target creatures is preserved and issues with soil and water quality are addressed as a result of the decreased use of pesticides [93]. Their distinct physicochemical characteristics enable the regulated release of active chemicals, guaranteeing a more successful method of pest control [92].

8 Nanotoxicity

Although NMs have been shown to have good benefits, they can also be phytotoxic based on the concentration, kind, exposure circumstances, and length of usage [94, 95]. The development of roots and stems is inhibited and the photosynthetic rate of plants is decreased by CuO-NPs, which are harmful to *H. sativum* [96]. When NPs change in plants, they harm the structure of the cells and lessen nutritional uptake and transport [97]. The germination percentage of ryegrass dropped by 6% after 30 days of exposure to PLA-NPs [98]. However, onions and wheat seeds treated with PS-NPs over 72 h, showed no change in germination percentage [98, 99]. Long-term exposure to polymer NPs can therefore reduce the rate of germination, mainly because the particles' blockage of pores will make it more difficult for the seedlings to absorb nutrients. ENMs were found to limit seed germination, root and shoot growth, and decrease chlorophyll and photosynthetic concentrations in numerous crops such as onion, radish, spinach, coriander, soybeans, barley, lettuce, wheat, cucumber, and rice [100, 101]. The smaller CuO-NPs (25 nm diameter) were more phytotoxic to soybeans [102]. Although there is contradictory evidence on the eco-toxicological effects of nanoparticles, most scientific research shows mild to average toxicity to terrestrial vegetation [103]. Exposure to metallic nanoparticles can cause oxidative damage in plants by triggering antioxidant defense systems and generating ROS (reactive oxygen species) [104]. The antioxidant defense system, which reduces free radicals, is composed of a variety of non-enzymatic and enzymatic antioxidants, including ascorbate peroxidase, superoxide dismutase, guaiacol peroxidase, glutathione reductase, and catalase [105]. Plants have established antioxidant capacity to mitigate the oxidative stress caused by nanoparticles [106]. Several NMs stimulate the antioxidant enzymes. Catalase is produced by CeO_2, Fe_2O_4, and CO_3O_4, while guaiacol peroxidase is produced by Fe_3O_4, CeO_2, MnO_2, CuO, and Au and Superoxide dismutase is produced by Pt, CeO_2, and fullerene [58]. NPs have the capacity to hinder seed germination, delay the growth and development of plants, and, on rare occasions, cause the plants to die through stimulating biochemical, physiological, and molecular mechanisms such as impairment of genomic DNA, bio-membranes and decrease in growth hormones, and photosynthetic rate, interferences in water transport and gene expression [5]. The application of fullerene NPs interfered with the energy and electron transport routes by inhibiting gene transcription [107]. However, certain nanoparticles enhanced the expression of several genes, including those connected to water channels and stress [58]. MWCNTs were implemented to upregulate the genes *NtLRX1*, *NNtPIP1*, and *CycB*, which control cell wall construction, water transport, and cell division, respectively [108].

Food safety, resilience to environmental change, and sustainable agricultural products are all greatly enhanced by the use of nanotechnology in farming [109]. CDs (Carbon dots) are an important nanomaterial in this context because of their special qualities, which include biocompatibility, adjustable photoluminescence, and low toxicity. CDs are suitable for use in agriculture because of their exceptional biocompatibility. Their capacity to accelerate photosynthesis and optimize solar energy usage, which raises the growth and output of crops. Additionally, the hydrophilicity of CDs promotes faster seed germination and plant water uptake, which benefits plant growth. Their antioxidant and antibacterial qualities also support plant health by reducing biotic and abiotic stressors. Additionally, CDs are used in agricultural sensors to facilitate the production of smart sensors and enable the recognition of pesticides and herbicides [110]. Furthermore, horticulture may benefit from nanotechnology by increasing fruit and vegetable production, quality, and shelf life. Crop growth, disease monitoring, and fertility can be improved with the use of targeted and effective nanosensors, nanopesticides, and nanofertilizers [111]. However, more investigation is required to comprehend the possible negative consequences of certain nanopesticides [109]. A revolutionary strategy with broad advantages for raising production, quality, and resilience in dealing with changing problems is the application of nanotechnology in agriculture.

9 Conclusions

This chapter discusses how NPs are absorbed and distributed by plants, as well as how they affect plant yield. Leaves absorb NPs through stomata and pores of stratum corneum, whereas roots absorb them through main roots, pores of cell walls, and damaged regions. These NPs are moved across various plant tissues via the apoplastic and symplastic channels. The process of absorption, movement and distribution is determined by the physical characteristics of NPs, including size, surface charge, and chemical composition. In addition, NPs have diverse impacts on plant development and stress response. NPs, in general, play a major role in effective and focused pest management, supporting long-term crop protection strategies. Finally, the impact of environmental variables on the stability and shape of NPs in natural settings should be considered in the real agricultural production process.

References

1. Wang X, Xie H, Wang P, Yin H (2023) Nanoparticles in plants: uptake, transport and physiological activity in leaf and root. Materials 16(8):3097. https://doi.org/10.3390/ma16083097
2. Jat SK, Bhattacharya J, Sharma MK (2020) Nanomaterial based gene delivery: a promising method for plant genome engineering. J Mater Chem B 8:4165–4175. https://doi.org/10.1039/D0TB00217H
3. Ahmed T, Masood HA, Noman M, Al-Huqail AA, Alghanem SM, Khan MM, Muhammad S, Manzoor N, Rizwan M, Qi X (2023) Biogenic silicon nanoparticles mitigate cadmium (Cd) toxicity in rapeseed (*Brassica napus* L.) by modulating the cellular oxidative stress metabolism and reducing Cd translocation. J Hazard Mater 459:132070. https://doi.org/10.1016/j.jhazmat.2023.132070
4. Li WQ, Qing T, Li CC, Li F, Ge F, Fei JJ, Peijnenburg WJGM (2020) Integration of subcellular partitioning and chemical forms to understand silver nanoparticles toxicity to lettuce (*Lactuca sativa* L.) under different exposure pathways. Chemosphere 258:127349. https://doi.org/10.1016/j.chemosphere.2020.127349
5. Usman M, Farooq M, Wakeel A, Nawaz A, Cheema SA, ur Rehman H, Ashraf I, Sanaullah M (2020) Nanotechnology in agriculture: current status, challenges and future opportunities. Sci Total Environ 721:137778. https://doi.org/10.1016/J.SCITOTENV.2020.137778
6. Khan I, Awan SA, Rizwan M, Hassan ZU, Akram MA, Tariq R, Brestic M, Xie W (2022) Nanoparticle's uptake and translocation mechanisms in plants via seed priming, foliar treatment, and root exposure: a review. Environ Sci Pollut Res 29(89823–89):833
7. Vega-Vásquez P, Mosier NS, Irudayaraj J (2020) Nanoscale drug delivery systems: from medicine to agriculture. Front Bioeng Biotechnol 8:79
8. Grillo R, Mattos BD, Antunes DR, Forini MML, Monikh FA, Rojas OJ (2021) Foliage adhesion and interactions with particulate delivery systems for plant nanobionics and intelligent agriculture. Nano Today 37:101078
9. Singh H, Sharma A, Bhardwaj SK, Arya SK, Bhardwaj N, Khatri M (2021) Recent advances in the applications of nano-agrochemicals for sustainable agricultural development. Environ Sci Process Impacts 23:213–239. https://doi.org/10.1039/D0EM00404A
10. Farooq MA, Hannan F, Islam F, Ayyaz A, Zhang N, Chen W, Zhang K, Huang Q, Xu L, Zhou W (2022) The potential of nanomaterials for sustainable modern agriculture: present findings and future perspectives. Environ Sci Nano 9:1926–1951. https://doi.org/10.1039/D1EN01124C
11. Kah M, Tufenkji N, White JC (2019) Nano-enabled strategies to enhance crop nutrition and protection. Nat Nanotechnol 14:532–540. https://doi.org/10.1038/S41565-019-0439-5
12. Noman M, Ahmed T, Shahid M, Nazir MM, Li D, Song F (2023) Nano-enabled crop resilience against pathogens: potential, mechanisms and strategies. J Adv Res. https://doi.org/10.1016/j.jare.2023.06.011
13. Pal A, Kaur P, Dwivedi N, Rookes J, Bohidar HB, Yang W, Cahill DM, Manna PK (2023) Clay-Nanocomposite based smart delivery systems: a promising tool for sustainable farming, ACS Agric. Sci Technol 3:3–16. https://doi.org/10.1021/acsagscitech.2c00140
14. Avellan A, Yun J, Zhang Y, Spielman-Sun E, Unrine JM, Thieme J, Li J, Lombi E, Bland G, Lowry GV (2019) Nanoparticle size and coating chemistry control foliar uptake pathways, translocation, and leaf-to-rhizosphere transport in wheat. ACS Nano 13:5291–5305
15. Hu P, An J, Faulkner MM, Wu H, Li Z, Tian X, Giraldo JP (2020) Nanoparticle charge and size control foliar delivery efficiency to plant cells and organelles. ACS Nano 14:7970–7986
16. Ha N, Seo E, Kim S, Lee SJ (2021) Adsorption of nanoparticles suspended in a drop on a leaf surface of *Perilla frutescens* and their infiltration through stomatal pathway. Sci Rep 11:11556
17. Li L, Luo Y, Li R, Zhou Q, Peijnenburg WJGM, Yin N, Yang J, Tu C, Zhang Y (2020) Effective uptake of submicrometre plastics by crop plants via a crack-entry mode. Nat Sustain 3:929–937

18. Lv J, Christie P, Zhang S (2019) Uptake, translocation, and transformation of metal-based nanoparticles in plants: recent advances and methodological challenges. Environ Sci Nano 6:41–59
19. Rajput V, Minkina T, Mazarji M, Shende S, Sushkova S, Mandzhieva S, Burachevskaya M, Chaplygin V, Singh A, Jatav H (2020) Accumulation of nanoparticles in the soil-plant systems and their effects on human health. Ann Agric Sci 65:137–143
20. Ali S, Mehmood A, Khan N (2021) Uptake, translocation, and consequences of nanomaterials on plant growth and stress adaptation. J Nanomater 2021:6677616
21. Larue C, Castillo-Michel H, Sobanska S, Trcera N, Sorieul S, Cécillon L et al (2014) Fate of pristine TiO_2 nanoparticles and aged paint-containing TiO_2 nanoparticles in lettuce crop after foliar exposure. J Hazard Mat 273:17–26. https://doi.org/10.1016/j.jhazmat.2014.03.014
22. Zhu J, Li J, Shen Y, Liu S, Zeng N, Zhan X, White JC, Gardea-Torresdey J, Xing B (2020) Mechanism of zinc oxide nanoparticle entry into wheat seedling leaves. Environ Sci Nano 7:3901–3913
23. Eichert T, Kurtz A, Steiner U, Goldbach HE (2008) Size exclusion limits and lateral heterogeneity of the stomatal foliar uptake pathway for aqueous solutes and water-suspended nanoparticles. Physiol Plant 134:151–160. https://doi.org/10.1111/j.1399-3054.2008.01135.x
24. Cai L, Cai L, Jia H, Liu C, Wang D, Sun X (2020) Foliar exposure of Fe_3O_4 nanoparticles on *Nicotiana benthamiana*: evidence for nanoparticles uptake, plant growth promoter and defense response elicitor against plant virus. J Hazard Mater 393:122415. https://doi.org/10.1016/j.jhazmat.2020.122415
25. El-Shetehy M, Moradi A, Maceroni M, Reinhardt D, Petri-Fink A, Rothen-Rutishauser B, Mauch F, Schwab F (2021) Silica nanoparticles enhance disease resistance in *Arabidopsis* plants. Nat Nanotechnol 16:344–353
26. Lian J, Liu W, Meng L, Wu J, Chao L, Zeb A, Sun Y (2021) Foliar-applied polystyrene nanoplastics (PSNPs) reduce the growth and nutritional quality of lettuce (*Lactuca sativa* L.). Environ Pollut 280:116978
27. Jia-Yi Y, Meng-Qiang S, Zhi-Liang C, Yu-Tang X, Hang W, Jian-Qiang Z, Ling H, Qi Z (2022) Effect of foliage applied chitosan based silicon nanoparticles on arsenic uptake and translocation in rice (*Oryza sativa* L.). J Hazard Mater 433:128781
28. Carlson C, Hussain SM, Schrand AM, Braydich-Stolle LK, Hess KL, Jones RL et al (2008) Unique cellular interaction of silver nanoparticles: size-dependent generation of reactive oxygen species. J Phys Chem B 112(13608–13):619. https://doi.org/10.1021/jp712087m
29. Torney F, Trewyn BG, Lin VSY, Wang K (2007) Mesoporous silica nanoparticles deliver DNA and chemicals into plants. Nat Nanotech 2:295–300. https://doi.org/10.1038/nnano.2007.108
30. Roberts A, Oparka KJ (2003) Plasmodesmata and the control of symplastic transport. Plant Cell Environ 26:103–124. https://doi.org/10.1046/j.1365-3040.2003.00950.x
31. Sattelmacher B (2001) The apoplast and its significance for plant mineral nutrition. New Phytol 149:167–192. https://doi.org/10.1046/j.1469-8137.2001.00034.x
32. Dietz KJ, Herth S (2011) Plant nanotoxicology. Trends Plant Sci 16:582–589. https://doi.org/10.1016/j.tplants.2011.08.003
33. Juárez-Maldonado A, Ortega-Ortíz H, Morales-Díaz AB, González-Morales S, Morelos-Moreno Á, Sandoval-Rangel A et al (2019) Nanoparticles and nanomaterials as plant biostimulants. Int J Molec Sci 20:162. https://doi.org/10.3390/ijms20010162
34. Hotze EM, Phenrat T, Lowry GV (2010) Nanoparticle aggregation: challenges to understanding transport and reactivity in the environment. J Environ Qual 39:1909–1924. https://doi.org/10.2134/jeq2009.0462
35. Sun X-D, Yuan X-Z, Jia Y, Feng L-J, Zhu F-P, Dong S-S, Liu J, Kong X, Tian H, Duan J-L et al (2020) Differentially charged nanoplastics demonstrate distinct accumulation in *Arabidopsis thaliana*. Nat Nanotechnol 15:755–760
36. Bosker T, Bouwman LJ, Brun NR, Behrens P, Vijver MG (2019) Microplastics accumulate on pores in seed capsule and delay germination and root growth of the terrestrial vascular plant *Lepidium sativum*. Chemosphere 226:774–781

37. Parkinson SJ, Tungsirisurp S, Joshi C, Richmond BL, Gifford ML, Sikder A, Lynch I, O'Reilly RK, Napier RM (2022) Polymer nanoparticles pass the plant interface. Nat Commun 13:7385
38. Sun L, Song F, Zhu X, Liu S, Liu F, Wang Y, Li X (2021) Nano-ZnO alleviates drought stress via modulating the plant water use and carbohydrate metabolism in maize. Arch Agron Soil Sci 67:245–259
39. Zhu J, Wang J, Zhan X, Li A, White JC, Gardea-Torresdey JL, Xing B (2021) Role of charge and size in the translocation and distribution of zinc oxide particles in wheat cells. ACS Sustain Chem Eng 9(11556–11):564
40. Zeng Y, Himmel ME, Ding SY (2017) Visualizing chemical functionality in plant cell walls. Biotechnol Biofuel 10:1–16. https://doi.org/10.1186/s13068-017-0953-3
41. Zhu ZJ, Wang H, Yan B, Zheng H, Jiang Y, Miranda OR et al (2012) Effect of surface charge on the uptake and distribution of gold nanoparticles in four plant species. Environ Sci Technol 46:12391–12398. https://doi.org/10.1021/es301977w
42. Liu M, Feng S, Ma Y, Xie C, He X, Ding Y, Zhang Z (2019) Influence of surface charge on the phytotoxicity, transformation, and translocation of CeO2 nanoparticles in cucumber plants. ACS Appl Mater Interf 11:16905–16913
43. Milewska-Hendel A, Zubko M, Stróz D, Kurczyńska EU (2019) Effect of nanoparticles surface charge on the *Arabidopsis thaliana* (L.) roots development and their movement into the root cells and protoplasts. Int J Molec Sci 20:1650. https://doi.org/10.3390/ijms20071650
44. Zhang H, Goh NS, Wang JW, Pinals RL, González-Grandío E, Demirer GS, Butrus S, Fakra SC, Del Rio FA, Zhai R et al (2022) Nanoparticle cellular internalization is not required for RNA delivery to mature plant leaves. Nat Nanotechnol 17:197–205
45. Sun H, Wang M, Wang J, Wang W (2022) Surface charge affects foliar uptake, transport and physiological effects of functionalized graphene quantum dots in plants. Sci Total Environ 812:151506
46. Spielman-Sun E, Avellan A, Bland GD, Tappero RV, Acerbo AS, Unrine JM et al (2019) Nanoparticle surface charge influences translocation and leaf distribution in vascular plants with contrasting anatomy. Environ Sci Nano 6:2508–2519. https://doi.org/10.1039/C9EN00626E
47. Larue C, Veronesi G, Flank AM, Surble S, Herlin-Boime N, Carrière M (2012) Comparative uptake and impact of TiO_2 nanoparticles in wheat and rapeseed. J Toxicol Environ Health A 75:722–734. https://doi.org/10.1080/15287394.2012.689800
48. Nadiminti PP, Dong YD, Sayer C, Hay P, Rookes JE, Boyd BJ et al (2013) Nanostructured liquid crystalline particles as an alternative delivery vehicle for plant agrochemicals. ACS Appl Mat Interf 5:1818–1826. https://doi.org/10.1021/am303208t
49. Avellan A, Schwab F, Masion A, Chaurand P, Borschneck D, Vidal V, Rose J, Santaella C, Levard C (2017) Nanoparticle uptake in plants: gold nanomaterial localized in roots of arabidopsis thaliana by x-ray computed nanotomography and hyperspectral imaging. Environ Sci Technol 51:8682–8691
50. Schwab F, Zhai G, Kern M, Turner A, Schnoor JL, Wiesner MR (2016) Barriers, pathways and processes for uptake, translocation and accumulation of nanomaterials in plants–critical review. Nanotoxicology 10:257–278. https://doi.org/10.3109/17435390.2015.1048326
51. Navarro DA, Bisson MA, Aga DS (2012) Investigating uptake of water-dispersible CdSe/Zn Squantum dot nanoparticles by *Arabidopsis thaliana* plants. J Hazard Mater 211:427–435. https://doi.org/10.1016/j.jhazmat.2011.12.012
52. Feng Y, Cui X, He S, Dong G, Chen M, Wang J et al (2013) The role of metal nanoparticles in influencing arbuscular mycorrhizal fungi effects on plant growth. Environ Sci Technol 47:9496–9504. https://doi.org/10.1021/es402109n
53. Cañas JE, Long M, Nations S, Vadan R, Dai L, Luo M et al (2008) Effects of functionalized and nonfunctionalized single-walled carbon nanotubes on root elongation of select crop species. Environ Toxicol Chem 27:1922–1931. https://doi.org/10.1897/08-117.1
54. Su Y, Ashworth VE, Geitner NK, Wiesner MR, Ginnan N, Rolshausen P et al (2020) Delivery, fate, and mobility of silver nanoparticles in citrus trees. ACSNano 14:2966–2981. https://doi.org/10.1021/acsnano.9b07733

55. Su Y, Ashworth V, Kim C, Adeleye AS, Rolshausen P, Roper C, White J, Jassby D (2019) Delivery, uptake, fate, and transport of engineered nanoparticles in plants: a critical review and data analysis. Environ Sci Nano 6:2311–2331
56. Miralles P, Church TL, Harris AT (2012) Toxicity, uptake, and translocation of engineered nanomaterials in vascular plants. Environ Sci Technol 46:9224–9239
57. Cornu JY, Bussiere S, Coriou C, Robert T, Maucourt M, Deborde C, Moing A, Nguyen C (2020) Changes in plant growth, Cd partitioning and xylem sap composition in two sunflower cultivars exposed to low Cd concentrations in hydroponics. Ecotoxicol Environ Saf 205:111145
58. Tripathi DK, Singh Shweta S, Singh S, Pandey R, Singh VP, Sharma NC, Prasad Dubey NK, Chauhan DK (2017) An overview on manufactured nanoparticles in plants: uptake, translocation, accumulation and phytotoxicity. Plant Physiol Biochemist 110:2–12. https://doi.org/10.1016/J.PLAPHY.2016.07.030
59. dela Rosa G, Vázquez-Núñez E, Molina-Guerrero C, Serafín-Muñoz AH, Vera-Reyes I (2021) Interactions of nanomaterials and plants at the cellular level: current knowledge and relevant gaps. Nanotechnol Environ Eng 6:7
60. Zhu J, Wang J, Chen R, Feng Q, Zhan X (2022) Cellular process of polystyrene nanoparticles entry into wheat roots. Environ Sci Technol 56:6436–6444
61. Azeem I, Adeel M, Ahmad MA, Shakoor N, Jiangcuo GD, Azeem K, Ishfaq M, Shakoor A, Ayaz M, Xu M et al (2021) Uptake and accumulation of nano/microplastics in plants: a critical review. Nanomaterials 11:2935
62. Xiong T, Zhang T, Xian Y, Kang Z, Zhang S, Dumat C, Shahid M, Li S (2021) Foliar uptake, biotransformation, and impact of CuO nanoparticles in *Lactuca sativa* L. var. *ramosa* Hort. Environ Geochem Health 43:423–439
63. Bueno V, Gao X, Rahim AA, Wang P, Bayen S, Ghoshal S (2022) Uptake and Translocation of a silica nanocarrier and an encapsulated organic pesticide following foliar application in tomato plants. Environ Sci Technol 56:6722–6732
64. Govea-Alcaide E, Masunaga SH, De Souza A, Fajardo-Rosabal L, Effenberger FB, Rossi LM et al (2016) Tracking iron oxide nanoparticles in plant organs using magnetic measurements. J Nano Res 18:305. https://doi.org/10.1007/s11051-016-3610-z
65. Ju M, Navarreto-Lugo M, Wickramasinghe S, Milbrandt NB, McWhorter A, Samia AC (2019) Exploring the chelation-based plant strategy for iron oxide nanoparticle uptake in garden cress (*Lepidium sativum*) using magnetic particle spectrometry. Nanoscale 11(18582–18):594. https://doi.org/10.1039/C9NR05477D
66. Bharti AS, Sharma S, Shukla N, Uttam KN (2018) Steady state and time resolved laser-induced fluorescence of garlic plants treated with titanium dioxide nanoparticles. Spectrosco Lett 51:45–54. https://doi.org/10.1080/00387010.2017.1417871
67. Mozhayeva D, Engelhard C (2020) A critical review of single particle inductively coupled plasma mass spectrometry–a step towards an ideal method for nanomaterial characterization. J Anal Atom Spectrom 43:5290–5294. https://doi.org/10.1039/C9JA00206E
68. Keller AA, Huang Y, Nelson J (2018) Detection of nanoparticles in edible plant tissues exposed to nano-copper using single-particle ICP-MS. J Nano Res 20:101. https://doi.org/10.1007/s11051-018-4192-8
69. Wojcieszek J, Jiménez-Lamana J, Bierla K, Asztemborska M, Ruzik L, Jarosz M et al (2019) Elucidation of the fate of zinc in model plants using single particle ICP-MS and ESI tandem MS. J Anal Atom Spectrom 34:683–693. https://doi.org/10.1039/C8JA00390D
70. Nath J, Dror I, Landa P, Vanek T, Kaplan-Ashiri I, Berkowitz B (2018) Synthesis and characterization of isotopically-labeled silver, copper and zinc oxide nanoparticles for tracing studies in plants. Environ Pollut 242:1827–1837. https://doi.org/10.1016/j.envpol.2018.07.084
71. Vaghar MS, Sayfzadeh S, Zakerin HR, Kobraee S, Valadabadi SA (2020) Foliar application of iron, zinc, and manganese nano chelates improves physiological indicators and soybean yield under water deficit stress. J Plant Nutr 43:2740–2756

72. Faizan M, Bhat JA, Chen C, Alyemeni MN, Wijaya L, Ahmad P, Yu F (2021) Zinc oxide nanoparticles (ZnO-NPs) induce salt tolerance by improving the antioxidant system and photosynthetic machinery in tomato. Plant Physiol Biochem 161:132–140
73. Choukri M, Abouabdillah A, Bouabid R, Abd-Elkader OH, Pacioglu O, Boufahja F, Bourioug M (2022) Zn application through seed priming improves productivity and grain nutritional quality of silage corn. Saudi J Biol Sci 29:103456
74. Armin M, Akbari S, Mashhadi S (2014) Effect of time and concentration of nano-Fe foliar application on yield and yield components of wheat. Int J Biosci 4:69–75
75. Tondey M, Kalia A, Singh A, Dheri GS, Taggar MS, Nepovimova E, Krejcar O, Kuca K (2021) Seed priming and coating by nano-scale zinc oxide particles improved vegetative growth, yield and quality of fodder maize (*Zea mays*). Agronomy 11:729
76. Gao G, Tester MA, Julkowska MM (2020) The use of high-throughput phenotyping for assessment of heat stress-induced changes in *Arabidopsis*. Plant Phenom 2020:3723916
77. Nadeem M, Khan R, Afridi K, Nadhman A, Ullah S, Faisal S, Ul Mabood Z, Hano C, Abbasi BH (2020) Green synthesis of cerium oxide nanoparticles (CeO_2 NPs) and their antimicrobial applications: a review. Int J Nanomed 15:5951–5961
78. Thabet SG, Sallam A, Moursi YS, Karam MA, Alqudah AM (2021) Genetic factors controlling TiO_2 nanoparticles stress tolerance in barley (*Hordeum vulgare*) during seed germination and seedling development. Funct Plant Biol 48:1288–1301
79. Elsheery NI, Helaly MN, El-Hoseiny HM, Alam-Eldein SM (2020) Zinc oxide and silicone nanoparticles to improve the resistance mechanism and annual productivity of salt-stressed mango trees. Agronomy 10:558
80. Miao Y, Luo X, Gao X, Wang W, Li B, Hou L (2020) Exogenous salicylic acid alleviates salt stress by improving leaf photosynthesis and root system architecture in cucumber seedlings. Sci Hortic 272:109577
81. Wang Z, Li H, Li X, Xin C, Si J, Li S, Li Y, Zheng X, Li H, Wei X et al (2020) Nano-ZnO priming induces salt tolerance by promoting photosynthetic carbon assimilation in wheat. Arch Agron Soil Sci 66:1259–1273
82. Lian J, Zhao L, Wu J, Xiong H, Bao Y, Zeb A, Tang J, Liu W (2020) Foliar spray of TiO_2 nanoparticles prevails over root application in reducing Cd accumulation and mitigating Cd-induced phytotoxicity in maize (*Zea mays* L.). Chemosphere 239:124794
83. Sharifan H, Ma X, Moore JM, Habib MR, Evans C (2019) Zinc Oxide Nanoparticles Alleviated the bioavailability of cadmium and lead and changed the uptake of iron in hydroponically grown lettuce (*Lactuca sativa* L. var. *Longifolia*). ACS Sustain. Chem Eng 7:16401–16,409
84. Derbalah A, Abdelsalam I, Behiry SI, Abdelkhalek A, Abdelfatah M, Ismail S, Elsharkawy MM (2022a) Copper oxide nanostructures as a potential method for control of zucchini yellow mosaic virus in squash. Pest Manag Sci 78(8):3587–3595. https://doi.org/10.1002/ps.7001
85. Derbalah A, Essa T, Kamel SM, Omara RI, Abdelfatah M, Elshaer A, Elsharkawy MM (2022b) Silver oxide nanostructures as a new trend to control strawberry charcoal rot induced by *Macrophomina phaseolina*. Pest Manag Sci 78:4638–4648
86. Kamel SM, Elgobashy SF, Omara RI, Derbalah AS, Abdelfatah M, El-Shaer A, Al-Askar AA, Abdelkhalek A, Abd-Elsalam KA, Essa T, Kamran M, Elsharkawy MM (2022) Antifungal activity of copper oxide nanoparticles against root rot disease in cucumber. J Fungi 8:911. https://doi.org/10.3390/jof8090911
87. Sumaira M (2022) Prospects of biosynthetically produced nanoparticles in biocontrol of pests and phytopathogens: a review. Int J Innov Sci Technol 4(2):552–563
88. do Espirito Santo Pereira A, Caixeta Oliveira H, Fernandes Fraceto L, Santaella C (2021) Nanotechnology potential in seed priming for sustainable agriculture. Nanomaterials 11:267. https://doi.org/10.3390/nano11020267
89. Dwivedi SA, Nameirakpam L (2022) The role of nanotechnology in insect pest management. In: Environmental management technologies: challenges and opportunities. CRC Press
90. Baliyarsingh B, Pradhan CK (2023) Prospects of plant-derived metallic nanopesticides against storage pests—a review. J Agric Food Res 14:100687. https://doi.org/10.1016/j.jafr.2023.100687

91. Kannan M, Mohan M, Elango K, Govindaraju K, Mani M (2022) Principles and application of nanotechnology in pest management. In: Mani M (ed) Trends in horticultural entomology. Springer, Singapore
92. Bahrulolum H, Nooraei S, Javanshir N, Tarrahimofrad H, Mirbagheri V, Easton A, Ahmadian G (2021) Green synthesis of metal nanoparticles using microorganisms and their application in the agrifood sector. J Nanobiotechnol. 19: 86 https://doi.org/10.1186/s12951-021-00834-3
93. Li Y, Zhang P, Li M, Shakoor N, Adeel M, Zhou P, Guo M, Jiang Y, Zhao W, Lou B, Rui Y (2023) Application and mechanisms of metal-based nanoparticles in the control of bacterial and fungal crop diseases. Pest Manag Sci. 79(1):21-36. doi: 10.1002/ps.7218. Epub 2022 Oct 21. PMID: 36196678
94. Verma SK, Das AK, Patel MK, Shah A, Kumar V, Gantait S (2018) Engineered nanomaterials for plant growth and development: a perspective analysis. Sci Total Environ 630:1413–1435. https://doi.org/10.1016/J.SCITOTENV.2018.02.313
95. Noori A, Donnelly T, Colbert J, Cai W, Newman LA, White JC (2020) Exposure of tomato (*Lycopersicon esculentum*) to silver nanoparticles and silver nitrate: physiological and molecular response. Int J Phytoremediation 22:40–51. https://doi.org/10.1080/15226514.2019.1634000
96. Fedorenko AG, Minkina TM, Chernikova NP, Fedorenko GM, Mandzhieva SS, Rajput VD, Burachevskaya MV, Chaplygin VA, Bauer TV, Sushkova SN et al (2021) The toxic effect of CuO of different dispersion degrees on the structure and ultrastructure of spring barley cells (*Hordeum sativum distichum*). Environ Geochem Health 43:1673–1687
97. Tian L, Shen J, Sun G, Wang B, Ji R, Zhao L (2020) Foliar application of SiO_2 nanoparticles alters soil metabolite profiles and microbial community composition in the pakchoi (*Brassica chinensis* L.) rhizosphere grown in contaminated mine soil. Environ Sci Technol 54:13137–13146
98. Boots B, Russell CW, Green DS (2019) Effects of microplastics in soil ecosystems: above and below ground. Environ Sci Technol 53(11496–11):506
99. Giorgetti L, Spano C, Muccifora S, Bottega S, Barbieri F, Bellani L, Castiglione MR (2020) Exploring the interaction between polystyrene nanoplastics and *Allium cepa* during germination: internalization in root cells, induction of toxicity and oxidative stress. Plant Physiol Biochem 149:170–177
100. Li Y, Cummins E (2021) A semi-quantitative risk ranking of potential human exposure to engineered nanoparticles (ENPs) in Europe. Sci Total Environ 778:146232. https://doi.org/10.1016/J.SCITOTENV.2021.146232
101. Noman EA, Al-Gheethi A, Al-Sahari M, Saphira Radin Mohamed RM, Crane R, Aziz NAA, Govarthanan M (2022) Challenges and opportunities in the application of bioinspired engineered nanomaterials for the recovery of metal ions from mining industry wastewater. Chemosphere:136165. https://doi.org/10.1016/J.CHEMOSPHERE.2022.136165
102. Yusefi-Tanha E, Fallah S, Rostamnejadi A, Pokhrel LR (2020) Particle size and concentration dependent toxicity of copper oxide nanoparticles (CuONPs) on seed yield and antioxidant defense system in soil grown soybean (*Glycinemax* cv. Kowsar). Sci Total Environ 715. https://doi.org/10.1016/J.SCITOTENV.2020.136994
103. Zheng Y, Chen X, Joseph ND, Zhang Y, Chen H, Gao B (2022) Occurrences and impacts of engineered nanoparticles in soils and groundwater. Emerg Contam Soil Groundw Syst:165–204. https://doi.org/10.1016/B978-0-12-824088-5.00005-7
104. Rico CM, Peralta-Videa JR, Gardea-Torresdey JL (2015) Chemistry, biochemistry of nanoparticles, and their role in antioxidant defense system in plants. Nanotechnol Plant Sci Nanoparticles Their Impact Plants 1–17. https://doi.org/10.1007/978-3-319-14502-0_1
105. Yin L, Colman BP, McGill BM, Wright JP, Bernhardt ES (2012) Effects of silver nanoparticle exposure on germination and early growth of eleven wetland plants. PLoS One 7:e47674. https://doi.org/10.1371/JOURNAL.PONE.0047674
106. Wei H, Wang E (2013) Nanomaterials with enzyme-like characteristics (nanozymes):next-generation artificial enzymes. Chem Soc Rev 42:6060–6093. https://doi.org/10.1039/C3CS35486E

107. Hossain Z, Mustafa G, Sakata K, Komatsu S (2016) Insights into the proteomic response of soybean towards Al_2O_3, ZnO, and Ag nanoparticles stress. J Hazard Mater 304:291–305. https://doi.org/10.1016/J.JHAZMAT.2015.10.071
108. Khodakovskaya MV, De Silva K, Biris AS, Dervishi E, Villagarcia H (2012) Carbon nanotubes induce growth enhancement of tobacco cells. ACS Nano 6:2128–2135. https://doi.org/10.1021/nn204643g
109. Wang D, Saleh NB, Byro A, Zepp R, Sahle-Demessie E, Luxton TP, Ho KT, Burgess RM, Flury M, White JC, Su C (2022) Nano-enabled pesticides for sustainable agriculture and global food security. Nat Nanotechnol 17:347–360. https://doi.org/10.1038/s41565-022-01082-8
110. Maholiya A, Ranjan P, Khan R, Murali S, Nainwal RC, Chauhan PS, Sathish N, Chaurasia JP, Srivastava AK (2023) An insight into the role of carbon dots in the agriculture system: a review. Environ Sci Nano 10:959–995. https://doi.org/10.1039/D2EN00954D
111. Rana RA, Siddiqui MN, Skalicky M, Brestic M, Hossain A, Kayesh E, Popov M, Hejnak V, Gupta DR, Mahmud NU, Islam T (2021) Prospects of nanotechnology in improving the productivity and quality of horticultural crops. Horticulturae 7:332. https://doi.org/10.3390/HORTICULTURAE7100332
112. Zhang P, Ma Y, Xie C, Guo Z, He X, Valsami-Jones E, Lynch I, Luo W, Zheng L, Zhang Z (2019) Plant species-dependent transformation and translocation of ceria nanoparticles. Environ Sci-Nano 6:60–67
113. Lopez-Vargas ER, Ortega-Ortiz H, Cadenas-Pliego G, Romenus KD, de la Fuente MC, Benavides-Mendoza A, Juarez Maldonado A (2018) Foliar application of copper nanoparticles increases the fruit quality and the content of bioactive compounds in tomatoes. Appl Sci 8:1020
114. Yang X, Pan H, Wang P, Zhao F-J (2017) Particle-specific toxicity and bioavailability of cerium oxide (CeO_2) nanoparticles to *Arabidopsis thaliana*. J Hazard Mater 322:292–300
115. Garcia-Lopez JI, Nino-Medina G, Olivares-Saenz E, Lira-Saldivar RH, Diaz Barriga-Castro E, Vazquez-Alvarado R, Rodriguez-Salinas PA, Zavala-Garcia F (2019) Foliar application of zinc oxide nanoparticles and zinc sulfate boosts the content of bioactive compounds in habanero peppers. Plants 8:254

Plant-Nanoparticle Interactions: Molecular Insights and Future Directions

Rahul Gogoi, Fung Swrangshee Daimari, Hridesh Harsha Sarma, Madhurjya Protim Borah, Abhisek Rath, Madhumita Barooah, and Sudipta Sankar Bora

Abstract The interaction between plants and nanoparticles (NPs) is a fast-evolving field of study with scope for future advancement in agriculture, environmental remediation, and biotechnology. Depending on their mode of action, nanoparticles may affect plant physiology and development both at the molecular and cellular levels as well as at the whole-plant level. They may help induce growth and better nutrient uptake, along with stress resistance, although there is considerable emerging disquiet with respect to their toxicity and environmental impact. This chapter thus dwells on the molecular basis of how nanoparticles interact with and are taken up by plants, their translocation, and biotransformation within plant tissues. We elaborate on how surface properties, dimensions, geometry, and chemical composition of nanoparticles modify plant responses, welcoming the latest insights from genomic, transcriptomic, and proteomic studies, and the last are focused on some aspects namely effects on plant hormone signaling, reactive oxygen species (ROS) generation, and gene expression regulation. We also discuss some current problems and challenges, such as the standardization of experimental protocols and the mechanisms of nanoparticle translocation in plants. Finally, we talk about future

R. Gogoi · M. Barooah
Department of Agricultural Biotechnology, Assam Agricultural University, Jorhat, Assam, India

F. S. Daimari
Department of Botany, Boroland University, Deborgaon, Assam, India

H. H. Sarma
Department of Agronomy, Assam Agricultural University, Jorhat, Assam, India

M. P. Borah
Department of Biosciences and Bioengineering, Indian Institute of Technology Jammu, Jammu and Kashmir, India

A. Rath
Department of Plant Pathology, Assam Agricultural University, Jorhat, Assam, India

S. S. Bora (✉)
Silchar Regional Forensic Science Laboratory, Ghungur, Silchar, Assam Under the Directorate of Forensic Science, Kahilipara, Guwahati, Assam, India

J.-T. Chen (ed.), *Nano-Delivering for Plant Genetic Engineering*, Smart Nanomaterials Technology, https://doi.org/10.1007/978-981-95-4165-2_10

directions, highlighting the practicalities of recombinant nanoparticles in enterprising agricultural technology and the responsible design of nanomaterials to minimize ecological impact. Thus, the chapter provides a general overview providing molecular understanding and guiding future research aiming at exploiting nanoparticles in plant science innovations.

Keywords Plant-nanoparticle interactions · Uptake and translocation · Reactive oxygen species (ROS) · Hormone signaling · Gene expression · Nanotechnology in plants

1 Introduction

Nanotechnology has recently emerged as a multi-faceted, highly promising tool with significant potential applications in agriculture, which could help guarantee food security on a global scale. The scientific community's concern about how this technology will affect plant biota as it develops has intensified alongside its rapid advancement [95]. Given that engineered nanoparticles (ENPs) or manufactured nanomaterials (MNMs), including nanopowders, nanocarriers, nanotubes, and nanosensors, significantly contribute to the burgeoning field of agricultural science, the remarkable physicochemical qualities of engineered nanoscale materials (ENMs) are explored in precision agriculture to (i) enhance the nutritional quality and agricultural production [69]; (ii) use targeted agrochemical delivery and controlled release [33]; (iii) protect crops against pathogens [70]; (iv) make them more resistant to environmental stresses [39]; and (v) assess soil quality and identify stresses utilizing advanced nanosensors [88]. Alongside their advantages for the targeted crop, it is also viable to dramatically minimize needless interaction with nontarget organisms and leaching into environmental waterways. Due to the technology's prevalence in various consumer goods—including biomedicine, cosmetics, and more—and its inevitable large-scale release into the atmosphere, scientists are increasingly concerned about its effects on other living biota [66]. Moreover, due to their unregulated distribution, MNMs exert both beneficial and detrimental impacts on plant communities, thus necessitating extensive future research. Furthermore, the best possible use of nanotechnology in the agricultural sector has been constrained by a lack of mechanistic understanding of how plants absorb, mobilize, and react biologically to nanoparticles. The main factors that determine the effects of these nanoparticles on plants are their type, size, composition, and specific surface area [8]. Additionally, various factors influence how plants respond to nanoparticles (NPs), including environmental conditions, the type of plant, its age and developmental stage, the amount of time exposed, as well as the characteristics of the NPs themselves, including their size, shape, structure, surface covering, stability, and disintegration [61]. Previous research has extensively examined both the positive (phytotoxic) and negative (nanoparticle) impacts on plants, along with their cumulative implications [108]. For example, titanium dioxide (TiO_2) nanoparticles reduce oxidative stress induced by ultraviolet B radiation [102], promote the conversion of inorganic nitrogen to organic form [106], aid in the assimilation and acquisition of nitrate, and increase the rate of photosynthesis [93]. TiO_2 NPs have the opposite effect on maize plants, slowing their water acquisition and transpiration rate. Similarly, carbon nanotubes (CNTs) are a type of NP that can penetrate the seed coat and help tomato seedlings acquire water, an essential component for germination [98]. Various techniques have been employed to synthesize nanoparticles, which exhibit diverse properties and demonstrate fascinating effects on plant biological systems [79]. Similarly, most studies show that NPs can affect plants' morphology and physiology, but there have been very few attempts to study how these MNMs affect plants at the genetic and physiological levels or to elucidate a comprehensive mode of action they employ. For instance, few research has demonstrated that certain nanoparticles, for example, cerium oxide (CeO_2) NPs and silver (Ag) NPs, can have a gene-level impact on the development of certain plant species, such as soybeans and onions, respectively [48]. Also, adding carbon (C) NPs to organic matter allowed the rice seedlings to flower for a longer period of time [55]. Zinc oxide (ZnO) nanoparticles are toxic to the cortical and epidermal cells of *Lolium perenne* seedlings [54] (Fig. 1).

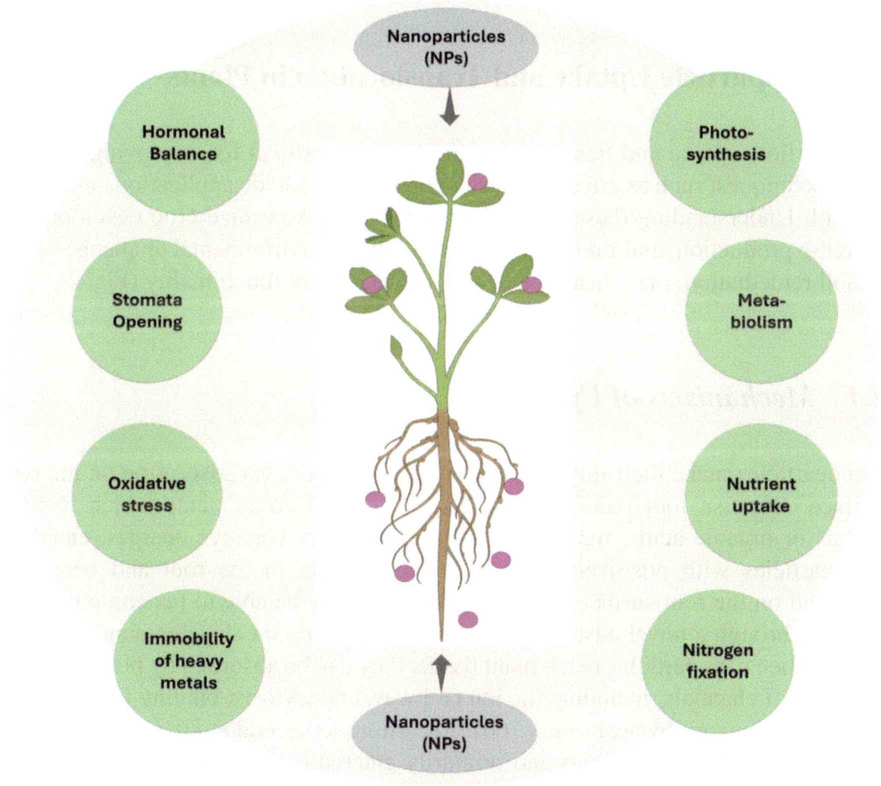

Fig. 1 Schematic representation of the multifaceted effects of nanoparticles (NPs) on plant physiology

This book chapter aims to achieve a comprehensive understanding of how nanoparticles interact with plant systems and how plants respond to these nanoparticles, specifically elucidating the signaling pathways, including nutrient flux, photosynthetic efficiency, and defense responses against oxidative stress, thereby establishing the interactive mechanisms is crucial [71]. In this context, omics studies, such as transcriptomics and proteomics, could evaluate the potential toxicity of nanomaterials in plants as well as concentrate on their effects on genes responsive to diverse biotic and abiotic stresses [35]. This would involve examining the influence on multiple cellular processes in plants, including hormonal signaling, transcriptional regulation and transcript processing, like changes to proteins that occur after translation [6]. On top of that, it will aid researchers in finding valuable information for developing creative tactics to increase agricultural sustainability and crop yields. With this information, genetic approaches can be developed to enhance growth characteristics, nutrient utilization efficiency, and crop stress tolerance. Moreover, knowledge gained from studying post-translational changes brought about by nanoparticles might direct the development of novel protein engineering strategies for agricultural advancement. Moreover, nanoparticles possess significant potential to transform agriculture and tackle global food security challenges.

2 Nanoparticle Uptake and Translocation in Plants

Nanoparticle uptake and translocation in plants are critical for improving agricultural techniques, such as effective nutrient delivery, pesticide application, and stress control. Understanding these processes allows us to maximize crop development, increase production, and find long-term solutions to environmental challenges such as soil remediation, plant health, and holistic ecosystem functionality (Fig. 2).

2.1 *Mechanisms of Uptake*

Nanoparticles make their initial contact with plant roots via adsorption on the root surface. Because root hairs can discharge chemical components which include mucus or organic acids, the surface of the root bears negative charges, enabling nanoparticles with positively charged to accumulate in the root and be easily absorbed on the root surface [60]. Nanoparticles may be able to penetrate the root column through a novel adsorption interface created by the development of lateral roots. When nanoparticles reach plant tissue, they can be absorbed by plant cells via a variety of channels, including the ion pathway, endocytosis, binding to cell membrane proteins, or physical injury [60]. According to several studies, nanoparticles were integrated by plant roots and primarily entered cells through the hydrophilic channel. However, nanoparticles' ability to enter cells through this route is severely restricted because of the tiny pore size [50]. *Catharanthus roseus* root cells had

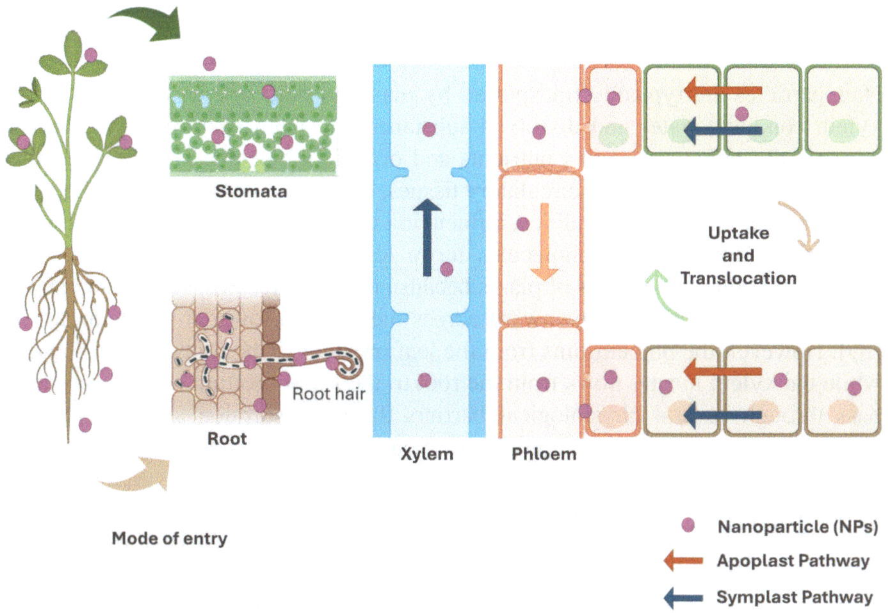

Fig. 2 Illustration of nanoparticle (NP) uptake and translocation in plants

absorbed carbon-based nanoparticles and nanotubes through endocytosis. Nanoparticles can also be absorbed by plants through interactions with transport proteins on the outer layer of the epidermis [57].

Nanoparticle attributes include particle size, chemical makeup, surface charge, and surface. According to current research, nanoparticle uptake in plants is determined by the nanoparticle's composition, plant species, and environment. Generally, there are two methods for solutes to get through the cuticle layer: hydrophilic for polar solutes and lipophilic for nonpolar solutes [21]. Material uptake via the cortex is size-dependent, with NPs smaller than 4.8 nm entering the leaf directly via the cuticular pathway [60]. The absorption activity of nanoparticles on plant leaves can be influenced by their characteristics, such as surface charge, surface modification, particle size, and chemical composition. The size exclusion restriction of nanoparticles in the blade uptake pathway has made nanoparticle size one of the most important factors to consider when examining NP absorption in the blade [50]. The primary target of foliar NP therapy is plant leaves. Plants with higher leaves have a waxy coating of cuticles that acts as a preliminary barrier to NP entry and shields the leaves from water loss [77].

Using transmission electron microscopy (TEM), several studies have demonstrated that NPs are absorbed by stomata and have been found in deeper tissues of a variety of plants, including *Arabidopsis thaliana*, *Allium porrum*, *Cucurbita pepo*, *Lactuca sativa*, and *Citrullus lanatus* [49]. The incorporation of NPs into leaf apoplasts via stomatal pathways results in extended movement across the vascular system.

2.2 Translocation in Plant Tissues

Nanoparticles are typically transported by plants via the xylem and phloem. The xylem conducts water and dissolved substances up from its roots to the leaves, whereas the phloem transfers nutrients and organic molecules in both directions. They can pass through these circulatory tissues, depending on their dimension, electrical charge, and surface qualities, influencing their dispersion and plant contact. It is generally assumed that molecules move long distances through the vascular (xylem and phloem) systems of plants because macromolecules, sugars, and photosynthates are transported downward across the phloem towards shoots and roots [59]. However, the phloem runs from the leaf or shoot to the roots (top to bottom), while the xylem mostly flows from the root to the shoot (bottom to top). NPs must pass through several physiological barriers during absorption and translocation before they may enter the xylem. The root cuticle, cortex, endodermis, epidermis, and Casparian strips are some examples of these root barriers. The vascular system is essentially non-circulatory, which means that material carried by the xylem or phloem cannot return to its original site [59]. The phloem system, thus, could be said to be in charge of NP intake and transmission from leaves to lower parts of the plant. According to a study, CuO NP (20–40 nm) moved through maize's xylem from root to shoot and through the phloem from shoot to root [99]. Since NPs can go from root to shoot or shoot to root through the xylem or phloem, respectively, the aforementioned results unequivocally support this theory. It is still unclear, though, how NPs move between the xylem and phloem. In the rhizosphere, the NPs and NP-products that were transported to the roots may be released, impacting the microbiota [60]. Researchers discovered that polystyrene nano plastic with a size of 93.6 nm could overcome the barrier on lettuce leaves and reach the phloem in plants utilizing TEM [53]. The nanoparticles are subsequently transported through the leaf to the root via vascular and phloem transport pathways. The dimension of the phloem sieve tube is significant, particles as small as 0.405 m are able to penetrate through the phloem [13].

The transport of nanoparticles in plants is influenced by their size, shape, and surface properties. Smaller particles, specialized forms, and surface alterations all influence absorption efficiency, transit through plant tissues, and contact with cellular structures. According to certain research, nanoparticles that are less than 50 nm are frequently delivered by the plastid pathway in plants, but the majority of nanoparticles around 50–200 nm are transported by the apoplast pathway [81]. The nanoparticles are subsequently transported via leaf into the root via the vascular system-phloem transport channels. Previous research discovered that metal nanoparticles that might be carried throughout the plant were smaller in size than nonmetal nanoparticles. This is because larger metal nanoparticles cannot be absorbed. Metal-based nanoparticles with sizes ranging from 30 nm to 50 nm were found in plant leaf phloem and other regions [100]. However, bigger nanoparticles consisting of nonmetal elements were frequently observed to be translocated inside plant leaves [18]. A new investigation looks into the uptake of four distinct kinds of NPs with sizes ranging from 24 nm to 47 nm in watermelon (*Citrullus lanatus* L.) through the foliar route. TEM examination revealed that smaller NPs were taken up by the leaves via stomatal channels and identified in the root and the shoots. They suggested using phloem sieve tubes to allow NPs to pass through shoots and reach roots [100]. The exterior characteristics of nanoparticles have a considerable influence on the endocytosis process. After entering cells, macromolecules can move to other cells via intercellular pores known as plasmodesmata. The transport of nanoparticles through the extracellular pathway is influenced by their surface charge. Since the cell wall is negatively charged due to the carboxyl groups in the pectin, positively charged nanoparticles will be drawn to the cell wall and find it simpler to move via the extracellular channel [12].

2.3 *Biotransformation of Nanoparticles*

Nanoparticle biotransformation in plants is the process by which plant enzymes or metabolites modify nanoparticles to improve their biocompatibility. This process improves nutrient availability, reduces toxicity, and promotes plant growth by enabling essential element uptake, boosting stress tolerance, and aiding in the detoxification of hazardous compounds. The movement and buildup of nanoparticles within plants trigger oxidative damage and genetic toxicity, and the interaction between nanoparticles and plants promotes the production of plant ROS [28]. ZnO NPs and the heavy metals Pb^{2+} and Cd^{2+} were used in a previous study to produce lettuce (*Lactuca sativa* L. var. *Longifolia*) in a hydroponic culture system and assess the levels of Pb and Cd in plant tissue [85]. ZnO NPs were observed to drastically reduce the buildup of lead and cadmium in lettuce roots by 81% and 49%, respectively. The amounts of Pb and Cd in lettuce shoots have decreased by 44% and 30%, respectively. Iron oxide nanoparticles at an amount of 100 mg/L lowered the yield of Bt-transgenic cotton while increasing hormone levels. Hormone levels are reduced in the roots of Bt-transgenic cotton [96]. Moreover, toxic metal ions are created and converted by metal-based nanoparticles in plants, which damage DNA and proteins and obstruct regular cell metabolism [68]. It has been observed that internalized Ag nanoparticles in wheat root tips can inhibit DNA synthesis, interfere with normal cell division, and result in chromosome abnormalities [1]. Additionally, onion cells subjected to PS NPs showed a sharp decline in their mitotic index along with chromosomal and nuclear abnormalities, confirming the genotoxicity of PS NPs and the resulting loss of genomic integrity [64].

3 Plant Reactions to Nanoparticles: Molecular Mechanisms

The effects of nanoparticles on stress physiology and plant development occur via a complex network of molecular pathways. Typically, these pathways manifest through hormone signaling, control of gene expression and reactive oxygen species

(ROS) regulation, which then provoke the plant's biochemical responses and physiological changes.

3.1 Interaction with Plant Hormone Signaling

Nanoparticles have emerged as modulators of plant growth across various physiological and developmental processes, including those governed by essential phytohormones such as auxins, cytokinins, gibberellins (GA), and abscisic acid (ABA). The study concluded that the transport and signaling functions of auxins were altered by silver nanoparticles, mainly due to the suppression of the PIN-FORMED (PIN) and AUX/IAA genes. Collectively, these alterations to the long-distance signal transduction system influence root elongation, lateral root formation, and the overall architecture of plants [4]. Exposure to TiO_2 and ZnO nanoparticles has caused both dicotyledonous and monocotyledonous plants to develop giant roots.

Cytokinin signaling would also respond to the therapy with nanoparticles. Some nanoparticles, especially carbon-based and ZnO NPs, contain the ability to increase the amount of cytokinin, resulting in cell division, shoot development, and inhibition of senescence. However, under such harsh conditions as drought or salinity, treatment with nanoparticles can limit the function of cytokinin pathways to concentrate the growth energy on stress tolerance [9]. The nanoparticle effect is likewise lacking in gibberellin: hormones directing the elongation process of the stem and germination of seeds. For example, TiO_2 NPs were reported to be able to impact both GA biosynthesis genes and hence differentiate the elongation of shoots in wheat seedlings [51].

Nano-priming also increased seedling growth and other physiological features. Abscisic acid (ABA), the primary hormone in abiotic stress response, was shown to rise following exposure to nanoparticles. ZnO and Se nanoparticles improve ABA level and activity of ABA-responsive genes such as RD29 and NCED, which in turn trigger stomatal closure and osmoprotectant production, thus enhancing plant tolerance to drought and salinity [45]. Thus, acknowledging NP's potential and its role in the hormone signaling pathway is an exciting technique to influence plant development and stress tolerance.

3.2 Generation of Reactive Oxygen Species (ROS)

A frequently reported physiological reaction to the introduction of nanoparticles that have been found across plant species is their creation of reactive oxygen species. These reactive oxygen species such as superoxide anions, hydrogen peroxide, and hydroxyl radicals are highly reactive and harmful to the plant when produced in excess. Reactive oxygen species (ROS) are seen as a crucial secondary messenger in the plant signaling and defense machinery [110]. Upon exposure of nanoparticles, the formation of ROS causes the subsequent activation of MAPK cascades. The mitogen-activated protein kinase cascade in stress tolerance serves to trigger the expression of stress-induced gene expression, which can lead to increased stress acclimation within the plant [65].

NPs typically trigger the plant's antioxidant defense mechanism at these sub-toxic amounts. Rice plants exposed to cerium oxide (CeO_2) and iron oxide (Fe_3O_4) nanoparticles showed increased activity of the antioxidant enzymes catalase (CAT), ascorbate peroxidase (APX), peroxidase (POD), and superoxide dismutase (SOD), as a result of this priming action, in addition to maize [72]. Redox equilibrium is re-established and excess ROS are detoxified by these enzymes functioning together. Whether plants suffer from the negative effects of ROS or acclimatize to stress depends on the balance between ROS production and ROS scavenging. For example, in *Pisum sativum* treated with salt stress, zinc oxide nanoparticles (NPs) greatly reduced lipid peroxidation and enhanced chlorophyll, largely via strengthening the antioxidative defense mechanism [34]. Recent studies have demonstrated that ROS created by NPs can cause histone modification and DNA methylation activities and also alter gene expression and plant stress memory over a long period. ROS are not merely the results of stress conditions but generate stress signaling and adaptation [37].

3.3 Gene Expression Regulation

NPs change the expression of genes in plants, affecting those related to growth, development, metabolism, and stress tolerance. Some basic changes in the gene expression terrain of plants treated by various nanoparticles were found in the few studies that examined the transcriptome of these plants. These included changes in the expression of a group of genes linked to signaling, hormone genes that switch and coordinate plant signaling, and genes that participate in the regulation of secondary metabolism [2]. To improve water use efficiency during drought, for example, ZnO and Se nanoparticles were expressed in soybeans, changing the expression levels of genes that respond to drought and other genes that are also involved in plant development mechanisms, such as GmRD20A, GmDREB2, and GmWRKY27 [14].

Apart from genes implicated in stress reactions, nanoparticles also alter genes related to basic metabolism, cellular photosynthesis, and nutrient transport. Using a proteomic analysis of kidney beans (*Phaseolus vulgaris*) treated with CeO_2 NPs, a study found up-regulated proteins related to ion transport, redox control, and energy metabolism that were suggestive of a metabolic shift toward adaptation [84]. Additionally, transcriptome analyses of wheat seedlings exposed to salt stress and pre-treated with Fe_3O_4 NPs showed that genes involved in the ABA and MAPK signaling pathways were upregulated, underscoring the role of NPs in protective responses to environmental stress [7].

The second type of nanoparticle regulation in plants takes place at the general regulatory level of the plant, in addition to the control of a single gene. Numerous omics studies that demonstrate how plants react to nanoparticles, including modifications in the activity of critical pathways including calcium signaling, transcription factors (WRKY, NAC, and MYB families), and epigenetic regulators, have also corroborated the aforementioned. In order to modify their entire metabolism and physiology, the plants react to the signals from the nanoparticles by expressing different genes and proteins. Plant resistance to biotic and abiotic stress may be improved by these effects [104].

4 Benefits of Nanomaterials in Agricultural Systems

4.1 Role of Nanomaterials in Promoting Root Growth, Water Retention, and Nutrient Use Efficiency

Nanotechnology is a cutting-edge subject that offers previously unheard-of answers to urgent problems in a variety of industries, including agriculture [78]. The use of nanotechnology in agriculture has paved the way for new developments in recent years, with the potential to boost crop resilience against environmental pressures, promote plant growth, and guarantee sustainable farming methods. Nanomaterials are distinguished from ordinary materials due to their high ratio of surface area to volume, distinctive physicochemical features, and nanoscale size. Nanomaterials are perfect for applications like nutrient delivery, promoting root growth, and water retention [56]. All phases of the development of plants, from seed germination to mature development, are significantly impacted by nanomaterials. Their distinct physicochemical characteristics allow them to interact with plants during crucial growth phases, improving nutrient metabolism, photosynthetic efficiency, and physiological processes. Because of these benefits, which frequently lead to increased biomass production and general plant health, nanotechnology holds promise as a solution to agricultural problems [24]. Nitric-oxide-releasing polymeric nanoparticles, for instance, have been demonstrated to enhance soybean seed germination and nodule growth, hence improving crop performance [80].

By improving the bioavailability of vital components in soil and enabling their movement inside plant tissues, nanomaterials improve nutrient uptake. For example, ZnO NPs increase the availability of Zn ions, which is necessary for protein synthesis and enzyme activity [40]. Similarly, Fe_2O_3 NPs provide an easily absorbed source of iron, which helps to alleviate iron deficiencies, especially in calcareous soils. These NPs guarantee that plants obtain enough nourishment for optimum growth and development by avoiding conventional obstacles to the absorption of nutrients, such as limited soil mobility and insolubility [87]. To improve nutrient uptake, nanomaterials work in concert with conventional farming methods like irrigation and fertilization. For instance, nanomaterials like ZnO NPs can improve plant nutrition and growth in nutrient-limited environments by increasing the availability of minerals like zinc when applied in combination with fertilizer [82]. Additionally, nanomaterials can be integrated with irrigation systems to improve water retention and nutrient solubility, ensuring more efficient use of fertilizers and water, thereby reducing overall input costs. These synergies help optimize nutrient management and resource utilization, promoting sustainable farming practices [58]. Nanomaterials boost crop productivity by enhancing the efficiency of agricultural commodities to allow for site-specific, regulated nutrient distribution and ensuring the least number of agri-inputs. Indeed, the application of nanotechnology in plant protection products has expanded quickly, which could lead to increased agricultural production. Furthermore, enabling faster plant adaptation to growing changes in climate factors—such as extreme temperatures, water scarcity, salinity, alkalinity, and pollution from the environment with hazardous metals—without endangering already delicate ecosystems is the main concern in agricultural production.

4.2 Pest and Disease Management

Nanopesticides (NPts) have attracted attention as a possible solution to these issues. According to the introduction, NPts possess various benefits compared to traditional pesticides, such as higher efficacy, lower toxicity, selectivity of effect, and controlled gradual release. Due to these properties, NPts are extremely favorable to be used in contemporary agricultural practices. NPts. They consist of stabilizing traditional pesticides in nanoscale sizes through the construction of carriers from them and employing nanoscale metal compounds to act as active components. This innovative approach reduces the negative impact of pesticides but enhances their effectiveness [15]. Nanopesticides are ideal for plant disease control due to their increased responsiveness, controlled absorption, and targeted delivery [107].

Some nanopesticides have been specifically engineered to employ targeted delivery systems to disperse them directly to areas where pests are located. This targeted approach maximizes the pesticide's effectiveness while minimizing its harmful effects on non-target organisms. Systems for controlled pesticide release can be engineered due to the inherent properties of nanomaterials. For instance, the response to environmental parameters such as temperature, pH, humidity, etc., allows for the control of pesticide release required. Besides enhancing pesticide effectiveness, this controlled release system reduces waste, leading to more efficient and sustainable pesticide use. For example, there are two main ways of employing silica nanoparticles as insecticides: spraying them onto crops to form a silica layer that inhibits the growth and penetration of insects and larvae and utilizing mesoporous silica nanoparticles to deliver commercial insecticides [52]. Direct use of silica is more lethal to adults and the eggs they produce because of its drying effect. Moreover, silica nanoparticles can also kill insects by clogging their stomata or trachea [91].

Where pesticides are concerned, the use of nanotechnology presents innovative solutions to conventional farming practices. There is the possibility of enhancing pesticide effectiveness considerably, as well as minimizing adverse environmental impacts, through the incorporation of nanotechnology into the formulation process. Preparations or products consisting of manufactured nanoparticles possessing biocidal activity are referred to as nano-pesticides [5]. There are some different nanopesticide types, such as metal-organic frameworks, lipids, and polymers. Considering pesticide release and delivery, these materials exhibit different processes. Polymer-based nano-pesticides' microstructures could be adjusted to closely regulate pesticide adhesion and release on plant surfaces [107]. Nevertheless, due to the solubility of the lipid layer, lipid-based nanopesticides can enhance durability and stability through the ability of gradual dissolution of pesticides in the water column [25]. In contrast, the highly ordered pore structure of metal-organic frameworks might facilitate the efficient loading of pesticides with slow release. It is established that a high number of nano-pesticide formulations have been more effective than their commercial equivalents [103].

4.3 *Application of Nanomaterials in Improving Drought, Salt, and Temperature Stress Tolerance*

Nanopesticides are effective at reducing stress because of their unique physicochemical properties and nanoscale size (Fig. 3). Through complex interactions with molecular and physiological systems that promote survival in harsh environments, nanomaterials help plants tolerate stress. To establish a strong foundation for stress resilience, these interactions include adjusting gene expression, controlling hormone balance, and modifying oxidative stress responses [113]. Nanomaterials assist plants in adapting to environmental stressors such as salinity, dryness, heavy metal pollution, and extremely high or low temperatures via controlling ROS levels, initiating signaling pathways, and promoting osmolyte accumulation [83]. Despite the obvious advantages of nanomaterials, there is still considerable concern about their long-term ecological effects. To properly assess their environmental impact, studies must be conducted on the ability of nanomaterials to persist in water and soil systems and the potential for them to bioaccumulate in plants and animals [47]. By comprehending these processes, it would be possible to minimize environmental hazards and phytotoxicity while optimizing agriculture's usage of nanotechnology.

4.4 *Nanomaterial-Based Sensors in Precision Agriculture for Soil and Crop Health Monitoring*

The development and application of nanosensors in precision farming have also greatly enhanced human control over the health of plants and soil, as well as quality and safety assurance [16]. Precision farming uses these sensors to assess and

Fig. 3 Overview of the beneficial impacts of nanoparticles (NPs) in agriculture

monitor environmental contamination, crop growth, soil conditions, illnesses, and the usage and penetration of agrochemicals. By offering a greater specific surface area, which is necessary for agricultural systems to be sustainable over the long run, nanomaterial engineering is an innovative study area that supports the growth of high-tech farming industries [73].

5 Potential Risks and Environmental Concerns

Although nanomaterials have a lot of potential to enhance farming methods, there are issues that need to be resolved due to their widespread use. As with any new technology, careful research is needed to determine how it will affect human health and the environment in the long run. Significant concerns include the persistence of nanomaterials in soil, their capacity for bioaccumulation, and their effects on organisms that are not their intended targets, including insects, aquatic life, and soil microbes [44]. To lessen these risks, biodegradable nanoparticles that decompose into non-toxic byproducts must be developed. Additionally, it is crucial to evaluate the adverse environmental effects of nanomaterials using a lifetime approach [3]. To handle the particular risks that nanomaterials present, regulatory frameworks need to be updated. This includes standardizing risk assessments and testing for environmental safety. These risks can be reduced by developing precise application systems like nanosensors and using green synthesis techniques that create biodegradable nanomaterials. Public awareness campaigns should be launched in tandem with these developments to guarantee the responsible and secure application of nanomaterials in farming [44]. Using nanomaterials in agriculture could revolutionize farming practices and offer sustainable solutions to boost crop growth and lessen stress. However, their widespread use presents serious health and environmental issues [38]. Unbalanced ecosystems, harm to organisms that are not the intended target, and contamination of soil and water are examples of unintended consequences that need to be carefully considered [95]. For nanomaterials to be used safely and sustainably, in-depth comprehension of their regulatory frameworks, risk mitigation strategies, and environmental behavior is required. Because of their special qualities, nanoparticles possess the promise of becoming an essential part of the majority of commercial goods, chemical fertilizers, healthcare, and drug development. As of right now, the toxicity of nanoparticles and their effects on human health are well understood. In addition to passing through biological membranes and organs, nanomaterials can enter the circulation by ingestion or inhalation. This might have genotoxic and biochemical effects [26]. Silver nanoparticles may trigger a variety of toxicities, such as DNA damage, developmental risk depending on the size of particles, toxicity to cells, and swelling. Silver nanoparticles may also trigger cellular damage in human lungs, which, because silver is released into the medium, can cause lung cancer [75]. This is the primary cause of the toxic effects of tiny particles of nano pesticides.

One major environmental concern is the buildup of nanomaterials in water systems and agricultural soils. Following application, nanomaterials have the ability to remain in soil matrices and change the microbial communities, nutrient availability, and soil structure [19]. For example, metallic nanoparticles (NPs) such as Ag and ZnO, which are beneficial for the development of plants, can accumulate over time and alter the soil's chemical and physical properties. These NPs have the potential to upset microbial ecosystems, which would have a detrimental effect on the microorganisms necessary for the decomposition of organic matter and the cycling of nutrients [10]. Additionally, certain nanoparticles may influence rhizosphere-associated bacteria linked to plants and promote advantageous microbial activity, enhancing the availability of nutrients and soil health [67]. Additionally, water bodies may become contaminated by runoff that contains nanomaterials, which could cause aquatic ecosystems to bioaccumulate nanoparticles. Through the food chain, this endangers human health in addition to aquatic life [62]. Nanomaterials can be harmful to the environment non-target organisms and phytotoxic to plants, depending on their concentration and composition. When plants are exposed to high quantities of metallic nanoparticles (NPs), such as Ag and Cu, they may suffer from oxidative stress, which slows growth and destroys cells [105]. In addition to their immediate effects on plants, nanoparticles can also affect higher trophic levels by entering the food chain through plant tissues and soil fauna [111]. In sensitive plant species, for example, carbon-based NPs prevent root elongation and seed germination, highlighting the necessity of species-specific toxicity evaluations. These unexpected outcomes emphasize how crucial it is to optimize dosage and thoroughly assess nanomaterial formulations in order to reduce ecological risks [74].

5.1 Phytotoxicity and Ecotoxicity

5.1.1 Toxicological Impacts on Plants and Non-target Organisms (Microbes, Insects, etc.)

Given that nanopesticides are a worldwide technology, nanoinformatics is a new and essential method for international harmonization of safety-by-design, environmental risk assessment, and Nano pesticide management in the direction of environmentally conscious growth and governance [17]. Based upon their kind, dimensions, quantity, and exposure route, these particles can have significant toxicological impacts on organisms alongside other non-target organisms, such as microbes and insects, primarily by interfering with the functioning of cells through mechanisms like oxidative stress, membrane damage, as well as disruption with vital functions. At high concentrations, they can also stunt plant growth, disrupt microbial communities, and kill insects [109].

In addition, the shape of the nanoparticle also plays a role in its toxicity; computer-based calculations revealed that both form and shape. NPS translocation across cell membranes can be accelerated by up to 60 times when charged [112]. Nanorod-shaped ZnO nanoparticles are more harmful to human lung epithelial cells than spherical ones. Likewise, while the gold nanoparticles serve the same purpose, their shapes vary, resulting in varying levels of toxicity. The sphere-shaped gold nanoparticles are more dangerous than the rod-shaped ones, perhaps because the spherical shape releases the beneficial chemical faster [20]. Human health is seriously endangered by nanopesticides that enter the targeted pathogen's habitat. These insecticides contain nanoparticles that can enter the body through the nose or respiratory system and travel to the lungs. The nasopharyngeal area has the ability to trap nanoparticles that are 10–20 nm in size [32]. If a particle has a comparable substance and crystalline form, its toxicity will increase with its size. It is possible for small nanoparticles to induce swelling, proliferation of cells, a condition called and eventually development of tumors [22]. Many kinds of nanoparticles increase the production of virus receptors in the body leading to excessive inflammation and hyperactivity of macrophages. But other nanoparticles, including SiO_2 and TiO_2, also lessen the expression of bacterial and viral receptors, which makes the body less resistant to germs and viruses [63].

According to a study, the following generation of earthworms responds to a solitary toxicant quicker than an amalgamation of poisonous substances [27]. In this study, two generations of earthworms were used to test a combination of chlorpyrifos in addition to zinc oxide (both bulk and nanoparticle) in both natural and artificial soil. The kind of soil, the size of the employed nanoparticles, and the earthworm generation all affected toxicity. They found that *Eisenia andrei* development and reproduction were more negatively impacted by zinc oxide nanoparticles in combination with chlorpyrifos compared to its application alone. On the other hand, nanoformulation has focused attention on the preparation of these formulations and research on their activities in the soil. Atrazine nano-formulation was found to be less toxic when applied to the soil of *Enchytraeus crypticus* (Oligochaeta) as compared to atrazine [30].

5.1.2 Impact on Soil Health and Ecosystem Balance

Since nanopesticides amplify the effect of common pesticides by virtue of the regulated release in a targeted pathogen environment, an increase in resistance to pathogens for pests and diseases together with a decline in biodiversity in addition to contamination of soil as well as water, has emerged following the exposure of up to 99.9% of nanopesticide to the surroundings where there will be an intended targeted pathogen. Barely a percent reaches the desired intended pathogen [15].

Many industrial, residential, agricultural, and biological applications are well suited for nano-enabled products made of a few nano-scale (1–100 nm) materials. Due to their special qualities, such as an extremely high surface area relative to volume ratio, greater reactivity, exterior potential, adjustable physical/chemical characteristics, biological manipulation, etc., in comparison with conventional salt and bulk materials, nanoparticles can and are preferred for larger applications [46]. Due to overproduction, inappropriate use, and misuse, nanoparticles have quickly been released into various compartments of the environment [31]. The effects of nanoparticles on soil organisms will depend on the complexity and characteristics of the soil. These include its capacity to act as a buffer, organic matter found in nature, the aggregation and immobilization of nanoparticles, their deposition, and the creation of environmental corona that encompasses nanoparticles [111]. Soil has been shown to negatively impact crop yields [76], microbes within the soil, and soil fertility [23].

The use of nanoparticles is at a level never before known globally, and nowadays, nanomaterials replace the normal materials used in almost all scientific disciplines. To prevent the sudden emission of NPs into the environment and its ecosystem, proper usage and disposal are very much in demand. These NPs can be transported between different food webs within an ecosystem after coming into contact with the environment, leading to bioaccumulation that could negatively impact its dependents [43]. One of the major pathways through which NPs could gain entry into various natural environments is via the aquatic environment. Only after making sure that there is an integrated study of different physicochemical properties of NPs and nano-bio interactions will it be confirmed if the nanomaterial is safe to be applied for other uses [94].

5.2 Long-Term Environmental Effects

There are still concerns about potential negative effects on human and environmental health due to the expanding uses of nanotechnology in numerous sectors of the global economy, including agriculture [41]. The intentional use of nanomaterials in agricultural operations may actually have unpredicted health effects [29, 42]. The exposure of humans and the environment to nanomaterial remnants in soil and crops is expected to rise in this scenario due to exposure pathways such as possible biological accumulation in the food chain and the environment. Thus, new advantages as well as new risks to the health of people and the environment could arise from the expected creative and improved activities of nano-enhanced applications. This perspective holds that sustainable agriculture requires the development of "green nanotechnology," a conceptual approach that strikes a balance between the benefits of nano-products to tackle ecological problems and the assessment and management of the possible hazards to safety, health, and the environment posed by nanoscale materials. It is also important to evaluate possible hazards associated with exposure to nanomaterials from a suitably customized life-cycle perspective. This means taking into account every step that could lead to the discovery of nano-solutions, such as field application, potential integration of food supplies, product disposal or recycling, and the possible impact of particular agro-system circumstances that could all affect the risk characterization and hazardous properties of nanomaterials [38].

6 Challenges and Current Issues

Studies investigating the long-term impacts of particular nanoparticles, like silver nanoparticles, on the elimination of phosphorus in artificial wetlands [36], and the change of zero-valent iron nanoparticles in soils [101], has made great strides. However, there is still a requirement for additional studies to examine the effects of various nanoparticles, including combinations of them, on the surrounding environment, microbial populations, and soil health. To completely recognise the wide-ranging environmental effects and ecotoxicity of exposure to nanoparticles in agricultural systems, thorough long-term investigations are essential, despite the fact that some research has focused on immediate impacts. To ensure the sustainability of using nanoparticles in agriculture, more research is also needed to assess how they affect non-target creatures, such as insects that are beneficial, pollinators, and soil organisms [89]. The fate and behaviour of nanoparticles in agricultural ecosystems are still not well understood. The transformation, transport, and destiny of nanoparticles in different types of soil should be the focus of research, with consideration given to soil properties, organic matter concentration, and irrigation techniques. However, maximising the way that nanoparticles interact with soil components will improve their application methods and make it easier to assess any possible environmental risks [90]. Additionally, the present research has demonstrated that plants are capable of uptaking nanoparticles; however, a more thorough comprehension is necessary to understand their accumulation patterns, translocation paths, and absorption mechanisms across different plant species. Moreover, more studies must examine the possible consequences of nanoparticle buildup in consumable plant components as well as the hazards linked to human ingestion [11]. Future studies must also look at how metal nanoparticles migrate, change, and settle in different soil particle sizes because they may have diverse effects on plants even at the same concentration.

However, there is a lack of uniformity and standardisation in the synthesis, characterisation, and reporting of nanoparticles used in agricultural research. This complicates the process of comparing and replicating research results from various prior studies. The reliability and repeatability of research findings can be improved by the creation of uniform procedures for nanoparticle synthesis, characterisation, and reporting [97]. Soil microbes like rhizobia and mycorrhizal fungi are essential to plant nutrition uptake and health, and nanoparticles may interact with them. However, how nanoparticles affect these interactions between plants and microbes still remains inadequately comprehended. To determine how nanoparticles influence soil microbiome composition and symbiotic relationships, additional studies are required [92].

Furthermore, it is critical to think about the socioeconomic ramifications and public perceptions about the application of nanoparticles in farming as research into this area advances. If nanoparticle-based technologies are to be effectively used in agriculture, it is essential to study their market acceptability, potential regulatory implications, and economic viability [86]. The potential benefits, risks, and limitations of nanoparticles for agricultural use will be better understood if these research gaps are filled. Additionally, it will pave the way for further study, into nanoparticles that can solve important problems with agricultural output, resource management, and environmental preservation in a way that is safe, effective, and long-lasting.

7 Conclusion and Future Prospects

Nowadays, the application of nanomaterials signifies a promising avenue towards a paradigm shift in sustainable agriculture, while also addressing food security issues and preserving environmental integrity. With the potential to completely transform farming methods on farms, our in-depth research has yielded important insights into the fundamental mechanisms and versatility of nanoparticles. More and more, plants and microbes, along with their extracts, are increasingly being used as natural green nano factories for future nanoscale technologies and the manufacturing of metal and metal oxide nanoparticles, offering enhanced control over their structural characteristics (shape, size, etc.) and broader applications compared to chemically synthesized counterparts. Controlled-release nanofertilizers enhanced nutrient management by facilitating accurate distribution and maximizing plant nutrient utilization efficiency, thereby reducing dependence on conventional fertilizers, minimizing environmental impacts, and adhering to sustainable agriculture principles. Moreover, metal nanoparticle-based nano-pesticides offer a sustainable alternative to chemical pesticides for disease control, lowering the environmental risks associated with them; the integration of nano-sensors enables real-time monitoring of crop health and environmental conditions, improving precision farming and optimizing resource utilization; the use of metal nanoparticles in nano-remediation methods holds great potential for improving soil and water quality, mitigating the negative impacts of pollutants in agricultural systems; and although these nanomaterials are considered transformative agents for plants, their possible long-term impacts on agricultural ecosystems, the surroundings, and human health are not well understood, exposing important research gaps that need to be addressed. Moreover, there is reason to be optimistic about the prospect of achieving sustainable agriculture through the interactions between plants and NPs, particularly in the areas of on-site pathogen detection, as well as the use of nanoformulations such as nano-pesticides, nano-fertilizers, nano-herbicides, and other nanofabricated materials. Long-term consequences, evaluations of nanotoxicity, ecotoxicity, and genomic deterioration, nanoparticle behavior in agroecosystems, plant absorption and translocation, and standardization of characterization techniques are therefore important areas that require more study. Furthermore, it is essential to elucidate the ramifications of the toxicity associated with nanoparticles/nanomaterials (NPs/NMs) in the environment, soil microbes and plant biosystems, including plant tissues, cells, and organelles like chloroplasts and mitochondria. Furthermore, the significance of transcriptome and proteomic contributions is further underscored by the pressing need to comprehend the genetic foundations of NP absorption and the response to stress mechanism. Furthermore, socioeconomic effects and public perception must be taken into account for agricultural solutions based on nanoparticles to be effectively embraced. Legislators still lack the necessary information to create a scientific regulatory framework for the commercialization and release of nanoparticles. Thus, we can lay the groundwork for safe and sustainable agricultural methods by fully understanding the benefits, drawbacks, and restrictions associated with metal nanoparticles in agriculture. With continued research and cooperation, we seek to develop solutions that improve nutrient management, reduce illness, increase agricultural output, and promote environmental sustainability.

References

1. Abdelsalam NR, Abdel-Megeed A, Ali HM et al (2018) Genotoxicity effects of silver nanoparticles on wheat (Triticum aestivum L.) root tip cells. Ecotoxicol Environ Saf 155:76–85. https://doi.org/10.1016/j.ecoenv.2018.02.069

2. Abideen Z, Hanif M, Munir N, Nielsen BL (2022) Impact of nanomaterials on the regulation of gene expression and metabolomics of plants under salt stress. Plants 11:691. https://doi.org/10.3390/plants11050691
3. Ahmed T, Noman M, Manzoor N et al (2021) Recent advances in nanoparticles associated ecological harms and their biodegradation: global environmental safety from nano-invaders. J Environ Chem Eng 9:106093. https://doi.org/10.1016/j.jece.2021.106093
4. Ahmed M, Decsi K, Tóth Z (2022) Different tactics of synthesized zinc oxide nanoparticles, homeostasis ions, and phytohormones as regulators and adaptatively parameters to alleviate the adverse effects of salinity stress on plants. Life 13:73. https://doi.org/10.3390/life13010073
5. Alam MW, Junaid PM, Gulzar Y et al (2024) Advancing agriculture with functional NM: "pathways to sustainable and smart farming technologies". Discov Nano 19:197. https://doi.org/10.1186/s11671-024-04144-z
6. Ali S, Mir RA, Tyagi A et al (2023) Chromium toxicity in plants: signaling, mitigation, and future perspectives. Plants 12:1502. https://doi.org/10.3390/plants12071502
7. Amirbakhtiar N, Ismaili A, Ghaffari M-R et al (2021) Transcriptome analysis of bread wheat leaves in response to salt stress. PLoS One 16:e0254189. https://doi.org/10.1371/journal.pone.0254189
8. An C, Sun C, Li N et al (2022) Nanomaterials and nanotechnology for the delivery of agrochemicals: strategies towards sustainable agriculture. J Nanobiotechnology 20:11. https://doi.org/10.1186/s12951-021-01214-7
9. Anjum S, Hashim M, Malik SA et al (2021) Recent advances in zinc oxide nanoparticles (ZnO NPs) for cancer diagnosis, target drug delivery, and treatment. Cancers (Basel) 13:4570. https://doi.org/10.3390/cancers13184570
10. Asadishad B, Chahal S, Akbari A et al (2018) Amendment of agricultural soil with metal nanoparticles: effects on soil enzyme activity and microbial community composition. Environ Sci Technol 52:1908–1918. https://doi.org/10.1021/acs.est.7b05389
11. Avellan A, Yun J, Morais BP et al (2021) Critical review: role of inorganic nanoparticle properties on their foliar uptake and in planta translocation. Environ Sci Technol 55:13417–13431. https://doi.org/10.1021/acs.est.1c00178
12. Azeem I, Adeel M, Ahmad MA et al (2021) Uptake and accumulation of nano/microplastics in plants: a critical review. Nanomaterials 11:2935. https://doi.org/10.3390/nano11112935
13. Bussières P (2014) Estimating the number and size of phloem sieve plate pores using longitudinal views and geometric reconstruction. Sci Rep 4:4929. https://doi.org/10.1038/srep04929
14. Chandrashekar HK, Singh G, Kaniyassery A et al (2023) Nanoparticle-mediated amelioration of drought stress in plants: a systematic review. 3 Biotech 13:336. https://doi.org/10.1007/s13205-023-03751-4
15. Chaud M, Souto EB, Zielinska A et al (2021) Nanopesticides in agriculture: benefits and challenge in agricultural productivity, toxicological risks to human health and environment. Toxics 9:131. https://doi.org/10.3390/toxics9060131
16. Chen YW, Lee HV, Juan JC, Phang S-M (2016) Production of new cellulose nanomaterial from red algae marine biomass Gelidium elegans. Carbohydr Polym 151:1210–1219. https://doi.org/10.1016/j.carbpol.2016.06.083
17. Côa F, Bortolozzo LS, Petry R et al (2020) Environmental toxicity of nanopesticides against non-target organisms: the state of the art. In: Nanopesticides. Springer, Cham, pp 227–279
18. de la Rosa G, Vázquez-Núñez E, Molina-Guerrero C et al (2021) Interactions of nanomaterials and plants at the cellular level: current knowledge and relevant gaps. Nanotechnol Environ Eng 6:7. https://doi.org/10.1007/s41204-020-00100-1
19. Dhanapal AR, Thiruvengadam M, Vairavanathan J et al (2024) Nanotechnology approaches for the remediation of agricultural polluted soils. ACS Omega 9:13522–13533. https://doi.org/10.1021/acsomega.3c09776
20. Egbuna C, Parmar VK, Jeevanandam J et al (2021) Toxicity of nanoparticles in biomedical application: nanotoxicology. J Toxicol 2021:1–21. https://doi.org/10.1155/2021/9954443

21. Eichert T, Goldbach HE (2008) Equivalent pore radii of hydrophilic foliar uptake routes in stomatous and astomatous leaf surfaces – further evidence for a stomatal pathway. Physiol Plant 132:491–502. https://doi.org/10.1111/j.1399-3054.2007.01023.x
22. Ettlinger R, Lächelt U, Gref R et al (2022) Toxicity of metal–organic framework nanoparticles: from essential analyses to potential applications. Chem Soc Rev 51:464–484. https://doi.org/10.1039/D1CS00918D
23. Fayiga A (2017) Nanoparticles in biosolids: effect on soil health and crop growth. Peertechz J Environ Sci Toxicol:059–067. https://doi.org/10.17352/pjest.000013
24. Fincheira P, Tortella G, Seabra AB et al (2021) Nanotechnology advances for sustainable agriculture: current knowledge and prospects in plant growth modulation and nutrition. Planta 254:66. https://doi.org/10.1007/s00425-021-03714-0
25. Fincheira P, Hoffmann N, Tortella G et al (2023) Eco-efficient systems based on nanocarriers for the controlled release of fertilizers and pesticides: toward smart agriculture. Nanomaterials 13:1978. https://doi.org/10.3390/nano13131978
26. Forest V (2022) Experimental and computational nanotoxicology—complementary approaches for nanomaterial hazard assessment. Nanomaterials 12:1346. https://doi.org/10.3390/nano12081346
27. García-Gómez C, Babín M, García S et al (2019) Joint effects of zinc oxide nanoparticles and chlorpyrifos on the reproduction and cellular stress responses of the earthworm Eisenia andrei. Sci Total Environ 688:199–207. https://doi.org/10.1016/j.scitotenv.2019.06.083
28. Ghosh I, Sadhu A, Moriyasu Y et al (2019) Manganese oxide nanoparticles induce genotoxicity and DNA hypomethylation in the moss Physcomitrella patens. Mutat Res Toxicol Environ Mutagen 842:146–157. https://doi.org/10.1016/j.mrgentox.2018.12.006
29. Gogos A, Knauer K, Bucheli TD (2012) Nanomaterials in plant protection and fertilization: current state, foreseen applications, and research priorities. J Agric Food Chem 60:9781–9792. https://doi.org/10.1021/jf302154y
30. Gomes SIL, Scott-Fordsmand JJ, Campos EVR et al (2019) On the safety of nanoformulations to non-target soil invertebrates – an atrazine case study. Environ Sci Nano 6:1950–1958. https://doi.org/10.1039/C9EN00242A
31. Gottschalk F, Nowack B (2011) The release of engineered nanomaterials to the environment. J Environ Monit 13:1145. https://doi.org/10.1039/c0em00547a
32. Gulati S, Kumar S, Jain S et al (2022) Toxicological perspectives and environmental risks of consumer nanoproducts. In: Handbook of consumer nanoproducts. Springer Nature, Singapore, pp 1253–1275
33. Guo H, White JC, Wang Z, Xing B (2018) Nano-enabled fertilizers to control the release and use efficiency of nutrients. Curr Opin Environ Sci Heal 6:77–83. https://doi.org/10.1016/j.coesh.2018.07.009
34. Hasanuzzaman M, Bhuyan MHM, Zulfiqar F et al (2020) Reactive oxygen species and antioxidant defense in plants under abiotic stress: revisiting the crucial role of a universal defense regulator. Antioxidants 9:681. https://doi.org/10.3390/antiox9080681
35. Hossain Z, Yasmeen F, Komatsu S (2020) Nanoparticles: synthesis, morphophysiological effects, and proteomic responses of crop plants. Int J Mol Sci 21:3056. https://doi.org/10.3390/ijms21093056
36. Huang J, Xiao J, Guo Y et al (2020) Long-term effects of silver nanoparticles on performance of phosphorus removal in a laboratory-scale vertical flow constructed wetland. J Environ Sci 87:319–330. https://doi.org/10.1016/j.jes.2019.07.012
37. Huang M, Wu Q, Jiang Z-H (2022) Epigenetic alterations under oxidative stress in stem cells. Oxid Med Cell Longev 2022:1–11. https://doi.org/10.1155/2022/6439097
38. Iavicoli I, Leso V, Beezhold DH, Shvedova AA (2017) Nanotechnology in agriculture: opportunities, toxicological implications, and occupational risks. Toxicol Appl Pharmacol 329:96–111. https://doi.org/10.1016/j.taap.2017.05.025
39. Jacobson A, Doxey S, Potter M et al (2018) Interactions between a plant probiotic and nanoparticles on plant responses related to drought tolerance. Ind Biotechnol 14:148–156. https://doi.org/10.1089/ind.2017.0033

40. Jan H, Shah M, Andleeb A et al (2021) Plant-based synthesis of zinc oxide nanoparticles (ZnO-NPs) using aqueous leaf extract of aquilegia pubiflora: their antiproliferative activity against HepG2 cells inducing reactive oxygen species and other in vitro properties. Oxid Med Cell Longev 2021. https://doi.org/10.1155/2021/4786227
41. Kah M (2015) Nanopesticides and nanofertilizers: emerging contaminants or opportunities for risk mitigation? Front Chem 3. https://doi.org/10.3389/fchem.2015.00064
42. Kah M, Beulke S, Tiede K, Hofmann T (2013) Nanopesticides: state of knowledge, environmental fate, and exposure modeling. Crit Rev Environ Sci Technol 43:1823–1867. https://doi.org/10.1080/10643389.2012.671750
43. Kahru A, Dubourguier H-C (2010) From ecotoxicology to nanoecotoxicology. Toxicology 269:105–119. https://doi.org/10.1016/j.tox.2009.08.016
44. Kaphle A, Navya PN, Umapathi A, Daima HK (2018) Nanomaterials for agriculture, food and environment: applications, toxicity and regulation. Environ Chem Lett 16:43–58. https://doi.org/10.1007/s10311-017-0662-y
45. Khalid MF, Iqbal Khan R, Jawaid MZ et al (2022) Nanoparticles: the plant saviour under abiotic stresses. Nanomaterials 12:3915. https://doi.org/10.3390/nano12213915
46. Khan I, Saeed K, Khan I (2019) Nanoparticles: properties, applications and toxicities. Arab J Chem 12:908–931. https://doi.org/10.1016/j.arabjc.2017.05.011
47. Khan ST, Adil SF, Shaik MR et al (2021) Engineered nanomaterials in soil: their impact on soil microbiome and plant health. Plants 11:109. https://doi.org/10.3390/plants11010109
48. Kumari M, Mukherjee A, Chandrasekaran N (2009) Genotoxicity of silver nanoparticles in Allium cepa. Sci Total Environ 407:5243–5246. https://doi.org/10.1016/j.scitotenv.2009.06.024
49. Larue C, Castillo-Michel H, Sobanska S et al (2014) Foliar exposure of the crop Lactuca sativa to silver nanoparticles: evidence for internalization and changes in Ag speciation. J Hazard Mater 264:98–106. https://doi.org/10.1016/j.jhazmat.2013.10.053
50. Li L, Luo Y, Li R et al (2020) Effective uptake of submicrometre plastics by crop plants via a crack-entry mode. Nat Sustain 3:929–937. https://doi.org/10.1038/s41893-020-0567-9
51. Li Y, Xi K, Liu X et al (2023) Silica nanoparticles promote wheat growth by mediating hormones and sugar metabolism. J Nanobiotechnology 21:2. https://doi.org/10.1186/s12951-022-01753-7
52. Li F, Hou Y, Chen L, Qiu Y (2025) Advances in silica nanoparticles for agricultural applications and biosynthesis. Adv Biotechnol 3:14. https://doi.org/10.1007/s44307-025-00067-7
53. Lian J, Liu W, Meng L et al (2021) Foliar-applied polystyrene nanoplastics (PSNPs) reduce the growth and nutritional quality of lettuce (Lactuca sativa L.). Environ Pollut 280:116978. https://doi.org/10.1016/j.envpol.2021.116978
54. Lin D, Xing B (2008) Root uptake and phytotoxicity of ZnO nanoparticles. Environ Sci Technol 42:5580–5585. https://doi.org/10.1021/es800422x
55. Lin S, Reppert J, Hu Q et al (2009) Uptake, translocation, and transmission of carbon nanomaterials in rice plants. Small 5:1128–1132. https://doi.org/10.1002/smll.200801556
56. Lin P-C, Lin S, Wang PC, Sridhar R (2014) Techniques for physicochemical characterization of nanomaterials. Biotechnol Adv 32:711–726. https://doi.org/10.1016/j.biotechadv.2013.11.006
57. Liu Q, Chen B, Wang Q et al (2009) Carbon nanotubes as molecular transporters for walled plant cells. Nano Lett 9:1007–1010. https://doi.org/10.1021/nl803083u
58. Liu Y, Zhou Y, Wang T et al (2019) Micro-nano bubble water oxygation: synergistically improving irrigation water use efficiency, crop yield and quality. J Clean Prod 222:835–843. https://doi.org/10.1016/j.jclepro.2019.02.208
59. Lough TJ, Lucas WJ (2006) Integrative plant biology: role of phloem long-distance macromolecular trafficking. Annu Rev Plant Biol 57:203–232. https://doi.org/10.1146/annurev.arplant.56.032604.144145
60. Lv J, Christie P, Zhang S (2019) Uptake, translocation, and transformation of metal-based nanoparticles in plants: recent advances and methodological challenges. Environ Sci Nano 6:41–59. https://doi.org/10.1039/C8EN00645H

61. Ma C, White JC, Zhao J et al (2018) Uptake of engineered nanoparticles by food crops: characterization, mechanisms, and implications. Annu Rev Food Sci Technol 9:129–153. https://doi.org/10.1146/annurev-food-030117-012657
62. MacCormack TJ, Goss GG (2008) Identifying and predicting biological risks associated with manufactured nanoparticles in aquatic ecosystems. J Ind Ecol 12:286–296. https://doi.org/10.1111/j.1530-9290.2008.00041.x
63. Mahon OR, Browe DC, Gonzalez-Fernandez T et al (2020) Nano-particle mediated M2 macrophage polarization enhances bone formation and MSC osteogenesis in an IL-10 dependent manner. Biomaterials 239:119833. https://doi.org/10.1016/j.biomaterials.2020.119833
64. Maity S, Chatterjee A, Guchhait R et al (2020) Cytogenotoxic potential of a hazardous material, polystyrene microparticles on Allium cepa L. J Hazard Mater 385:121560. https://doi.org/10.1016/j.jhazmat.2019.121560
65. Majeed Y, Zhu X, Zhang N et al (2023) Harnessing the role of mitogen-activated protein kinases against abiotic stresses in plants. Front Plant Sci 14. https://doi.org/10.3389/fpls.2023.932923
66. Mathur P, Chakraborty R, Aftab T, Roy S (2023) Engineered nanoparticles in plant growth: phytotoxicity concerns and the strategies for their attenuation. Plant Physiol Biochem 199:107721. https://doi.org/10.1016/j.plaphy.2023.107721
67. Mgadi K, Ndaba B, Roopnarain A et al (2024) Nanoparticle applications in agriculture: overview and response of plant-associated microorganisms. Front Microbiol 15. https://doi.org/10.3389/fmicb.2024.1354440
68. Moazzami Farida SH, Karamian R, Albrectsen BR (2020) Silver nanoparticle pollutants activate oxidative stress responses and rosmarinic acid accumulation in sage. Physiol Plant 170:415–432. https://doi.org/10.1111/ppl.13172
69. Munir N, Hanif M, Dias DA, Abideen Z (2021) The role of halophytic nanoparticles towards the remediation of degraded and saline agricultural lands. Environ Sci Pollut Res 28:60383–60405. https://doi.org/10.1007/s11356-021-16139-9
70. Munir N, Gulzar W, Abideen Z et al (2023) Nanotechnology improves disease resistance in plants for food security: applications and challenges. Biocatal Agric Biotechnol 51:102781. https://doi.org/10.1016/j.bcab.2023.102781
71. Mustafa G, Komatsu S (2021) Plant proteomic research for improvement of food crops under stresses: a review. Mol Omi 17:860–880. https://doi.org/10.1039/D1MO00151E
72. Nelson B, Johnson M, Walker M et al (2016) Antioxidant cerium oxide nanoparticles in biology and medicine. Antioxidants 5:15. https://doi.org/10.3390/antiox5020015
73. Panpatte DG, Jhala YK, Shelat HN, Vyas RV (2016) Nanoparticles: the next generation technology for sustainable agriculture. In: Microbial inoculants in sustainable agricultural productivity. Springer India, New Delhi, pp 289–300
74. Park S, Ahn Y-J (2016) Multi-walled carbon nanotubes and silver nanoparticles differentially affect seed germination, chlorophyll content, and hydrogen peroxide accumulation in carrot (Daucus carota L.). Biocatal Agric Biotechnol 8:257–262. https://doi.org/10.1016/j.bcab.2016.09.012
75. Parveen R, Maiti PK, Murmu N, Datta A (2021) Preparation of serum capped silver nanoparticles for selective killing of microbial cells sparing host cells. Sci Rep 11:11610. https://doi.org/10.1038/s41598-021-91031-7
76. Pittol M, Tomacheski D, Simões DN et al (2017) Macroscopic effects of silver nanoparticles and titanium dioxide on edible plant growth. Environ Nanotechnol Monit Manag 8:127–133. https://doi.org/10.1016/j.enmm.2017.07.003
77. Pollard M, Beisson F, Li Y, Ohlrogge JB (2008) Building lipid barriers: biosynthesis of cutin and suberin. Trends Plant Sci 13:236–246. https://doi.org/10.1016/j.tplants.2008.03.003
78. Pramanik P, Krishnan P, Maity A et al (2020) Application of nanotechnology in agriculture. IntechOpen, London, pp 317–348
79. Prasad R, Pandey R, Barman I (2016) Engineering tailored nanoparticles with microbes: quo vadis? WIREs Nanomed Nanobiotechnol 8:316–330. https://doi.org/10.1002/wnan.1363

80. Preisler AC, do Carmo GC, da Silva RC et al (2024) Improving soybean germination and nodule development with nitric oxide-releasing polymeric nanoparticles. Plants 14:17. https://doi.org/10.3390/plants14010017
81. Raliya R, Saharan V, Dimkpa C, Biswas P (2018) Nanofertilizer for precision and sustainable agriculture: current state and future perspectives. J Agric Food Chem 66:6487–6503. https://doi.org/10.1021/acs.jafc.7b02178
82. Rashid MI, Shah GA, Sadiq M et al (2023) Nanobiochar and copper oxide nanoparticles mixture synergistically increases soil nutrient availability and improves wheat production. Plants 12:1312. https://doi.org/10.3390/plants12061312
83. Rehman A, Khan S, Sun F et al (2024) Exploring the nano-wonders: unveiling the role of nanoparticles in enhancing salinity and drought tolerance in plants. Front Plant Sci 14. https://doi.org/10.3389/fpls.2023.1324176
84. Salehi H, Chehregani Rad A, Raza A, Chen J-T (2021) Foliar application of CeO_2 nanoparticles alters generative components fitness and seed productivity in bean crop (Phaseolus vulgaris L.). Nanomaterials 11:862. https://doi.org/10.3390/nano11040862
85. Sharifan H, Ma X, Moore JM et al (2019) Zinc oxide nanoparticles alleviated the bioavailability of cadmium and lead and changed the uptake of iron in hydroponically grown lettuce (Lactuca sativa L. var. Longifolia). ACS Sustain Chem Eng 7:16401–16409. https://doi.org/10.1021/acssuschemeng.9b03531
86. Shelar A, Nile SH, Singh AV et al (2023) Recent advances in nano-enabled seed treatment strategies for sustainable agriculture: challenges, risk assessment, and future perspectives. Nano-Micro Lett 15:54. https://doi.org/10.1007/s40820-023-01025-5
87. Soliemanzadeh A, Fekri M (2021) Effects of green iron nanoparticles on iron changes and phytoavailability in a calcareous soil. Pedosphere 31:761–770. https://doi.org/10.1016/S1002-0160(21)60035-8
88. Srivastava AK, Dev A, Karmakar S (2018) Nanosensors and nanobiosensors in food and agriculture. Environ Chem Lett 16:161–182. https://doi.org/10.1007/s10311-017-0674-7
89. Suazo-Hernández J, Arancibia-Miranda N, Mlih R et al (2023) Impact on some soil physical and chemical properties caused by metal and metallic oxide engineered nanoparticles: a review. Nanomaterials 13:572. https://doi.org/10.3390/nano13030572
90. Takeshita V, Munhoz-Garcia GV, Pereira AES et al (2023) Radiometric strategy to track nanopesticides: an important approach to understand the fate, mechanisms of action and toxicity. TrAC Trends Anal Chem 165:117156. https://doi.org/10.1016/j.trac.2023.117156
91. Thabet AF, Boraei HA, Galal OA et al (2021) Silica nanoparticles as pesticide against insects of different feeding types and their non-target attraction of predators. Sci Rep 11:14484. https://doi.org/10.1038/s41598-021-93518-9
92. Tian H, Kah M, Kariman K (2019) Are nanoparticles a threat to mycorrhizal and rhizobial symbioses? A critical review. Front Microbiol 10. https://doi.org/10.3389/fmicb.2019.01660
93. Tighe-Neira R, Reyes-Díaz M, Nunes-Nesi A et al (2020) Titanium dioxide nanoparticles provoke transient increase in photosynthetic performance and differential response in antioxidant system in Raphanus sativus L. Sci Hortic (Amsterdam) 269:109418. https://doi.org/10.1016/j.scienta.2020.109418
94. Tortella GR, Rubilar O, Durán N et al (2020) Silver nanoparticles: toxicity in model organisms as an overview of its hazard for human health and the environment. J Hazard Mater 390:121974. https://doi.org/10.1016/j.jhazmat.2019.121974
95. Usman M, Farooq M, Wakeel A et al (2020) Nanotechnology in agriculture: current status, challenges and future opportunities. Sci Total Environ 721:137778. https://doi.org/10.1016/j.scitotenv.2020.137778
96. Van Nhan L, Ma C, Rui Y et al (2016) The effects of Fe2O3 nanoparticles on physiology and insecticide activity in non-transgenic and bt-transgenic cotton. Front Plant Sci 6. https://doi.org/10.3389/fpls.2015.01263
97. Vega-Vásquez P, Mosier NS, Irudayaraj J (2020) Nanoscale drug delivery systems: from medicine to agriculture. Front Bioeng Biotechnol 8:79. https://doi.org/10.3389/fbioe.2020.00079

98. Verma SK, Das AK, Gantait S et al (2019) Applications of carbon nanomaterials in the plant system: a perspective view on the pros and cons. Sci Total Environ 667:485–499. https://doi.org/10.1016/j.scitotenv.2019.02.409
99. Wang Z, Li N, Zhao J et al (2012) CuO nanoparticle interaction with human epithelial cells: cellular uptake, location, export, and genotoxicity. Chem Res Toxicol 25:1512–1521. https://doi.org/10.1021/tx3002093
100. Wang W-N, Tarafdar JC, Biswas P (2013) Nanoparticle synthesis and delivery by an aerosol route for watermelon plant foliar uptake. J Nanoparticle Res 15:1417. https://doi.org/10.1007/s11051-013-1417-8
101. Wang C, Liu X, Li J et al (2021a) Copper nanoclusters promote tomato (Solanum lycopersicum L.) yield and quality through improving photosynthesis and roots growth. Environ Pollut 289:117912. https://doi.org/10.1016/j.envpol.2021.117912
102. Wang J, Li M, Feng J et al (2021b) Effects of TiO2-NPs pretreatment on UV-B stress tolerance in Arabidopsis thaliana. Chemosphere 281:130809. https://doi.org/10.1016/j.chemosphere.2021.130809
103. Wang N, Wang B, Wan Y et al (2023) Alginate-based composites as novel soil conditioners for sustainable applications in agriculture: a critical review. J Environ Manage 348:119133. https://doi.org/10.1016/j.jenvman.2023.119133
104. Xuan L, Ju Z, Skonieczna M et al (2023) Nanoparticles-induced potential toxicity on human health: applications, toxicity mechanisms, and evaluation models. MedComm 4. https://doi.org/10.1002/mco2.327
105. Yan A, Chen Z (2019) Impacts of silver nanoparticles on plants: a focus on the phytotoxicity and underlying mechanism. Int J Mol Sci 20:1003. https://doi.org/10.3390/ijms20051003
106. Yang F, Hong F, You W et al (2006) Influences of nano-anatase TiO2 on the nitrogen metabolism of growing spinach. Biol Trace Elem Res 110:179–190. https://doi.org/10.1385/BTER:110:2:179
107. Yin J, Su X, Yan S, Shen J (2023) Multifunctional nanoparticles and nanopesticides in agricultural application. Nanomaterials 13:1255. https://doi.org/10.3390/nano13071255
108. Younis SA, Kim K-H, Shaheen SM et al (2021) Advancements of nanotechnologies in crop promotion and soil fertility: benefits, life cycle assessment, and legislation policies. Renew Sustain Energy Rev 152:111686. https://doi.org/10.1016/j.rser.2021.111686
109. Yousef HA, Fahmy HM, Arafa FN et al (2023) Nanotechnology in pest management: advantages, applications, and challenges. Int J Trop Insect Sci 43:1387–1399. https://doi.org/10.1007/s42690-023-01053-z
110. Yu Z, Li Q, Wang J et al (2020) Reactive oxygen species-related nanoparticle toxicity in the biomedical field. Nanoscale Res Lett 15:115. https://doi.org/10.1186/s11671-020-03344-7
111. Zhang P, Guo Z, Zhang Z et al (2020) Nanomaterial transformation in the soil–plant system: implications for food safety and application in agriculture. Small 16. https://doi.org/10.1002/smll.202000705
112. Zhang X, Ma G, Wei W (2021) Simulation of nanoparticles interacting with a cell membrane: probing the structural basis and potential biomedical application. NPG Asia Mater 13:52. https://doi.org/10.1038/s41427-021-00320-0
113. Zhao W, Wu Z, Amde M et al (2023) Nanoenabled enhancement of plant tolerance to heat and drought stress on molecular response. J Agric Food Chem 71:20405–20418. https://doi.org/10.1021/acs.jafc.3c04838

Nanoparticle-Mediated Nucleic Acid Delivery in Plant Cells: Molecular Insights and Applications

Amitesh Bhattacharyya, Anirneeta De, Shreya Sarkar, Biswajit Pramanik, and Sandip Debnath

Abstract The ever-increasing global population and intensifying effects of climate change have significantly strained food security worldwide. At this eleventh hour of utmost need, plant genetic engineering emerges as a critical solution, offering resilience, improved yields, and enhanced nutritional quality. However, conventional gene delivery methods such as *Agrobacterium*-mediated transformation, electroporation, and biolistics face multiple limitations including species-specific barriers, low efficiency, and tissue damage. In this context, nanoparticle (NP)-mediated nucleic acid delivery has gained immense attention as a transformative alternative. Due to their tunable physico-chemical properties, NPs facilitate precise, efficient, and minimally invasive delivery of genetic materials, including DNA, RNA, siRNA, and CRISPR-Cas components, across diverse plant species. Moreover, NP-mediated systems enable both transgenic and non-transgenic applications, addressing regulatory and biosafety concerns. This chapter delves into the molecular mechanisms governing NP acquisition, transport, and intracellular trafficking within plant systems, alongside recent advancements in employing various types of NPs such as gold nanoparticles (AuNPs), carbon nanotubes (CNTs), mesoporous silica nanoparticles (MSNs), and chitosan-based carriers. Special emphasis is given to NP-based delivery's role in virus resistance, crop improvement, and sustainable agriculture. Although challenges regarding toxicity, scalability, and environmental safety remain, continued interdisciplinary research promises to position NP-mediated delivery as a cornerstone technology for next-generation plant biotechnology.

A. Bhattacharyya · S. Sarkar · B. Pramanik
Biotechnology Unit, Crop Improvement Division, ICAR-Central Research Institute for Jute and Allied Fibres, Barrackpore, West Bengal, India

A. De
School of Agriculture, Swami Vivekananda University, Barrackpore, West Bengal, India

S. Debnath (✉)
Department of Genetics and Plant Breeding, Palli Siksha Bhavana (Institute of Agriculture), Visva-Bharati, Sriniketan, West Bengal, India
e-mail: sandip.debnath@visva-bharati.ac.in

J.-T. Chen (ed.), *Nano-Delivering for Plant Genetic Engineering*, Smart Nanomaterials Technology, https://doi.org/10.1007/978-981-95-4165-2_11

Keywords CRISPR-Cas · Gene expression · Nanoparticles (NPs) · Nucleic acid delivery · Plant cell · Virus resistance

1 Introduction

The ever-exploding population and the escalating impacts of climate change have threatened global food security in the last few decades. Plant genetic engineering stands out to be a superior critical approach surpassing other conventional strategies followed to date. It has become an essential tool in this contemporary era of agriculture, allowing the development of crops with enhanced nutritional quotient, resilience, and productivity [1]. Although the applicability of genetic engineering in plants has limitations while considering conventional transformation methodologies which are *Agrobacterium*-mediated, electroporation-mediated, or associated with biolistic gene delivery, and protoplast transfection, nanoparticles (NPs)-mediated nucleic acid delivery supposed to be a path-breaking approach gaining popularity surpassing these constraints. This approach offers unique advantages in efficiency, precision, and versatility, opening new avenues for both transgenic and non-transgenic plant improvement strategies [2].

Plant genetic engineering, in wholesome, answers all the queries regarding urgent global issues inclusive of food insecurity, climatic adversities, and erosion of cultivable land. A simple modification in genetic architecture in a plant body can result in improved yield, enhanced stress tolerance, ameliorated nutritional content, and boosted resistance to biotic stressors [3]. Traditional protocols of plant transformation methods (e.g. *Agrobacterium tumefaciens*-mediated transfer, electroporation-mediated gene delivery, biolistics, protoplast fusion, etc.) have long been recruited to introduce alien genes into plant genomes. In order to insert DNA vectors, Aminomethyltransferase (AMT) gene circumvents the cellular defense through the plant's innate immunological response [4]. Due to *Agrobacterium*'s limited host-specificity, the majority of plant species are resistant to ATM, leading to restricted applicability in genetic transformation in a wide range of plant species [5]. Furthermore, *Agrobacterium*-dependent plant regeneration protocols are labor-intensive, time-consuming, and have limited reproducibility. According to Liu et al. [3], viral vectors used in genetic engineering usually produce temporary expressions that are not inheritable, which limits their usefulness in attempts at persistent transformation. Biolistic (gene gun) methods, though applicable to a broader range of species, frequently result in tissue damage due to high-velocity particle bombardment that might lead to multiple gene insertions, random integration, or gene silencing due to rearrangements. This unpredictability can jeopardize the stability and expression of transgenes in addition to decreasing transformation efficiency [6]. On the other hand, electroporation demands precisely calibrated parameters, possessing poor efficiency and high cell mortality, particularly due to the fragile nature of plant protoplasts and species-specific preparation protocols [1]. Moreover, regeneration of whole plants from protoplasts turns out to be a major impediment for a group of crops. Most of these conventional protocols are often intrusive, expensive, and labor-intensive, which restricts their applicability and scalability in large-scale agricultural contexts [7]. Furthermore, effective intracellular delivery is particularly difficult due to the strong physical barriers present on plant cell walls and cuticles. Delivery efficiency is further decreased by environmental degradation of nucleic acids caused by UV radiation, rain, or nucleases, particularly for RNA-based applications [6]. These limitations, comprehensively, underscore the urgent need for alternative gene delivery systems that are efficient, species-independent, non-damaging, and capable of transiently stable transformation. This has led to growing interest in NP-mediated nucleic acid delivery, which is potent enough to overcome many of these restraints through precise, and minimally invasive mechanisms.

Nanotechnology outshines these conventional methodologies as a premium transformative tool in plant biotechnology. NPs, owing to their tiny size (1–100 nm), larger surface area, and adjustable properties, are being explored for introducing nucleic acids e.g., DNA, RNA, siRNA, and CRISPR-Cas components directly into the phyto-cellular system [3]. NP-mediated delivery offers numerous advantages such as species-independent delivery, safeguarding nucleic acids against deterioration, minimal tissue damage, and the possibility of tailored release in particular cellular compartments. Additionally, plant nanobiotechnology has paved the door for non-transgenic applications, which address biosafety and regulatory problems by implementing genome editing or temporary expression systems without the integration of foreign DNA. This present chapter elaborately discusses the potential of NP-based delivery protocols to transform plant genetic engineering for precision and sustainable agriculture.

2 NP-Mediated Nucleic Acid Delivery: An Overview

NPs serve as vectors or carriers with the capability of transporting nucleic acids across plant cell walls and membranes, allowing either stable or transient gene expression. Gold NPs (AuNPs), lipid-based carriers, carbon nanotubes, mesoporous silica, polymeric particles, and magnetic NPs have been established to deliver genetic cargo effectively and with minimal tissue damage [3]. It has also been demonstrated that both magnetic nanoparticle-based transformation (magnetofection) and carbon dot-mediated delivery can boost transformation efficiency across a broad spectrum of plant species avoiding the need and drawbacks of conventional approaches [3].

Species-non-specific compatibility always bears a major advantage of NP-mediated gene delivery over other methodologies. Unlike *Agrobacterium*-based systems, which have host-range restrictions, NPs can be designed to interact with a variety of plant species and tissues, including both monocots and dicots [6]. Moreover, these systems can be used in non-transgenic applications which reduces biosafety issues by allowing transient gene expression, RNA interference (RNAi), or CRISPR/Cas9-based genome editing without integrating foreign DNA permanently. Beyond transformation, NPs can shield nucleic acids from environmental degradation by UV radiation, enzymatic digestion, and moisture, while ensuring their timely and targeted release inside plant cells. This makes them particularly suited for topical applications such as spray-induced gene silencing or foliar delivery of siRNA/dsRNA for pathogen control [6].

Crucially, nanotechnology can also be incorporated into the domain of precision agriculture and sustainable farming, offering tools for nutrient delivery, disease detection, and stress management [7]. These platforms are anticipated to become more important in the next-generation plant biotechnology research studies as they deliver more focused, effective and eco-friendly protocols of genetic engineering. Basically, NP-mediated nucleic acid introduction, with its adjustable qualities, protective capabilities, and compatibility with both transgenic and non-transgenic approaches, offers a revolutionary transformative alternative over the traditional transformative approaches. Continued advancements in this field promise more efficient, precise, and environmentally safe genetic interventions, paving the way for the next generation of stress-resilient and high-yielding crops.

2.1 Principles of NP-Mediated Nucleic Acid Delivery

NP-associated delivery systems represent a groundbreaking advancement in plant genetic engineering, enabling precise, minimally invasive, and efficient delivery of DNA, RNA, and gene-editing tools.

This innovative technique hinges on three key principles:

1. Type and design of NPs.
2. Type of nucleic acids delivered.
3. Intracellular uptake and transport mechanism.

The efficacy, specificity, and safety of NP-mediated delivery are defined by these principles, establishing the foundation for sustainable next-generation plant biotechnology solutions at a high scale across a wide range of crop species.

2.1.1 Types of NPs Used in Gene Delivery

NPs serve as promising carriers for nucleic acid delivery due to their customizable size, surface properties, and ability to protect genetic cargo. Various types of NPs have been developed for efficient and species-independent delivery of genetic materials and gene-editing tools in plants (Fig. 1).

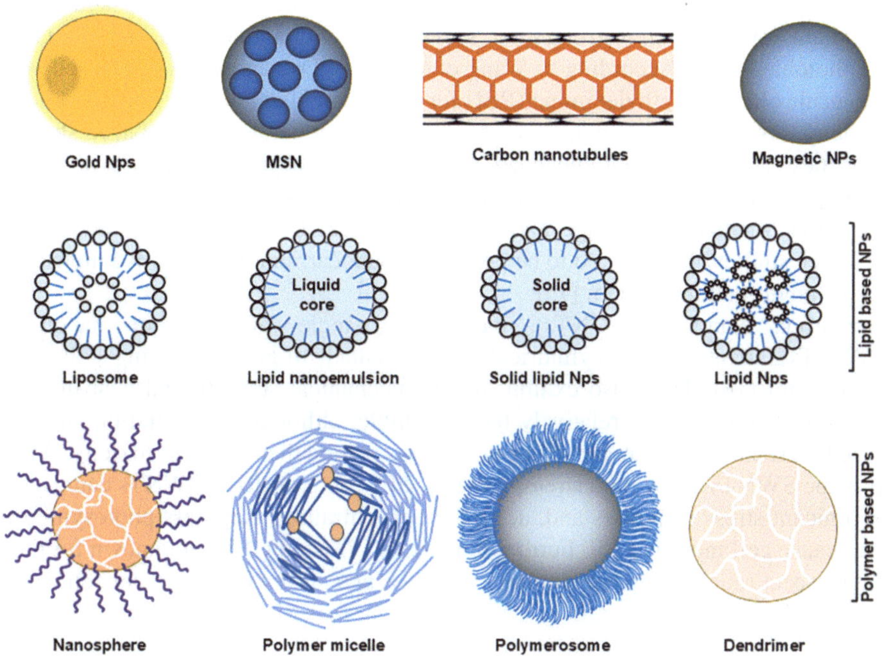

Fig. 1 Types of NPs used in plant genetic material delivery

Gold NPs (AuNPs)

AuNPs are often used in plant biotechnology for their simplicity of surface functionalization, biocompatibility and capacity to bind with nucleic acids through thiol or amine linkages. Gold carriers like gold nanosphere, gold nanorods, and gold nanoclusters have been used for sustained and efficient delivery of nucleic acids in the last few decades. Successful grafting of thiolated nucleic acids to their carriers has been found without degradation of nuclease in payloads. Also, these NPs were observed to contain large amounts of cargo for being highly adjustable in nature. Compared to macro gold particles, AuNP administration caused less harm to plant tissue and showed a greater DNA delivery efficiency of about 40% [8]. Moreover, to mention, AuNPs are mostly used in foliar spray and biolistic delivery due to their stability and minimal toxicity.

Carbon-Based Nanomaterials

Carbon dots (CDs) or carbon-based microscopic NPs, with less than 10 nm in diameter, are categorized into three groups, namely graphene quantum dots (GQDs), carbon nanodots (CNDs), and polymer dots (PDs), on the basis of their chemical composition [9]. CDs, synthesized from biomass or organic compounds like amino acids and sugars, are considered as non-toxic alternatives to heavy metal-based quantum dots. They emit strong fluorescence, greater quantum yield, and resistance to photobleaching. According to Wang et al. [10], CDs can be functionalized with molecules like Polyethylenimine (PEI) for nucleic acid delivery despite increasing toxicity to some extent. Studies in wheat, mung bean, lettuce, and tobacco showed that these crops exhibited enhanced growth and photosynthesis upon exposure to CDs [11–14]. High CD concentrations promoted dose-dependent phytotoxicity, including membrane damage and oxidative stress. Despite a bunch of handful research studies, CDs proved to be promising nanocarriers for bio-active molecules in the phyto-cellular system [15].

Carbon Nanotubes (CNTs), particularly single-walled carbon nanotubes (SWNTs), are needle-like cylindrical structures made of graphite having a diameter of 1–3 nm [16]. They also exhibit unique mechanical strength and near-infrared fluorescence. CNTs are relatively biocompatible, although concentration and surface functionalization are the major two factors behind their levels of toxicity. While pristine SWNTs cause mild stress in plants, PEI-functionalized SWNTs can trigger significant stress responses and, at higher concentrations, suppress photosynthesis and cause cell mortality [6]. Furthermore, both CDs and CNTs are known for their high aspect ratio, strong penetration capability, and ability to cross plant cell walls.

Mesoporous Silica NPs (MSNs)

MSNs are an additional promising nanocarrier for drug- and gene delivery due to their large surface area, adjustable pore size, and biocompatibility [17]. Initially, their negative surface charge in hydrophilic environments hampered the process of delivery of genetic materials, but this has been resolved through surface modifications with cationic polymers like polylysine, polyarginine, and polyethyleneimine [6]. Although nucleic acids commonly bind to the NP surface, more recent techniques allow encapsulation of siRNA within the mesopores, offering protection and controlled release. Though MSNs are gaining immense interest in the field of phytobiotechnology, only a few studies conducted by Chang et al. [18], Hajiahmadi et al. [19] have so far demonstrated their successful use in introducing plasmid DNA into phyto-cellular system.

Chitosan-Based NPs

Chitosan, a biocompatible and low-cost polymer derived from chitin, forms positively charged NPs under relatively acidic conditions, making it suitable for nucleic acid delivery [20]. These NPs allow controlled cargo release in agricultural applications, protecting DNA or dsRNA from degrading. Chitosan has also shown antibacterial, antifungal, and antiviral activity, enhanced stress resistance, and productivity when administered topically or as foliar spraying. Despite having limited research in plant systems, studies in insects revealed that chitosan facilitated endosomal escape via the proton sponge effect, as reported by Komarova et al. [6], which might be generalized to plants despite the paucity of evidence in the same. However, NP performance depends heavily on chitosan quality and homogeneity, and its full biosafety in plants still remains to be established.

Magnetic NPs (MNPs)

Magnetofection, another promising technique for delivering nucleic acids into plant cells using MNPs, and an external magnetic field enables stable transient gene expression, especially through pollen magnetofection, where MNPs coated with DNA are introduced into pollen grains, irrespective of regeneration [21]. Studies in maize [22], tobacco [23], and okra [24] showed the ideal ratio of DNA: MNP and magnetic field settings to achieve successful transformation. While magnetofection study is largely confined to pollen, its potential for protoplasts and seeds, among other plant tissues, is being investigated. Optimizing NP properties, magnetic field strength, and exposure time are the major keys behind its broader application in plant biotechnology.

2.1.2 Nucleic Acid Types Delivered by NPs

NP-based delivery systems' one chief benefit over others is their ability to carry a wide range of nucleic acid cargo, including plasmid DNA (pDNA), messenger RNA (mRNA), small interfering RNA (siRNA), double-stranded RNA (dsRNA), antisense oligonucleotides (ASOs), and even CRISPR-Cas components. The intended use, such as precise genome editing, gene silencing, or temporary gene expression, determines the type of nucleic acid to be carried. NPs protect these molecules from environmental degradation as well as facilitate targeted intracellular delivery and timely release, making them superior alternatives to traditional transformation tools [3].

DNA (pDNA and Linear DNA)

pDNA is widely used for gene expression and transformation studies. NPs such as CNTs, MSNs, and chitosan have successfully introduced pDNA into plant cells for transient or stable expression [3]. Through magnetofection, particularly using Fe_3O_4 coated with PEI, pDNA was bound via electrostatic attraction and was directed into plant cells using a magnetic field. This method has enabled efficient transformation in species like *Brassica napus*, *Nicotiana tabacum*, and *Gossypium hirsutum*, facilitating both transient and stable expression of genes such as GUS, EGFP, and bar bypassing tissue culture [3, 23, 25]. Parallelly CNTs and CDs have emerged as effective pDNA-delivery vehicles. Single-walled and multi-walled CNTs have been used to introduce pDNA like pGreen0029 into tobacco leaf explants, protoplasts, and callus cells, demonstrating transient YFP and stable *nptII* expression [26]. Functionalization with arginine or PEI enhanced DNA-carrying capacity and cellular uptake. In several research studies, DNA was conjugated with NPs and sprayed directly onto leaves, which eventually eliminated the requirement for tissue culture [6].

RNA

mRNA

mRNA delivery is beneficial for expressing proteins transiently without altering the genome. Nanocarriers like lipid NPs and CDs have shown promise for delivering mRNA in both plant and animal cells for gene therapy [3].

siRNA and dsRNA

siRNAs are commonly used in RNAi to silence target genes by degrading complementary mRNA through the RISC complex after Dicer-like protein processing [27]. Unlike traditional viral vector methods, NPs offer a non-transgenic and species-independent alternative. Carbon-based nanocarriers have delivered siRNA effectively, achieving >90% efficiency in knocking down target mRNAs such as GFP and ROQ1 in *Nicotiana benthamiana* with effects lasting up to 3 days after a single infiltration [28].

Graphene oxide NPs (GONs) loaded with siRNA achieved 97.2% silencing within 24 h, demonstrating high cellular uptake and barrier penetration [29]. PEI-functionalized AuNPs were used to deliver siRNA targeting NPR1 in *Arabidopsis*, causing an 80% reduction in gene expression [30]. Other than these clay nanosheets, and polymeric NPs are being explored for topical siRNA delivery [6]. Further to indicate, this is especially vital for non-transgenic pest and disease control.

CRISPR-Cas Components

CRISPR/Cas systems can be delivered in multiple forms, such as pDNA expressing Cas9 and sgRNA, mRNA (Cas9) with sgRNA, or ribonucleoprotein complexes (RNPs). NPs protect these sensitive cargos and improve editing efficiency by avoiding degradation and facilitating cellular uptake. Delivery of RNPs also minimizes off-target effects and avoids foreign DNA integration, making it suitable for non-GMO genome editing. Reports on nanocarrier-mediated CRISPR-Cas9-based plant genome editing are very few. The primary cause behind this discrepancy is the CRISPR-Cas9 system's encoding DNA plasmid's size, which is around 15 kb. This is significantly larger than the pCAMBIA-series reporter plasmids that are often utilized in nano-delivery investigations [3].

2.2 Mechanisms of Uptake and Delivery

The success of NP-based gene delivery in plants relies on the design of NPs, the nature of cargo and the mechanism by which these nanocarriers interact with plant cells. Unlike animal cells, plant cells possess multiple physical and chemical barriers, specifically cuticles, rigid cell walls and cell membranes. In-depth knowledge of uptake pathways, transport mechanisms and intracellular release strategies is, therefore, very crucial to optimize delivery efficiency to ensure successful gene expression or silencing [3].

2.2.1 Interaction with Phyto-cellular Barriers

The pectin-rich plant cell wall is a major physical barrier to NP-based gene delivery due to its small pore size (5–20 nm) [31]. The size, surface chemistry and surface charge of NPs influence their interaction with these barriers. While the precise mechanism of NP uptake is still a mystery, endocytosis, channel-operated transport, lipid membrane exchange and pore formation are reported to be the possible pathways [32]. Cationic NPs are often more effective due to their electrostatic interaction with negatively charged cell walls, facilitating ion exchange and penetration [7]. An unusual case was reported by Zhang et al. [33] where AuNPs, without entering the cell, enabled siRNA delivery by serving as external reservoirs where it intercalated with the plant cells, protecting siRNA from degradation. NPs are likely to be translocated via the xylem and phloem after proper introduction into plant tissues. Hence, pests feeding on phloem sap (*e.g.* aphids) are targeted when NPs are carried over through phloem. Likewise, while targeting pests feeding on xylem sap, NPs carried over by xylem are to be used [31]. However, there are concerns about undesired NP accumulation in the food chain upon foliar spraying on leaves, flowers and fruits. For instance, Fe_3O_4 MNPs at high concentrations harm mycorrhizal fungi, disrupt symbiosis, and reduce key soil enzymes, indicating environmental and biodiversity risks [34].

2.2.2 Acquisition Pathways

There are two primary uptake mechanisms followed by plants for the acquisition of NPs.

Passive Uptake

It is a non-invasive method where NPs enter plant cells via natural openings such as stomata, root hairs, or symplastic pathways without any use of external force [35]. NPs ranging from 10 to 40 nm in diameter are considered optimal for passive transformation. CDs and oxides of Zn, Au, Ce, and Ti can easily penetrate roots and leaves and move through vascular tissues. Pal et al.[36] further confirmed that CDs successfully delivered pDNA in Nipponbare rice root tips through soaking, indicating passive uptake. Similarly, CNTs enabled high-efficiency pDNA delivery into protoplasts with 76% transformation via passive entry, without compromising cell viability [37]. Passive uptake takes place when methods like high-pressure spraying, abrasion, or foliar infiltration occur. This method is more effective for smaller and neutrally charged particles [6].

Active Uptake

Active acquisition involves energy-dependent processes such as endocytosis, channel-mediated transport, and lipid membrane exchange. Lipid exchange envelope penetration (LEEP) offers a kinetic model for NP entry, influenced by particle size and zeta potential [38]. Despite these assurances, low endocytic rates, and the structural complexity of plant cell walls, cause hindrance in active delivery in plants [37]. In specialized setups like protoplast transformation, active uptake mechanisms, often supported by cationic polymers or engineered carriers, enhanced precise delivery and transgene expression [1].

2.2.3 Intracellular Transport and Cargo Release

The nucleic acid must be carefully released from NP after internalization in a controlled manner. This could be accomplished by stimuli-responsive linkers that are sensitive to light, temperature, or enzymatic activity; biodegradable coatings that decompose gradually; or pH-sensitive release systems, in which acidic conditions within vesicles cause dissociation [3]. Attaching nuclear localization signals (NLS) or using NP architectures that make organelle targeting easier are common ways to help CRISPR/Cas RNPs target the nucleus [6].

3 Molecular Insights into NP-Plant Cell Interactions

3.1 Phyto-Cellular Envelop-NPs Interaction

NPs are present in abundant form all over the environment because of their natural sources as well as due to the results of geological catastrophes. These include dynamic volcanic structures, forest fires, as well as the shoot that is formed during the thermal oxidation process. The proliferation of NPs in the environment is more due to the amplification of advancements in nanotechnology because the synthesis of NPs is either done intentionally or they are released inadvertently as byproducts of various industrial processes including, metal smelting, soldering, vulcanization, plasma cutting, high power generation unit, cooking, grilling, and also use of laser emitting office equipment [39]. Recent trend indicates that nanomaterials (NMs) are inherently and increasingly utilized across a wide range of industries spanning from healthcare and electronic applications to the food industry, cosmetology, and even in the agricultural sector. This exponential growth of NPs has contributed heavily towards the accumulation of NMs in the environment, however, its behavior, transformation, and moreover, extended impacts are still poorly understood. While numerous studies have explored and intercepted the potential benefits of these microparticles, their detrimental effects on human health and plant systems remain largely ambiguous and under investigation. To study the harmful impact of these NPs on ecosystems, an emerging discipline of science, Nanotoxicology, has made substantial progress in recent years. Current research not only analyzes the toxicity of NMs in living beings, but also investigates their acquisition, transport, and distribution throughout plant systems at various organizational levels *i.e.* from organs and tissues to individual cells.

The physicochemical properties of NPs, which affect their reactivity, cellular penetration, and transportation throughout the plant systems, can be significantly altered by their composition. However, the diversity of effect is contrasting, but different plants exhibit multiple reactions with the same kind of NPs. The rising application of NPs in agriculture, biotechnology, and environmental remediation has exerted revived attention in understanding the complex interaction among these NPs with plant systems. One of the most delicate interplays in this relationship is the plant cell wall and the plasma membrane, which acts as the selective barriers to inhibit the entry and movement of externally derived particles including the NPs, within the plant's internal environment. A better understanding of the universal model of NP-plant interaction and the dual nature of the system is the chief prerequisite behind getting a broader knowledge of this relationship. Firstly, these NPs vary widely in their chemical composition, coating material, size, shape and even surface charge, all of which combine to influence their behavior and interaction potential. Plant systems, on the other hand, during this interaction exhibit species-specific, developmental stage-specific, and even organ-specific responses toward the competency of these NPs which are further influenced by multiple natural variables like light, temperature, and nutrient availability. Therefore, to get an overall understanding of how these NPs cross the boundaries, it is essential to study the underlying principles and the multiple entities that are playing a major role in governing this interaction. Plants, being sessile, comprise multiple organs like leaves, roots, stems, etc. and their response towards the delivery of NPs is crucial to understand. For instance, regarding the interaction between NPs and leaf cells, it is very common to include that NPs can communicate with any leaf cell via the stomatal or cuticular pathway (Fig. 2). Now, as the cuticle comprises a waxy layer, which is hydrophobic in nature it limits the movement of large NPs across the cell. Compared to this cuticular pathway, which is generally used for the uptake of small NPs, the stomatal pathway dominates at the forefront of plant foliar uptake of NPs (Fig. 2). Stomata, predominantly present in the epidermis of leaf cells, comprise guard cells which are thick internally and thin externally forming a bulging-like apparatus for

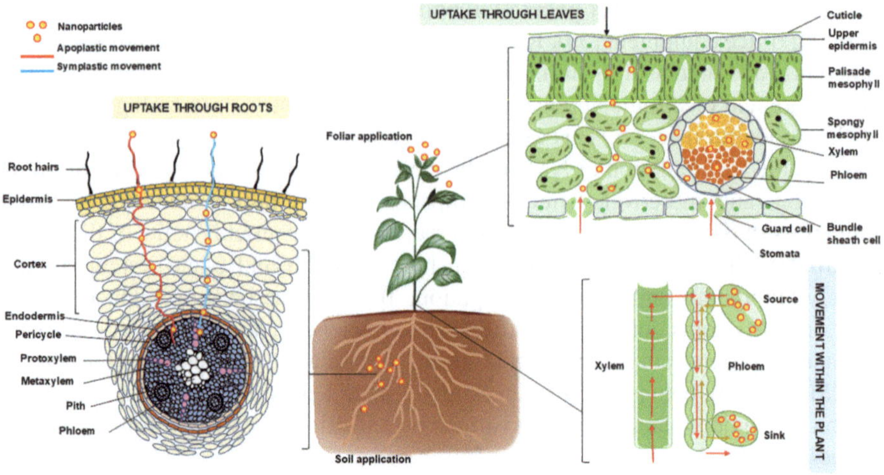

Fig. 2 Schematic diagram representing NP uptake through roots and leaves, and their movement through vascular tissues

the gaseous movement required for photosynthesis. The diameter of the stomatal pore during its open state is around 100 μm^2, which is enough for the passing of NPs. Therefore, maintaining this opening and avoiding the closure of the stomatal pore is one of the rate-limiting factors for the efficient delivery of NPs through the leaf cells. After crossing the epidermis layer of leaf cells via stomatal and cuticular pathways, NPs are further transported throughout the whole plant via phloem. Among the vascular tissues, the xylem is responsible for carrying water and minerals from the basal portion of the plant to its apical end via the transpiration-mediated mechanism, while the phloem carries the photosynthates from the leaves to different parts of the plant via its loading and unloading process. Hence, it is crucial for NPs to reach phloem for their overall movement throughout the plant system. They generally reach the phloem via two routes; the first one is from mesophyll cells (palisade and spongy mesophyll cells) to the phloem, and the other is moving through the cell wall spaces and intercellular gaps between mesophyll cells to the phloem, which is devoid of cytoplasm and other specialized structures. In the former pathway, plasmodesmata connect the mesophyll cell with the bundle sheath cell and eventually mediate the movement of NPs to phloem cells, while in the latter one, NPs squeeze through the gaps between mesophyll cells and reach the bundle sheath cell, and from there directly move towards phloem (Fig. 2). Other than the movement through leaf cells, NPs can also take the entry route through the root system. For efficient assimilation of NPs into the root system, multiple subsidiaries like the role of root exudates, root epidermis, Casparian strips, and even the rate of transpiration should be taken into account. Root exudates being the first line of barrier within the root cells, have the capability to allow the movement of both low molecular weight (*e.g.* amino acids, sugars, phenolic compounds and secondary metabolites) and high molecular weight (mainly proteins) compounds through itself.

Beyond root exudates, the root epidermis forms the second layer. Movement of NPs through the root epidermis is naturally through apoplastic and symplastic pathways. Like a series of boundaries, NPs now face root endodermis which comes after the epidermis and cortex, respectively. And within endodermis, it Casparian strip brings out a major difference. Casparian strip, being made up of suberin and lignin, forms an impenetrable wall towards the movement of NPs. Although multiple studies have been carried out to show the possibility of NPs crossing the Casparian strip, when NPs actually face the strips, their movement halts due to plasmodesmata's inability to play its role, leading to the closure of the symplastic pathway. All these biological events result in the pro-activation of the apoplastic pathway, which still carries the movement through intercellular gaps and finally towards the stele (Fig. 2). After crossing all these impediments, NPs finally reach the xylem, and transpiration rate is one of the limiting factors for the acquisition of NPs into roots in this section. Along with transpiration, both pressure potential and water potential equally contribute to maintaining the transpiration rate, which further regulates the xylem loading of NPs. Although a detailed intra-phyto-cellular delivery protocol of NPs has been discussed so far, the scenario from outside to inside of the cell is still incomplete without knowing about the interaction of NPs with cell walls and plasma membranes. Therefore, in this context, the first arena of discussion will be the cell wall, which forms the first or primary boundary for the NPs to enter within the plant system.

Contrary to all other cells, plant cells come with a distinct cell wall. Being the first shield, it surrounds the plasma membrane and provides the mechanical stability and tensile capacity to protect it from osmotic shocks and stress, although, for a particle to penetrate, it must be at its appropriate size to pass through the pores or the spaces within the cell wall. As per studies, plant cell wall porosity is roughly around 13–20 nm while it varies with plant species as well as development stage [32]. Besides, the plant cell wall consists of major components like cellulose microfibrils, pectins, hemicellulose, structural glycoprotein, and phenolic compounds like lignin. Among these constituents, pectins have the ability to modulate cell wall porosity and regulate apoplastic fluxes. Even a slight shift in pH, for example, acidic to basic and vice versa, can induce the capability of cell wall modifying enzymes like polygalacturonase (Pgs) and endo-1,4-β-glucanase (Egase), which can facilitate the structural configuration of the whole porous matrix. Hence, changes in cell wall structures due to chemical components, pH change and enzymatic mechanisms have the potential to expand as well as loosen the cell wall matrix. This can even allow NPs with large sizes to pass through the matrix pore by passive transport. Therefore, designing an NP while keeping the size in its appropriate formulation is very crucial. Despite all these, it is clear that the porosity of the cell wall is not the only criterion that can effectively describe the efficiency of NPs to cross the rigid cell wall. Rather, this interaction is slightly complex beyond expectation. Another property that equally plays a vital role is the electrochemical gradient. As it stands, plant cell wall has both charge and membrane potential that make it more resistant to the entry of exogenous particles. The electrical charge in a plant cell generally varies, with roughly around −40 to −110 millivolts (mV), and this charge is sufficient for any plant to uptake or resist penetrating any charged NPs from its surroundings. At a glance, a positively charged cell wall will always attract any negatively charged NPs, causing them to stick to the outer wall of the cell, not necessarily internalizing them. In their research study, Avellan et al. [40] found that negatively charged AuNPs were distributed inside the roots of *Arabidopsis* while the positively charged AuNP were not detected. This showed the retainment of positively charged NPs into the outer wall by the negatively charged cell wall. Another contradictory work conducted by a group of researchers showed that positively charged NPs can transverse the wall more efficiently [41].

3.2 Endocytosis and Intracellular Trafficking

After crossing the primary level of cell barriers, for its smooth internalization, NPs need to cross another barrier, the plasma membrane. Till now, the interaction between NPs and plasma membrane is limited but to get a vivid understanding of this interaction is essential. The plasma membrane is single-layered, mainly composed of lipids, fatty acids, and proteins. The protein is embedded within the lipid layer, either tightly with multiple strong bonds (termed intrinsic proteins) or hanging on the termini of the fatty layer (termed extrinsic proteins). Referred to animal cells, which are without cell walls the most common mechanism of internalization is clathrin-mediated or endosome-mediated endocytosis in some cases. As discussed earlier, the size of NMs overlaps with the pore diameter of the cell wall, allowing them to reach the plasma membrane directly. Therefore, endocytosis is not the universal mechanism for the uptake of NPs from outside. However, several studies have shown that the protoplast of plant cells can penetrate particles with a size less than 1 μm through endocytosis, suggesting that NPs that are engulfed by the process of endocytosis have strict particle size selectivity (Fig. 3). Depending on size and shape, NMs can also be assimilated with the aid of transporters or channels present on the membrane to gain entry within the cell. For instance, AgNPs are capable of crossing the mechanically gated ion channel having a pore size of around 3 nm [42]. There are also some other non-endocytosis-mediated mechanisms, among which, the lipid exchange envelope penetration method remains a significant one. This mechanism is based on a non-endocytic route by which nanomaterials enter the cell. It is generally applicable for long and high-aspect ratio NPs, for example, nanorods, nanotubes, etc. In this method, unlike being swallowed by the cell, the nonmaterial integrates itself with that of the lipid membrane. To some extent, this process involves agitation or exchange of lipid particles, which hampers the integrity of the membrane structure. Apart from this, there are Quantum dots also, which are responsible for creating tiny holes or nanopores within the bi-layer membrane of the cell. These pores facilitate the movements of charged ions or molecules to pass through this minute aperture, thus disrupting the cell's natural membrane barrier. Another aforementioned factor that also requires a praiseworthy mention while maintaining the membrane compactness is the electrochemical gradient, also known as the membrane potential. The typical membrane potential

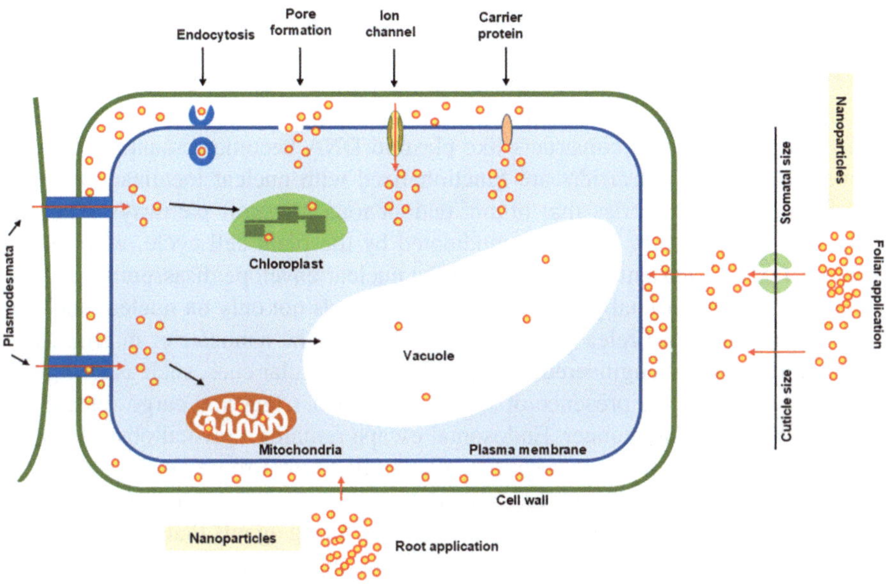

Fig. 3 Introduction of external NPs through endocytosis, ion-gated channels, cell wall carrier proteins into the cytosol and their internal movement through plasmodesmata

across the plasma membrane of a plant cell is −120 mV, which provides the essential activation force for the movement of charged NPs.

Post entry through cellular barriers, NPs infiltrate into cytoplasmic space, the gateway for all essential intracellular processes. Among various types of NPs that are designed, the majority act as the cargo for biomolecules such as DNA, RNA, or any formulated small peptides, which serve not only as simple carrier molecules but also are active mediators of genetic transformations. The ultimate goal of mediating such delivery lies in the regulation of gene expression, to silence or express any crucial gene to modify the biochemical pathways or trigger any desirable physiological or developmental outcomes in the plant system. However, to bring any functional result, it must not only reach within the nucleus, rather it should also be interpreted with the transcriptional or post-transcriptional machinery. This brings to the avenue of understanding how delivered nucleic acids are expressed and regulated and how their interaction controls the entire molecular dogma within the plant cell.

3.3 Gene Expression and Regulation

Following the successful intracellular delivery of nucleic acid-loaded NPs into plant cells, a cascade of finely regulated biological processes governs their eventual expression and functional integration. Within the cytoplasm, these complexes often rely on the plant cell's cytoskeletal infrastructure, namely actin filaments, and microtubules, for intracellular trafficking. This guided movement is orchestrated through vesicle-mediated transport and molecular motors, directing the cargo toward the nucleus. Given the stringent size restrictions imposed by the nuclear pore complex, which limits passive diffusion to molecules under ~40 kDa, the active nuclear import of large constructs like plasmid DNA becomes essential. To overcome this, many nanocarriers are functionalized with nuclear localization signals (NLS) or surface moieties that mimic endogenous transport pathways. The efficiency of nuclear entry is further modulated by the plant cell cycle, with mitotic phases offering transient windows when the nuclear envelope disassembles, allowing easier access. Crucially, gene expression depends not only on nuclear entry but also on the controlled release of nucleic acids from the nanocarrier matrix. Many delivery systems are engineered to respond to intracellular cues, such as changes in pH, redox state, or the presence of specific enzymes, triggering cargo release in a compartment-specific manner. Endosomal escape remains a pivotal obstacle, without successful evasion of the degradative endo-lysosomal route, nucleic acids risk enzymatic degradation. To address this, delivery systems have incorporated viral-inspired elements like fusogenic peptides and buffering agents that induce osmotic rupture through the proton sponge effect, promoting efficient cytoplasmic dispersion. Inside the nucleus, transgene expression is governed by the interaction between delivered DNA and host transcriptional machinery. The choice of promoter, whether constitutive (*e.g.* CaMV 35S) or inducible, determines the temporal and spatial specificity of transcription. However, along with promoter activity, chromatin accessibility, histone modifications, and nuclear architecture also influence transcriptional efficiency. Epigenetic features, including DNA methylation and chromatin remodeling, can either facilitate or repress transgene expression depending on the genomic context of integration. At the post-transcriptional level, mRNAs undergo essential modifications such as 5′-capping, splicing, and polyadenylation, processes that shape their stability and translational readiness. The delivery of RNA molecules, particularly siRNAs or synthetic mRNAs, engages the plant's innate gene regulatory pathways, notably RISC [43]. This post-transcriptional gene silencing (PTGS) can either enhance specificity or unintentionally degrade functional transcripts. Codon optimization, RNA structural refinement, and UTR engineering are employed to improve translation rates and mRNA longevity. Yet, plant translational machinery may still detect foreign sequences as aberrant, leading to repression or immune activation. A persistent obstacle in transgene-based strategies is epigenetic silencing, often triggered by high expression levels or transgene repeats. Such silencing may stem from homology-dependent gene silencing (HDGS), DNA methylation, or histone modifications. Furthermore, position effects from random genomic integration led to variable expression profiles depending on the local chromatin environment. Strategies involving matrix attachment regions (MARs), insulators, and silencing suppressors are being explored to circumvent these issues. Ultimately, nucleic acid-induced modulation of gene expression can trigger feedback responses within the plant's signaling networks. Alterations in transcription factor levels, stress response pathways, or hormonal balances may reshape metabolic and developmental processes. Thus, integrating systems-level analyses becomes essential to fully predict and manage the phenotypic consequences of such precise molecular interventions.

4 Applications of NPs-Mediated Nucleic Acid Delivery in Plant Cells

4.1 Genetic Engineering for Crop Improvement

Among the achievements towards continuing the efforts of sustainable development, genetic engineering is the centered one. Leveraging genetic engineering for agricultural crop improvement, and natural product synthesis are certain pillars towards achieving the aim of sustainability. In the past three decades, it has been a revolution for the biotechnology sector, with the amalgamation of genome editing and sequencing tools, which in turn has innovated plant synthetic biology and bioengineering. In agriculture, genetic engineering has the immense potential to transform crop production by improving the yield as well as the nutritional quality while providing resistance to multiple factors like herbicides, pests, diseases, and even abiotic stresses like drought and heat. Still, while considering genetically engineered mechanisms, it is basically the gene level adjustment or manipulation that can be achieved either by inserting a foreign gene to encode our desired trait of interest or targeted gene suppression by harboring genome editing methodologies or RNAi silencing mechanism, although these novelly acquired technologies also come with multiple limitations which have been discussed in-depth later in this chapter. Among the most accepted protocols for transformation, there is the *Agrobacterium*-mediated transformation protocol which is highly efficient in inserting any heterologous gene for concurrent expression but it is also limited to few plant species because though it is efficient in showing its efficacy in nuclear genome, it faces a severe breakdown while transforming chloroplast or mitochondrial genome. Among the other renowned gene delivery protocols, there are biolistic DNA delivery and electroporation methods. In biolistic mode, it basically involves deploying AuNPs which adhere to the nucleic acid (DNA) within the plant cell employing a highly pressurized gene gun. Due to such rapid propulsion force, it transverses into both the cell wall and plasma membrane simultaneously, positioning the micro compounds into the specific region of the chromosome. To carry out this process, a large concentration of DNA, sophisticated equipment and cautious handling expertise is required. On the other hand, electroporation leverages the medium of an electric field to disrupt the plasma membrane of plant cells, although sometimes this strong electric field is potent enough to create discrepancies in the recombinant gene of interest. The main hindrance lies in the rigid cell wall which is rich in cellulose, xylan as well as pectin which forms multiple lines of barrier for an entry of genes. To get rid of these shortcomings, protoplast isolation and chemical transfection method using Polyethylene glycol (PEG) enables the incorporation of

DNA directly into the plant's protoplast, but the challenge remains in protoplast-mediated whole plant regeneration, which comes with a minimum transformation rate to carry out the further downstream procedures. Various studies have emphasized the potency of NPs across a diverse selection of plant species, *viz. Nicotiana benthamiana, Triticum aestivum, Gossypium hirsutum, Viscia fava*, and *Cannabis sativa*. These observational results open a summary that NPs are designed in such a way to transfer multiple biomolecules like DNA, RNA, or even protein in a way without regard to species. To date, a range of NPs has been evaluated and used for modifying any desired traits or in transformation including hybrid NPs, carbon nanotubes, metal NPs, quantum dots, titanium and graphene oxide NPs, small peptide-based NPs and nanohydrogels (Table 1).

Table 1 Effects of diversified NPs on various physiological parameters of different crops

Crop	Types of NPs	NPs	Delivery method	Physiological effects	References
Mustard (*Brassica juncea*)	Metal oxide	AuNPs	Foliar spraying	Enhancement in yield attributing characters and elevation in sugar content	Arora et al. [44]
Sorghum (*Sorghum bicolor*)	Metal oxides	AgNPs	Indirect uptake via agar medium	Cessation of plant growth	Lee et al. [45]
			Acquisition via roots	Enhanced plant growth	
Wheat (*Triticum aestivum*)	Metal oxides	Fe_2O_3 NPs	Passive uptake through seed coat while germination	Higher germination ability upon lower concentration of NPs	Feizi et al. [46]
Onion (*Alieum cepa)*	Metal oxide	ZnO NPs	Passive uptake through seed coat while germination	Higher germination ability upon lower concentration of NPs	Raskar and Laware [47]
Wheat (*Triticum aestivum*)	Metallic oxides	CeO_2 NPs	Acquisition via roots	Improvement in multiple agronomic traits such as plant height, chlorophyll content, number of tillers, peduncles length, and number of internodes, and overall biomass	Rico et al. [48]
Chickpea (*Cicer arietinum*)	Metallic oxide nanostructures	CuO NPs	*In vitro* exposure via ½MS medium	Stunted shoot and root, higher production of ROS	Nair and Chung [49]

(continued)

Table 1 (continued)

Crop	Types of NPs	NPs	Delivery method	Physiological effects	References
Mung bean (*Vigna radiata*)	Metal oxides	TiO_2 NPs	Foliar spray	Enhancement of overall plant length, significant increase in chlorophyll content and soluble leaf protein content	Raliya et al. [50]
Alfalfa (*Medicago sativa*)	Metallic oxides	K_2SO_4 NPs	Acquisition via roots	Improved resistance again salt stress	El-Sharkawy et al. [51]
Watermelon (*Citrullus lanatus*)	Metallic oxides	CuO NPs	Foliar spray	Resistance against *Fusarium oxysporum*	Elmer et al. [52]
Rice (*Oryza sativa*)	Bio-based NPs	ZnO NPs	Seed priming	Enhanced seed-physiological and yield-attributing agronomic traits	Itroutwar et al. [53]
Tomato (*Solanum lycopersicum*)	Hybrid NPs	Cu + QACs NPs	Foliar spray	Resistance against *Xanthomonas perforans*,	Ozcan et al. [54]

Ag: silver, Au: gold, CeO_2: cerium oxide, Cu: copper, CuO: copper oxide, Fe_2O_3: iron oxide, K_2SO_4: potassium sulphate, MS: Murashige and Skoog, NPs: nanoparticles, QACs: quaternary ammonium compounds, TiO_2: titanium oxide, ZnO: zinc oxide

4.2 *Virus Resistance and Plant Defense*

As previously stated, NPs, being a typical small particulate material having a size within the range of 1–100 nm, have been categorized in multiple forms based on their chemical properties. They can be categorized into metallic and non-metallic NPs depending upon their surface composition, forms, densities, and reactivity with respect to adsorption properties and redox reactions. Other than this, NPs also possess various inherent properties which makes them crucial in multiple avenues of biological sciences. The properties that make these compounds at the forefront include their chemical reactivity, being optically active in nature, small size having access to large surface area, mechanical durability and strength, and inertness. Due to this, the application of NPs or nano compounds has been leveraged in numerous sectors like fuel cells, energy retention technologies, frequency tunable devices, and even in molecular electronics and photonics systems. But in recent decades, NPs have received widespread interest and attention within the biological sciences community. Nanobiotechnology, an interdisciplinary field combining nanotechnology with biotechnology, has provided the way for innovative solutions from healthcare to agriculture. These small-scale compounds have shown a sparkling result in biosensing technology, pesticide and herbicide methodology, advancing image technology, molecular validation, and even cancer treatment strategies. Nanomedicine, another field of leveraging these NPs having antimicrobial, antiviral, and anticancer activity, has opened new avenues for the diagnosis, treatment, and prevention of various diseases. These promising outcomes are not only limited to *in vitro* medical applications but now there is also growing interest from the agriculture domain, giving rise to the field of agro-nanotechnology. In agricultural systems, nanotechnology provides multiple benefits reducing the dependency on the use of chemical fertilizers, increasing the uptake of water and nutrient utilization, providing plants protection against various arrays of plant pathogens and viruses, improving the efficiency of fertilizers, and enhancing crop productivity through effective pest and nutrient management [7].

Plant viruses are one of the serious causative agents that lead to some disastrous problems in agriculture, and it is very crucial to take strong measures to control viral infection. Basic farming practices like utilization of infected plant material unknowingly, growing the same crop throughout a large area (monoculture farming), lack of pest control, insufficient crop rotation, or mechanical transmission often help in the spreading of phyto-viral infection. It is estimated that more than half of plant diseases are caused by viruses, and so far, nearly around 900 plant viruses have been elucidated which is directly or indirectly affecting a huge number of crops. The intensity of viral infection in plants depends on the environmental condition, type of viral strain, host organism, host plant type, multiplication rate, and concentration. When a plant gets affected, it usually shows multiple symptoms like rolling of leaf, molting, wilting, leaf drying, fruit deformities, tissue death, stunted growth, and eventually death. Since viral infection on plants is beyond cure, the preliminary strategy for managing them is prevention and early intervention, but due to influential groundbreaking research, NPs have shown immense potential due to their antibacterial, antiviral, and antifungal agents, mostly in the animal system but increasingly in the plant system too. Among these, both metallic and non-metallic NPs have been fruitful in their application, with metallic NPs being at the forefront. Regarding the current status of virus mitigation, plant viruses remain a persistent threat to the global agriculture ecosystem, with some infestations causing devastating damage mainly in resource-limited regions. Even the economic consequences are significant, with high annual loss due to compromising loss of crop yield and quality. So, effective management practices to reduce plant viral infection have become an urgent priority. In protected cultivated systems, the use of greenhouse chambers is crucial, where efforts are made to mitigate virus transmission by focusing on the management of insect vectors. Use of synthetic compounds like acephate, pymetrozine, imidacloprid, etc., and biocontrol agents are widely employed to reduce vector populations as many of the plant viruses depend upon these vectors for their life cycle from survival to replication and even dissemination. Infected plants or residual plant debris, which acts as a potential virus reservoir should also be removed. Utilizing reflective mulches, colored traps, or mineral oils can also interfere with vector landing and virus transmission. On the laboratory level, both highly advanced methodologies and conventional technologies can be implemented to address plant-viral infections. Conventional methods include meristem tip culture, cryotherapy, thermotherapy, and chemotherapy. While these strategies reduce the viral content in plant tissues, they are primarily limited to *in vitro* settings and not always fully scalable for large agricultural practices. Contrastingly, utilizing advanced techniques provides more targeted interventions. Analytical techniques like RNA silencing mechanisms, further including RNA interference (RNAi) provide gene-specific suppression of viruses through implementation of double-stranded RNA (dsRNA) mediated pathways (Fig. 4). This strategy can also be executed on insect vectors offering a dual tone of benefits. In plants, RNAi-based techniques have received popularity for viral transmission control, pest resistance, and functional genomics. However, double-stranded RNA molecules face tremendous complications in assimilation and distribution throughout tissues because they are intrinsically unstable, and are vulnerable to RNase disintegration. Over and above to enhance foliar or root uptake and enable widespread translocation throughout the plant, NPs can shield dsRNA from decomposition. One of the most successful applications of NPs used in RNAi distribution is layered double hydroxide (LDH) nanosheets [55], which were made available as BioClay. Upon exposure to plant moisture or enzymes, these nanosheets ligate with dsRNA molecules and facilitate slow liberation, inhibiting gene activity for a prolonged period of time. BioClay-mediated RNAi methodology [56] has facilitated successful pathogen repression in crops such as tomato and lettuce for more than 20 days after a single application. However, the effectiveness of these research frameworks must be validated under field conditions to ensure long-term durability. Other genetic strategies,

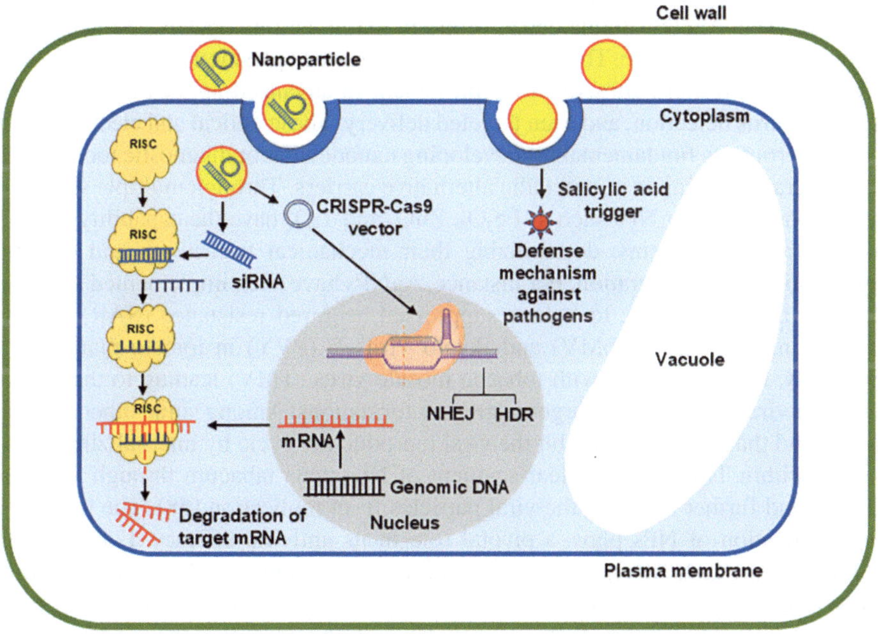

Fig. 4 Overview of NP-mediated gene silencing, defense mechanism, and CRISPR-based genome editing in plants

such as the development of transgenic plants, employing virus-tolerant and resistant varieties, and employing lower virus strains for cross-protection have also been carried out in recent years. Chemotherapy, whose application is precisely more employed in animal virology, has also been adapted to use in plant tissue culture. Compounds such as ribavirin, acyclovir, and azidothymidine have been executed against the infestations of plant viruses.

In plant systems, metallic NPs have emerged as a promising solution, mainly as antiviral agents against multiple odds due to their physicochemical and biological properties. They have an added advantage over the use of non-metallic NPs, particularly in terms of antiviral efficacy, cellular interactions, and multidimensionality. Firstly, metals like silver (Ag), gold (Au), copper (Cu), and zinc (Zn) possess internal antiviral and antimicrobial attributes. These metals have the ability to inhibit the viral replication mechanism, impede the capsid integrity, or disrupt the viral genetic material through systemic mechanisms like generation of Reactive oxygen species (ROS) and release of ions. Such type of inherent functionality is not present in non-metal-based NMs, limiting their viral applications unless it is modified exogenously. Secondly, metallic NPs are very minute in their size but with a large surface area. This makes them easily bind to the viral components, enabling them to easily internalize themselves within cells and potentially inactivate viruses at the site of infection. Now these metallic NPs are very potent in generating ROS as stated above via the chemical redox reaction which further provides them with extramural adaptability to damage the viral components. This redox-driven antiviral mechanism makes the metallic NPs more effective in viral inhibition as compared to the non-metallic ones. Another fundamental property that metallic NPs have is their being optically active in nature. They exhibit localized surface plasmon resonance (LSPR) which makes it more attractive to deploy them in advanced applications like biosensing, virus detection, and even targeted delivery. These optical and electrochemical properties are fundamental for developing nanodevices or diagnostic tools which are generally lacking in nonmetallic alternative carriers. Through multiple studies it is demonstrated that NPs such as Fe_3O_4, ZnO, and TiO_2 have the capability to bind with viral coat proteins, destabilizing their mechanical framework and causing aggregation or disintegration. For instance, AgNPs have been implemented successfully as antiviral agents to induce suppressed acquired resistance (SAR) against Tomato mosaic virus (ToMV) and Potato Virus Y (PVY) in tomato plants [57]. Similarly, Fe_3O_4 interacts with tobacco mosaic virus (TMV) leading to the breakdown of viral integrity and larger aggregate formations. Among some other studies, it is found that CeO_2 NPs inhibit the viral reproduction cycle by internalizing themselves within TMV induced leaf segment of Nicotiana tabacum through vascular tissues and further stopping the viral particles from replication [58]. The timing of the application of NPs plays a pivotal role in its antiviral efficacy. For instance, foliar spray of TiO_2 NPs in *Vicia faba* plants after inoculation with broad bean strain virus (BBSV) shows a dramatic reduction in viral concentration as well as infection intensity [59]. This observation paves an array of understanding that post-infection treatment allows the NPs to interact directly at the site of viral repositories and possibly also interfere with viral systemic spreads. Eventually, the activity of NPs as immunosuppression agents or antiviral agents depends sometimes on the timing of their deployment, *i.e.*, whether they are administered before, during, or after the viral inoculation. Treating before inoculation serves to provide a unique direction toward plants' immune response leading to a systemic resistance mechanism, while the post-inoculation strategy is more directed toward counteracting the viral constituents and restricting their proliferation. In parallel, NPs being in their multifaceted form induce systemic resistance in plants by mimicking pathogen-associated molecular patterns (PAMPs) and inducing plant immune responses. When it interacts with plant cells, it directly activates the salicylic acid (SA) dependent signaling pathway (Fig. 4), which further prolongs the systemic acquired resistance (SAR). Not only the SAR pathway but also jasmonic acid (JA)- and ethylene-mediated pathways are also involved in generating induced systemic resistance (ISR). This cascade of signaling events leads to the up-regulation of multiple genes associated with its pathogen resistance. Along with that, the production of secondary metabolites and antioxidant enzymes like superoxide dismutase, catalase, peroxidase, etc. further strengthen the plant's cellular defense by detoxifying ROS, which is able to disrupt the plant's own cellular membrane integrity. In a study when the scientists implemented the combination of CuNPs and chitosan polyvinyl alcohol hydrogels (Cs-PVA), they found that it elevates the expression analysis of JA pathway-related genes in tomato under abiotic stress [60]. Finally, NPs comprise a novel and sustainable eco-benign method to plant viral disease management, generating both prompt viral reduction and enduring development of plant immunity, making them an important component of integrated plant protection strategy.

4.3 Overview of Other Potential Applications

Plant transgenic technology is evolving day by day and it has significantly improved the grain quality, crop yield, size of the crop, and secondary metabolite production, by enhancing fundamental plant biological systems. As studied earlier, the plant cell wall being the first barrier of defense against any exogenous molecules, it is tough to incorporate any foreign DNA within the nucleus of the cell. Plant improvement strategy began long ago with crop breeding methodologies with the cultivation of wild plant species to improve crop productivity and nutrient richness. The Green Revolution of 1950, with the incorporation of a new crop breeding strategy has generated high-yielding varieties of semi-dwarf wheat and rice and is glorified as one of the finest examples of how plant breeding can be employed to boost the yield value to cope with scarcity of food for an exponentially growing population, but breeding procedures do have some limitations like it demands more time and labor power; it is nonselective at the genetic level; it can reduce plant fitness due to incorporation of undesirable traits and many more. Chemical agents and various irradiation-based technologies have been employed to introduce beneficial traits within the plant system but due to randomly occurring mutation [61], non-specific alterations and sometimes accumulation of excess nucleotides hinder the process.

Due to expanding demand and the introduction of new procedures to lower the liability of conventional breeding techniques, newly arrived genetic engineering technologies like *Agrobacterium*-mediated transformation or the use of tools like particle bombardment have enhanced the delivery method as compared to the conventional breeding one. With this, a broad range of desirable traits have been introduced, whose transformation efficacy through desirable phenotypic characteristics can also be checked, but the problem still persists, *i.e.* incorporation of the target gene is randomly taking place within the genome which can lead to undesirable outcomes. To provide a solution for this much-needed accumulation, researchers in the last decade have revolutionized the way of targeting in a more specific manner, by introducing nuclease-based genome editing methods, such as ZFNs (Zinc Finger Nucleases), TALEN (Transcription Activator Like Effector Nucleases) and CRISPR (Clustered Regularly Interspaced Short Palindromic Repeats) which is also associated with a nuclease known as Cas, and collectively these all are target specific, rapid, precise and moreover, they provide long-standing solutions to incorporate desired new traits into specific locations in the genome [62]. In the field of plant or agricultural biotechnology, CRISPR-Cas tools hold immense potential in the domain of transformation efficacy and functional genomics research, by incorporating new desirable traits within crop genomes and making them resilient against both abiotic and biotic stresses [63]. However, this methodology also faces some challenges while conducting an experimental design as well as several barriers. Most of the challenges include a lack of proper transformation efficiency in plant tissue and cell culture, depending on the nature of generation, somatic mosaicism, and lack of information about the nature of edits in plants at the ploidy level and in propagation methods. Even introducing CRISPR constructs within the plant system faces multiple global regulatory rules and ethical considerations. NMs on the other hand, possess unique characteristics to penetrate the plant cell wall and its subsidiary tissues and it has the capability to reach both cellular and subcellular compartments. Beyond this ability, they can even facilitate the targeted cargo delivery and controlled release while keeping the genetic material intact, protecting them from any enzymatic degradation. NPs generally act in a species-independent manner which makes them compatible and accessible across diverse plant systems. Although NPs can be an effective solution for CRISPR-Cas-mediated genome editing in plants (Fig. 4), there is still some space for improvements required in the field of nanotechnology.

This innovative technology still persists with several drawbacks despite having exceptional advantages over previously discovered methodologies. One of the major concerns in target-specific edits of genetic engineering is to induce somatic embryogenesis, where germline transformation is not possible. Even within the same species, different genotypes react differently during the tissue culture process. Additionally, delivery of CRISPR reagents is limited within the cell's internal environment by crossing the cell wall barrier of plant cells. As stated earlier, the plant cell wall, mainly composed of extracellular matrix and polysaccharides, is literally tough to penetrate due to its selectivity in nature. An alternate way to get rid of this problem is by the use of protoplasts, which are basically plant cells having their cell wall removed by enzymatic mechanisms. However, regenerating a whole healthy plant is quite difficult from protoplast and it is constrained in many of the plant species. Leveraging the potential of particle bombardment to keep the cell wall of the plant intact tends to possess another way out, and microscopic particles coated with CRISPR constructs are shot into the cell at a higher speed through this protocol, although sometimes this can act in a false positive manner, where the cells get damaged permanently and even the efficiency of gene editing through this method is low. Before delving into its whole purpose, another shortcoming that can hinder the procedure is the selection of appropriate single guide RNA (sgRNA) for sequentially working the genome modification procedure. As this sgRNA will guide the Cas nuclease to the exact region of the plant's DNA, it can initiate the restriction procedure. If this whole directional procedure collapses, then Cas nuclease will make a cut in an undesired location which leads to off-target editing and might lead to further conspiracies [64]. Even though CRISPR tools are truly implemented to make site-specific, precise, heritable changes within a crop plant without causing any unwanted mutations, a lack of knowledge about the functionalities of most of the plant genes remains a backdrop. Therefore, precise knowledge about the targeted genes and their mode of action at molecular, biochemical pre- and post-transcriptional levels is a major pre-requisite for the study ahead, which will lead to transcriptomic analysis and several omics-assisted approaches at proteomic and metabolomics levels. These layers of information are all interconnected, and any desired changes made at the gene level will manifest through the entire network of cellular function.

Nanotechnology will provide a longstanding solution by enabling the efficient delivery of nucleotide bio-molecules into a diverse range of plant species. Beyond just delivery, they can even maintain the targeted release, protecting the cargo from degradation and at a broader aspect even allow real-time imaging of biomolecules as cargo inside the plant cell harboring their engineered fluorescent properties. However, delivery of DNA, RNA, or proteins into plant cells, using NPs for CRISPR/Cas genome editing remains challenging due to the significantly larger size and charge of the plasmid, which makes them sometimes inappropriate to load and deliver effectively. Similarly, Cas9 constructs delivery as RNP poses severe complications due to its large size, strong charge density and instability in the ionic environment. To get rid of these shortcomings, researchers nowadays are exploring multiple objectives like the covalent attachment of Cas9 RNPs to NPs with chemical bonds that can break once inside the cell due to its reducing environment. The employment of enzymatically cleavable linkers that break when exposed to certain plant enzymes is known as one of the probable solutions. Nanotechnology, on the other hand, can overcome the drawbacks of germline regeneration by enabling direct genetic insertion of plant germline cells, which can lead to the development of natural offspring bypassing tissue culture. Editing the male gametophyte (pollen) is one promising concept that might lead to the formation of edited plants just by fertilizing a plant with the altered male gametophyte [65]. This is challenging in several aspects because pollen is delicate with a hard exterior shell. Hence, to introduce any new modification without any harm to the pollen cells is really tough.

Because of their strength and small size, NPs like CNTs may be able to transfer CRISPR constructs into pollen, particularly through naturally thinner regions of the pollen surface. To create NPs that can efficiently penetrate the pollen without damaging it, more study is required in this prospect. Leveraging the potential of NPs and using them to transport CRISPR-gRNA constructs directly to the reproductive organs of flowers, including the pollen and ovules (female gametophytes) can be suggested as an alternative approach. Gene editing in many floral sections, including the developing embryo, may be possible using this method.

5 Challenges and Future Perspectives

Nanoparticle-oriented delivery systems provide superior transformative potential in the vast area of plant biotechnology, with the capability to overcome several practical and theoretical challenges in order to ensure the proper feasibility of this advanced tool. The following section delves into numerous key limitations with an emphasis on future directions, ranging from toxicity concerns to commercialization pipeline, aiming to provide a holistic overview of the future research paves forward.

5.1 Toxicity and Biocompatibility of NPs

Nanotechnology-assisted gene introductory systems gain greater eagerness despite having concerns regarding the toxicity and biocompatibility of NPs. The core issue lies in the interaction between engineered NPs and unpredictably complex phyto-cellular machinery. Numerous previous research studies stated that NPs, especially those having heavy metal cores or non-degradable shells, can result in oxidative stress generation, destabilizing cell membranes, and even cellular death in NP-treated plants. Metal-based NPs based on Ag and ZnO are popularly known for their antimicrobial properties but may inadvertently disrupt phyto-cellular physiological processes upon indiscriminate usage [66].

Plant-specific responses towards NPs add another layer of complexity to its applicability. Metal oxide-associated NPs formulations may exhibit phytotoxic effects in crops like rice (*Oryza sativa*), maize (*Zea mays*), etc. [67]. This efficiently makes universal solutions elusive and underlines the importance of species-specific biocompatibility testing. Furthermore, the majority of the toxicity studies, to date, concentrate on prompt post-treatment physiological alteration, often sidelining long-term or transgenerational impacts that could manifest in growth, fertility, or gene expression patterns, although further extensive studies need to be chalked out for future research. Therefore, mitigation of these concerns requires suggesting recent research studies to be highly focussed on the development of biodegradable or bio-derived NPs, *viz.* those based on chitosan, lignin, or cellulose. These organic carriers offer greater compatibility with phyto-cellular systems with limited environmental accumulation as well as magnification. Functionalization of NPs with biocompatible coatings like polyethylene glycol (PEG), plant peptides, or polysaccharides can also elevate compatibility and reduce cytotoxicity. Hence, the formulation of truly non-toxic, highly efficient NPs remains a research frontier in the field of biotechnology and toxicology requiring interdisciplinary collaboration among plant scientists, chemists, biochemists, plant physiologists, and toxicologists.

5.2 *Scalability and Efficiency of NP-Mediated Delivery*

The rise of laboratory-level success stories requires the simultaneous addition of NP-mediated delivery systems with widely scalable, field-ready technologies, which is practically far from straightforward. Most studies utilized techniques like syringe infiltration, vacuum infiltration, or microinjection, which are impractical on a commercial scale. Furthermore, NP synthesis methods such as hydrothermal treatment or chemical vapor deposition are time-demanding, resource-intensive, and often lack consistency across the entire process. On the other hand, scalability demands robust efficiency in delivery protocols spanning across an array of environmental conditions, which introduce hurdles in maintaining NP behavior in non-sterile, outdoor, open environments. Several factors like ambient temperature, optimal humidity, pedological chemistry, and even the soil-associated microbiome can impact the efficiency of NP acquisition and internalization. Despite these factors, the ability to uniform delivery of nucleic acids across multiple plant tissues and developmental stages poses another potential impediment.

The reproducibility of delivery outcomes is closely linked with the physicochemical properties of NPs, such as shape, size, surface charge, and functional groups [68]. Moreover, transfection efficiency varies drastically with even a slight variation in the said parameters. Hence, this research field requires standardized protocols for NP characterization as well as confirmation of delivery. Some promising strides have been procured in this direction. Spray-based delivery systems and root-absorption models are now being explored as alternatives to injection-based methods. These tools, combined with advanced precision agricultural technologies like drone-enabled spraying, nanofabrication, etc. offer a glimmer of hope for scalable deployment. However, comprehensively evaluative field trials, cost-benefit analyses, and eco-risk assessments must accompany these innovative strategies to establish their practical viability.

5.3 *Future Research Directions and Opportunities*

The area of NP-mediated gene delivery is still in its infancy, followed by the rich tapestry of research opportunities across inter-scientific disciplines. Integration of computational modeling along with machine learning or deep learning for designing and predicting NP behavior in plants tends to be a promising avenue in this regard. Simulation of NP-cellular interactions helps to dig the path for researchers for pre-screenable delivery with improved efficacy and pre-synthesis safety, which leads to better time and resource management. Moreover, multi-omics approaches including proteomics, transcriptomics, and metabolomics are being extensively employed for exploring in-depth aspects of cellular response while exposed to NPs. Such comprehensive studies can illuminate several unexplored biochemical pathways, detect biologically sourced markers for gaining successful delivery, and unveil novel mechanisms of internalization and endosomal escape. These findings integrated with rational designing of next-generation NPs eventually result in smart delivery systems capable of reacting to internal or external stimuli *viz.* temperature, pH, and light, which provide spatiotemporal control over gene expression. Implementation of NPs for *in situ* genome editing with CRISPR-Cas offers another potential direction to this innovative research field. Unlike conventional transformation protocols, NP carriers can introduce gene-editing tools transiently into the cell, effectively reducing the risk of off-targets. This co-aligns well with regulatory trends favoring non-GMO approaches, especially in countries with strict biotechnology laws.

Inclusive exploration into NP delivery in recalcitrant species, especially resistant to *Agrobacterium* or biolistic methods, could greatly expand the scope of plant biotechnology. For instance, fiber crops, woody perennials, and legumes often present challenges for genetic transformation [69, 70]. Therefore, customizing NP properties for species-specific acquisition could deliver viable alternatives for such phytosystems. Potential interdisciplinary collaborations, particularly between plant scientists, materialistic engineers, and regulatory bodies, are crucially important for these opportunities to materialize into tangible innovations in this field.

5.4 Potential Commercial Applications and Industrial Perspectives

The commercial potential of NP-mediated nucleic acid delivery is immensely renowned, particularly in the context of an ever-expanding global population, ecological turbulence, and the utmost need for sustainable agriculture. By enabling precise, transient, and target-oriented gene expression, NPs can revolutionize multifaceted aspects of crop advancement, from stress resistance to yield enhancement as well as nutritional biofortification.

NPs with their immense potential to bypass regulatory backdrops associated with transgenic crops possess an added advantage over that of conventional protocols [71]. Since many NP-linked transformations do not involve stable integration of foreign DNA, they may not fall under existing GMO regulations in several jurisdictions, which reveals a bunch of opportunities for rapid deployment and public acceptance, especially in regions wary of genetic alterations. Alternatively, industrial interest is steadily expanding through the intensive keenness of biotechnology firms, agrochemical companies, and seed corporations with active investment in the advancement of NP platforms for gene delivery in phyto-cellular systems.

Additionally, the specialization of NP formulations to target specific crops or traits allows for personalized agricultural solutions. Such precision can not only intensify productivity but also minimize the requirement for resource-associated inputs, aligning with the aims of developing climate-resilient plants.

From the development of NP-mediated nucleic acid introduction inside plant cells to owning a commercial protocol finds its entire path with potential impediments including patentability, intellectual property disputes, lack of regulatory clarity, etc. [72]. Moreover, the cost of NP synthesis, particularly if not scalable, may deter adoption in resource-limited settings. It is, therefore, imperative that future research includes economic feasibility assessments, stakeholder engagement, and policy discussions to ensure inclusive and sustainable deployment.

6 Conclusion

The advent of NP-mediated nucleic acid delivery has ushered in a new era in plant molecular biology. This technology represents an immense shift from conventional gene transfer methods, such as *Agrobacterium*-mediated transformation and biolistics, towards more precise, detailed, efficient, and adaptable approaches, although, following all pioneering technologies, it comes with both unparalleled potential and formidable challenges. In the course of this chapter, fundamental principles, molecular mechanisms, and practical applications of NP-mediated gene delivery in plant systems were thoroughly studied. NPs enable non-integrative, species-independent nucleic acid introduction, allowing researchers to bypass conventional bottlenecks associated with plant transformation.

Several types of NPs, ranging from carbon-based to polymeric and metallic, possess their own sets of fors and againsts. Molecular acquisition pathways, including endocytosis and plasmodesmata trafficking, offer new biological insights into the foreign introduction of genetic materials through the phyto-cellular barrier. An array of multifaceted challenges including issues regarding toxicity, delivery consistency, eco-safety, and scalability remain to be resolved, to date. Commercialization of this tool still demands wider applicability towards mitigating these issues for further adaptation outside laboratory environments. Nonetheless, the recent surge in interdisciplinary research, advances in biocompatible materials, and the growing synergy with genome editing technologies mark significant progress.

The future of NP-mediated gene delivery appears both stimulative as well as transformative. The incorporation of AI-based NP designing, real-time delivery monitoring, and responsive carrier systems are pushing the biotechnological era towards a new age of precision biotechnology. These innovations could allow for real-time detection of gene expression in response to environmental stimuli, ushering in a form of dynamic, adaptive agriculture. Moreover, the alignment of this technology with regulatory frameworks favoring non-transgenic interventions creates a fertile ground for rapid commercial adoption. For developing countries, particularly those in Asia and Africa, where agricultural challenges are deeply intertwined with socio-economic factors, such innovative approaches could redefine food security strategies. From an academic perspective, this technology enables exploration into fundamental plant biology queries, *viz.* genetic regulation, epigenetic inheritance, and developmental plasticity, by providing tools for precise molecular interrogation.

In a nutshell, NP-associated phyto-cellular nucleic acid introduction is not merely a technological advancement, it is essentially a paradigm shift in the general concept of gene transfer, plant breeding, and sustainability in the agricultural field. However, its journey from lab to land will require conscientious collaboration, transparent policy frameworks, and inclusive innovation that takes into account the needs of both smallholder farmers and industrial stakeholders. Moving towards recommendations, the establishment of open-source platforms for NP synthesis and characterization protocols, global regulatory harmonization for transient genetic interventions, and capacity-building initiatives offer democratized access to this technology. With thoughtful development and ethical stewardship, NP-mediated nucleic acid delivery is poised to play a central role in the future of food systems, climate resilience, and agricultural equity.

Authors' Contribution BP, SD conceived the idea; AB, AD, BP collected data and prepared the basic manuscript; SS incorporated the pictorial diagrams; AB added the table; BP edited and corrected the manuscript; BP, SD revised and finalized the manuscript. All the authors read and approved the manuscript.

References

1. Shivashakarappa K, Marriboina S, Dumenyo K, Taheri A, Yadegari Z (2025) Nanoparticle-mediated gene delivery techniques in plant systems. Front Nanotechnol 7:1516180
2. Yong J, Wu M, Carroll BJ, Xu ZP, Zhang R (2024) Enhancing plant biotechnology by nanoparticle delivery of nucleic acids. Trends Genet 40(4):352–363
3. Liu S, Zheng Y, Pan L, Wang W, Li Y, Liu Z, Zhang X (2025) Nanodelivery of nucleic acids for plant genetic engineering. Discov Nano 20(1):31
4. Kumar S, Raj SK, Sharma AK, Varma HN (2012) Genetic transformation and development of Cucumber mosaic virus resistant transgenic plants of *Chrysanthemum morifolium* cv. Kundan. Sci Hortic 134:40–45
5. Su W, Xu M, Radani Y, Yang L (2023) Technological development and application of plant genetic transformation. Int J Mol Sci 24(13):10646
6. Komarova T, Ilina I, Taliansky M, Ershova N (2023) Nanoplatforms for the delivery of nucleic acids into plant cells. Int J Mol Sci 24(23):16665
7. Pramanik B, Sar P, Bharti R, Gupta RK, Purkayastha S, Sinha S et al (2023) Multifactorial role of nanoparticles in alleviating environmental stresses for sustainable crop production and protection. Plant Physiol Biochem 201:107831
8. Zhi H, Zhou S, Pan W, Shang Y, Zeng Z, Zhang H (2022) The promising nanovectors for gene delivery in plant genome engineering. Int J Mol Sci 23(15):8501
9. Lopes RC, Rocha BG, Maçôas EM, Marques EF, Martinho JM (2022) Combining metal nanoclusters and carbon nanomaterials: opportunities and challenges in advanced nanohybrids. Adv Colloid Interface Sci 304:102667

10. Wang Q, Zhang C, Shen G, Liu H, Fu H, Cui D (2014) Fluorescent carbon dots as an efficient siRNA nanocarrier for its interference therapy in gastric cancer cells. J Nanobiotechnol 12:1–12
11. Li W, Zheng Y, Zhang H, Liu Z, Su W, Chen S et al (2016) Phytotoxicity, uptake, and translocation of fluorescent carbon dots in mung bean plants. ACS Appl Mater Interfaces 8(31):19939–19945
12. Sai L, Liu S, Qian X, Yu Y, Xu X (2018) Nontoxic fluorescent carbon nanodot serving as a light conversion material in plant for UV light utilization. Colloids Surf B: Biointerfaces 169:422–428
13. Wang H, Zhang M, Song Y, Li H, Huang H, Shao M et al (2018) Carbon dots promote the growth and photosynthesis of mung bean sprouts. Carbon 136:94–102
14. Zheng Y, Xie G, Zhang X, Chen Z, Cai Y, Yu W et al (2017) Bioimaging application and growth-promoting behavior of carbon dots from pollen on hydroponically cultivated Rome lettuce. ACS Omega 2(7):3958–3965
15. Mohammadinejad R, Dadashzadeh A, Moghassemi S, Ashrafizadeh M, Dehshahri A, Pardakhty A et al (2019) Shedding light on gene therapy: carbon dots for the minimally invasive image-guided delivery of plasmids and noncoding RNAs—a review. J Adv Res 18:81–93
16. Liu Z, Robinson JT, Tabakman SM, Yang K, Dai H (2011) Carbon materials for drug delivery & cancer therapy. Mater Today 14(7–8):316–323
17. Niculescu VC (2020) Mesoporous silica nanoparticles for bio-applications. Front Mater 7:36
18. Chang FP, Kuang LY, Huang CA, Jane WN, Hung Y, Hsing YIC, Mou CY (2013) A simple plant gene delivery system using mesoporous silica nanoparticles as carriers. J Mater Chem B 1(39):5279–5287
19. Hajiahmadi Z, Shirzadian-Khorramabad R, Kazemzad M, Sohani MM (2019) Enhancement of tomato resistance to Tuta absoluta using a new efficient mesoporous silica nanoparticle-mediated plant transient gene expression approach. Sci Hortic 243:367–375
20. Román-Doval R, Torres-Arellanes SP, Tenorio-Barajas AY, Gómez-Sánchez A, Valencia-Lazcano AA (2023) Chitosan: properties and its application in agriculture in context of molecular weight. Polymers 15(13):2867
21. Park CW, Choi JY, Son YJ, Kim DH, Li H, Liang W et al (2024) Magnetofected pollen gene delivery system could generate genetically modified Cucumis sativus. Hortic Res 11(8):uhae179
22. Wang ZP, Zhang ZB, Zheng DY, Zhang TT, Li XL, Zhang C et al (2022) Efficient and genotype independent maize transformation using pollen transfected by DNA-coated magnetic nanoparticles. J Integr Plant Biol 64(6):1145–1156
23. George ACS (2018) Magnetofection of tobacco protoplasts: a novel mechanism for plant transformation. Doctoral dissertation, Murdoch University
24. Farooq N, Ather L, Shafiq M, Nawaz-ul-Rehman MS, Haseeb M, Anjum T et al (2022) Magnetofection approach for the transformation of okra using green iron nanoparticles. Sci Rep 12(1):16568
25. Cao SS, Xue J, Chen XQ, Wang QN, Zhang XH, An XH (2020) Establishment of Chlorella ellipsoidea transformation system mediated by magnetic nanoparticles. Jiangsu J Agric Sci 36:306–311
26. Burlaka OM, Pirko YV, Yemets AI, Blume YB (2015) Plant genetic transformation using carbon nanotubes for DNA delivery. Cytol Genet 49(6):349–357
27. Jat SK, Bhattacharya J, Sharma MK (2020) Nanomaterial based gene delivery: a promising method for plant genome engineering. J Mater Chem B 8(19):4165–4175
28. Demirer GS, Zhang H, Goh NS, Pinals RL, Chang R, Landry MP (2020) Carbon nanocarriers deliver siRNA to intact plant cells for efficient gene knockdown. Sci Adv 6(26):eaaz0495
29. Li S, Li J, Du M, Deng G, Song Z, Han H (2022) Efficient gene silencing in intact plant cells using siRNA delivered by functional graphene oxide nanoparticles. Angew Chem Int Ed 61(40):e202210014
30. Lei WX, An ZS, Zhang BH, Wu Q, Gong WJ, Li JM, Chen WL (2020) Construction of gold-siRNA NPR1 nanoparticles for effective and quick silencing of NPR1 in *Arabidopsis thaliana*. RSC Adv 10(33):19300–19308

31. Wang JW, Grandio EG, Newkirk GM, Demirer GS, Butrus S, Giraldo JP, Landry MP (2019) Nanoparticle-mediated genetic engineering of plants. Mol Plant 12(8):1037–1040
32. Wu H, Li Z (2022) Nano-enabled agriculture: how do nanoparticles cross barriers in plants? Plant Commun 3(6):100346
33. Zhang H, Goh NS, Wang JW, Pinals RL, González-Grandío E, Demirer GS et al (2022) Nanoparticle cellular internalization is not required for RNA delivery to mature plant leaves. Nat Nanotechnol 17(2):197–205
34. Cao J, Feng Y, Lin X, Wang J, Xie X (2017) Iron oxide magnetic nanoparticles deteriorate the mutual interaction between arbuscular mycorrhizal fungi and plant. J Soils Sediments 17:841–851
35. Sembada AA, Lenggoro IW (2024) Transport of nanoparticles into plants and their detection methods. Nanomaterials 14(2):131
36. Pal G, Ingole KD, Yavvari PS, Verma P, Kumari A, Chauhan C, & Vemanna RS. (2024) Exogenous application of nanocarrier-mediated double-stranded RNA manipulates physiological traits and defence response against bacterial diseases. Molecular Plant Pathology, 25(1), e13417
37. Demirer GS, Zhang H, Matos JL, Goh NS, Cunningham FJ, Sung Y et al (2019) High aspect ratio nanomaterials enable delivery of functional genetic material without DNA integration in mature plants. Nat Nanotechnol 14(5):456–464
38. Lew TTS, Wong MH, Kwak SY, Sinclair R, Koman VB, Strano MS (2018) Rational design principles for the transport and subcellular distribution of nanomaterials into plant protoplasts. Small 14(44):1802086
39. Kurczyńska E, Godel-Jędrychowska K, Sala K, Milewska-Hendel A (2021) Nanoparticles—plant interaction: what we know, where we are? Appl Sci 11(12):5473
40. Avellan A, Schwab F, Masion A, Chaurand P, Borschneck D, Vidal V et al (2017) Nanoparticle uptake in plants: gold nanomaterial localized in roots of Arabidopsis thaliana by X-ray computed nanotomography and hyperspectral imaging. Environ Sci Technol 51(15):8682–8691
41. Cunningham FJ, Goh NS, Demirer GS, Matos JL, Landry MP (2018) Nanoparticle-mediated delivery towards advancing plant genetic engineering. Trends Biotechnol 36(9):882–897
42. Perozo E, Cortes DM, Sompornpisut P, Kloda A, Martinac B (2002) Open channel structure of MscL and the gating mechanism of mechanosensitive channels. Nature 418(6901):942–948
43. Kryovrysanaki N, James A, Tselika M, Bardani E, Kalantidis K (2021) RNA silencing pathways in plant development and defense. Int J Dev Biol 66(1–2–3):163–175
44. Arora S, Sharma P, Kumar S, Nayan R, Khanna PK, Zaidi MGH (2012) Gold-nanoparticle induced enhancement in growth and seed yield of *Brassica juncea*. Plant Growth Regul 66:303–310
45. Lee WM, Kwak JI, An YJ (2012) Effect of silver nanoparticles in crop plants Phaseolus radiatus and *Sorghum bicolor*: media effect on phytotoxicity. Chemosphere 86(5):491–499
46. Feizi H, Moghaddam PR, Shahtahmassebi N, Fotovat A (2013) Assessment of concentrations of nano and bulk iron oxide particles on early growth of wheat (*Triticum aestivum L.*). Annu Res Rev Biol 3(4):814–824
47. Raskar SV, Laware SL (2014) Effect of zinc oxide nanoparticles on cytology and seed germination in onion. Int J Curr Microbiol Appl Sci 3(2):467–473
48. Rico CM, Lee SC, Rubenecia R, Mukherjee A, Hong J, Peralta-Videa JR, Gardea-Torresdey JL (2014) Cerium oxide nanoparticles impact yield and modify nutritional parameters in wheat (*Triticum aestivum L.*). J Agric Food Chem 62(40):9669–9675
49. Nair PMG, Chung IM (2015) Changes in the growth, redox status and expression of oxidative stress related genes in chickpea (*Cicer arietinum L.*) in response to copper oxide nanoparticle exposure. J Plant Growth Regul 34:350–361
50. Raliya R, Biswas P, Tarafdar JC (2015) TiO_2 nanoparticle biosynthesis and its physiological effect on mung bean (*Vigna radiata L.*). Biotechnol Rep 5:22–26
51. El-Sharkawy MS, El-Beshsbeshy TR, Mahmoud EK, Abdelkader NI, Al-Shal RM, Missaoui AM (2017) Response of alfalfa under salt stress to the application of potassium sulfate nanoparticles. Am J Plant Sci 8(8):1751–1773

52. Elmer W, De La Torre-Roche R, Pagano L, Majumdar S, Zuverza-Mena N, Dimkpa C et al (2018) Effect of metalloid and metal oxide nanoparticles on Fusarium wilt of watermelon. Plant Dis 102(7):1394–1401
53. Itroutwar PD, Govindaraju K, Tamilselvan S, Kannan M, Raja K, Subramanian KS (2020) Seaweed-based biogenic ZnO nanoparticles for improving agro-morphological characteristics of rice (*Oryza sativa L.*). J Plant Growth Regul 39:717–728
54. Ozcan A, Young M, Lee B, Liao YY, Da Silva S, Godden D, Colee J, Huang Z, Mendis HC, Campos MGN, Jones JB, Freeman JH, Paret ML, Tetard L, Santra S (2021) Copper-fixed quat: a hybrid nanoparticle for application as a locally systemic pesticide (LSP) to manage bacterial spot disease of tomato. Nanoscale Adv 3(5):1473–1483
55. Mitter N, Worrall EA, Robinson KE, Li P, Jain RG, Taochy C et al (2017) Clay nanosheets for topical delivery of RNAi for sustained protection against plant viruses. Nat Plants 3(2):1–10
56. Niño-Sánchez J, Sambasivam PT, Sawyer A, Hamby R, Chen A, Czislowski E et al (2022) BioClay™ prolongs RNA interference-mediated crop protection against Botrytis cinerea. J Integr Plant Biol 64(11):2187–2198
57. Noha K, Bondok AM, El-Dougdoug KA (2018) Evaluation of silver nanoparticles as antiviral agent against ToMV and PVY in tomato plants. Sciences 8(1):100–111
58. Eugene K, Zholobak N (2016) Antiviral activity of cerium dioxide nanoparticles on Tobacco mosaic virus model. In: Proceedings of the topical issues of new drugs development, Kharkiv, Ukraine, 21
59. Elsharkawy MM, Derbalah A (2019) Antiviral activity of titanium dioxide nanostructures as a control strategy for broad bean strain virus in faba bean. Pest Manag Sci 75(3):828–834
60. Hernández-Hernández H, Juárez-Maldonado A, Benavides-Mendoza A, Ortega-Ortiz H, Cadenas-Pliego G, Sánchez-Aspeytia D, González-Morales S (2018) Chitosan-PVA and copper nanoparticles improve growth and overexpress the SOD and JA genes in tomato plants under salt stress. Agronomy 8(9):175
61. Pramanik B, Debnath S (2024) Induced mutagenesis for Gene Pool expansion with special attention on pulse crops. In: Kumar N (ed) Plant mutagenesis and crop improvement. CRC Press, Boca Raton, pp 58–83
62. Bharti DR, Pramanik B, Paul A, Debnath S (2025) CRISPR-based precision breeding for crop improvement: current achievements and future directions. In: Chen JT (ed) Plant stress tolerance. CRC Press, Boca Raton, pp 118–138
63. Das A, Mahanta M, Pramanik B, Purkayastha S (2023) Genetically modified crops and crop species adapted to global warming in dry regions. In: Naorem A, Machiwal D (eds) Enhancing resilience of dryland agriculture under changing climate: interdisciplinary and convergence approaches. Springer Nature Singapore, Singapore, pp 385–409
64. Sar P, Pramanik B, Debnath S (2022) CRISPR/Cas: a new horizon in crop improvement. In: Debnath S, Manjula IK, Shah KR, Bhattacharjee S (eds) Futuristic trends in biotechnology. IIP International Publishers, Chikkamagaluru, pp 9–27
65. Farinati S, Draga S, Betto A, Palumbo F, Vannozzi A, Lucchin M, Barcaccia G (2023) Current insights and advances into plant male sterility: new precision breeding technology based on genome editing applications. Front Plant Sci 14:1223861
66. Ehsan M, Waheed A, Ullah A, Kazmi A, Ali A, Raja NI et al (2022) Plant-based bimetallic silver-zinc oxide NPs: a comprehensive perspective of synthesis, biomedical applications, and future trends. Biomed Res Int 2022(1):1215183
67. Yang Z, Chen J, Dou R, Gao X, Mao C, Wang L (2015) Assessment of the phytotoxicity of metal oxide NPs on two crop plants, maize (*Zea mays* L.) and rice (*Oryza sativa* L.). Int J Environ Res Public Health 12(12):15100–15109
68. Yusuf A, Almotairy ARZ, Henidi H, Alshehri OY, Aldughaim MS (2023) NPs as drug delivery systems: a review of the implication of NPs' physicochemical properties on responses in biological systems. Polymers 15(7):1596
69. Choudhury A, Rajam MV (2021) Genetic transformation of legumes: an update. Plant Cell Rep 40:1813–1830

70. Wang P, Si H, Li C, Xu Z, Guo H, Jin S, Cheng H (2025) Plant genetic transformation: achievements, current status and future prospects. Plant Biotechnol J 23:1–25
71. Rani N, Kumari K, Hooda V (2025) The role of NPs in transforming plant genetic engineering: advancements, challenges and future prospects. Funct Integr Genomics 25(1):1–22
72. dos Santos Fonseca LM, Machado BAS, Oliveira FO, de Jesus Santos JR, da Silva JW, Hodel KVS et al (2024) An overview on recent patents and technologies on NPs for nucleic acid delivery. Expert Opin Ther Pat 34(3):171–186

Nano-delivering RNAi in Plants: Opportunities and Challenges

Piyush Bania, Dyuti Ballav Chongder, Ankan Das, Bodhisattya Pal, and Sandip Debnath

Abstract The application of RNA interference (RNAi) in plant biotechnology holds immense potential for crop improvement, pest management, and climate resilience. However, the efficacy of RNAi-based strategies is hindered by the biological barriers of plants, including the cuticle, cell wall, and apoplastic defences, which limit the efficiency of conventional delivery methods. This chapter explores the transformative role of nanotechnology in overcoming these challenges through innovative, non-transgenic delivery systems. Recent advances in nanocarrier design, such as lipid-based nanoparticles, pH-responsive polymeric systems, mesoporous silica nanoparticles, and biomimetic platforms inspired by plant exosomes or viral capsids, offer unprecedented precision in RNAi payload delivery. Surface functionalization strategies, including ligand-mediated targeting and stimuli-responsive coatings (e.g., light- or enzyme-triggered release), enhance tissue-specific uptake while minimizing off-target effects. Novel insights into nanoparticle trafficking mechanisms, such as symplastic transport via plasmodesmata and vascular systemic delivery, further enable whole-plant gene silencing. Breakthroughs in RNAi payload engineering, such as chemically stabilized siRNA and dual-function carriers co-delivering RNAi with micronutrients, address extracellular degradation and enhance agroecological compatibility. Despite these advancements, scalability challenges, regulatory uncertainties, and unresolved biosafety concerns underscore the need for interdisciplinary collaboration. Emerging solutions like AI-driven nanocarrier optimization, UV-resistant coatings, and "safe-by-design" frameworks aim to bridge lab-to-field translation gaps. This chapter synthesizes the latest inno-

P. Bania · D. B. Chongder · A. Das
Palli Siksha Bhavana (Institute of Agriculture), Visva-Bharati, Sriniketan, West Bengal, India

B. Pal
Crop Improvement Division, ICAR—Central Research Institute for Jute and Allied Fibres, Barrackpore, West Bengal, India

S. Debnath (✉)
Department of Genetics and Plant Breeding, Palli Siksha Bhavana (Institute of Agriculture), Visva-Bharati, Sriniketan, West Bengal, India
e-mail: sandip.debnath@visva-bharati.ac.in

J.-T. Chen (ed.), *Nano-Delivering for Plant Genetic Engineering*, Smart Nanomaterials Technology, https://doi.org/10.1007/978-981-95-4165-2_12

vations in nano-enabled RNAi delivery, emphasizing their potential to revolutionize precision agriculture while navigating technical, environmental, and ethical frontiers. By uniting nanotechnology, synthetic biology, and sustainable practices, these advances pave the way for resilient, high-yielding crops in an era of global food insecurity.

Keywords RNA interference (RNAi) · Nanocarriers · Non-transgenic delivery · Targeted delivery · Sustainable agriculture

1 Introduction

The convergence of nanotechnology and RNA interference (RNAi) has emerged as a transformative approach to crop protection, offering unprecedented opportunities to develop sustainable, species-specific pest management strategies while addressing the fundamental challenges that limit widespread agricultural adoption. Spray-induced gene silencing (SIGS) represents a revolutionary non-transgenic alternative to traditional chemical pesticides, where double-stranded RNA (dsRNA) molecules are topically applied to plants to silence essential genes in pests and pathogens without requiring genetic modification of crops [1]. However, the inherent instability of naked RNA molecules under environmental conditions, degradation by UV light, rainfall, and ubiquitous nucleases, has necessitated the development of sophisticated nanocarrier systems to protect RNA cargo and facilitate efficient cellular uptake. Nanocarriers, including layered double hydroxide (LDH) clay particles, carbon dots, chitosan nanoparticles, and liposomes, have demonstrated remarkable success in extending RNA half-life from minutes to days while enabling controlled release and enhanced penetration through plant cell walls [2]. The BioClay™ formulation, utilizing LDH nanosheets to encapsulate dsRNA, has shown particular promise by providing protection against fungal pathogens for up to 30 days compared to just hours for naked dsRNA [3]. Carbon dots, with their ultra-small size (<10 nm), effectively bypass cell wall barriers and demonstrate over 50% protection against RNase III degradation while facilitating efficient siRNA delivery into plant cells [4]. Despite these technological advances, significant challenges remain, including manufacturing costs, regulatory complexity across jurisdictions, environmental fate assessment, and the need for comprehensive safety evaluations regarding non-target organism impacts. The integration of nanotechnology with RNAi delivery systems not only addresses RNA stability and delivery challenges but also enables precision agriculture approaches that minimize environmental impact while maximizing crop protection efficacy, positioning nano-enabled SIGS as a cornerstone technology for sustainable food production in the face of growing global food security demands.

2 Nanocarrier Platforms for RNAi Delivery in Plants

The efficacy of RNAi highly depends on the efficient delivery of the RNA molecules into the targeted sites. Nanocarriers are nano-sized delivery systems (typically ranging from 1 to 1000 nm in at least one dimension) that encapsulate or bind therapeutic cargo [5]. These are specially engineered nanoscale vehicles designed to transport RNAi molecules, such as small interfering RNA (siRNA), microRNA (miRNA), and short hairpin RNA (shRNA), safely and efficiently to target cells. Their multifaceted role is critical for overcoming several major challenges inherent to RNA-based therapeutics [4, 6–8].

2.1 Emerging Nanocarriers (Lipid, Polymeric, Inorganic, Biomimetic) and Their Unique Advantages

2.1.1 Lipid Nanocarriers

Lipid nanocarriers, which include liposomes, lipid nanoparticles (LNPs), and nanostructured lipid carriers (NLCs), are especially intended to improve miRNA delivery. Lipid nanoparticles (LNPs) have attracted interest for their biocompatibility, biodegradability, and ability to successfully contain RNA. These carriers bind to RNA by electrostatic interactions, increasing the RNA payload's stability against nuclease destruction [9]. LNPs have high RNA encapsulation efficiency, which is critical for increasing the amount of RNA supplied to target cells [10]. Furthermore, they may be adjusted to increase circulation times and target specificity, making them appropriate for systemic delivery in plants. The capacity of certain lipid carriers to allow endosomal escape contributes to increased RNA bioavailability following absorption. Furthermore, lipid-based formulations have a lower risk of immunogenic reactions than other nanocarrier types, improving their safety profile for agricultural applications [11, 12].

2.1.2 Polymeric Nanocarriers

Polymeric nanocarriers have distinct benefits due to their versatility and ability to be modified for individual delivery requirements [13]. Among the most common are chitosan and polyethyleneimine (PEI), which have high biocompatibility and low toxicity. Polymeric carriers may be constructed to provide controlled release, allowing the given siRNA to have long-term effects in target tissues, which is useful for pest management solutions. Furthermore, the structural modification of polymers enables increased cellular uptake and stability under different environmental circumstances, which addresses one of the primary obstacles in RNAi application [11]. Guanylated polymers, for example, have high stability and may effectively protect

RNA against enzymatic destruction, especially in alkaline environments like those found in the insect stomach [14]. Other examples include the well-explored nanocarrier, Chitosan, which has been employed in various plants, including tobacco, spinach, etc. and can be readily mixed with other forms of nanocarriers like carbon-based nanocarriers or SWNTs (single-walled nanotubes) [14, 15]. Overall, polymeric nanocarriers represent a robust platform for addressing practical constraints in RNA delivery while minimizing toxicity.

2.1.3 Inorganic Nanocarriers

Carbon-based nanoparticles and layered double hydroxide nanoparticles are examples of inorganic nanocarriers that provide distinct benefits [8] due to their nanoscale stability and tunability. Inorganic carriers frequently have a large surface area, which facilitates the attachment of RNA molecules and increases their stability in biological settings [16]. For example, clay nanosheets have shown efficient distribution capabilities by accessing intact plant leaf cells and offering long-term protection against environmental elements that can damage RNA, such as nucleases and UV radiation [17]. Furthermore, inorganic nanoparticles may be engineered to regulate the release of RNA, allowing for tailored activity where and when needed. However, the potential toxicity of some inorganic components must be carefully examined to ensure they do not negatively affect plant development or non-target creatures [18, 19].

In another study, *Solanum lycopersicum* plants were treated with dsRNA encoding genes required for *Fusarium oxysporum* to colonize plants effectively. LDH/dsRNA complexes measuring 30–90 nm were administered to tomato plants in three ways: foliar spraying, petiole adsorption, and root dipping in solution. It was established that all three techniques were effective: they offered greater resistance to *F. oxysporum* and avoided the development of severe symptoms such as crown and root rot for up to 60 days on the monitored plants [20].

Another example includes combating tomato yellow leaf curl virus (TYLCV) infection by a combination of pDNA encoding three artificial microRNAs against TYLCV was loaded into LDH and sprayed onto *S. lycopersicum* or *N. benthamiana* plants. Pre-treatment with pDNA/LDH increases plant resistance to TYLCV infection [21].

2.1.4 Biomimetic Nanocarriers

Biomimetic nanocarriers use biological materials' natural features to enhance the distribution and effectiveness of RNA molecules. These carriers improve the biocompatibility and targeted distribution capability of nanoparticles by covering them with plant or animal cell membranes [22]. For example, membranes generated from red blood cells or cancer cells have inherent homing characteristics, enabling efficient targeting of tumour areas or particular vascular regions inside plants. These

Table 1 Different types of nanocarriers used for delivery of RNAi

Sl no.	Nanoparticle	Plant	Remark	References
1	Carbon nanotubes	*Nicotiana benthamiana*	siRNA was delivered into the leaves	Demirer et al. [25, 26]
2	Lignin-based nanocarriers	Grapevine	Demonstrated targeted fungicide delivery with enzyme-responsive release, minimizing off-target effects	Yong et al. [20]
3	Au-MSN	*Nicotiana tabacum*	siRNA was delivered into the leaves and also achieved codelivery of the gene and nutrient	Yong et al. [20]
4	Chitosan	Potato	Used to improve resistance against biotic and abiotic stress	Mujtaba et al. [5]
5	LDH	*Vigna unguiculata*	LDH complexed with dsRNA provided systemic RNAi-based viral protection after topical application, lasting up to 30 days	Mujtaba et al. [5]
6	Guanidinium-containing disulfide nanoparticles	Rice and *Arabidopsis*	Silencing of key genes to enhance salt tolerance	Lin et al. [27]

systems can escape immune reaction and enhance the half-lives of their payloads in circulation, allowing for improved delivery efficiency [23]. The capacity of biomimetic nanocarriers to undergo surface alterations gives researchers the freedom to develop responders that can adjust to the dynamic settings encountered during RNA delivery [16, 24] (Table 1).

2.2 *Innovations in Nanocarriers like Chitosan-PLGA Hybrids for pH-Responsive Release or Plant-Derived Exosomes for Biocompatibility*

2.2.1 Chitosan-PLGA Hybrids for pH-Responsive Release

Chitosan and poly (lactic-co-glycolic acid) (PLGA) hybrids have garnered considerable attention in the design of pH-responsive delivery systems, particularly for the controlled release of small interfering RNA (siRNA) and other RNA interference (RNAi) therapeutics. These hybrids combine the advantageous properties of chitosan, a biocompatible and biodegradable polysaccharide, with PLGA, an FDA-approved polymer, to create effective delivery vehicles for RNAi applications. The incorporation of chitosan into PLGA nanoparticles enhances their biocompatibility and responsiveness to varying pH environments. Chitosan is a cationic polysaccharide with primary amine groups that allow it to form electrostatic interactions with negatively charged RNA molecules, such as siRNA, thereby facilitating encapsulation. Chitosan is applied as a thin layer over the PLGA nanoparticle and the siRNA is loaded onto this hybrid and then the siRNA-loaded nanoparticle is sprayed on the cells or cell culture. When the siRNA is delivered via PLGA core nanoparticles coated with chitosan, it protects the siRNA from nuclease degradation and facilitates the fast release of siRNA into the cytosolic portion. This innovation is equally beneficial for both RNAi delivery and delivery of CRISPR/Cas9 complex into the cells [28]. The major characteristics of Chitosan/PLGA hybrid are [28]:

- **Biocompatibility**: Both chitosan and PLGA are biodegradable, reducing potential toxicity and immunogenicity upon administration.
- **pH Sensitivity**: Chitosan exhibits solubility at low pH (acidic conditions), whereas PLGA is stable across a broader range of pH levels. This property allows for pH-triggered release mechanisms, wherein RNA payloads can be released more effectively in acidic environments, such as those found in tumour tissues or endosomes (Fig. 1).

2.2.2 Plant-Derived Exosomes for Biocompatibility

Plant-derived exosomes, also known as exosome-like nanoparticles (ELNs), have emerged as promising nanocarriers for various therapeutic applications. These extracellular vesicles possess unique properties that enhance their potential for the transport of small RNAs, proteins, and therapeutic agents [29].

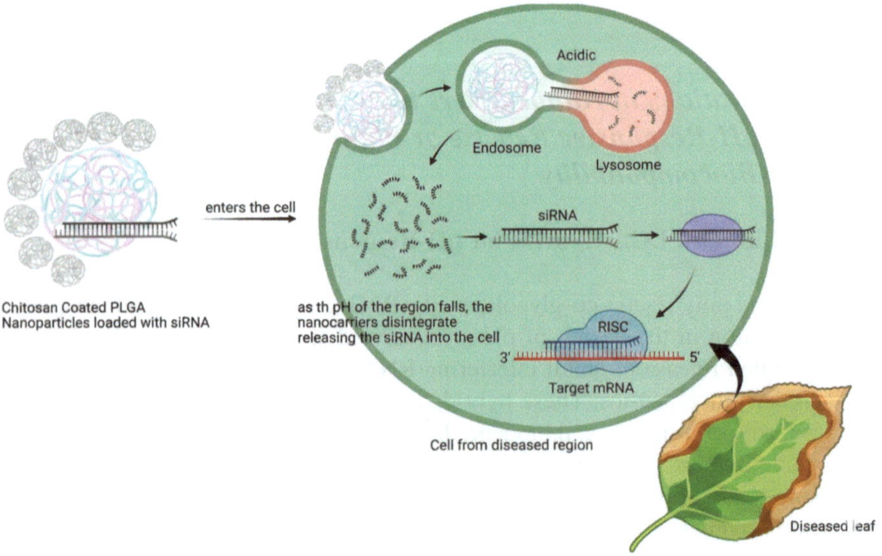

Fig. 1 Chitosan-PLGA hybrid as nanocarriers for pH-responsive release, a way of controlled and site-specific delivery of RNAi in plants

Characteristics of Plant-Derived Exosomes

Exosomes are nanoscale membrane vesicles with a diameter range from 30 to 100 nm that are secreted by almost every kind of cell. They can transmit a variety of signaling molecules, including nucleic acids (mainly mRNA and microRNA), functional proteins, and lipids. Owing to their small size of exosomes, they can escape from the various microbes present on the plant surface, steadily carry and deliver the cargo, and pass through cell walls to target cells [9].

These tools are extracellular vesicles that look like exosomes and migrate across kingdoms to silence organisms and are also naturally present in organisms. For example, extracellular vesicles are used by the host cell to transport sRNAs into fungal cells during interactions between *Arabidopsis* and *Botrytis cineraria* [30]

Plant-derived exosomes are nanosized vesicles ranging from 30 to 500 nm and originate from different plant tissues. Their composition typically includes lipids, proteins, and nucleic acids, such as microRNAs (miRNAs) and messenger RNA (mRNA) [9]. Their key characteristics are:

- **Biocompatibility**: ELNs are derived directly from plants, which makes them biologically compatible with human cells, reducing the risk of immune reactions.
- **Stability**: They exhibit high stability in various environments, including physiological and gastrointestinal conditions, which is critical for therapeutic applications.
- **Natural Cargo**: ELNs naturally encapsulate bioactive molecules, including phytochemicals and bioactive RNAs, which can be harnessed for therapeutic purposes [31].

2.3 *Biomimetic Designs (e.g., Viral Capsid-Mimicking Nanoparticles) and Inorganic MSNs with Tunable Pore Sizes for Controlled Payload Release*

Biomimetic designs for nanoparticles, particularly viral capsid-mimicking nanoparticles, draw inspiration from the natural efficiency of viruses in delivering genetic material into host cells. These nanoparticles closely resemble the structure of viral capsids, which facilitates the efficient encapsulation and delivery of nucleic acids such as DNA and RNA. By mimicking the inherent geometric and biochemical properties of viruses, these biomimetic nanoparticles can achieve higher rates of cellular uptake compared to conventional delivery systems. Their design enables them to undergo specific interactions with cell receptors, resulting in improved targeting of diseased tissues and enhanced therapeutic outcomes.

The advantages of viral capsid-mimicking nanoparticles are pronounced in their ability to traverse biological barriers. The natural entry mechanisms of viruses, such as endocytosis, allow these engineered nanoparticles to deploy their payloads effectively within the cytosol of target cells. Furthermore, extensive modification options exist, allowing for the incorporation of targeting ligands or therapeutic agents directly onto the nanoparticle surface, enhancing specificity and reducing off-target effects. This versatility makes viral capsid-mimicking nanoparticles a highly precise option for delivering therapeutic payloads, particularly in applications involving RNA interference (RNAi) [32].

2.3.1 Inorganic MSNs with Tunable Pore Sizes

Mesoporous silica nanoparticles (MSNs) have a high potential as a nanomaterial due to their enormous surface area, variable pore size, and biocompatibility. As previously stated, the main disadvantage of using nanoparticles is the successful entry of a cargo-carrier complex or cargo into the protoplast [33]. Plant cells may internalize MSNs functionalized with various chemical compounds (N-trimethoxysilylpropyl-N,N,N-trimethylammonium chloride, for example) and release pDNA cargo intracellularly. MSNs are also quite stable and may be functionalized by a variety of chemical changes, including decorating with cationic polymers such as polylysine, polyarginine, polyethyleneimine, and so on. MSNs were shown to be relevant to many plant species (*N. tabacum, S. lycopersicum, A. thaliana, Zea mays*) [20] and plant sections (protoplasts, suspension culture BY-2, leaves, roots) [21, 34].

3 Surface Functionalization and Targeting Strategies

Surface functionalization refers to the chemical modification of nanoparticle surfaces to enhance their interaction with biological systems. This process enables the inclusion of specific ligands, polymers, or other functional groups that promote cellular uptake, increase stability, and enable effective release profiles for encapsulated RNA molecules. By altering the physicochemical properties of a nanocarrier, scientists can optimize the interactions between the carrier and its target cells, facilitating a targeted delivery that maximizes therapeutic effects while minimizing off-target toxicity.

The effectiveness of RNAi delivery systems greatly relies on the ability to specifically target the cells of interest. Targeting capitalizes on the enhanced permeability and retention (EPR) effect, or involves the use of ligands that bind to specific receptors overexpressed on the surface of target cells. This strategy often employs antibodies, peptides, or small molecules that bind with high affinity to receptor targets, facilitating receptor-mediated endocytosis of the nanocarrier and subsequent release of the RNA payload inside the target cells.

3.1 *Enhancing Tissue-Specific Uptake Through Surface Modifications in Nanocarriers*

The development of advanced delivery systems, particularly for the effective delivery of biologics such as small interfering RNAs (siRNAs), relies heavily on enhancing tissue-specific uptake. A vital approach to achieving this goal is through surface modifications of nanocarriers. Techniques such as the incorporation of cell-penetrating peptides (CPPs) and carbohydrate ligands have shown great promise in improving the targeting efficiency of these nanocarriers, leading to greater therapeutic efficacy and reduced side effects.

3.1.1 Cell-Penetrating Peptides (CPPs)

Cell-penetrating peptides are short, typically cationic peptides that facilitate the internalization of various molecular cargoes, including nucleic acids and proteins, into cells [35]. Several essential aspects contribute to CPPs' ability to increase tissue-specific absorption. First, CPPs' positive charge allows them to engage electrostatically with the negatively charged components of the cell membrane, which promotes cellular interaction and uptake. When CPPs come into contact with the plasma membrane, they can induce either direct penetration or receptor-mediated endocytosis, thereby circumventing the lipid bilayer barrier, which normally prevents bigger molecules from entering [36, 37].

There have been various studies on CPPs derived from endogenous proteins in plant systems. In addition to the Tat-derived peptide, an 18-aa pVEC peptide from the murine vascular endothelial cadherin protein was shown to enter wheat mesophyll protoplasts and other plant organs [37, 38]. Many CPPs are derived from viral capsid proteins [39].

One of the most widely utilized instances is the Brome Mosaic Virus (BMV) capsid protein, which can easily penetrate the protoplast of mesophyll cells as well as barley and Arabidopsis roots. The BMV CPP has RNA-binding capability, allowing for the easy intracellular transport of both BMV RNA and in vitro-synthesized dsRNA into barley protoplasts. CPP sequences from barley yellow dwarf virus (BYDV), tobacco curly shoot virus (TCSV), and bean yellow dwarf virus (BeYDV) were also found to be identical. As a result, such viral-derived natural CPPs might be used as potential dsRNA delivery agents in plant cells [40] (Fig. 2).

One of the directions of CPP application is the organelle-specific delivery of cargo. And these have been demonstrated to be effective peptide-based chloroplast transformation for three species: tobacco, rice, and kenaf [40].

Moreover, CPP-modified nanoparticles can also overcome some common barriers, such as low bioavailability and rapid clearance. This is particularly beneficial in the context of RNAi therapeutics, where effective cellular uptake and delivery to the cytosol are critical for triggering the desired gene silencing effects [35].

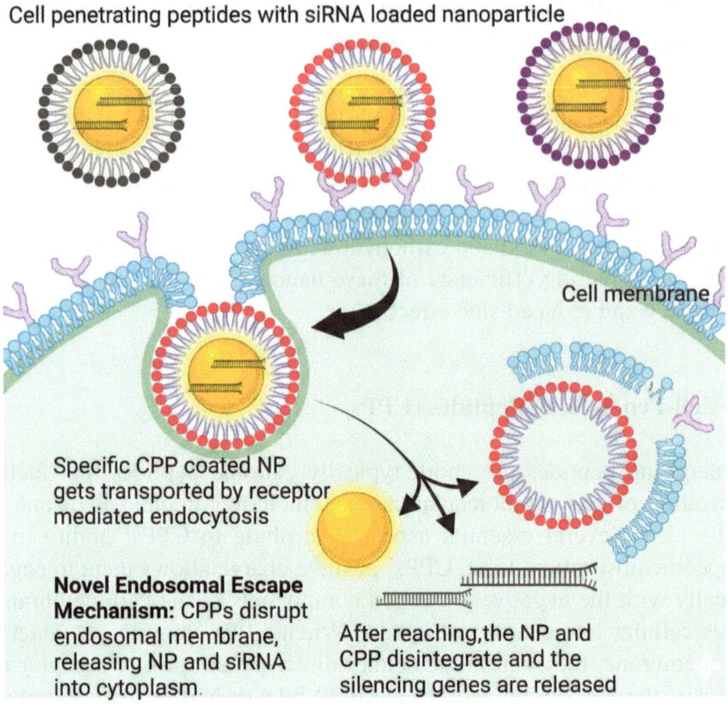

Fig. 2 Modification of nanoparticle (NP) surface with CPP resulting in tissue-specific uptake

3.1.2 Carbohydrate Ligands

Surface functionalization with carbohydrate ligands is an alternative but equally successful technique for improving tissue-specific absorption. Many biological processes rely on interactions between cell surface carbohydrates (glycans) and carbohydrate-binding proteins (lectins). Carbohydrates can be used as targeting moieties because of their capacity to precisely interact with receptors on the surface of target cells, which are overexpressed in certain pathological situations. Mannose, galactose, and other sugar residues, for example, can be used to target cells that express the appropriate receptors, increasing the specificity and efficacy of medicinal delivery systems [41].

The mechanism underpinning this targeting method is receptor-mediated endocytosis, which occurs when carbohydrate ligands bind to their relevant receptors and the complete nanocarrier complex is internalized into the cell [42]. This approach not only promotes increased absorption but also reduces the systemic toxicity commonly associated with bulk medication delivery strategies. Researchers use carbohydrates to create nanocarriers that traverse through the complicated biological environment, ensuring gene delivery to the appropriate tissues or cells while reducing exposure to healthy tissue [43].

3.2 *Stimuli-Responsive Coatings (e.g., Light-Activated Nanocarriers, ROS-Triggered RNA Release)*

The advent of nanotechnology has opened new avenues for RNAi delivery systems, particularly through the use of stimuli-responsive coatings. These innovative coatings allow for controlled release of genes in response to specific environmental triggers, thereby enhancing the therapeutic efficacy while minimizing side effects. Notably, light-activated nanocarriers and reactive oxygen species (ROS)-triggered systems have emerged as promising strategies in the context of targeted therapy [44, 45].

3.2.1 Light-Activated Nanocarriers

Light-activated nanocarriers utilize photochemical reactions to trigger cargo release in response to light irradiation. This mechanism provides unique advantages, including precise spatial and temporal control over cargo delivery, making it exceedingly suitable for applications requiring heightened therapeutic efficacy within specific regions of interest [44].

Typically, light-responsive systems incorporate photosensitive materials that change their properties upon exposure to light. For example, nanocarriers can be designed using azobenzene or spiropyran molecules, which undergo conformational changes when activated by specific wavelengths of light [46]. This conformational alteration can lead to the cleavage of chemical bonds, resulting in the release of encapsulated cargos at the target site. Additionally, the use of near-infrared (NIR) light is particularly advantageous because it offers deeper tissue penetration with minimal damage to surrounding healthy tissues [47].

3.2.2 ROS-Triggered RNA Release

Reactive oxygen species (ROS) are highly reactive molecules that are often overproduced in disease states, as during fungal and bacterial infections. The unique microenvironment of these diseases presents an opportunity to exploit ROS as a trigger for cargo release. Nanocarriers engineered with ROS-responsive coatings can undergo controlled degradation or release upon encountering elevated ROS levels, which are characteristic of pathological conditions [45].

Typically, these ROS-responsive systems are constructed using materials that contain ROS-sensitive linkers, such as thioketals or boronic esters, which can cleave in response to ROS-induced oxidative conditions [48]. For instance, nanoparticles that encapsulate small interfering RNA (siRNA) can be designed with a thioketal linker that breaks down under oxidative stress, resulting in the release of the RNA payload precisely at the site of heightened ROS levels [49]. This targeted release mechanism not only enhances the therapeutic effect of RNA-based treatments, such as RNAi approaches, but also minimizes off-target effects associated with conventional delivery methods.

3.3 Ligand-Mediated Targeting and Enzyme-Responsive Systems That Degrade Only in Specific Plant Tissues, Reducing Off-Target Effects

3.3.1 Ligand-Mediated Targeting Using Pectin-Binding Motifs

One of the most difficult obstacles in developing plant-based medication delivery systems is penetrating the plant cell wall, which is a complicated structure that acts as a barrier to foreign chemicals. Ligand-mediated targeting makes use of specialized ligands that can bind to plant tissue receptors or components, allowing medicinal medicines to be taken up and delivered more effectively. Pectin-binding motifs are particularly intriguing in this context since they occur naturally in plant tissues and may precisely target the cell wall's pectin component [50].

Pectins are key polysaccharides that make up the majority of the plant cell wall and are essential for structural stability. Researchers can increase the affinity of nanocarriers for plant cells by designing them with pectin-binding motifs. This focused method not only improves cell wall penetration but also enhances internalization via receptor-mediated endocytosis. Furthermore, the biocompatibility of pectin derivatives decreases the likelihood of negative interactions within plant systems, which aligns with environmental sustainability goals by employing natural components found in plant tissues [51].

For example, research has shown that nanoparticles embedded with pectin-binding motifs KH9-OEP34 coated on a nanocarrier have better accumulation in plant chloroplasts, resulting in increased organelle internalization of cargo such as siRNAs. This technique guarantees that therapeutic chemicals are delivered precisely to target tissues, enhancing effectiveness while limiting exposure to non-target tissues. Thagun et al. [52] successfully developed a peptide-based transformation platform by combining the functions of a chloroplast-targeting peptide (CTP; KH9-OEP3412) and a cell-penetrating peptide (CPP) to deliver DNA molecules across cell boundaries and then mediate transgene expression in plastids [52].

3.3.2 Enzyme-Responsive Systems for Selective Degradation

Building on the basis of ligand-mediated targeting, enzyme-responsive systems provide a novel way for fine-tuning cargo delivery by guaranteeing that therapeutic molecules are only released in response to specified enzymatic triggers. This method is very useful for eliminating off-target impacts, which improves the safety profile of these devices [53].

Enzyme-responsive nanocarriers are designed to dissolve or release their payloads when exposed to certain enzymes found in disease states or plant tissues. For example, pathogen-secreted lignin-degrading enzymes or certain tissue circumstances might stimulate the release of encapsulated therapeutic medicines. By using this natural enzymatic activity, the site and time of RNAi distribution may be regulated [54].

The inclusion of such systems into nanocarriers may be demonstrated by the use of lignin-based coating materials that break down in the presence of ligninolytic enzymes linked to fungal infections. This method guarantees that encapsulated nucleic acids are only released in sick or damaged tissues, avoiding unwanted exposure to healthy plant parts [55]. Furthermore, the selective breakdown of the carrier promotes therapeutic efficacy by providing high local concentrations of active substances at the site of infection or damage while maintaining the plant's general health [53].

4 Mechanisms of Cellular Uptake and Systemic Transport

The nanotechnology application opens up new opportunities for precise and species-independent delivery of biomolecules., Engineered nanoparticles (NPs) have a clear advantage in breaking through plant cell walls and membrane barriers due to their physicochemical properties, which are adaptable. Recent developments have shown their potential to disperse functional genetic material, such as siRNA and CRISPR constructs, throughout entire plants.

4.1 Physicochemical Characteristics of Nanoparticles and Cellular Absorption

4.1.1 Entry Based on Size

Although NPs smaller than 50 nm can passively enter the cell wall, the pore size of the cell wall, which typically ranges from 5 to 20 nm, can momentarily enlarge under stress or with particle contact [56, 57]. Nanoparticles smaller than 50 nm can passively diffuse through cell wall pores or exploit transient apertures to enter the plasma membrane. Furthermore, smaller NPs are more mobile and effective due to better tissue penetration, less aggregation, and less steric hindrance.

4.1.2 . Interactions Between Surface Charge and Cell Wall

The surface charge of nanoparticles is important when they interact with the negatively charged plant cell wall and plasma membrane. Positively charged NPs (like chitosan-coated, cationic liposomes) strongly attract the negatively charged heads of pectin and phospholipids, increasing their adhesion and uptake [58]. This property enhances endocytosis, membrane fusion, or direct penetration, especially for lipid and polymer-based carriers.

4.2 Two Intracellular and Systemic Transport Mechanisms

4.2.1 Transportation of Nanoparticles Through Plasmodesmata

They can move between cells through the symplast, which is the networked cytoplasm of neighboring plant cells. Plasmodesmata (PD) are cytoplasmic channels that facilitate molecular trafficking and intercellular communication. The exclusion limit of Plasmodesmata usually ranges within 1 and 2 nm, so that it can seldom reach 50 nm [59]. Some specially made nanoparticles (such as flexible, PEGylated NPs) can sometimes disassemble due to the pore size of Plasmodesmata. This cell-to-cell transfer made the transfer of siRNA to the phloem carriers.

4.2.2 Transport of Vascular by Way of the Phloem

To attain complete systemic distribution, nanoparticles must enter the circulatory system, particularly the phloem. The phloem transports small molecules, including sugars, hormones, RNAs, and others, from source tissues to sink tissues. Nanoparticles can enter the zones of phloem loading, also into partner cells, either by imitating natural mobile molecules (carbohydrates or RNA motifs) or by utilizing the pathway of symplast movement [60, 61]. Nanoparticles can go long distances to reach roots, all apex regions, and reproductive organs after entering the phloem. This systemic transport pathway is very important for RNAi-based genome editing, the regulation of hormones, and controlling pest technologies. It should function outside of the original treatment site.

4.3 Mechanistic Perspectives on Trafficking of Nanoparticles

4.3.1 Avoiding Conventional Endocytosis

Despite the fact that endocytosis is a known route for nanoparticle entry, a number of advanced nanocarriers can get past endocytic constraints. Lipid nanoparticles, carbon nanotubes (CNTs), and DNA origami have been shown to be able to transport genetic material directly into the cytoplasm without getting trapped in endosomes [62, 63]. This direct cytosolic distribution ensures better functionality of cargo such as siRNA, which may degrade or become inactive in endosomes.

Additionally, avoiding endocytosis reduces the energy load on the cell, which is critical for metabolic processes.

4.3.2 Systemic RNAi Engineered Nanocarriers

RNA interference (RNAi) has the potential to silence genes in plants, but delivery is very challenging due to poor absorption and RNA degradation. Layered double hydroxide (LDH) clay nanosheets protect dsRNA molecules from nucleases and release them gradually and continuously on leaf surfaces [64, 65]. There is a claim in 2020 by Diemer et al. that RNA functionalized carbon nanotubes have shown serious systemic activity by genuinely silencing the genes which are located across distant organs. According to Table 2, these nanocarriers give a comparatively safer environment, as well as a non-transgenic alternative to Agrobacterium for the modification of traits and regulation of genes (Fig. 3).

Table 2 A comparative analysis of nanoparticles and Agrobacterium-mediated methods

Features	Nanoparticles	Agrobacterium
Special range	Broadly, mostly independent of species	Mostly seen in dicots
Integration	Transgene-free possible	Mostly seen locally
RNAi delivery	Good, effective, and systematic delivery	Not up to the mark, systematic delivery
Regulatory concerns	In the case of the transient method, it is lower	For genetically modified nature, it's higher
Delivery area	Local and systemic	Mostly local

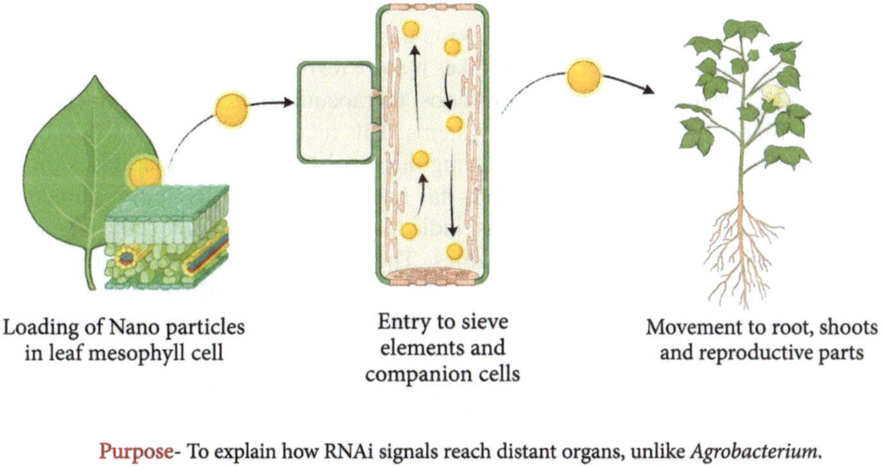

Fig. 3 RNAi nanoparticles' long-distance transport via the phloem pathway

5 RNAi Payload Engineering for Enhanced Stability and Release

One of the major obstacles to the effective application of RNAi technology is the inherent instability of RNA molecules in biological environments. Due to their susceptibility to degradation by nucleases and environmental factors, unmodified RNA molecules experience significant challenges when delivered in vivo. To combat this issue, various chemical modifications have been introduced in RNAi payload engineering, aimed at improving the stability and bioavailability of these therapeutic agents.

In addition to enhancing stability, achieving controlled release of RNAi payloads is crucial for maximizing therapeutic efficacy while minimizing off-target effects. Controlled release systems can be engineered to deliver RNA therapeutics precisely when and where they are needed, significantly improving the therapeutic response.

One promising approach is the development of smart nanocarriers that respond to specific stimuli in the target environment. Enzyme-responsive systems, pH-sensitive materials, etc., are designed to release their RNA payloads selectively in the presence of specific enzymes prevalent in certain tissues or disease states.

5.1 Chemical RNA Modifications to Resist RNases

Chemical changes are an important method in RNAi payload engineering for enhancing the stability of RNA molecules against RNase destruction. One of the most noticeable changes is the addition of 2′-fluoro modifications. These alterations entail replacing the hydroxyl (-OH) group at the ribose sugar's 2′ position with a fluorine atom. This change not only boosts the RNA molecule's binding affinity to its complementary target, but it also makes it more resistant to RNase action. Studies have revealed that 2′-fluoro alterations boost therapeutic effectiveness by increasing the half-life of siRNAs in biological systems. This increased stability enables long-term gene silencing effects while reducing the requirement for regular doses [66].

The efficacy of 2′-fluoro alterations has been further investigated in numerous therapeutic applications, with results indicating a lower immunogenic response compared to unmodified RNA molecules. The changes boost nuclease resistance while keeping the structural features necessary for optimal RNAi action. These chemical alterations make it easier to use RNA therapies in a variety of biological settings by increasing their pharmacokinetic qualities [67].

Another novel change is the incorporation of 5′ ethylene-bridged nucleic acids (ENAs). ENAs feature a covalent bond between the ribose sugar's 2′ oxygen and 4′ carbon, which locks the sugar in a rigid C3′-endo (north) conformation that mirrors the native RNA structure, stabilizing the duplex created with complementary RNA. This change includes the creation of ethylene bridges between the nucleotides, which stabilizes their conformation and reduces vulnerability to exonucleases. ENAs outperform conventional nucleotides in terms of binding affinity and stability, leading to increased RNAi effectiveness. The innate resistance to enzymatic degradation provides for a longer therapeutic window and greater intracellular storage of the RNA payload [68].

Another novel change is the incorporation of 5′ ethylene-bridged nucleic acids (ENAs). ENAs feature a covalent bond between the ribose sugar's 2′ oxygen and 4′ carbon, which locks the sugar in a rigid C3′-endo (north) conformation that mirrors the native RNA structure, stabilizing the duplex created with complementary RNA [69]. This change includes the creation of ethylene bridges between the nucleotides, which stabilizes their conformation and reduces vulnerability to exonucleases. ENAs outperform conventional nucleotides in terms of binding affinity and stability, leading to increased RNAi effectiveness. The innate resistance to enzymatic degradation provides for a longer therapeutic window and greater intracellular storage of the RNA payload [70].

5.2 Dual-Function Nanocarriers Co-delivering RNAi and Micronutrients (e.g., Zn, Si) for Synergistic Effects

In the pursuit of sustainable agricultural practices and increased crop production, the study of dual-function nanocarriers has emerged as a new way to co-deliver RNA interference (RNAi) agents and important micronutrients, notably zinc (Zn) and silicon (Si). These revolutionary nanocarriers not only allow for the targeted delivery of RNA molecules designed to silence certain genes associated with pests or illnesses, but they also provide essential micronutrients that may improve plant resistance and growth. This synergistic dual delivery approach improves the effectiveness of RNAi while also treating nutritional shortages, increasing overall plant health and production [71].

Different nanomaterials can be used to develop hybrid systems with improved gene and micronutrient delivery capabilities. For example, alginate-based hydrogels may be utilized to embed metal oxide nanoparticles such as ZnO and CuO, allowing for micronutrient release while simultaneously aiding gene transport. Nanofertilizers are created by embedding zinc oxide (ZnO) and copper oxide (CuO) nanoparticles within biopolymer matrices such as alginic acid hydrogels, which act as dual-function carriers. These nanofertilizers gradually release important micronutrients like Zn and Cu into the soil, conditioning it to retain moisture and promote nutrient absorption. Though largely focused on nutrition release, alginate-based nanocarriers show potential for gene transfer applications due to their biocompatibility and capacity to preserve cargo [71].

MSNs adorned with AuNPs were employed to transfer plasmid DNA and proteins into plant cells using biolistic particle bombardment. This technique provides a mechanism to successfully distribute genetic material, while the silica component may be functionalized for nutrient encapsulation, implying potential for dual gene and micronutrient administration [33].

5.3 Self-Amplifying RNAi Systems Encapsulated in Nanoparticles and RNase Inhibitor-Loaded Carriers, Which Prolong RNA Stability in Harsh Extracellular Environments

The development of self-amplifying RNA interference (RNAi) systems encapsulated in nanoparticles represents a significant leap forward in the field of molecular therapeutics, particularly in combating diseases in both agricultural and medical contexts. These innovative RNAi systems not only leverage the natural ability of RNA to act as a potent regulator of gene expression, but they also employ sophisticated delivery mechanisms that significantly enhance RNA stability in the harsh extracellular environments typically encountered during delivery. By integrating RNase inhibitors into carrier formulations, the effectiveness and viability of these RNAi systems can be maximized, opening new avenues for applications in gene therapy and crop protection [72].

5.3.1 Self-Amplifying RNAi Systems

Self-amplifying RNAi systems (saRNA) use viral replicon technology to improve cellular absorption and long-term production of RNA molecules. What distinguishes saRNA from other RNA delivery techniques is its inherent capacity to reproduce once inside the host cell. This replication causes larger amounts of RNA molecules, which amplifies the desired silencing impact on target genes [73]. The capacity to self-replicate needs an encapsulating technique that ensures the saRNA's stability in the systemic environment, allowing it to reach the target location before degradation occurs [74].

Nanoparticles are frequently used in encapsulation strategies because they protect the RNA payload from enzymatic degradation, particularly by ribonucleases (RNases) present in the extracellular milieu. These nanoparticle systems can be composed of lipids, polymers, or inorganic materials, and each offers unique advantages in terms of stability, biocompatibility, and release dynamics. Lipid nanoparticles (LNPs) are particularly [75].

5.3.2 RNase Inhibitor-Loaded Carriers

Incorporating RNase inhibitors into delivery methods increases RNA stability. RNase inhibitors are substances that may bind to RNases and limit their action, protecting the integrity of RNA molecules throughout passage through the different tissues. The inclusion of RNase inhibitors not only increases the half-life of encapsulated RNA in extracellular conditions but also enhances systemic circulation timings, allowing RNAi systems to reach target cells more efficiently [76].

The combination of nanoparticle encapsulation with RNase inhibitors produces an extremely effective platform for RNA delivery. This combination strategy allows for synergy, effectively establishing a strong barrier against the many components that contribute to RNA breakdown. Furthermore, the sustained release capabilities of many nanoparticle systems guarantee that a larger concentration of RNA remains at the target region for a prolonged length of time, which is critical for delivering the intended therapeutic effects [76].

6 Overcoming Plant-Specific Delivery Barriers

Exogenous dsRNA or siRNA applications for RNA interference-based gene silencing have grown in popularity due to the non-requirement of any genetic modifications. However, RNA administered to leaves must travel through the apoplast, cell wall, cuticle, and external stressors before it can reach its target cells. Novel nanoparticle (NP) compositions addressed these issues. This review examines the plant-specific delivery barriers and nanoparticle-based engineering methods. That improves the distribution of RNA, systemic movement, also robustness in the field.

6.1 Hindrance to Plant RNAi Delivery

- **Cuticle that is hydrophobic in nature**: According to Schreiber [77], the plant cuticle is a waxy, lipid-enriched coating that guards aerial tissues, prohibits RNA, and hydrophilic nanoparticles. That leads to poor adhesion and penetration on the leaf surface.
- **Electrostatic repulsion and exclusion of cell wall size**: Due to its negative charge and size exclusion limit, which should be less than 20–50 nm, the cellulose, hemicellulose, and pectin networks that form the cell wall under the cuticle. It keeps larger or anionic nanocarriers from reaching the plasma membrane [78].
- **RNases that cause cancer**: The apoplast's wall and intercellular spaces contain RNases that degrade exposed RNA strands, reducing the efficiency of delivery [64, 65].
- **Aspects of the environment**: While UV light speeds up the breakdown of RNA, rain can wash off formulations applied topically, reducing exposure time and absorption.

6.2 Design Strategies for Nanoparticles

6.2.1 Improving Surface Retention and Cuticle Penetration of Cellulose Nanocrystals (CNCs)

CNCs are rod-shaped and range in size from 5 to 20 nm. They also have hydroxyl-rich shells for modification and a large surface area. They can be covered by a surfactant or cationized to improve the adherence to the hydrophobic cuticle layer. CNCs facilitate the release should be controlled and loading of large Biomolecules [79].

- **Coatings made by Chitosan**: Chitosan nanoparticles are biodegradable and positively charged in nature. They assist in penetration by attaching to waxy cuticles electrostatically [80].
- **Nanoparticles modified by Surfactants**: For creating Nano particles more efficiently spread and penetrate surface waxes, surfactants like lecithin or Tween are being used recently.

6.2.2 Nanocarriers for Cell Wall Mobility

- **Double Hydroxides in Layers (LDHs)**: These clay-based 2D nanomaterials can pass through cell wall apertures because of their thin, flexible structure (~20–50 nm) and positive charge, which helps them bind and protect dsRNA [64, 65].
- **Nanostructures of DNA**: Tetrahedrons and nanocages are examples of programmable DNA origami structures that offer modular, malleable carriers that pass through cell walls and release siRNA directly, avoiding endocytosis [59].
- **LNPs, or lipid nanoparticles**: Malleable LNPs can directly transport RNA into the cytoplasm by fusing with plasma membranes and passing through flexible cell wall holes [62, 63].

6.2.3 Keep Protected Against the Extracellular RNases

- **Electrostatics encapsulating**: Intercalation of LDH and chitosan-dsRNA complexes attaches to RNA. It also prevents it from being harmed by enzymes [64, 65].
- **PEGylation**: Attachment of a steric PEG chain improves stability of RNA and also systemic transport by blocking RNase access [60, 61].
- **Hydrogel embedding**: Resistance against RNAse activity and controlled release are ensured by encapsulation in hydrogel nano matrices [81].

6.2.4 Enhancements in Environmental Stability

- **UV-Protection Coatings**: Lignin-Based Polymers Lignin nanoparticles, which are 40–200 nm in size, have antioxidative properties and naturally absorb UV light. Modified lignin coatings have been shown to block UV-A/B rays and shield embedded molecules [82].
- **UV-blocking inorganic**: Nanosized ZnO or TiO_2 enhances light protection and decreases ROS production when co-encapsulated in lignin nanocarriers (ACS reviews).
- **Rainfast formulations**: While adhesive polymers like PVA, xanthan gum, and latex binders enhance leaf adherence, hydrophobic or wax-compatible coatings help embed NPs into the cuticle. These formulations resist simulated rainfall for over a week [83].

6.3 A Comprehensive Formulation Approach

Through the use of lignin or TiO_2-based UV-resistant coatings, chitosan or CNCs for cuticle adherence, DNA nanocarriers or LDHs for RNA protection and cell wall passage, and polymeric adhesives for rainfastness. We can construct a robust delivery truck that: stays on leaf surfaces, overcomes structural barriers, protects the RNA payload, and keeps working outside (Fig. 4).

7 Scalability and Environmental Considerations

7.1 Industrial-Scale Synthesis Methods

Effective delivery systems that can shield the RNA molecules from deterioration while guaranteeing cellular uptake and continuous activity are essential for the successful field application of RNA interference [84]. Depending on the target organism and application technique, the global agricultural sector needs production scales of 2–10 g of dsRNA per hectare [85]. This emphasizes the need for reliable, scalable, and cost-effective manufacturing techniques that can move from laboratory-scale synthesis to industrial-scale production while maintaining regulatory compliance and consistent product quality. In vivo bacterial fermentation systems, utilizing carrier-state cells with the ϕ6 RNA polymerase complex, offer a scalable solution for industrial dsRNA production, achieving virtually unlimited yield and enabling continuous synthesis of dsRNA up to 4.0 kb in length [86]. According to recent optimization studies, bacterial dsRNA production can produce 53.3 μg/mL in fed-batch bioreactor operations, while flask cultures only produce 6.2 μg/mL [85]. T7 RNA polymerase-based in vitro transcription (IVT) provides more control over

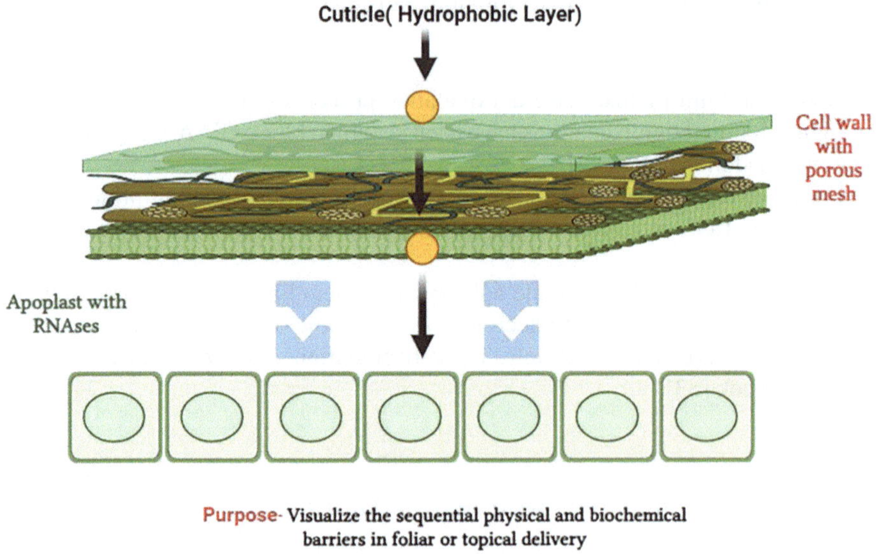

Fig. 4 Plant barriers hindering RNAi delivery

product purity but has scalability issues. Recent developments, like modified T7-68 polymerase variants and feed strategies that control UTP levels, have significantly lowered dsRNA by-products while preserving high transcriptional yields [87, 88].

7.1.1 Microfluidic Manufacturing Systems

With its ability to precisely control particle size, shape, and composition and enable high-throughput manufacturing appropriate for industrial applications, microfluidic technology has become a game-changing platform for the production of scalable nanoparticles [89]. The Silicon Scalable Lipid Nanoparticle Generation (SCALAR) platform, which integrates 256 mixing units on a single 100-mm silicon chip and achieves production rates of up to 17 L/h, represents a breakthrough in the manufacturing of microfluidic nanoparticles [90]. Microfluidic systems are excellent for GMP-compliant industrial nanoparticle production because of their excellent solvent compatibility and sterilizability made possible by silicon and glass construction, scalability through parallelization without changing individual unit parameters, and highly consistent particle properties due to millisecond-scale controlled mixing [90–92].

7.1.2 Continuous Flow Manufacturing

Flash nanoprecipitation methods using confined impingement jet (CIJ) and multi-inlet vortex mixers (MIVM) have been successfully scaled for industrial use, allowing for the quick mixing of polymer solutions with anti-solvents to create uniformly sized nanoparticles at production rates that are profitable [91]. Inline sonication methods have been specifically developed for large-scale polymeric nanoparticle production, achieving rates of 84 g/h (approximately 2 kg/day) for PLGA nanoparticles while consistently maintaining particle sizes of 150 ± 50 nm and polydispersity indices below 0.2 [93].

7.1.3 Quality by Design Implementation

Implementing Quality by Design (QbD) principles is essential for the successful scale-up of RNAi nanocarrier manufacturing. QbD principles offer a methodical framework for identifying and controlling critical quality attributes (CQAs) through the assessment of critical material attributes (CMAs) and critical process parameters (CPPs). Establishing a Quality Target Product Profile (QTPP) that details the ideal particle size (50–200 nm), >80% encapsulation efficiency, environmental stability, targeted release kinetics, and biocompatibility is important component of QbD. Particle size and homogeneity, dsRNA loading effectiveness, storage stability, controlled release, and the lack of hazardous residues are all included in the CQAs for RNAi nanocarriers. Full batch traceability, automated feedback systems for continuous production, statistical control of critical variables, and real-time monitoring through Process Analytical Technology (PAT) are all necessary for effective process control strategies [94, 95].

7.2 *Eco-Toxicity of Non-degradable Nanoparticles*

Non-biodegradable nanoparticles, which are usually between 1 and 100 nm in size, can withstand degradation in biological and environmental systems and endure for months or even years. These nanoparticles, which include materials like titanium dioxide, zinc oxide, silver nanoparticles, carbon nanotubes, fullerenes, and quantum dots, retain structural integrity in contrast to their biodegradable counterparts [96, 97]. Because they are chemically inert, resistant to enzymatic degradation, stable against oxidation and hydrolysis, and have the ability to bioaccumulate within organisms and food chains, non-degradable nanoparticles exhibit environmental persistence [98]. Synthetic nanoparticles have an incredibly long half-life in biological systems; in animal models, some studies have found persistence for up to 2 years. Particle properties like high surface area and reactivity (found in smaller particles), photochemical activity (found in TiO_2, ZnO, and fullerenes under light exposure), and metal content—which can drive ROS production via Fenton-type reactions—all have a significant impact on the main mechanism of nanoparticle eco-toxicity, which is the generation of reactive oxygen species (ROS), which results in oxidative stress [99].

In addition to causing mechanical damage and ROS-induced lipid peroxidation at the plasma membrane, which results in changed fluidity and permeability, non-biodegradable nanoparticles also affect other cellular components and functions. By affecting the electron transport chain, they impair mitochondrial function by depolarizing the membrane and producing less ATP. As these nanoparticles build up in lysosomes, they change pH, inhibit enzymes, and destabilize lysosomal membranes. They also cause DNA damage through strand breaks, oxidative lesions, and chromosomal mutations [100].

Studies have shown that even low concentrations (0.5 mg/L) of fullerenes (C_{60}, C_{70}) can cause oxidative damage in fish brains and gills, including lipid peroxidation and glutathione depletion. Fullerenes were among the first engineered nanoparticles to be examined for ecotoxicity [101]. Furthermore, studies on freshwater ecosystems show that fullerenes can upset ecological balance by changing community structure and predation rates, even when there is no acute toxicity [102]. The ecotoxicity of carbon nanotubes varies with their structure; single-walled carbon nanotubes (SWCNTs) are generally more toxic than multi-walled carbon nanotubes (MWCNTs). Along with the potential for bioaccumulation and food chain transfer, they have been demonstrated to cause respiratory toxicity in aquatic organisms, including gill damage and altered ventilation rates [103]. One of the most widely produced and researched nanomaterials, titanium dioxide (TiO_2) nanoparticles exhibit eco-toxicity through concentration-dependent inhibition of algal growth, bioaccumulation and biomagnification in aquatic food chains (with biomagnification factors >1), and increased toxicity as a result of photocatalytic activity under ultraviolet light [104]. Because of their solubility and release of Zn2+ ions, zinc oxide (ZnO) nanoparticles are extremely toxic, causing oxidative stress, cellular damage, microbial community disruption, genotoxic effects, and inflammation in a variety of organisms [105].

7.3 Biodegradable Alternatives to Conventional Nanocarriers

Protein- and starch-based nanocarriers are extremely promising green nanotechnology platforms for delivering RNA interference (RNAi) in plants. They provide sustainable substitutes for synthetic materials while preserving superior functional performance, biocompatibility, and biodegradability. Quaternized starch (Q-starch), chemically modified with quaternary ammonium groups, serves as an effective starch-based platform for siRNA delivery due to its cationic nature, enabling efficient electrostatic complexation with negatively charged RNA molecules [106, 107]. At low N/P ratios (as low as 2:1), these nanoparticles form stable ~30 nm complexes that provide strong protection against nuclease degradation and sustained release for up to 30 days [107]. In optimized formulations, Q-starch systems have shown high gene silencing efficiency, achieving 85–90% knockdown. Endocytosis is the mechanism by which cells are taken up; 15% of siRNA complexes are able to get past endosomes and initiate RNA interference in the cytoplasm [107]. Carboxymethyl starch (CMS) is a promising platform for nucleic acid delivery, offering pH-responsive release and high encapsulation efficiency of up to 87%, making it suitable for targeted RNAi applications [108]. Self-assembled hydroxyethyl starch-cholesterol nanoparticles have shown promise in co-delivering siRNA and other therapeutic agents [109].

Among protein-based carriers, cationized bovine serum albumin (cBSA) has demonstrated remarkable efficacy in silencing plant genes, with 60–70% knock-down of constitutively expressed genes in poplar and tobacco. By delivering RNAi signals over long distances, cBSA allows systemic gene silencing when applied locally to leaf petioles or shoot bases. Additionally, it works with constitutive and inducible gene targets, making it a flexible tool for plant functional genomics [110]. Gelatin, a collagen-derived polymer that is both biocompatible and biodegradable, is a useful vehicle for delivering nucleic acids. The mechanical stiffness of cationized gelatin nanoparticles has been found to be a critical component of RNA interference efficiency. An ideal elastic modulus of 0.87 MPa allows for maximum siRNA uptake and gene silencing, particularly in suspension cells that are difficult to transfect [111]. Casein-Based Nanocarriers, Silk fibroin nanoparticles, and ferritin nanocages are other promising protein-based carriers. Starch- and protein-based nanocarriers have transformed spray-induced gene silencing (SIGS) by enabling effective, non-transgenic, topical delivery of dsRNA for sustainable plant protection. This approach has achieved over 90% efficacy against fungal pathogens, provided virus resistance lasting over 20 days, and enhanced drought tolerance through targeted gene silencing [112].

Green chemistry-synthesised zinc oxide nanoparticles (ZnO-NPs) containing plant extracts from basil (*Ocimum basilicum*) and mint (*Mentha spicata*) show good biocompatibility and can be functionalized for effective nucleic acid delivery [113]. Layered double hydroxide (LDH) nanosheets represent a significant advancement in green nanotechnology for RNAi delivery, as studies have shown that non-toxic, degradable LDH clays can stably load dsRNA, resist wash-off, enable sustained release, and remain detectable on sprayed leaves for up to 30 days post-application [18, 19, 64, 65]. The cationic nature of chitosan, a biodegradable green polymer made from the chitin of crustacean shells, makes it ideal for RNAi delivery because it allows for efficient binding with negatively charged RNA molecules while maintaining environmental safety [114, 115].

7.4 Green Nanotechnology Approaches

The increasing pressures of population growth and climate change on modern agriculture have sparked interest in the use of green, biologically derived nanoparticles from plants, fungi, or microbes, which provide a more cost-effective and ecologically friendly alternative to conventional chemical and physical nanoparticle synthesis [116]. For the delivery of RNA interference, plant virus-like particles (VLPs) have become one of the most promising green nanotechnology platforms. Naturally occurring as defective viral forms, virus-like particles (VLPs) are engineered or naturally occurring nanostructures that resemble viruses in form but lack the necessary viral genetic material for replication [117]. A promising VLP platform, Cowpea

Chlorotic Mottle Virus (CCMV) permits RNA loading before reassembly at neutral pH, thereby preventing degradation of encapsulated dsRNA while permitting controlled release [118]. Tobacco Mosaic Virus (TMV)-derived VLPs, with their elongated structure, are well-suited for carrying larger RNA cargo and can be engineered to simultaneously display multiple functional proteins [119].

8 Regulatory and Biosafety Challenges in Nano-enabled RNAi Delivery

Systemic distribution of double-stranded RNA (dsRNA) via nanoparticles, along with RNA-based tropical sprays are very good alternative to traditional pesticides as well as for transgenic crops. These platforms promise an ample amount for boosting output, reducing stress levels, and safeguarding crops. However, the lack of a clear regulatory framework and a lack of knowledge about long-term ecological effects significantly hinder large-scale deployment [58, 120].

8.1 Inadequate Regulations for Nano-agriproducts

8.1.1 Ambiguity Nature and Fragmentation While Regulating

The RNAi-based goods and nanomaterials are nowadays subject to variable regulation throughout the global. There are some guidelines for nanoparticle safety testing, like though it has been published by the Organization for Economic Co-operation and Development (OECD) [121], they are not specific to the complex hybrid systems which employed in RNAi delivery or applications of agrochemicals.

- **United States**: Nanoparticle carriers frequently gap which are differential in nature, in spite of the EPA's regulation of dsRNA products under FIFRA [122].
- **European Union**: Under REACH, Nanocarriers in agrochemicals are governed as "novel materials" though they lack RNAi-specific protocols.
- **China and India**: There are no such agricultural-specific nanotechnology biosafety rules that are comprehensive in nature, though there is regulatory interest.

8.1.2 Lack of Frameworks for Composites Risk Assessment

Nano-enabled RNAi sprays combine genetic materials (dsRNA) with chemical carriers (like carbon nanotubes, chitosan, and LDHs). These hybrids are not compatible with current frameworks designed for either genetic constructions or conventional insecticides. Standard testing of toxicity might miss interactions that are either synergistic in nature or extended exposure [123].

8.2 *Nanoparticles' Biosafety Concerns in Agroecosystems*

- **Accumulation of nanoparticles in soil and food chains**: When sprayed on crops, nanoparticles may remain in the soil after spraying on the crops. After that, they either accumulate in roots or move up the trophic levels through interactions with microbes or herbivores.
- **Hindrance to the soil microbiology**: Positively charged nanoparticles (e.g, silver and ZnO) have been shown to alter the enzyme activity. It also reduces microbial biomass and disrupts the nitrogen cycle ([124]; Simonin).
- **Risks of bioaccumulation**: Uncertain long-term effects may result from engineered NPs moving from plants to insects and predators. According to certain research, earthworms and bees may bioaccumulate [125].
- **Transfer to trophic**: According to Keller et al. [126], LDHs and carbon-based NPs have the potential to build up in plant tissues. It can also make its way to consumers. Beyond avery much toxicity, these hazards occur in chronic exposure studies and life cycle assessment (LCA) methodologies [127].

8.3 *Techniques for Observation and Reduction*

- **Real-time environmental monitoring with nanosensors**: To allay concerns regarding the permanence and safety of nanoparticles, real-time monitoring is essential. Nanosensors that can track the rate of NP degradation, mobility, and environmental release, and uptake by nontarget organisms include quantum dot or FRET-based fluorescence devices. For example, in Arabidopsis, Scientists can successfully track the foliar absorption as well as degradation of carbon dot-labeled dsRNA [128].
- **Nanoparticles that are "safe by design"**: Recently, scientists have been able to create nanocarriers that are biocompatible and biodegradable and self-limit toxicity rather than retrofitting safety after development. Biodegradable cellulose nanocrystals and lignin nanoparticles can be propagated from agricultural waste [129]. A marine biopolymer called chitosan, which encourages the growth of beneficial microorganisms, is enzymatically degradable in soil [130].

8.4 *Moving Towards Coordinated Risk Management*

- **Initiatives for Regulatory Harmonization**—Nano-specific testing recommendations have been proposed by international collaborations, including the OECD Working Party on Manufactured Nanomaterials (WPMN) and the International Council for Nanotechnology (ICON). But they must be integrated with: Risk evaluations unique to RNA interference (such as off-target silencing), Models of environmental behavior for RNA-NP hybrids. It might be possible to create a specific "nano-RNAi risk assessment module" to serve as a guide for developers and regulators.

- **Ethical Considerations and Public Perception**—In addition to technical regulation, public trust is necessary for the implementation of nano-enabled RNAi. Independent biosafety audits, labeling, and open communication will all aid in avoiding opposition like that observed in the GMO controversy [131].

9 Conclusion and Future Perspectives

The field of RNA interference (RNAi) delivery in plants has reached a critical inflection point, propelled by significant innovations in nanoscale carrier systems. The strategic integration of stimuli-responsive nanocarriers and biomimetic design principles constitutes a transformative approach for enhancing the precision, efficiency, and tissue-specificity of genetic modulation. This paradigm directly addresses the fundamental delivery barriers unique to plant systems, offering pathways to minimize off-target effects and maximize therapeutic impact. Stimuli-responsive nanocarriers, engineered for payload release upon encountering specific environmental cues such as pH gradients, enzymatic activity, reactive oxygen species (ROS) fluctuations, or light exposure, provide essential spatiotemporal control. This targeted activation enhances efficacy, exemplified by light-triggered systems enabling cellular-level precision and ROS-mediated deployment within stressed tissues. Complementing these advances, biomimetic strategies leverage natural cellular uptake pathways, optimizing carrier internalization and cargo release efficiency within complex plant architectures.

Future progress necessitates focused development across several interconnected areas. Central to advancement is deeper cross-disciplinary convergence, integrating expertise from nanotechnology, synthetic biology, computational modeling, and artificial intelligence. AI-driven design pipelines hold particular promise for rapidly optimizing nanoparticle parameters, such as size, surface functionalization, and release kinetics, to overcome plant-specific delivery challenges, accelerating the translation of foundational research into scalable agricultural applications. Equally critical are advances in intelligent closed-loop delivery systems capable of autonomous responsiveness. Such next-generation platforms would release RNAi payloads exclusively upon detecting predefined biotic or abiotic stimuli, such as specific pathogen biomarkers or herbivore elicitors, enabling precise, on-demand crop protection. This self-regulating approach promises to revolutionize sustainable agriculture by drastically reducing dependence on broad-spectrum chemical pesticides while minimizing ecological disruption. Concurrently, pioneering work in nanobionic integration offers a paradigm-shifting strategy for enhancing climate resilience. Embedding functionalized nanoparticles within plant systems, designed to sense environmental pressures like drought, salinity, or disease, could activate targeted endogenous RNAi pathways or deliver protective agents directly. This proactive enhancement of crop adaptability directly addresses the global imperative for food security amidst increasingly volatile climatic conditions.

In conclusion, the continued refinement of stimulus-responsive nanotechnologies and biomimetic delivery platforms illuminates a robust trajectory for RNAi-based interventions in plant science. Prioritizing interdisciplinary collaboration while vigorously pursuing the development of intelligent closed-loop systems and nanobionic integrations will be pivotal in overcoming current delivery limitations. This concerted effort promises to catalyze the development of high-yielding, environmentally resilient crops, ultimately supporting the establishment of sustainable agricultural systems and enhanced global food security. The pursuit of these innovations heralds a new era of precision plant biotechnology firmly grounded in both efficacy and ecological stewardship.

References

1. Chen C, Imran M, Feng X, Shen X, Sun Z (2025) Spray-induced gene silencing for crop protection: recent advances and emerging trends. Front Plant Sci 16:1527944
2. Hu T, Gu Z, Williams GR, Strimaite M, Zha J, Zhou Z et al (2022) Layered double hydroxide-based nanomaterials for biomedical applications. Chem Soc Rev 51(14):6126–6176
3. Vaddoriya HK, Shelar VB, Pampaniya A, Panwala A (2024) "BioClay™": one step toward the sustainable and novel plant protection method. Int J Econ Plants 11(2):142–146
4. Ray P, Sahu D, Aminedi R, Chandran D (2022) Concepts and considerations for enhancing RNAi efficiency in phytopathogenic fungi for RNAi-based crop protection using nanocarrier-mediated dsRNA delivery systems. Front Fungal Biol 3:977502. https://doi.org/10.3389/ffunb.2022.977502
5. Mujtaba M, Wang D, Carvalho LB, Oliveira JL, Pereira ADES, Sharif R, Jogaiah S, Paidi MK, Wang L, Ali Q, Fraceto LF (2021) Nanocarrier-mediated delivery of miRNA, RNAI, and CRISPR-CAS for plant protection: current trends and future directions. ACS Agric Sci Technol 1(5):417–435. https://doi.org/10.1021/acsagscitech.1c00146
6. Christiaens O, Petek M, Smagghe G, Taning CNT (2020) The use of nanocarriers to improve the efficiency of RNAi-based pesticides in agriculture. In: Dubey AK, Kumar A, Kumar V (eds) Nanopesticides: from research and development to mechanisms of action and sustainable use in agriculture. Academic, New York, pp 49–68. https://doi.org/10.1016/B978-0-12-819831-4.00003-2
7. Christiaens O, Petek M, Smagghe G, Taning CNT (2020) The use of nanocarriers to improve the efficiency of RNAi-based pesticides in agriculture. In: Fraceto LF, de Castro SSVL, Grillo R, Ávila D, Caixeta Oliveira H, Lima R (eds) Nanopesticides. Springer, Cham. https://doi.org/10.1007/978-3-030-44873-8_3
8. Xing Y, Jiang H, Cai L (2025) Engineered nanotransporters for efficient RNAi delivery in plant protection applications. J Integr Plant Biol 67(5):1223–1245. https://doi.org/10.1111/jipb.13887
9. Alghuthaymi MA, Ahmad A, Khan Z, Khan SH, Ahmed FK, Faiz S, Nepovimova E, Kuča K, Abd-Elsalam KA (2021) Exosome/liposome-like nanoparticles: new carriers for CRISPR genome editing in plants. Int J Mol Sci 22(14):7456. https://doi.org/10.3390/ijms22147456
10. Tenchov R, Bird R, Curtze AE, Zhou Q (2021) Lipid nanoparticles—from liposomes to mRNA vaccine delivery, a landscape of research diversity and advancement. ACS Nano 15(11):16982–17015. https://doi.org/10.1021/acsnano.1c04996

11. Chandrasekaran R, Seetharaman PK, Danaraj J, Rajiv P, Abd-Elsalam KA (2021) Polymer and lipid-based nanoparticles to deliver RNAi and CRISPR systems. Elsevier eBooks, Amsterdam, pp 635–659. https://doi.org/10.1016/b978-0-12-821910-2.00016-3
12. Qiu M, Glass Z, Chen J, Haas M, Jin X, Zhao X, Rui X, Ye Z, Li Y, Zhang F, Xu Q (2021) Lipid nanoparticle-mediated codelivery of Cas9 mRNA and single-guide RNA achieves liver-specific in vivo genome editing of Angptl3. Proc Natl Acad Sci 118(10):e2020401118. https://doi.org/10.1073/pnas.2020401118
13. Pascoli M, Lopes-Oliveira PJ, Fraceto LF et al (2018) State of the art of polymeric nanoparticles as carrier systems with agricultural applications: a minireview. Energy Ecol Environ 3:137–148. https://doi.org/10.1007/s40974-018-0090-2
14. Scarpin D, Nerva L, Chitarra W, Moffa L, D'Este F, Vuerich M, Filippi A, Braidot E, Petrussa E (2023) Characterisation and functionalisation of chitosan nanoparticles as carriers for double-stranded RNA (dsRNA) molecules towards sustainable crop protection. Biosci Rep 43(11):BSR20230817. https://doi.org/10.1042/bsr20230817
15. Kwak SY, Lew TTS, Sweeney CJ et al (2019) Chloroplast-selective gene delivery and expression in planta using chitosan-complexed single-walled carbon nanotube carriers. Nat Nanotechnol 14:447–455. https://doi.org/10.1038/s41565-019-0375-4
16. Mat Jalaluddin NS, Asem M, Harikrishna JA, Ahmad Fuaad AAH (2023) Recent progress on nanocarriers for topical-mediated RNAi strategies for crop protection—a review. Molecules 28(6):2700. https://doi.org/10.3390/molecules28062700
17. Mostafa M, Ahmed FK, Alghuthaymi M, Abd-Elsalam KA (2021) Inorganic smart nanoparticles: a new tool to deliver CRISPR systems into plant cells. Elsevier eBooks, Amsterdam, pp 661–686. https://doi.org/10.1016/b978-0-12-821910-2.00036-9
18. Yong J, Wu M, Zhang R, Bi S, Mann CWG, Mitter N, Carroll BJ, Xu ZP (2022) Clay nanoparticles efficiently deliver small interfering RNA to intact plant leaf cells. Plant Physiol 190(4):2187–2202. https://doi.org/10.1093/plphys/kiac430
19. Yong J, Wu M, Zhang R, Bi S, Mann CW, Mitter N et al (2022) Clay nanoparticles efficiently deliver small interfering RNA to intact plant leaf cells. Plant Physiol 190(4):2187–2202
20. Yong J, Wu M, Carroll BJ, Xu ZP, Zhang R (2024) Enhancing plant biotechnology by nanoparticle delivery of nucleic acids. Trends Genet 40(4):352–363. https://doi.org/10.1016/j.tig.2024.01.005
21. Komarova T, Ilina I, Taliansky M, Ershova N (2023) Nanoplatforms for the delivery of nucleic acids into plant cells. Int J Mol Sci 24(23):16665. https://doi.org/10.3390/ijms242316665
22. Yang F, Liu Y, Jing X, Liu X, Wang G, Jian-Ping A, Huo H, Zhang Z, Wang X, Zhang S, You C (2023) A biomimetic nanoparticle for PDNA delivery and expression in plant cells in a pH-dependent manner. ACS Agric Sci Technol 3(8):631–641. https://doi.org/10.1021/acsagscitech.3c00068
23. Islam MS, Ahmed MR, Noman M, Zhang Z, Wang J, Lu Z, Cai Y, Ahmed T, Li B, Wang Y, Golam Sarwar AKM, Wang J (2025) Integrating RNA interference and nanotechnology: a transformative approach in plant protection. Plants 14(6):977. https://doi.org/10.3390/plants14060977
24. Zhang Z, Luo H, Zhang X, Yang R, Yan S, Yang Q, Yang J (2024) Extracellular vesicles mimetic design of membrane chimeric nanovesicles for DSRNA delivery in spray-induced gene silencing for crop protection. ACS Nano 18(47):32468–32480. https://doi.org/10.1021/acsnano.4c06282
25. Demirer GS, Silva TN, Jackson CT, Thomas JB, Ehrhardt DW, Rhee SY, Landry MP (2020) Nanotechnology to advance CRISPR–Cas genetic engineering of plants. Nat Nanotechnol 15(7):541–548. https://doi.org/10.1038/s41565-020-0703-x
26. Demirer GS, Zhang H, Goh NS, Pinals RL, Chang R, Landry MP (2020) Carbon nanocarriers deliver siRNA to intact plant cells for efficient gene knockdown. Sci Adv 6(26). https://doi.org/10.1126/sciadv.aaz0495
27. Lin S, Zhang Q, Bai S, Yang L, Qin G, Wang L, Wang W, Cheng C, Zhang D, Lu C, Yuan J, Li J, Yang H, Gu X, Han X (2025) Beyond species and spatial boundaries: enabling long-

distance gene silencing in plants via guanidinium-siRNA nanoparticles. Plant Biotechnol J 23(4):1165–1177. https://doi.org/10.1111/pbi.14575

28. Srivastav A, Gupta K, Chakraborty D et al (2022) Efficiency of chitosan-coated PLGA nanocarriers for cellular delivery of siRNA and CRISPR/Cas9 complex. J Pharm Innov 17:180–193. https://doi.org/10.1007/s12247-020-09496-4
29. Bai C, Liu J, Zhang X, Li Y, Qin Q, Song H, Yuan C, Huang Z (2024) Research status and challenges of plant-derived exosome-like nanoparticles. Biomed Pharmacother 174:116543. https://doi.org/10.1016/j.biopha.2024.116543
30. Akuma P, Okagu OD, Udenigwe CC (2019) Naturally occurring exosome vesicles as potential delivery vehicle for bioactive compounds. Front Sustain Food Syst 3. https://doi.org/10.3389/fsufs.2019.00023
31. Feng H, Yue Y, Zhang Y, Liang J, Liu L, Wang Q et al (2024) Plant-derived exosome-like nanoparticles: emerging nanosystems for enhanced tissue engineering. Int J Nanomedicine 19:1189–1204. https://doi.org/10.2147/IJN.S448905
32. He J, Yu L, Lin X, Liu X, Zhang Y, Yang F, Deng W (2022) Virus-like particles as nanocarriers for intracellular delivery of biomolecules and compounds. Viruses 14(9):1905. https://doi.org/10.3390/v14091905
33. Cai Y, Liu Z, Wang H, Meng H, Cao Y (2023) Mesoporous silica nanoparticles mediate SiRNA delivery for long-term multi-gene silencing in intact plants. Adv Sci 11(9):e2301358. https://doi.org/10.1002/advs.202301358
34. Okuda A, Futaki S (2022) Protein delivery to cytosol by cell-penetrating peptide bearing tandem repeat penetration-accelerating sequence. In: Langel Ü (ed) Cell penetrating peptides. Methods in molecular biology, vol 2383. Humana, New York, NY. https://doi.org/10.1007/978-1-0716-1752-6_18
35. Liu BR, Chiou S-H, Huang Y-W, Lee H-J (2022) Bio-membrane internalization mechanisms of arginine-rich cell-penetrating peptides in various species. Membranes 12(1):88. https://doi.org/10.3390/membranes12010088
36. Durzyńska J, Przysiecka Ł, Nawrot R, Barylski J, Nowicki G, Warowicka A, Musidlak O, Goździcka-Józefiak A (2015) Viral and other cell-penetrating peptides as vectors of therapeutic agents in medicine. J Pharmacol Exp Ther 354(1):32–42. https://doi.org/10.1124/jpet.115.223305
37. Liu BR, Chen C-W, Huang Y-W, Lee H-J (2023) Cell-penetrating peptides for use in development of transgenic plants. Molecules 28(8):3367. https://doi.org/10.3390/molecules28083367
38. Taylor RE, Zahid M (2020) Cell penetrating peptides, novel vectors for gene therapy. Pharmaceutics 12(3):225. https://doi.org/10.3390/pharmaceutics12030225
39. Qi X, Droste T, Kao CC (2010) Cell-penetrating peptides derived from viral capsid proteins. Mol Plant-Microbe Interact 24(1):25–36. https://doi.org/10.1094/mpmi-07-10-0147
40. Ji Q, Kowalski KP, Golenberg EM, Chung SH, Barker ND, Bickford WA, Gong P (2025) Cell-penetrating peptide-mediated delivery of gene-silencing nucleic acids to the invasive common reed Phragmites australis via foliar application. Plants 14(3):458. https://doi.org/10.3390/plants14030458
41. Maguregui A, Abe H (2020) Developments in siRNA modification and ligand conjugated delivery to enhance RNA interference ability. Chembiochem 21(13):1808–1815. https://doi.org/10.1002/cbic.202000009
42. Kumar V, Turnbull WB (2023) Targeted delivery of oligonucleotides using multivalent protein-carbohydrate interactions. Chem Soc Rev 52(4):1273–1287. https://doi.org/10.1039/d2cs00788f
43. Das PR, Sherif SM (2020) Application of exogenous DSRNAs-induced RNAI in agriculture: challenges and triumphs. Front Plant Sci 11:946. https://doi.org/10.3389/fpls.2020.00946
44. Ji J, Wu T, Zhang Y, Feng F (2019) Light-controlled in vitro gene delivery using polymer-tethered spiropyran as a photoswitchable photosensitizer. ACS Appl Mater Interfaces 11(17):15222–15232. https://doi.org/10.1021/acsami.8b22505

45. Wang J, He X, Shen S, Cao Z, Yang X (2018) ROS-sensitive cross-linked polyethylenimine for red-light-activated siRNA therapy. ACS Appl Mater Interfaces 11(2):1855–1863. https://doi.org/10.1021/acsami.8b18697
46. Giménez VMM, Arya G, Zucchi IA, Galante MJ, Manucha W (2021) Photo-responsive polymeric nanocarriers for target-specific and controlled cargo delivery. Soft Matter 17(38):8577–8584. https://doi.org/10.1039/d1sm00999k
47. Xiong Q, Lim Y, Li D, Pu K, Liang L, Duan H (2019) Photoactive nanocarriers for controlled delivery. Adv Funct Mater 30(2):1903896. https://doi.org/10.1002/adfm.201903896
48. Liang C, Chang J, Jiang Y, Liu J, Mao L, Wang M (2019) Selective RNA interference and gene silencing using reactive oxygen species-responsive lipid nanoparticles. Chem Commun 55(56):8170–8173. https://doi.org/10.1039/c9cc04517a
49. Zheng N, Xie D, Zhang Z, Kuang J, Zheng Y, Wang Q, Li Y (2019) Thioketal-crosslinked: ROS-degradable polycations for enhanced in vitro and in vivo gene delivery with self-diminished cytotoxicity. J Biomater Appl 34(3):326–338. https://doi.org/10.1177/0885328219845081
50. Lowry GV, Giraldo JP, Steinmetz NF, Avellan A, Demirer GS, Ristroph KD, Wang GJ, Hendren CO, Alabi CA, Caparco A, Da Silva W, González-Gamboa I, Grieger KD, Jeon S, Khodakovskaya MV, Kohay H, Kumar V, Muthuramalingam R, Poffenbarger H et al (2024) Towards realizing nano-enabled precision delivery in plants. Nat Nanotechnol 19(9):1255–1269. https://doi.org/10.1038/s41565-024-01667-5
51. Narváez-Barragán DA, Tovar-Herrera OE, Guevara-García A, Serrano M, Martinez-Anaya C (2022) Mechanisms of plant cell wall surveillance in response to pathogens, cell wall-derived ligands and the effect of expansins to infection resistance or susceptibility. Front Plant Sci 13:969343. https://doi.org/10.3389/fpls.2022.969343
52. Thagun C, Chuah J, Numata K (2019) Targeted gene delivery into various plastids mediated by clustered cell-penetrating and chloroplast-targeting peptides. Adv Sci 6(23):1902064. https://doi.org/10.1002/advs.201902064
53. Yousefiasl S, Zare I, Ghovvati M, Ghomi M (2023) Enzyme-responsive materials: properties, design, and applications. In: ACS symposium series, pp 203–229. https://doi.org/10.1021/bk-2023-1436.ch009
54. Zhou H, Chen DS, Hu CJ, Hong X, Shi J, Xiao Y (2023) Stimuli-responsive nanotechnology for RNA delivery. Adv Sci 10(36):e2303597. https://doi.org/10.1002/advs.202303597
55. Anjum F, Lienemann PS, Metzger S, Biernaskie J, Kallos MS, Ehrbar M (2016) Enzyme responsive GAG-based natural-synthetic hybrid hydrogel for tunable growth factor delivery and stem cell differentiation. Biomaterials 87:104–117. https://doi.org/10.1016/j.biomaterials.2016.01.050
56. Etxeberria E, Gonzalez P, Pozueta-Romero J, Etxeberria M (2016) Evidence for the existence of two different size exclusion limits in plant cell walls. Plant Signal Behav 11(3):e1141874. https://doi.org/10.1080/15592324.2016.1141874
57. Nair R, Varghese SH, Nair BG, Maekawa T, Yoshida Y, Kumar DS (2010) Nanoparticulate material delivery to plants. Plant Sci 179(3):154–163. https://doi.org/10.1016/j.plantsci.2010.04.012
58. Kah M, Kookana RS, Gogos A, Bucheli TD (2018) A critical evaluation of nanopesticides and nanofertilizers against their conventional analogues. Nat Nanotechnol 13(8):677–684. https://doi.org/10.1038/s41565-018-0131-1
59. Zhang D, Demirer GS, Zhang H, Ye T, Goh NS, Aditham AJ et al (2022) Engineering DNA nanostructures for plant gene silencing. Nat Protoc 17:409–428. https://doi.org/10.1038/s41596-021-00645-4
60. Lv J, Christie P, Zhang S (2019) Uptake, translocation, and transformation of metal-based nanoparticles in plants: recent advances and methodological challenges. Environ Sci Nano 6(1):41–59. https://doi.org/10.1039/C8EN00645G
61. Lv J, Christie P, Zhang S (2019) Uptake, translocation, and transformation of metal-based nanoparticles in plants: recent advances and methodological challenges. Environ Sci Nano 6(1):41–59. https://doi.org/10.1039/C8EN00877A

62. Demirer GS, Zhang H, Goh NS, Gonzalez-Grandio E, Landry MP (2019) Carbon nanotube–mediated DNA delivery without transgene integration in intact plants. Nat Protoc 14(8):2954–2971. https://doi.org/10.1038/s41596-019-0209-1
63. Demirer GS, Zhang H, Goh NS, Pinals RL, Chang R, Landry MP, Yang D (2019) Carbon nanotube–mediated DNA delivery without transgene integration in intact plants. Nat Protoc 14(8):2954–2971. https://doi.org/10.1038/s41596-019-0208-1
64. Mitter N, Worrall EA, Robinson KE, Li P, Jain RG, Taochy C et al (2017) Clay nanosheets for topical delivery of RNAi for sustained protection against plant viruses. Nat Plants 3(2):16207. https://doi.org/10.1038/nplants.2016.207
65. Mitter N, Worrall EA, Robinson KE, Xu ZP, Carroll BJ (2017) Induction of virus resistance by topical application of double-stranded RNA in plants. Plant Biotechnol J 15(7):877–888. https://doi.org/10.1111/pbi.12678
66. Howard JD, Beghyn M, Dewulf N, De Vos Y, Philips A, Portwood D, Kilby PM, Oliver D, Maddelein W, Brown S, Dickman MJ (2022) Chemically modified dsRNA induces RNAi effects in insects in vitro and in vivo: a potential new tool for improving RNA-based plant protection. J Biol Chem 298(9):102311. https://doi.org/10.1016/j.jbc.2022.102311
67. Chen S, Le BT, Chakravarthy M, Kosbar TR, Veedu RN (2019) Systematic evaluation of 2′-fluoro modified chimeric antisense oligonucleotide-mediated exon skipping in vitro. Sci Rep 9(1):6078. https://doi.org/10.1038/s41598-019-42523-0
68. Halder K, Chaudhuri A, Abdin MZ, Majee M, Datta A (2022) RNA interference for improving disease resistance in plants and its relevance in this clustered regularly interspaced short palindromic repeats-dominated era in terms of dsRNA-based biopesticides. Front Plant Sci 13:885128. https://doi.org/10.3389/fpls.2022.885128
69. Prall W, Ganguly DR, Gregory BD (2023) The covalent nucleotide modifications within plant mRNAs: what we know, how we find them, and what should be done in the future. Plant Cell 35(6):1801–1816. https://doi.org/10.1093/plcell/koad044
70. Morita K, Takagi M, Hasegawa C, Kaneko M, Tsutsumi S, Sone J, Ishikawa T, Imanishi T, Koizumi M (2003) Synthesis and properties of 2′-O,4′-C-ethylene-bridged nucleic acids (ENA) as effective antisense oligonucleotides. Bioorg Med Chem 11(10):2211–2226. https://doi.org/10.1016/s0968-0896(03)00115-9
71. Zhi H, Zhou S, Pan W, Shang Y, Zeng Z, Zhang H (2022) The promising nanovectors for gene delivery in plant genome engineering. Int J Mol Sci 23(15):8501. https://doi.org/10.3390/ijms23158501
72. UF Innovate—Building business on innovation (n.d.). https://ufinnovate.technologypublisher.com/tech/Self-Amplifying_RNA_for_Gene_Silencing_in_Plants. Accessed 4 May 2025
73. Blakney AK, McKay PF, Hu K, Samnuan K, Jain N, Brown A, Thomas A, Rogers P, Polra K, Sallah H, Yeow J, Zhu Y, Stevens MM, Geall A, Shattock RJ (2021) Polymeric and lipid nanoparticles for delivery of self-amplifying RNA vaccines. J Control Release 338:201–210. https://doi.org/10.1016/j.jconrel.2021.08.029
74. Vallet T, Vignuzzi M (2025) Self-amplifying RNA: advantages and challenges of a versatile platform for vaccine development. Viruses 17(4):566. https://doi.org/10.3390/v17040566
75. Bera K, Rojas-Gómez RA, Mukherjee P, Snyder CE, Aksamitiene E, Alex A, Spillman DR, Marjanovic M, Shabana A, Johnson R, Hood SR, Boppart SA (2024) Probing delivery of a lipid nanoparticle encapsulated self-amplifying mRNA vaccine using coherent Raman microscopy and multiphoton imaging. Sci Rep 14(1). https://doi.org/10.1038/s41598-024-54697-3
76. Muhammad T, Zhang F, Zhang Y, Liang Y (2019) RNA interference: a natural immune system of plants to counteract biotic stressors. Cells 8(1):38. https://doi.org/10.3390/cells8010038
77. Schreiber L (2010) Transport barriers made of cutin, suberin and associated waxes in the plant cuticle: a review. Trends Plant Sci 15(10):546–553. https://doi.org/10.1016/j.tplants.2010.06.002
78. Carpita NC, Gibeaut DM (1993) Structural models of primary cell walls in flowering plants: consistency of molecular structure with the physical properties of the walls during growth. Plant J 3(1):1–30. https://doi.org/10.1046/j.1365-313X.1993.03010001.x

79. Putro IGNE, Wijaya CJ, Ismadji S, Gunawan S (2021) A review of lignocellulosic-derived nanoparticles for drug delivery applications. Molecules 26(3):676. https://doi.org/10.3390/molecules26030676
80. Zhang X, Zhang X, Yang Y, Ma Y, Wu M (2016) Chitosan/dsRNA nanoparticles loaded on leaf surfaces to silence genes in the agricultural pest Spodoptera litura. Pestic Biochem Physiol 134:30–36. https://doi.org/10.1016/j.pestbp.2016.04.007
81. Ghosh D, Roy P, Dey D, Das R, Ghosh S (2021) Hydrogel-based nanocarriers for controlled delivery of RNAi in plants: an emerging biotechnological platform. Biomacromolecules 22(5):2207–2217. https://doi.org/10.1021/acs.biomac.1c00122
82. Jahl A, Kath M, Stracke R, Müller M, Kaul H-P (2021) Lignin-based nanocarriers for UV-protective agrochemical delivery: preparation, characterization and performance. J Agric Food Chem 69(6):1802–1811. https://doi.org/10.1021/acs.jafc.0c07234
83. Ma X, He Y, Wang M, Liu X (2022) Rainfastness enhancement of nanopesticide formulations: a review of water-resistant coating strategies. Pest Manag Sci 78(9):3857–3868. https://doi.org/10.1002/ps.6893
84. Chen A, Halilovic L, Shay JH, Koch A, Mitter N, Jin H (2023) Improving RNA-based crop protection through nanotechnology and insights from cross-kingdom RNA trafficking. Curr Opin Plant Biol 76:102441. https://doi.org/10.1016/j.pbi.2023.102441
85. da Rosa J, Viana AJC, Ferreira FRA, Koltun A, Mertz-Henning LM, Marin SRR et al (2024) Optimizing dsRNA engineering strategies and production in E. coli HT115 (DE3). J Ind Microbiol Biotechnol 51:kuae028
86. Aalto AP, Sarin LP, van Dijk AA, Saarma M, Poranen MM, Arumäe U, Bamford DH (2007) Large-scale production of dsRNA and siRNA pools for RNA interference utilizing bacteriophage phi6 RNA-dependent RNA polymerase. RNA (New York, N.Y.) 13(3):422–429. https://doi.org/10.1261/rna.348307
87. Miller M, Alvizo O, Baskerville S, Chintala A, Chng C, Dassie J et al (2024) An engineered T7 RNA polymerase for efficient co-transcriptional capping with reduced dsRNA byproducts in mRNA synthesis. Faraday Discuss 252:431–449
88. Ziegenhals T, Frieling R, Wolf P, Göbel K, Koch S, Lohmann M, Baiersdörfer M, Fesser S, Sahin U, Kuhn AN (2023) Formation of dsRNA by-products during in vitro transcription can be reduced by using low steady-state levels of UTP. Front Mol Biosci 10:1291045. https://doi.org/10.3389/fmolb.2023.1291045
89. Shen Y, Gwak H, Han B (2024) Advanced manufacturing of nanoparticle formulations of drugs and biologics using microfluidics. Analyst 149(3):614–637
90. Shepherd SJ, Han X, Mukalel AJ, El-Mayta R, Thatte AS, Wu J et al (2023) Throughput-scalable manufacturing of SARS-CoV-2 mRNA lipid nanoparticle vaccines. Proc Natl Acad Sci U S A 120(33):e2303567120
91. Shepherd SJ, Issadore D, Mitchell MJ (2021) Microfluidic formulation of nanoparticles for biomedical applications. Biomaterials 274:120826. https://doi.org/10.1016/j.biomaterials.2021.120826
92. Shepherd SJ, Warzecha CC, Yadavali S, El-Mayta R, Alameh MG, Wang L, Weissman D, Wilson JM, Issadore D, Mitchell MJ (2021) Scalable mRNA and siRNA lipid nanoparticle production using a parallelized microfluidic device. Nano Lett 21(13):5671–5680. https://doi.org/10.1021/acs.nanolett.1c01353
93. Operti MC, Bernhardt A, Sincari V, Jager E, Grimm S, Engel A, Hruby M, Figdor CG, Tagit O (2022) Industrial scale manufacturing and downstream processing of PLGA-based nanomedicines suitable for fully continuous operation. Pharmaceutics 14(2):276. https://doi.org/10.3390/pharmaceutics14020276
94. Panigrahi D, Sahu PK, Swain S, Verma RK (2021) Quality by design prospects of pharmaceuticals application of double emulsion method for PLGA loaded nanoparticles. SN Appl Sci 3:1–21
95. Troiano G, Nolan J, Parsons D, Van Geen Hoven C, Zale S (2016) A quality by design approach to developing and manufacturing polymeric nanoparticle drug products. AAPS J 18(6):1354–1365

96. Gupta RK, Guha P, Srivastav PP (2024) Investigating the toxicological effects of nanomaterials in food packaging associated with human health and the environment. J Hazard Mater Lett 5:100125
97. Mrowiec B (2025) Nanowaste in the aquatic environment—threats and risk countermeasures. Desalin Water Treat 322:101112
98. Nthunya LN, Mosai AK, López-Maldonado EA, Bopape M, Dhibar S, Nuapia Y et al (2025) Unseen threats in aquatic and terrestrial ecosystems: nanoparticle persistence, transport and toxicity in natural environments. Chemosphere 382:144470
99. Yu Z, Li Q, Wang J, Yu Y, Wang Y, Zhou Q, Li P (2020) Reactive oxygen species-related nanoparticle toxicity in the biomedical field. Nanoscale Res Lett 15(1):115
100. Rahimi S, Chen Y, Zareian M, Pandit S, Mijakovic I (2022) Cellular and subcellular interactions of graphene-based materials with cancerous and non-cancerous cells. Adv Drug Deliv Rev 189:114467
101. Ponte S, Moore EA, Border CT, Babbitt CW, Tyler AC (2019) Fullerene toxicity in the benthos with implications for freshwater ecosystem services. Sci Total Environ 687:451–459. https://doi.org/10.1016/j.scitotenv.2019.05.362
102. Moore EA, Babbitt CW, Connelly SJ, Tyler AC, Rogalskyj G (2019) Cascading ecological impacts of fullerenes in freshwater ecosystems. Environ Toxicol Chem 38(8):1714–1723
103. Kobayashi N, Izumi H, Morimoto Y (2017) Review of toxicity studies of carbon nanotubes. J Occup Health 59(5):394–407. https://doi.org/10.1539/joh.17-0089-RA
104. Li M, Zhang Y, Feng S, Zhang X, Xi Y, Xiang X (2022) Bioaccumulation and biomagnification effects of nano-TiO2 in the aquatic food chain. Ecotoxicology (London) 31(6):1023–1034. https://doi.org/10.1007/s10646-022-02572-0
105. Fujihara J, Nishimoto N (2024) Review of zinc oxide nanoparticles: toxicokinetics, tissue distribution for various exposure routes, toxicological effects, toxicity mechanism in mammals, and an approach for toxicity reduction. Biol Trace Elem Res 202(1):9–23. https://doi.org/10.1007/s12011-023-03644-w
106. Amar-Lewis E, Buaron N, Chintakunta R, Benafsha C, Goldbart R, Traitel T et al (2021) Quaternized starch-based composite nanoparticles for siRNA delivery to tumors. ACS Appl Nano Mater 4(2):2218–2229
107. Amar-Lewis E, Cohen L, Chintakunta R, Benafsha C, Lavi Y, Goldbart R et al (2024) Elucidating siRNA cellular delivery mechanism mediated by quaternized starch nanoparticles. Small 20(51):2405524
108. Adeli F, Pourmadadi M, Abdouss M, Rahdar A, Fathi-karkan S, Samandari SS, Díez-Pascual AM (2024) Controlled curcumin delivery via carboxymethyl starch-modified gamma alumina nanoparticles in a polyethylene glycol-based hydrogel. J Appl Polym Sci 141(37):e55946
109. Wu Z, Tang Y, Liu Y, Chen Z, Feng Y, Hu H et al (2025) Co-delivery of fucoxanthin and twist siRNA using hydroxyethyl starch-cholesterol self-assembled polymer nanoparticles for triple-negative breast cancer synergistic therapy. J Adv Res 70:463–479. https://doi.org/10.1016/j.jare.2024.04.017
110. Sun H, Kalluri A, Tang D, Ding J, Zhai L, Gu X et al (2024) Engineered dsRNA–protein nanoparticles for effective systemic gene silencing in plants. Hortic Res 11(4):uhae045
111. Zhang W, Han B, Lai X, Xiao C, Xu S, Meng X et al (2020) Stiffness of cationized gelatin nanoparticles is a key factor determining RNAi efficiency in myeloid leukemia cells. Chem Commun 56(8):1255–1258
112. Chen Y, Jiang S, Wang Y, Zhang F, Wang L, Zhao L et al (2024) Small-sized starch nanoparticles for efficient penetration of plant cells. Chem Commun 60(95):14113–14116
113. Khan F, Shariq M, Asif M, Siddiqui MA, Malan P, Ahmad F (2022) Green nanotechnology: plant-mediated nanoparticle synthesis and application. Nanomaterials (Basel) 12(4):673. https://doi.org/10.3390/nano12040673
114. Lallana E, Rios de la Rosa JM, Tirella A, Pelliccia M, Gennari A, Stratford IJ et al (2017) Chitosan/hyaluronic acid nanoparticles: rational design revisited for RNA delivery. Mol Pharm 14(7):2422–2436

115. Liew WJM, Alkaff SA, Leong SY, Yee MZL, Hou HW, Czarny B (2024) Cell membrane- and extracellular vesicle-coated chitosan methacrylate-tripolyphosphate nanoparticles for RNA delivery. Int J Mol Sci 25(24):13724. https://doi.org/10.3390/ijms252413724
116. Bhandari G, Dhasmana A, Chaudhary P, Gupta S, Gangola S, Gupta A, Rustagi S, Shende SS, Rajput VD, Minkina T, Malik S, Slama P (2023) A perspective review on green nanotechnology in agro-ecosystems: opportunities for sustainable agricultural practices & environmental remediation. Agriculture (Basel) 13(3):668. https://doi.org/10.3390/agriculture13030668
117. Pushko P, Pumpens P, Grens E (2013) Development of virus-like particle technology from small highly symmetric to large complex virus-like particle structures. Intervirology 56(3):141–165
118. Ramos-Carreño S, Giffard-Mena I, Zamudio-Ocadiz JN, Nuñez-Rivera A, Valencia-Yañez R, Ruiz-Garcia J et al (2021) Antiviral therapy in shrimp through plant virus VLP containing VP28 dsRNA against WSSV. Beilstein J Org Chem 17(1):1360–1373
119. Eber FJ, Eiben S, Jeske H, Wege C (2015) RNA-controlled assembly of tobacco mosaic virus-derived complex structures: from nanoboomerangs to tetrapods. Nanoscale 7(1):344–355
120. Heinemann JA, Agapito-Tenfen SZ, Carman JA (2013) A comparative evaluation of the regulation of GM crops or products containing dsRNA and suggested improvements to risk assessments. Environ Int 55:43–55. https://doi.org/10.1016/j.envint.2013.02.010
121. OECD (2021) Test guidelines for the testing of chemicals: guidelines for nanomaterials. OECD Publishing, Paris. https://www.oecd.org/chemicalsafety/nanosafety/
122. Roberts AF, Devos Y, Lemgo GNY, Zhou X (2015) Biosafety research for non-target organism risk assessment of RNAi-based GE plants. Front Plant Sci 6:958. https://doi.org/10.3389/fpls.2015.00958
123. Atha DH, Wang H, Petersen EJ, Cleveland D, Holbrook RD, Jaruga P et al (2012) Copper oxide nanoparticle mediated DNA damage in terrestrial plant models. Environ Sci Technol 46(3):1819–1827. https://doi.org/10.1021/es202660k
124. Rico CM, Majumdar S, Duarte-Gardea M, Peralta-Videa JR, Gardea-Torresdey JL (2011) Interaction of nanoparticles with edible plants and their possible implications in the food chain. J Agric Food Chem 59(8):3485–3498. https://doi.org/10.1021/jf104517j
125. Servin A, Elmer W, Mukherjee A, de la Torre-Roche R, Hamdi H, White JC, Bindraban P (2015) A review of the use of engineered nanomaterials to suppress plant disease and enhance crop yield. J Nanopart Res 17(2):92. https://doi.org/10.1007/s11051-015-2881-3
126. Keller AA, McFerran S, Lazareva A, Suh S (2013) Global life cycle releases of engineered nanomaterials. J Nanopart Res 15:1692. https://doi.org/10.1007/s11051-013-1692-4
127. Nel A, Xia T, Mädler L, Li N (2006) Toxic potential of materials at the nanolevel. Science 311(5761):622–627. https://doi.org/10.1126/science.1114397
128. Wang Q, Zhang S, Wang X et al (2020) Tracking carbon-dot-labeled RNAi molecules in plant cells. Nano Res 13(10):2772–2780. https://doi.org/10.1007/s12274-020-2887-2
129. Frangville C, Rutkevičius M, Richter AP, Velev OD, Stoyanov SD, Paunov VN (2012) Fabrication of environmentally biodegradable lignin nanocarriers for the delivery of active compounds. ACS Sustain Chem Eng 1(3):295–302. https://doi.org/10.1021/sc300140h
130. Li J, Zeng Z, Zhang J, Zhao L, Cheng Y (2020) Biodegradable chitosan-based nanoparticles for sustainable agriculture and food packaging applications. J Clean Prod 276:123219. https://doi.org/10.1016/j.jclepro.2020.123219
131. Bawa AS, Anilakumar KR (2013) Genetically modified foods: safety, risks and public concerns–a review. J Food Sci Technol 50(6):1035–1046. https://doi.org/10.1007/s13197-012-0899-1

Nanoparticles for Organelle-Targeted Gene Delivery in Plants

A. K. M. Aminul Islam, Nor Anis Nadhirah Md. Nasir, Suresh Kadaru, Prakash I. Gangashetty, and Sadhana Singh

Abstract Plant genetic engineering is a potential tool for the improvement of plants for productivity, stress tolerance, and climate resilience. Conventional plant genetic engineering methods of gene transfer such as Agrobacterium-facilitated, gene gun, electroporation, viral transfection, and PEG-mediated transfection have several limitations. Organelle-targeted nanoparticle-based gene transfer in plants shows huge potential in plant transformation through genetic engineering for their sustainability and resilience. Nanoparticles may impact plants both positively and negatively, and promote plant growth and development through precise delivery, enhanced protection of genetic materials, uptake of nutrient elements, stable transfer of genes, phytoremediation, and stress tolerance. Nanoparticle-based plant genetic engineering enables plant trait improvement, enhanced stress tolerance, production of new biomolecules, environmental remediation, etc. This chapter discusses the application of nanoparticles in organelle organelle-targeted delivery of

A. K. M. Aminul Islam (✉)
Department of Genetics and Plant Breeding, Gazipur Agricultural University, Gazipur, Bangladesh
e-mail: aminulgpb@bsmrau.edu.bd

Nor Anis Nadhirah Md. Nasir
Faculty of Civil Engineering & Technology, Universiti Malaysia Perlis, Arau, Perlis, Malaysia

S. Kadaru
Medium Maturity Rice Breeding Program, International Rice Research Institute (IRRI), Los Baños, Laguna, Philippines
e-mail: s.kadaru@irri.org

P. I. Gangashetty
International Crops Research Institute for the Semi-Arid Tropics (ICRISAT), Hyderabad, India
e-mail: Prakash.Gangashetty@icrisat.org

S. Singh
Department of Biochemistry, School of Sciences, U. P. Rajarshi Tandon Open University, Prayagraj, India

J.-T. Chen (ed.), *Nano-Delivering for Plant Genetic Engineering*, Smart Nanomaterials Technology, https://doi.org/10.1007/978-981-95-4165-2_13

genetic materials into plants focusing on methods, advantages, applications, and challenges of this approach.

Keywords Crop improvement · Genetic engineering · Genetic transformation · Nanomaterials · Nanotechnology · Stress tolerance

1 Introduction

Organelle-targeted gene delivery in plants comprises transporting foreign DNA or RNA into the cell organelles of the plant cells. This targeted transportation of genetic materials into cell organelles allows manipulation of their genomes which assists improvement of photosynthetic activity, respiration, and production of biomolecules in plant systems [1].

1.1 Plant Cellular Organelles

Plant cell organelles are intracellular, membrane-bound structures embedded in the cytosol with specific functions. Every organelle has its function and plays a vital role in the development of plants (Table 1). Plant cell organelles include the nucleus, mitochondria, chloroplasts, Golgi body, ribosomes, endoplasmic reticulum, lysosomes, and vesicles (Fig. 1). Among the organelles nucleus, mitochondria, and chloroplast contain genetic materials (DNA). The genetic material in the cell organelles serves as an essential component for the improvement of plant traits [5, 6]. Therefore, the advancement of structural and functional investigation of cell organelles of the plant and their utilization come to the forefront of biologists, biotechnologists, geneticists, and plant breeders.

Table 1 Plant cell organelles carrying genetic materials and their specific function

Organelles	Functions	References
Nucleus	• Command center of the cell • A storehouse of genetic materials • Contains both DNA and RNA, DNA in the form of chromosomes	Cooper [2], Deslandes and Rivas [3]
Mitochondria	• Responsible for producing energy through cellular respiration • Keep the cell full of energy • Contain circular DNA	Millar [4], Wang et al. [5]
Chloroplast	• Work as a site of photosynthesis • Convert light energy into chemical energy • Contain DNA and ribosomes	Song et al. [6]

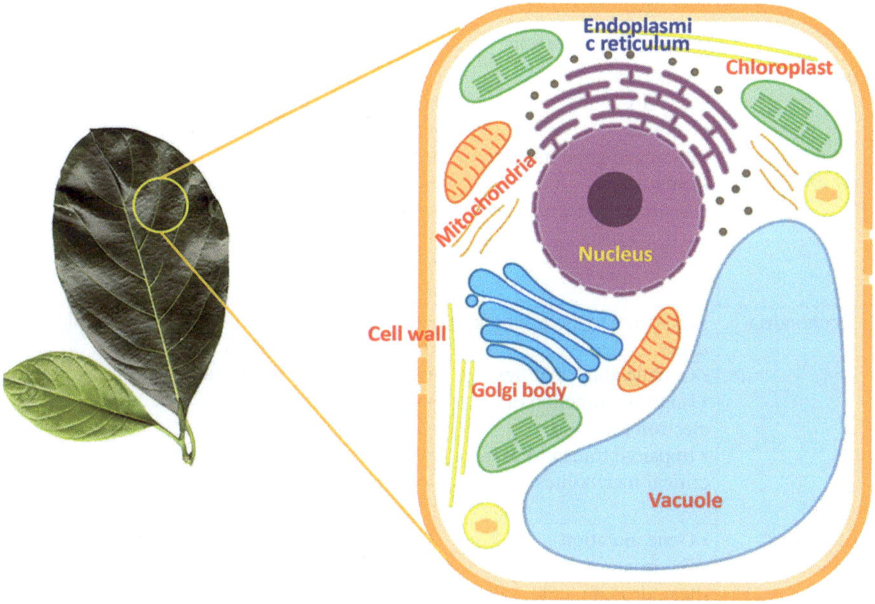

Fig. 1 Plant cell organelles (nucleus, chloroplasts, mitochondria, endoplasmic reticulum, Golgi body, ribosomes, vesicles, etc.), bound by cell wall

1.2 Organelle Dysfunctions

Organelle dysfunction in plants creates numerous problems by altering gene expression, impairing the growth-development process and stress responses [7]. Common organelle dysfunction in plants are mutations in mitochondria and plastids. Mitochondrial or plastid dysfunction affects the function of RecG protein, an enzyme with helicase, translocase, and ATPase activity [8] which plays a vital role in the replication, repair, and stability of genetic materials or DNA or organelle genomes. RecG plays an important role in DNA replication, repair, and maintain the stability of organelle genomes. The dysfunction of different plant cell organelles is presented in Table 2.

1.3 Nanoparticles

A nanoparticle is an ultrafine particle having a size that varies from 1 to 100 nm in diameter. Nanoparticles possess distinct physio-chemical properties (colloidal properties) compared to larger counterparts, micro or macroparticles [9, 10]. It can be solid, colloidal, or layered with a core, shell, or surface and invisible to the naked eye due to its nanostructure. Nanomaterials can be organic, inorganic, or

Table 2 Causes and consequences of plant cell organelle dysfunctions

Organelles	Causes of dysfunction	Consequence of dysfunctions	
Nucleus	• Mutation in genetic materials • Stresses • Pathogen-gene interaction, symbiosis • Nuclear squeezing	• Alter gene expression • Disrupt nuclear trafficking • Errors in transcription and translation • Affects protein transportation • Affects defense mechanism • Impaired plant development	Deslandes and Rivas [3]
Mitochondria	• Exposure to environmental stress • Genetic abnormalities • Improper function of electron transport • Impaired transport of critical metabolites	• Redox balance, overproduction of ROS • Affects energy production • Oxidative stress • mtDNA mutation • Male sterility, embryo lethality, chlorotic phenotypes	Busi et al. [11]
Chloroplast	• Gene mutation • Environmental stresses (heat, drought, cold, salt, heavy metals) • Pathogen infection • Mitochondrial dysfunction • Nutrient deficiencies	• Reduced photosynthesis, stunted growth • Reduced chlorophyll production, chlorosis • Chloroplast degradation • Cell cycle defect • Affects sugar metabolism, accumulation of ROS • Constitutive activation of stress-responsive genes	Hudik et al. [12], Mondal et al. [13]
Endoplasmic reticulum	• Protein misfolding • Oxidative stress • Calcium imbalance	• Disrupt protein synthesis, folding, and packaging	Manghwar and Li [14]
Vacuole	• Disruption of membrane integrity • Genetic mutation • Stress responses	• Affects nutrient storage, waste disposal, and cell growth • Programmed cell death • Affect plant survival	Jiang et al. [15], Acuna et al. [16]
Golgi body	• Genetic mutation • Cellular stress • Some diseases	• Impaired protein transport • Cell death • Neurodegenerative disease	Iacobescu et al. [17]

carbon-based based, have high surface area, which makes them sensitive and useful for various applications in fields such as agriculture, medicine, environment, and engineering [18]. Nanoparticles significantly impact plants both positively and negatively reliant on the category, concentration, and plant species. It can promote plant growth and development through enhanced uptake of nutrient elements, phytoremediation, and stress tolerance [19]. Nanoparticles also induce toxicity and interrupt plant development disrupting photosynthesis, cell integrity, altered gene expression, and DNA damage.

2 Targeting the Organelles in Plants

Targeting the cell organelles in plants is critical for any biological, genetic engineering, or biotechnological research in plants [20]. It comprises directing proteins or other biomolecules in the chloroplast or mitochondria for understanding and manipulating plant cellular functions. Targeting organelle involves the delivering of specific proteins or other biomolecules to plant organelles which is achieved through various mechanisms viz., targeting peptides, electrostatic interaction, and nanoparticles, facilitates precise and effective transport to the target organelle [21]. It is used for targeted gene expression, genome modification, and bioengineering in a specific organelle. Genes can be delivered to chloroplast to improve photosynthetic activity, to mitochondria to improve respiration, and to the nucleus for various plant genetic engineering. Targeting organelles can help in manipulating metabolic pathways, improve crop yield, stress tolerance, efficiency of molecule delivery, understanding of cell functions, organelle interaction, and transport of specific proteins into multiple organelles. Organelle targeting faces challenges like selectivity in biomolecule delivery to specific organelle, lack of reliable delivery system, and knowledge of organelle dynamics. Therefore, the implication of organelle targeting is a vital process for plant metabolism, bioengineering, and defense mechanisms. Development of new delivery methods for organelle targeting enfolds potential application in plant science.

3 Organelle Targeted Gene Delivery in Plants

Organelle-targeted delivery of genetic materials in plants denotes accurate delivery of DNA or RNA into specific plant organelles for the purpose of modification of genetic materials. Organelle-targeted gene delivery into the nucleus, chloroplast, and mitochondria provides several benefits such as modification of cellular metabolism, development of new biomolecules, enhance stress tolerance, improvement of plant traits, enhance plant regeneration [22]. Organelle-targeted delivery of genes in plants is a propitious field with enormous potential to transform agriculture and plant genetic engineering for sustainable and resilient food systems.

3.1 Nucleus-Targeted Gene Delivery

The nucleus is the key target of any alteration of genetic materials in the plant as the majority of genetic materials are contained by the nucleus in the form of DNA in chromosomes [20]. Nucleus-targeted gene delivery method aims to introduce genetic materials directly to the cell nucleus bypassing the cytoplasm.

3.2 Mitochondria-Targeted Gene Delivery

Mitochondria-mediated delivery of genes refers to the process of delivering genetic materials to mitochondria of the plant cells [23]. The mitochondrial gene delivery technique is used to improve traits like stress resistance and alter metabolic processes by targeting the mitochondria of the plant cell. Nanoparticle-based mitochondrial gene delivery system involves overcoming the challenges of crossing membrane barriers to deliver the genes into the mitochondria which opens a new era of plant genetic engineering.

3.3 Chloroplast Targeted Gene Delivery

Chloroplast-mediated gene delivery in plants denotes the method of transferring genetic materials into the chloroplast of the plant cells. This gene-delivery technique is used to improve organelle-responsible photosynthesis [24]. Nanoparticle-based chloroplast gene delivery system differs from the traditional system by offering advantages like high-level transgene expression and potential for transgene containment [25]. This technique overcame the challenges of crossing membrane barriers to deliver the genes into the chloroplast and open a new era of plant genetic engineering.

3.4 Golgi Apparatus Targeted Gene Delivery

It is reported that the Golgi apparatus is involved in gene delivery in plants. This involvement is often explored in plant genetic engineering and the development of strategies for introducing foreign genetic materials into the plant cell organelles [26] though the primary function of the Golgi apparatus in the cell is to process and package proteins. Some plant viruses exploit the Golgi body for the replication and spread of their own genetic materials. Researchers are investigating whether and how the Golgi body could be utilized in the targeted gene delivery in plants. Golgi's role in generating and trafficking vesicles could potentially be utilized to deliver gene-containing cargo directly into the plant cell.

4 Nanoparticles Used for Organelle-Targeting

Nanoparticles (NPs) are emerging as a multipurpose and useful tool for targeting plant cell organelles because of their tiny size, high surface-to-volume ratio, and transportation ability of various biomolecules and genetic materials into the organelles [21]. Nanoparticles are designed with specific characteristics like size, shape, surface modification, and composition for facilitating effective transportation and NP-targeted organelle interaction. Several types of nanomaterials are explored for organelle targeting, for example, liposomes, mesoporous silica nanoparticles (MSNs), carbon nanotubes (CNTs), polymeric nanoparticles, metals and magnetic nanoparticles, dendrimers, etc.

4.1 Liposomes

Liposome nanoparticle is a tiny and sphere-shaped vesicle, made of lipid bilayers with polar and non-polar head and tail groups, respectively. These liposomes are capable of encapsulating the genetic materials, and other biomolecules, and deliver effectively to the targeted specific organelles [27]. This nanoparticle is also used to deliver nutrients into the targeted organelles and can be designed to target delivery to the mitochondria and the chloroplast. Research is also ongoing with liposomes nanoparticles to target other organelles like endoplasmic reticulum, Golgi apparatus, etc.

4.2 Polymeric Nanoparticles

This type of nanoparticle is made from synthetic polymers (polyethylene glycol) or natural polymers (cellulose or proteins) having flexibility in size (1–1000 nm), shape, and modified surface for organelle targeting [28]. Polymeric nanoparticles are grouped into two-groups, (i) nanosphere—particle solid core, and (ii) nanocapsules—particle with liquid core surrounded by polymer membrane. Polymeric nanoparticles are designed to deliver nucleic acids or small molecules to targeted organelles for various purposes in plants. The efficiency of the delivery depends on the size, shape, charge, and functionality of the nanoparticles [22].

4.3 Mesoporous Silica Nanoparticles (MSNs)

It is a versatile nanomaterial with a porous structure with particle sizes ranging from 30 to 300 nm. MSNs are known for their tunable size, surface properties, and porous structure [29]. Pore dimensions of MSNs vary from 2 to 50 nm, surface of the particle easily be modified with numerous functional groups [30]. MSNs are used in organelle-targeted gene delivery in plants for their exceptional properties for example big surface area, small pore size, and biocompatibility. In plants, MSNs are used to target specific organelles particularly mitochondria, nucleus, and cytosol for material delivery and studying organelle functions. MSNs are used for delivering protein molecules to targeted cellular compartments.

4.4 Carbon Nanotubes (CNTs)

Carbon nanotubes are made from carbon with nano diameter and cylindrical structure, rolled up as graphene sheets exhibit extraordinary mechanical (strength, elasticity, density), electrical, and thermal properties [31]. Carbon nanotubes are grouped into single-walled (SWCNTs: 0.2–2.0 nm) or multi-walled (MWCNTs) carbon nanotubes based on their graphene layers. CTNs are discovered as a powerful tool for targeting cell organelles in plants particularly mitochondria for enhanced growth and metabolic engineering [20]. It allows more precise delivery of genes and other biomolecules to targeted organelles.

4.5 Metals and Magnetic Nanoparticles

Metal nanoparticles are made from gold, silver, zinc, titanium, platinum, iron, etc., or oxide, hydroxide, sulfides, phosphates, fluorides, and chlorides of these metals [20]. Metallic nanoparticles are used to deliver gene cargoes in plants with great success. Metallic nanoparticles are of two types based on their composition: metal oxide nanoparticles and gold nanoparticles [32]. Metal nanoparticles are very enchanting materials used for various purposes including the delivery of biomolecules, environmental remediation, and catalysts in various reactions. Gold (Au), silver (Ag), and platinum (Pt) nanoparticles (NPs) are widely used for organelle targeting [21]. The magnetic nanoparticles are composed of magnetic metals such as iron, nickel, and cobalt and are manipulated by magnetic fields. Commonly used magnetic nanoparticles are iron oxide nanoparticles such as magnetite (Fe_3O_4) and maghemite (γ-Fe_2O_3) due to their stability, biocompatibility, and magnetic field sensitivity [33].

4.6 Lipid-Based Nanoparticles

Lipid-based nanoparticles (LNPs) are increasingly used for organelle-targeted materials delivery in plants due to their versatility, biocompatibility, controlled release system, and high delivery efficiency [21]. LNPs efficiently deliver various molecules like genes, proteins, and nutrients into specific cellular organelles which leads to new applications in plant biotechnology and plant health improvement [20].

Research on LNPs continues to address the challenges like improvement of targeting efficiency, reduce environmental impact, and make the methods cost-effective to expand the applications of this technology in plant science and agriculture.

4.7 *Dendrimers*

Dendrimers are hyperbranched tree-like synthetic molecules having three-dimensional polymer structures with very low polydispersity, high compatibility, predictable biodistribution, and features of cell-membrane interaction [34]. Dendrimer nanoparticles can carry multiple targeting ligands on the surface and be used to deliver genetic materials to targeted organelles. Due to high versatility dendrimers have been tiling way to the front in gene delivery to targeted organelles and monitoring diseases [35].

5 Methods for Organelle-Targeted Gene Delivery

Several methods are used in organelle-targeted gene delivery in plants such as peptide-based targeting sequences, nanoparticles, viral vectors, protein-based, and nucleic acid-based targeting systems [36]. Organelle-targeted gene delivery methods involve delivering genetic materials into specifically targeted organelles in the plant cells to enhance gene expression and biomolecule transfer [20]. The organelles are the nucleus, mitochondria, chloroplast, etc.

There are several methods of gene delivery into cell organelles (particularly the nucleus) which are categorized as physical through particle bombardment and electroporation, biological through *Agrobacterium*-mediated gene transfer and viral transfection, and chemical through PEG-mediated transfection. Recently nanoparticle-mediated gene delivery is getting more attention due to its precise and enhanced entry, and stable delivery of genes into targeted organelles. The nanoparticle-based method delivers genetic materials into cell organelles using nanomaterials as carriers [32]. This method is an attractive alternative to ATM because of its diffusive and non-invasive nature and not only delivers genes in the nucleus but is also successfully used in other targeted cell organelles like mitochondria and plastid [37]. Nanoparticles are designed to encapsulate genetic materials (DNA and RNA or proteins) and deliver to specific targeted organelles with improved efficiency and safety. Various methods are used for organelle-targeted gene delivery as discussed in the following.

5.1 Peptide-Based Methods

Peptide, a short sequence of amino acids, could be attached to a specific protein to deliver it to specific organelles. Peptide-based systems provide promising pathways for targeted gene delivery to specific organelles within the cell. The unique properties of peptides enhance targeting, protect DNA, and facilitate transmembrane transport of the molecules. To enhance targeting. N-terminal and C-terminal extensions of peptides are used to target proteins to the mitochondria, plastid, and endoplasmic reticulum [38]. The N-terminal extension is added to the beginning of the protein and the C-terminal extension is added to the end of the protein, the earlier one is frequently used in targeting organelles while the latter one is less frequently used. N-terminal extension influences protein folding and packaging while C-terminal extension influences protein stability. These two extensions are used as valuable tools to modify protein behavior and target them to specific organelle locations. Peptides can be engineered to attach to specific organelle surfaces to facilitate their entry and localization. For example, mitochondria-targeting peptides deliver genes into mitochondria, and chloroplast-targeting peptides deliver genes into chloroplasts. Cytcox/KaibA-Mic [23], Peptide/pDNA Complexes [20], and Fusion Peptides [39] are unique examples of peptide-based systems for organelle-targeted gene delivery.

5.2 Protein-Based Method

Protein-based methods for organelle targeting gene delivery involve protein engineering to carry and deliver genetic materials to specific organelles. In this method, a protein is used to carry directly the genetic materials or used as a carrier for nanoparticles, or through cell-penetrating peptides or vesicular transport [38]. For direct protein targeting, proteins are designed to attach specific amino acid sequences, signal peptides, that carry genetic materials directly to the specific organelle [40]. In the case of protein-based nanoparticles, a protein used as a carrier of nanoparticles encapsulated the genes to deliver targeted organelle. Cell-penetrating peptides (CPPs) are short peptides, that contain many basic and hydrophobic amino acids that can improve cellular uptakes. The side chain of basic amino acid of CPPs interacts with the phosphate group of the cell membrane and disrupts cell membranes due to its hydrophobic properties and enables intracellular transportation of genetic materials. For vesicular transport, proteins are delivered to endosomes and then deliver the gene to the targeted organelles. Protein-based gene delivery is a powerful tool and results in the precise delivery of genetic materials to the targeted organelles [36].

5.3 Nanoparticles

Nanoparticle-based gene delivery is an efficient and safe method for introducing genetic materials to cells. Various nanoparticle-based methods or platforms are explored for organelle-targeted gene delivery. The platform includes polymeric nanoparticles, metal nanoparticles, magnetic nanoparticles, and lipid nanoparticles, each platform with its advantages. These platforms are designed with specific sizes, shapes, and surface areas to carry and protect genetic materials (DNA or RNA) in the process of targeted delivery to the organelle [32]. Nanoparticles exhibit a large surface area that allows nanoparticles as cargo of genetic materials in organelle-targeted delivery. Nanoparticle cargo can be introduced into plant cells through the methods listed in Table 3.

5.4 Viral Vectors

Due to the natural ability of viruses to enter the cell, viral vectors are the most potential delivery vehicles and are commonly used to carry genetic materials into targeted organelles. Viral vectors are modified to target specific organelles to express proteins that bind to the surface of the organelle. For example, adeno-associated viruses (AAVs) have been successfully used to deliver genes into mitochondria [52]. Targeted molecules are incorporated into the viral cascade and direct the virus to the specific organelles viz., mitochondria, chloroplasts, etc. It also allows CRISPR/Cas system delivery into cell organelles [53].

6 Advantages of Nanoparticles for Organelle-Targeted Gene Delivery

Nanoparticles offer various advantages for organelle-targeted gene delivery in plants including precise delivery, specificity, efficient delivery, cargo protection, protection of genetic materials, enhanced plant regeneration, reduction of off-target delivery, etc. The nano size, large surface area, and biocompatibility make nanoparticles a versatile delivery transport of selected materials to targeted organelles. Nanoparticles can overcome cell wall barriers for their tunable properties. It has good safety profiles and makes them suitable for safe applications.

Table 3 Lists of organelle targeted cargo delivery into plant cell with process of delivery, targeted tissues, positive and negative aspects associated with each delivery method

Methods	Goods needed	Process insertion	Targeted tissue	Advantages	Disadvantages	References
Syringe infiltration	Needleless syringe (size: 1 mL)	A needleless syringe is used to incorporate nanoparticle cargo encapsulated with genetic materials into mature intact leaves	Leaves	Ability to deliver a single gene across an entire leaf or multiple genes into different regions of a single leaf. It has broad applications in plant biotechnology	Limited to a narrow range of plant species and appropriate entirely to leaf tissues	Demirer et al. [41], Kaur et al. [42]
Vacuum infiltration	Vacuum desiccator	Infiltration-mediated permeation pathway is used to deliver nanocarrier complex. Placing tissue in the infiltration solution and mixing with gene-containing cargo complex, incubating a specific period at the desired concentration in the large syringe	Seedlings, seeds, embryos, leaves	Gene can deliver a wide range of plant cells, no special equipment is needed	Require optimization of infiltration time, and chance of tissue damage due to vacuum pressure	Wang et al. [43], Bian et al. [44]
Biolistic delivery	Gene gun	Specific mechanical pressure is used to deliver nanoparticles carrying genetic materials to the organelles	Seeds, cotyledon, tissues, callus	Biolistic gene guns don't have species constraints, applicable to a variety of plants	Gene gun instruments and spermidine are costly, risk of tissue damage, and a chance of undesired integration of genetic materials	Liu et al. [45], Cunningham et al. [46]

Magnetofection	Magnetic stirrer	Exogenous genetic materials are incorporated into plant cell organelle under a magnetic field, and nanoparticles coated with genetic materials incubated with plant cells, apply a magnetic field for nucleic acid uptake under a magnetic field	Pollen grain, protoplast	A highly efficient method, to generate more genetically modified stable lines	Tissue culture skill requires poor regeneration risk, optimization of the duration of the magnetic field is difficult, and limited tissue range	Wang et al. [43], Levengood et al. [47]
Ultrasound based delivery	Sonicator bath	Plant parts are mixed with nanoparticles carrying genetic materials in a sonicator bath containing buffer solution, and exposed to ultrasound waves in different frequencies for a definite period	Protoplast, intact plant cells	Straight forward, cost-effective, enhanced efficiency of cargo delivery, possible to apply in a wide range of tissues	Require optimization of the ultrasound wave, frequencies, and intensity, the chance of tissue damage, and risk of cell disruption due to lengthy exposure duration	Anami et al. [48], Sanzari et al. [49], Levengood et al. [47]
Spray method	Spray bottle or atomizer	The innovative approach, integrates genetic materials through the application of spray with transformation solution, prepared through blending genetic materials with carrier solution	Plant surface	Simple, cost-effective, efficient method for large-scale application, non-invasive, applicable for a wide range of plant species	Nanoparticle accumulation in plant cells and soils during large-scale application, negative impact on beneficial bacteria residing in the soil, tissue toxicity	Cunningham et al. [46], Rank and Koch [50]
Passive diffusion/ cellular uptake	Specific instruments not required	Nanoparticles compatible with the cell or tissue of the plant designed with desired genetic materials, incubate with protoplast or root or leaves, allowing them for passive diffusion to deliver nanoparticles encapsulated genetic materials into the organelles	Protoplast, seedlings, whole plant	Simple, no instruments or external force is required, cost-effective, no risk of tissue damage as no external force applies	Optimization of incubation duration requires optimization of the incubation period, applicable exclusively to nanoparticles with dimensions <30 nm	Wang et al. [43], Sembada and Lenggoro [51]

6.1 Precise or Targeted Delivery

Precise gene delivery using nanoparticles is a promising approach that enables the efficient delivery of genetic materials to the targeted organelles. Nanoparticles attach genetic materials or genes on the surface and carry them to targeted organelles viz., nucleus, mitochondria, and plastids precisely. Ligand-mediated targeting of nanoparticles enhances the internalization of delivery systems through interaction between cell surface receptors and the nanoparticles.

6.2 Enhanced Delivery Efficiency

Nanoparticle offers several advantageous properties including a high surface area to volume ratio for efficient loading of genetic materials and a high ability to attach targeting elements that enhance delivery. Nanoparticles enhance gene delivery to targeted organelle by protecting genetic materials and facilitating its entry as a carriers into the specific organelles [54]. The uptake of nanoparticles by cells can be improved by using sonication and other methods.

6.3 Protection of Genetic Materials and Cargo

Nanoparticles are commonly used to protect genetic materials and cargo during targeted delivery to the cell organelles. It can protect genetic materials or the cargo from degradation or denaturation and ensure effective delivery. This is because nanoparticles form a shield by encapsulating genetic materials and protecting them from enzymatic degradation or other environmental factors.

6.4 Potential for Multi-gene Delivery

Nanoparticle is a promising carrier of multi-gene delivery to the targeted organelles within plant cells. It can be engineered to accommodate multiple genes equipped with targeting ligands for efficient delivery of the genes to specific cellular compartments. Some nanoparticles can accommodate multiple genes in the cargos enabling delivery of multiple genes simultaneously. Nanoparticle-based gene delivery to targeted organelles paves the way for novel gene transport to plants for its improvements and disease tolerance.

6.5 Enhanced Plant Regeneration

Nanoparticle-mediated gene delivery methods enhanced the regeneration of transformed plants which is a crucial stage in plant genetic engineering. Nanoparticles can enhance plant regeneration by efficient and targeted gene delivery to subcellular organelles through overcoming limitations (cell wall barriers) of traditional methods, improving transformation efficiency (high rate of transformation, compatibility with biological systems), and facilitating the development of improved plant varieties.

6.6 Effective Transportation of CRISPR/Cas9 System

Nanoparticles are highly efficient for transporting CRISPR/Cas9 system in organelle-targeted gene delivery in plants due to their ability to encapsulate and protect the CRISPR/Cas9 complexes [55]. Transportation is also enhanced by its cellular uptake, ability to protect CRISPR/Cas9 cargo from degradation, and targeting ability of specific cell organelles. Nanoparticles can easily deliver large-size cargos like plasmids and proteins. Nanoparticles are versatile in their preparation and composition which allows customization of their size, shape, surface chemistry, and functionality to optimize delivery and target specificity.

6.7 Reduced Off-Target Effects

Nanoparticle-based gene delivery methods target specific organelles in the cell and reduce the unintended consequences in other organelles within the cell. To reduce off-target effects, optimized nanoparticle design (size, composition, and surface properties) and control the release mechanism in response to specific stimuli (light, pH, enzymatic activity) for minimizing non-specific interactions and ensuring the controlled release of the gene cargo to the desired location [20]. Optimizing nanoparticle design, highly specific and efficient organelle-targeted gene delivery is possible by minimizing off-target effects and maximizing the delivery efficiency.

7 Applications Organelle-Targeted Gene Delivery in Plants

Organelle-targeted gene delivery in plants is a promising field with enormous potential to modify specific cellular compartments for crop trait improvement, enhanced yield, increased disease resistance, and longer shelf lives [20]. Organelle-targeted gene delivery in plants offers a way to produce biopharmaceuticals, develop novel crop varieties, and enhance plant regeneration through bioengineering and molecular farming, contributing to more sustainable and resilient agriculture [56]. In molecular farming, plants can be engineered for the production of valuable biomolecules (antibodies, vaccines) through nanoparticle-based organelle-targeted gene delivery. Organelle-targeted gene delivery also helps to understand the process and functions of the cellular organelles inside the plant cell.

7.1 Plant Genetic Engineering

Nanoparticle-mediated gene delivery methods are widely used to deliver genes to targeted organelles in plants for the genetic transformation of different plant species [57]. Due to its reduced toxicity in plant systems, nanoparticles create potential prospects in the engineering of plant cells through organelle-targeted gene delivery. Plant genetic transformation based on nanoparticle-mediated gene delivery is an important tool for manipulating regulatory genes in biotechnology-based crop breeding and promoting genotype-independent plant transformation. Nanoparticles can penetrate easily into the plant cell wall without any external forces, paving the way for efficient plant transformation through plant genetic engineering [46].

7.2 Mitochondrial Disorders

Nanoparticle-based organelle targeted gene delivery can be a promising tool to control mitochondrial disorders like male sterility, reduced ATP production, and oxidative stress in plants by introducing functional components directly to the mitochondria. Mitochondrial disorders in plants also impact chloroplast, disrupt plant metabolism, affect photosynthesis, and even trigger cell death pathways [58]. Effective delivery of the targeted gene into the mitochondria can reverse the disorder by enabling cells to produce essential components of the mitochondria. This approach also helps to correct genetic defects, improve energy production, and enhance plant resilience to various stresses.

7.3 Plant Traits Improvements

New traits can be introduced in plants through the application of organelle-targeted gene delivery, influencing the plant phenotypes. The application of organelle-targeted gene delivery can modify plant traits such as photosynthetic efficiency, nutrient uptake, stress tolerance, disease tolerance, and synthesis of new biomolecules leading to the improvement of yield and quality [20]. Organelle-targeted gene delivery facilitates faster and efficient crop breeding allowing rapid introduction of desirable traits into new varieties [59].

7.4 Develop Stress and Disease Resistance in Plants

The introduction of resistant genes into the organelles particularly mitochondria or chloroplast confers the resistance capacity of plants to stresses [56]. The plant stresses include drought, salinity, heat, cold, and specific pathogens or insects. Agricultural practices need to reduce chemical pesticide use to enhance crop productivity and sustainability [60]. Resistance to diseases or stress in plants can improve through plant genetic engineering by targeting specific cell organelles.

7.5 Develop New Biomolecules in Plants

Plant organelles, particularly chloroplast engineering through new gene introduction can produce pharmaceutically important biomolecules, biofuels, and other industrial chemicals. Nanoparticle-based gene delivery into the targeted organelle helps to optimize the metabolic pathway of plant cells and produce desired biomolecules with greater control [20].

7.6 Understanding Cellular Process

Organelle-targeted gene delivery in plants can provide insights into the role of organelles in different cellular processes such as metabolism, plant development process, and stress responses [61]. Nanoparticle-based gene delivery system by targeting organelles allows for the investigation of organelle-specific functions and interaction, and provides valuable insights into plant biology.

7.7 Enhance Transformation and Plant Regeneration

Nanoparticles and other carriers of extraneous molecules into plant cell organelles lead to improve the efficiency of plant regeneration in the process of transformation through plant genetic engineering. Organelle-targeted gene delivery in plants particularly chloroplast and mitochondria improve plant regeneration by enhancing transformation efficiency [32]. Nanoparticles-based organelle targeted gene delivery in plants allows manipulation of organelles' genomes for the development of improved functionalities and crop traits. It also overcomes limitations of traditional methods such as Agrobacterium-mediated genetic transformation in plants [20]. Nanoparticles-based organelle targeted gene delivery in plants expand the range of plant species and genotypes.

7.8 Precise Genetic Modification

Nanotechnology-based organelle targeted gene delivery is widely used for precise modification of genetic materials and successful transformation of mitochondrial and chloroplast genomes in higher plants [62]. Nanoparticles act as vectors to deliver genes to specific organelles with enhanced efficiency and target accuracy by overcoming plant cell wall barriers [63]. This approach allows the finetuning of plant metabolism and functions at the organelle level.

7.9 Effective Transportation of DNA, RNA and Biomolecules

Nanoparticles are effective carrier vehicles for DNA, RNA, and other biomolecules to targeted organelles for their size, surface-to-volume ratio, functional properties, target release etc. Application of nanotechnology-based organelle targeted gene delivery allows precise and efficient transport of genetic materials (DNA and RNA) or biomolecules to specific organelles, enabling targeted genetic modification for the improvement of plant traits and production of novel biomolecules [64].

8 Challenges of Nanoparticle Mediated Gene Delivery

8.1 Stable Genetic Transformation

The major challenge in nanoparticle-mediated gene delivery for stable genetic transformation is the persistent integration of delivered genes into the host organelles [61]. Numerous results showed nanoparticle-based transformations are transient, meaning that introduced genes are not incorporated in stable conditions and are finally missing or silenced [55]. Nanoparticle-based gene delivery faces challenges in achieving stable and long-lasting genetic transformation. The challenge embraces nanoparticle stability, cellular uptakes, and integration of the DNA or RNA organelle surface. These challenges limit the effectiveness of organelle-targeted gene delivery in plant genetic transformation.

8.2 Toxicity and Biodegradability

Nanoparticle-based organelle targeted gene delivery faces challenges related to toxicity and biodegradability [65]. Toxicity is induced due to long-term accumulation of nanoparticles in the cell organelles and low biodegradability. Nanoparticle triggers toxic effects in the cell organelles through cell death, ROS (reactive oxygen species) generation, and cellular process disruption. Designing nanoparticles with degradable properties at an appropriate rate is the crucial challenge of this system [55]. Therefore, the development of nanoparticle-based gene delivery systems requires nontoxic, biodegradable, biocompatible nanoparticles with enhanced delivery efficiency.

8.3 Optimizing Delivery and Regeneration

Optimizing nanoparticle-based gene delivery systems and the regeneration efficiency of genetically transformed cells are major challenges in organelle-targeted delivery of genetic materials. The challenges include maximizing nanoparticle uptake, minimizing cell damage, protocol optimization for plant regeneration after transformation for a wide range of plant species, ensuring stable and efficient gene expression, etc. [46]. Therefore, overcoming these challenges requires improved nanoparticle design, optimized delivery systems, and regeneration protocols.

8.4 Combination with Other Technologies

Combining nanoparticle-based gene delivery systems with other methods faces numerous challenges including cell wall barriers, stable gene expression, precise targeting of the organelles, risk of toxicity, off-target effects, and large-scale applications [66]. These challenges become prominent when nanoparticle-based gene delivery methods integrate with modern crop improvement technologies like speed breeding, CRISPR/Cas-based genome editing, RNA interference, etc. Additionally, there are challenges in delivering nanoparticle-carrying genetic materials to specific cell organelles and optimizing the interaction with plant defense mechanisms.

8.5 Regulatory Frameworks

Challenges related to safety, efficacy, and regulatory hurdles are also faced in nanoparticle-based gene delivery to targeted plant cell organelles [67]. Further challenges include safeguarding the intellectual property rights and public health issues of genetically transformed plants through nanoparticle-based gene delivery. Due to their smaller size, nanoparticles exhibit unique physiochemical properties that result in different biological interactions, causing critical regulatory risk [68]. Therefore, regulatory frameworks need to consider the potential risk from the long-term effects of nanoparticle accumulation in plant cell organelles.

9 Conclusion

Nanoparticles offer a promising method for organelle-targeted gene delivery to specific cell organelles of plants, overcoming the barriers of traditional methods. The tiny size, large surface area, biocompatibility, and ability to pass through cell walls make them ideal for delivering genetic materials to cell organelles such as the nucleus, mitochondria, and chloroplast. Advancement of nanotechnology paved the way for more precise and efficient gene delivery with enhanced protection of genetic materials to the targeted organelles, enabling plant genetic engineering for various applications line plant traits improvement, enhanced stress tolerance, production of new biomolecules, environmental remediation, etc. In-depth safety measures related to the impact of nanoparticles on plant health and the environment are required for the complete utilization of nanomaterial-based gene delivery in plants. Further research and development in nanotechnology is necessary to unlock the full potential of nanomaterial-based gene delivery in plants for addressing the challenges of global food security and sustainability.

References

1. Alberts B, Johnson A, Lewis J, Raff M, Roberts K, Walter P (2002) Chloroplasts and photosynthesis—molecular biology of the cell, 4th edn. Garland Science, New York. https://www.ncbi.nlm.nih.gov/books/NBK26819/. Accessed 12 May 2025
2. Cooper GM (2000) The nucleus. In: The cell: a molecular approach, 2nd edn. Sinauer Associates, Sunderland, MA. https://www.ncbi.nlm.nih.gov/books/NBK9845
3. Deslandes L, Rivas S (2011) The plant cell nucleus: a true arena for the fight between plants and pathogens. Plant Signal Behav 6(1):42–48. https://doi.org/10.4161/psb.6.1.13978
4. Millar AH (2014) Plant mitochondria biogenesis and cellular function. In: Assmann S, Liu B (eds) Cell biology. Springer, New York, NY. https://doi.org/10.1007/978-1-4614-7881-2_13-1
5. Wang J, Xu G, Ning Y, Wang X, Wang GL (2022) Mitochondrial functions in plant immunity. Trends Plant Sci 27(10):1360–1385. https://doi.org/10.1016/j.tplants.2022.04.007
6. Song Y, Feng L, Alyafei MAM, Jaleel A, Ren M (2021) Function of chloroplasts in plant stress responses. Int J Mol Sci 22(24):13464. https://doi.org/10.3390/ijms222413464
7. Wallet C, Le Ret M, Bergdoll M, Bichara M, Dietrich A, Gualberto JM (2015) The RECG1 DNA translocase is a key factor in recombination surveillance, repair, and segregation of the mitochondrial DNA in Arabidopsis. Plant Cell 27(10):1–19. https://doi.org/10.1105/tpc.15.00680
8. Lloyd RG, Rudolph CJ (2016) 25 years on and no end in sight: a perspective on the role of RecG protein. Curr Genet 62:827–840. https://doi.org/10.1007/s00294-016-0589-z

9. Khan I, Saeed K, Khan I. (2019). Nanoparticles: properties, applications and toxicities. Arab J Chem. 12(7): 908–931. https://doi.org/10.1016/j.arabjc.2017.05.011.
10. Sarker S. D., and Nahar L. (2017). Application of nanotechnology in phytochemical research. pharm Sci. 23(3): 170–171. https://doi.org/10.15171/PS.2017.25
11. Busi MV, Gomez-Lobato ME, Araya A, Gomez-Casati DF (2011) Mitochondrial dysfunction affects chloroplast functions. Plant Signal Behav 6(12):1904–1907. https://doi.org/10.4161/psb.6.12.18050
12. Hudik E, Yoshioka Y, Domenichini S, Bourge M, Soubigout-Taconnat L, Mazubert C, Yi D, Bujaldon S, Hayashi H, De Veylder L, Bergounioux C, Benhamed M, Raynaud C (2014) Chloroplast dysfunction causes multiple defects in cell cycle progression in the Arabidopsis crumpled leaf mutant. Plant Physiol 166(1):152–167. https://doi.org/10.1104/pp.114.242628
13. Mondal R, Antony S, Roy S, Kumar Chattopadhyay S (2022) Programmed cell death (PCD) in plant: molecular mechanism, regulation, and cellular dysfunction in response to development and stress. IntechOpen, London. https://doi.org/10.5772/intechopen.97940
14. Manghwar H, Li J (2022) Endoplasmic reticulum stress and unfolded protein response signaling in plants. Int J Mol Sci 23(2):828. https://doi.org/10.3390/ijms23020828
15. Jiang YT, Yang LH, Ferjani A, Lin WH (2021) Multiple functions of the vacuole in plant growth and fruit quality. Mol Hortic 1:4. https://doi.org/10.1186/s43897-021-00008-7
16. Acuna E, Ndlovu E, Molaeitabari A, Shahina Z, Dahms TES (2023) Carvacrol-induced vacuole dysfunction and morphological consequences in Nakaseomyces glabratus and Candida albicans. Microorganisms 11(12):2915. https://doi.org/10.3390/microorganisms11122915
17. Iacobescu GL, Corlatescu AD, Popa M, Iacobescu L, Cirstoiu C, Orban C. Exploring the implications of golgi apparatus dysfunction in bone diseases. Cureus. 2024 Mar 26;16(3): e56982. https://doi.org/10.7759/cureus.56982.
18. Salata OV (2004) Applications of nanoparticles in biology and medicine. J Nanobiotechnol 2(1):1–6. https://doi.org/10.1186/1477-3155-2-3
19. Joudeh N, Linke D (2022) Nanoparticle classification, physicochemical properties, characterization, and applications: a comprehensive review for biologists. J Nanobiotechnol 20:262. https://doi.org/10.1186/s12951-022-01477-8
20. Law SSY, Miyamoto T, Numata K (2023) Organelle-targeted gene delivery in plants by nanomaterials. Chem Commun (Camb) 59(47):7166–7181. https://doi.org/10.1039/d3cc00962a
21. Soukar J, Peppas NA, Gaharwar AK (2025) Organelle-targeting nanoparticles. Adv Sci (Weinh) 12(7):e2411720. https://doi.org/10.1002/advs.202411720
22. Wang JW, Grandio EG, Newkirk GM, Demirer GS, Butrus S, Giraldo JP, Landry MP (2019) Nanoparticle-mediated genetic engineering of plants. Mol Plant 12(8):1037–1040. https://doi.org/10.1016/j.molp.2019.06.010
23. Abe N, Fujita S, Miyamoto T, Tsuchiya K, Numata K (2023) Plant mitochondrial-targeted gene delivery by peptide/DNA micelles quantitatively surface-modified with mitochondrial targeting and membrane-penetrating peptides. Biomacromolecules 24(8):3657–3665. https://doi.org/10.1021/acs.biomac.3c00391
24. Kumar AU, Ling APK (2019) Gene introduction approaches in chloroplast transformation and its applications. J Genet Eng Biotechnol 19(1):148. https://doi.org/10.1186/s43141-021-00255-7
25. Kwak SY, Lew TTS, Sweeney CJ, Koman VB, Wong MH, Bohmert-Tatarev K, Snell KD, Seo JS, Chua NH, Strano MS (2019) Chloroplast-selective gene delivery and expression in planta using chitosan-complexed single-walled carbon nanotube carriers. Nat Nanotechnol 14(5):447–455. https://doi.org/10.1038/s41565-019-0375-4
26. Dupree P, Sherrier DJ (1998) The plant Golgi apparatus. Biochim Biophys Acta Mol Cell Res 1404(1–2):259–270. https://doi.org/10.1016/S0167-4889(98)00061-5
27. Li M, Du C, Guo N, Teng Y, Meng X, Sun H et al (2019) Composition design and medical application of liposomes. Eur J Med Chem 164:640–653. https://doi.org/10.1016/j.ejmech.2019.01.007

28. Zielińska A, Carreiró F, Oliveira AM, Neves A, Pires B, Venkatesh DN, Durazzo A, Lucarini M, Eder P, Silva AM, Santini A, Souto EB (2020) Polymeric nanoparticles: production, characterization, toxicology and ecotoxicology. Molecules 25(16):3731. https://doi.org/10.3390/molecules25163731
29. Xu B, Li S, Shi R, Liu H (2023) Multifunctional mesoporous silica nanoparticles for biomedical applications. Sig Transduct Target Ther 8:435. https://doi.org/10.1038/s41392-023-01654-7
30. Paul S, Hmar EBL, Pathak H, Sharma HK (2022) An overview on nanocarriers. In: Kumar L, Pathak YY (eds) Micro and nano technologies, nanocarriers for drug-targeting brain tumors. Elsevier, Amsterdam, pp 145–204. https://doi.org/10.1016/B978-0-323-90773-6.00004-X
31. Mehdipour-Ataei S, Aram E (2023) High-end applications of unsaturated polyester composites. In: Thomas S, Chirayil CJ (eds) Applications of unsaturated polyester resins. Elsevier, Amsterdam, pp 421–439. https://doi.org/10.1016/B978-0-323-99466-8.00009-5
32. Shivashakarappa K, Marriboina S, Dumenyo K, Taheri A, Yadegari Z (2025) Nanoparticle-mediated gene delivery techniques in plant systems. Front Nanotechnol 7:1516180. https://doi.org/10.3389/fnano.2025.1516180
33. Altammar KA (2023) A review on nanoparticles: characteristics, synthesis, applications, and challenges. Front Microbiol 14:1155622. https://doi.org/10.3389/fmicb.2023.1155622
34. Abbasi E, Aval SF, Akbarzadeh A, Milani M, Nasrabadi HT, Joo SW, Hanifehpour Y, Nejati-Koshki K, Pashaei-Asl R (2014) Dendrimers: synthesis, applications, and properties. Nanoscale Res Lett 9:247. http://www.nanoscalereslett.com/content/9/1/247
35. Cruz A, Barbosa J, Antunes P, Bonifácio VDB, Pinto SN (2023) A glimpse into dendrimers integration in cancer imaging and theranostics. Int J Mol Sci 24(6):5430. https://doi.org/10.3390/ijms24065430
36. Abe N, Numata K (2025) Peptide-mediated gene and protein delivery systems to plant mitochondria for modifying mitochondrial functions. Polym J 57:57–68. https://doi.org/10.1038/s41428-024-00973-y
37. Campos G, Chialva C, Miras S, Lijavetzky D (2021) New technologies and strategies for grapevine breeding through genetic transformation. Front Plant Sci 12:767522. https://doi.org/10.3389/fpls.2021.767522
38. Chuah JA, Matsugami A, Hayashi F, Numata K (2016) Self-assembled peptide-based system for mitochondrial-targeted gene delivery: functional and structural insights. Biomacromolecules 17(11):3547–3557. https://doi.org/10.1021/acs.biomac.6b01056
39. Watanabe K, Odahara M, Miyamoto T, Numata K (2021) Fusion peptide-based biomacromolecule delivery system for plant cells. ACS Biomater Sci Eng 7(6):2246–2254. https://doi.org/10.1021/acsbiomaterials.1c00227
40. Er S, Laraib U, Arshad R, Sargazi S, Rahdar A, Pandey S, Thakur VK, Díez-Pascual AM (2021) Amino acids, peptides, and proteins: implications for nanotechnological applications in biosensing and drug/gene delivery. Nanomaterials 11(11):3002. https://doi.org/10.3390/nano11113002
41. Demirer GS, Zhang H, Goh NS, Pinals RL, Chang R, Landry MP (2020) Carbon nanocarriers deliver siRNA to intact plant cells for efficient gene knockdown. Sci Adv 6:eaaz0495. https://doi.org/10.1126/sciadv.aaz0495
42. Kaur M, Manchanda P, Kalia A, Ahmed FK, Nepovimova E, Kuca K, Abd-Elsalam KA (2021) Agroinfiltration mediated scalable transient gene expression in genome edited crop plants. Int J Mol Sci 22:10882. https://doi.org/10.3390/ijms221910882
43. Wang ZP, Zhang ZB, Li XL, Zhang C, Yin LF, Yu R, Wei J, Wu Z (2020) Efficient and genotype independent maize pollen transfection mediated by magnetic nanoparticles. https://doi.org/10.21203/rs.3.rs-97984/v1
44. Bian Y, Zhang X, Chen C, Wang G, Zhang P, Liu G, Ma F, Bao Z (2024) Establish an efficient inoculation system of tomato yellow leaf curl virus to assist tomato resistance breeding. Sci Hortic 324:112567. https://doi.org/10.1016/j.scienta.2023.112567
45. Liu W, Rudis MR, Cheplick MH, Millwood RJ, Yang J-P, OndzighiAssoume CA, Montgomery GA, Burris KP, Mazarei M, Chesnut JD, Stewart CN Jr (2020) Lipofection-mediated genome

editing using DNA-free delivery of the Cas9/gRNA ribonucleoprotein into plant cells. Plant Cell Rep 39:245–257. https://doi.org/10.1007/s00299-019-02488-w
46. Cunningham FJ, Goh NS, Demirer GS, Matos JL, Landry MP (2018) Nanoparticle-mediated delivery towards advancing plant genetic engineering. Trends Biotechnol 36:882–897. https://doi.org/10.1016/j.tibtech.2018.03.009
47. Levengood H, Zhou Y, Zhang C (2024) Advancements in plant transformation: from traditional methods to cutting-edge techniques and emerging model species. Plant Cell Rep 43(11):273. https://doi.org/10.1007/s00299-024-03359-9
48. Anami S, Njuguna E, Coussens G, Aesaert S, Van Lijsebettens M (2013) Higher plant transformation: principles and molecular tools. Int J Dev Biol 57:483–494. https://doi.org/10.1387/ijdb.130232mv
49. Sanzari I, Leone A, Ambrosone A (2019) Nanotechnology in plant science: to make a long story short. Front Bioeng Biotechnol 7:120. https://doi.org/10.3389/fbioe.2019.00120
50. Rank AP, Koch A (2021) Lab-to-field transition of RNA spray applications—how far are we? Front Plant Sci 12:755203. https://doi.org/10.3389/fpls.2021.755203
51. Sembada AA, Lenggoro IW (2024) Transport of nanoparticles into plants and their detection methods. Nanomaterials 14:131. https://doi.org/10.3390/nano14020131
52. Jang Y-h, Lim K-i (2018) Recent advances in mitochondria-targeted gene delivery. Molecules 23(9):2316. https://doi.org/10.3390/molecules23092316
53. Li HR, Harb M, Heath JE, Trippett JS, Shapiro MK, Szablowski JO (2024) Engineering viral vectors for acoustically targeted gene delivery. Nat Commun 15:4924. https://doi.org/10.1038/s41467-024-48974-y
54. Azevedo C, Macedo MH, Sarmento B (2018) Strategies for the enhanced intracellular delivery of nanomaterials. Drug Discov Today 23(5):944–959. https://doi.org/10.1016/j.drudis.2017.08.011
55. Piperno A, Sciortino MT, Giusto E, Montesi M, Panseri S, Scala A (2021) Recent advances and challenges in gene delivery mediated by polyester-based nanoparticles. Int J Nanomedicine 16:5981–6002. https://doi.org/10.2147/IJN.S321329
56. Chung YH, Church D, Koellhoffer EC, Osota E, Shukla S, Rybicki EP, Pokorski JK, Steinmetz NF (2021) Integrating plant molecular farming and materials research for next-generation vaccines. Nat Rev Mater 7:372–388. https://doi.org/10.1038/s41578-021-00399-5
57. Yan T, Hou Q, Wei X, Qi Y, Pu A, Wu S, An X, Wan X (2023) Promoting genotype independent plant transformation by manipulating developmental regulatory genes and/or using nanoparticles. Plant Cell Rep 42:1395–1417. https://doi.org/10.1007/s00299-023-03037-2
58. Sánchez AL, Luelmo SH, Izquierdo Y, López B, Cascón T, Castresana C (2021) Mitochondrial stress induces plant resistance through chromatin changes. Front Plant Sci 12:704964. https://doi.org/10.3389/fpls.2021.704964
59. Yong J, Wu M, Carroll BJ, Xu ZP, Zhang R (2024) Enhancing plant biotechnology by nanoparticle delivery of nucleic acids. Trends Genet 40:352–363. https://doi.org/10.1016/j.tig.2024.01.005
60. Mora C, McKenzie T, Gaw IM, Dean JM, Von Hammerstein H, Knudson TA, Setter RO, Smith CZ, Webster KM, Patz JA, Franklin EC (2022) Over half of known human pathogenic diseases can be aggravated by climate change. Nat Clim Chang 12:869–875. https://doi.org/10.1038/s41558-022-01426-1
61. Desai N (2012) Challenges in development of nanoparticle-based therapeutics. AAPS J 14(2):282–295. https://doi.org/10.1208/s12248-012-9339-4
62. Su W, Xu M, Radani Y, Yang L (2023) Technological development and application of plant genetic transformation. Int J Mol Sci 24(13):10646. https://doi.org/10.3390/ijms241310646
63. Osmani Z, Wang L, Xiao W, Kulka M (2024) Nanomaterials as tools in plant transformation: a protoplast-centric perspective. Plant Nano Biol 10:100100. https://doi.org/10.1016/j.plana.2024.100100

64. Byun MJ, Lim J, Kim SN, Park DH, Kim TH, Park W, Park CG (2022) Advances in nanoparticles for effective delivery of RNA therapeutics. Biochip J 16(2):128–145. https://doi.org/10.1007/s13206-022-00052-5
65. Sharma S, Parveen R, Chatterji BP (2021) Toxicology of nanoparticles in drug delivery. Curr Pathobiol Rep 9(4):133–144. https://doi.org/10.1007/s40139-021-00227-z
66. Demirer GS, Landry MP (2017) Delivering genes to plants. Chem Eng Prog 113(4):40–45. https://escholarship.org/uc/item/2hp0j2g2
67. Souto EB, Blanco-Llamero C, Krambeck K, Kiran NS, Yashaswini C, Postwala H, Severino P, Priefer R, Prajapati BG, Maheshwari R (2024) Regulatory insights into nanomedicine and gene vaccine innovation: safety assessment, challenges, and regulatory perspectives. Acta Biomater 180:1–17. https://doi.org/10.1016/j.actbio.2024.04.010
68. De Jong WH, Geertsma RE, Borchard G (2022) Regulatory safety evaluation of nanomedical products: key issues to refine. Drug Deliv Transl Res 12:2042–2047. https://doi.org/10.1007/s13346-022-01208-4

Ethical Considerations and Limitations of Nano-delivering Genetic Engineering in Agriculture and Food Systems

Luis Alejandro Martínez-Chávez, Mariana Yalid Hernández-Ramírez, Emmanuel Aguirre-García, Daniel Mendoza Jiménez, Edgardo Ulises Esquivel Naranjo, Bertha Isabel Carvajal Gámez, José Antonio Cervantes-Chávez, Amanda K. Rico-Chávez, Alexandro Escobar-Ortíz, Ana A. Feregrino-Pérez, and Karen Esquivel Escalante

Abstract This chapter explores the ethical concerns and limitations of using nanotechnology to deliver genetic engineering in agriculture and food systems. Nanotechnology can improve gene delivery methods, such as CRISPR/Cas9, by making them more efficient and precise, helping crops become more resilient to pests, drought, and environmental stress. However, this technology also raises significant ethical issues, including potential risks to human health, the environment, and biodiversity. Topics discussed include public acceptance, equity in access to nanotechnology, respect for cultural and religious values, and concerns about long-term effects on food security. The chapter also highlights scientific and technical challenges, such as potential toxicity, environmental accumulation of nanoparticles, and unknown long-term impacts. Regulatory and policy gaps are examined, showing the need for stronger global frameworks to ensure safe use. And while nano-enabled genetic engineering holds promise for sustainable agriculture, it must be used responsibly. Future research should focus on understanding risks, improving safety, and ensuring fair access for all communities. Ethical and ecological consid-

L. A. Martínez-Chávez · M. Y. Hernández-Ramírez · E. Aguirre-García · A. K. Rico-Chávez · A. Escobar-Ortíz · A. A. Feregrino-Pérez · K. E. Escalante (✉)
División de Investigación y Posgrado, Facultad de Ingeniería, Universidad Autónoma de Querétaro, Santiago de Querétaro, Querétaro, Mexico

D. M. Jiménez · E. U. E. Naranjo · J. A. Cervantes-Chávez
Laboratorio de Microbiología Molecular, Cuerpo Académico Biotecnología y Sanidad Vegetal, Facultad de Ciencias Naturales, Universidad Autónoma de Querétaro, Santiago de Querétaro, Querétaro, Mexico

B. I. C. Gámez
Laboratorio de Proteogenómica y Diagnóstico Molecular. Cuerpo Académico Salud Animal y Microbiología Ambiental, Facultad de Ciencias Naturales, Universidad Autónoma de Querétaro, Santiago de Querétaro, Querétaro, Mexico

J.-T. Chen (ed.), *Nano-Delivering for Plant Genetic Engineering*, Smart Nanomaterials Technology, https://doi.org/10.1007/978-981-95-4165-2_14

erations must guide technology development to avoid harm and ensure benefits are shared equitably across societies.

Keywords Nanomaterials · Agriculture · Food systems · Nano-delivering genetic engineering · Ethics · Regulatory frameworks · Toxicity · Environmental issues

1 Introduction

The ethical importance of addressing societal, environmental, and ecological implications of nano-delivering genetic engineering in agriculture and food systems is significant. This technology holds promise for enhancing agricultural productivity and sustainability but also poses potential risks that must be carefully managed.

For instance, nanotechnology significantly enhances genetic engineering in plants by improving the delivery of genes and the CRISPR-Cas9 system. This integration addresses several challenges in plant biotechnology, such as species dependency, low transformation efficiency, and potential tissue damage [1–3]. Specifically, nanotechnology helps overcome critical barriers in CRISPR applications, such as low editing efficiency and challenges in germline transformation, thus unlocking the full potential of CRISPR in plant genetic engineering [4, 5]. The use of nanoparticles allows for the delivery of larger Cas9 enzymes and complex guide RNAs, enabling the editing of previously inaccessible genetic regions [6, 7].

Nano-delivery systems significantly contribute to improving crop traits by enhancing drought resistance, pest resilience, and yields. They offer precise, efficient, and sustainable agricultural solutions, addressing the challenges of climate change and food security. These technologies hold promise for revolutionizing agriculture by optimizing resource use and minimizing environmental impact [8–11].

Nevertheless, the ethical importance of addressing societal, environmental, and ecological implications of nano-delivering genetic engineering in agriculture and food systems should be taken care of as soon as possible. This technology holds promise for enhancing agricultural productivity and sustainability but also poses potential risks that must be carefully managed [12]. Nanotechnology can improve crop yields and nutritional quality, addressing food security challenges posed by a growing global population [10]. However, public acceptance and regulatory frameworks are crucial to ensure these technologies are safely integrated into food systems [13]. The integration of nanotechnology in agriculture can lead to economic benefits by reducing waste and improving efficiency. However, the cost-effectiveness and accessibility of these technologies for small-scale farmers need consideration [4]. There are concerns about the long-term environmental impacts of nanomaterials, including their effects on soil and water quality, and biodiversity [12, 13]. The potential toxicity of nanomaterials to plants, animals, and humans is a critical concern. Understanding bio-nano interactions is essential to prevent adverse ecological impacts [14].

Addressing the ethical implications of nano-delivering genetic engineering in agriculture involves balancing the benefits of enhanced productivity and sustainability with the potential risks to society, the environment, and ecosystems [15]. Ensuring public acceptance, developing robust regulatory frameworks, and conducting thorough risk assessments are essential steps in responsibly advancing these technologies. In this chapter, some of these topics are presented to have a global situation panorama to have a starting point of how to overcome the ethical considerations and limitations of using nanotechnology in agriculture.

2 Ethical Considerations

The ethical implications of any endeavor, particularly in fields such as biotechnology, agriculture, and food production, are multifaceted. The subsequent sections explore various dimensions of ethical considerations that must be considered to ensure that practices are equitable, sustainable, and culturally sensitive [16].

The integration of nanotechnology with genetic engineering in agriculture plays a crucial and complex ethical view that confronts traditional perceptions of food production and feeding. The main ethical consideration is the principle of beneficence "the obligation to act for the benefit of others" [17]. Researchers keep up that nano-delivering genetic engineering has the potential to substantially enhance crop yields, and nutritional content, and recover resistance to pests and environmental stresses. These benefits could be provided to address global food security challenges and upgrade public health outcomes. Nevertheless, this notion must be balanced next to the ethical imperative of non-maleficence—the duty to avoid causing harm. The novelty and complexity of these technologies suggest that potential long-term risks to human health and the environment are not yet absolutely understood, originating an ethical dilemma in weighing potential advantages against unknown risks [17, 18].

Another critical ethical factor is the principle of autonomy, specifically which links consumer choice and informed consent. The use of nano-delivering genetic engineering in food production increases questions about customers to identify what is in their food and how it was produced. There is an ethical requirement for transparency and clear classification to allow individuals to make informed decisions about the foods they consume. Nevertheless, this principle is complicated because of the challenges in successfully communicating complex scientific information to the public and the potential for distortion of the information which influences public perception. Additionally, if nano-genetically engineered foods become established in the food supply, it may limit the practical capacity of consumers to opt out, potentially interfering with their autonomy [19].

Lastly, the ethical principle of respect for nature and the environment must be considered. The use of nanotechnology represents a significant intervention in natural biological processes, raising philosophical doubts about the restrictions of human alteration of nature. Although these technologies may offer keys to environmental challenges like climate change adaptation, there are likewise troubles with negative impacts on biodiversity and ecosystem balance. The risk of unplanned consequences, such as the formation of "superweeds" or interruptions to soil microbiomes, presents ethical challenges in balancing technological progress with the conservation of natural systems. This principle also intersects with cultural and religious values that residence importance on the concept of naturalness in food, adding another layer of complexity to ethical deliberations and public acceptance of these technologies in agriculture and food systems [20].

2.1 Public Perception and Acceptance

The connection between nanotechnology and genetic engineering in agriculture and food production displays a complex landscape of ethical attention, specifically when it comes to public perception and acceptance. As these new technologies promise to reform food production, increase crop yields, and in theory address global food security challenges; they also raise significant ethical questions and worries among consumers, policymakers, and stakeholders in the food supply chain. The first ethical consideration around nano-delivering genetic engineering in agriculture is the question of safety and long-term health impacts. While researchers argue that this knowledge can produce more nutritious and robust crops, there are valid concerns about potential unforeseen effects on human health and the environment. The new nature of combining nanoparticles with genetic modification incomes is partial long-term data on their effects when consumed or released into environments. Unfortunately, it generates ethical dilemmas for policymakers and regulators tasked with equilibrating potential benefits upon unknown risks [21, 22].

Public insight into these technologies is often prejudiced by a complex interplay of scientific thought, cultural values, and emotional responses to the idea of "tampering" with nature or food [20]. Many consumers express discomfort about consuming foods that have been genetically modified or treated with nanomaterials, yet scientific consent proposes they are safe. This highlights the ethical challenge concerning public sentiment and consumer choice while also promoting potentially beneficial technological improvements. Clearness and informed consent appear as critical ethical considerations in this context. For public acceptance to be important, consumers must have access to accurate information about how nano-delivering genetic engineering is being applied to their food supply [22]. However, linking complex scientific concepts to the public without oversimplification or fearmongering exhibits significant challenges. An extra ethical consideration is the potential to strengthen existing inequalities in the global food system. While these technologies have the hypothetical view of increasing crop yields and creating resilient food sources, there is unease about who will have access to and power over these improvements. If the profits in agriculture are primarily gained by larger agrobusiness corporations or wealthy nations, it could further downgrade small-scale farmers and food-insecure residents [18, 19].

Environmental ethics has a crucial character in the dialogue surrounding nano-delivering genetic engineering in agriculture. These technologies may propose solutions to challenges like pest resistance and climate adaptation, but there are worries about negative impacts on biodiversity and ecosystem balance. The perception of "naturalness" and its relation to food is another ethical thought that influences public perception and acceptance of nano-genetic engineering in agriculture. Many consumers award high value on "natural" or "organic" foods, recognizing them as healthier and more ethically produced. The use of nanoparticles and genetic modifications confronts traditional notions of what represents natural food. This introduces philosophical questions about the ethics of technological intrusion in food production and the magnitude to which "naturalness" should be conducted in agricultural practices [19, 23].

2.2 *Equity and Accessibility*

Equity and accessibility are fundamental ethical views, as upgrades in technology should not overly benefit one group over another. Making certain that disregarded communities have access to advanced solutions (healthcare, agriculture, or food systems) is crucial. Programs should aim to reject barriers that inhibit equitable access, such as socioeconomic category, geography region, and education system [19]. This also implies supporting fair valuing models and establishing programs that specifically target disadvantaged populations to guarantee that all could benefit from technological advances. The mixing of nanotechnology and genetic engineering in agriculture and food systems depicts significant hypothetical benefits but also raises critical ethical fears involving equity and accessibility. As these advanced technologies arise, it is essential to deliberate their suggestions for social justice, global food security, and the moderate allocation of both benefits and hazards across diverse populations [20].

One of the key ethical considerations around equity in nano-delivering genetic engineering is the potential aggravation of existing unfairness in the global food system. While these technologies promise increased crop yields, enhanced nutritional content, and improved resistance to environmental stresses, there are valid concerns about who will have access to these advancements. The development and accomplishment of nano-genetic engineering techniques require substantial financial investments, advanced scientific infrastructure, and specialized skills. This creates a risk that the benefits of these technologies will be extremely available to prosperous nations, large agribusiness corporations, and industrial-scale farming operations [16, 24].

Small-scale agriculturalists, particularly in developing countries, may look at significant difficulties in accessing nano-genetic engineering technologies. These difficulties could bring high costs, a lack of technical knowledge, and limited access to the essential equipment and materials. Even if small farmers are incapable of getting these new technologies, they may become less competitive in global markets, potentially directing to additional consolidation of agricultural production in the roles of large organizations. This scenario creates ethical worries about the impact on rural livelihoods, traditional farming practices, and food sovereignty in developing regions [23].

The main part of intellectual property presents an extra critical ethical challenge in making certain equitable access to nano-delivering genetic engineering in agriculture. Patents and other forms of intellectual property protection on genetically engineered seeds, nanoparticles, or related technologies could limit access for farmers who cannot provide licensing fees. This could create a system where the benefits of agricultural advancements are mostly reaped by those who can pay them, instead of those looking for the based-on need or potential impact on food security [22, 25].

Ethical considerations to offer the global distribution of potential hazards associated with nano-genetic engineering in agriculture. If this knowledge is principally confirmed and executed in developing countries owed to less stringent regulations or the promise of increased yields, it could lead to a situation where exposed populations bear an unequal problem of any unexpected negative consequences. This raises inquiries about the ethical responsibilities of technology developers, multinational corporations, and monitoring bodies in ensuring that the consequences and benefits of these technologies are rightfully distributed on a global scale [21, 26].

Availability of information and decision-making processes regarding nano-delivering genetic engineering is an extra fundamental ethical consideration. Impartial employment of these technologies, almost all sponsors (small farmers, native communities, and customers from diverse socioeconomic families) should have the prospect of participating in discussions and decisions about how these technologies are advanced, regulated, and arranged. However, the complex and technical nature of nano-genetic engineering can create obstacles to deep participation for those without concentrated scientific knowledge [27, 28].

The promise for nano-delivering genetic engineering to keep on global challenges like climate change variation and food security enhances another layer to equity regards. In case these technologies can clearly influence solving necessary issues that strangely affect susceptible populations, there may be an ethical reason for giving precedence to their development and ensuring prevalent accessibility. For instance, drought-resistant crops enhanced through nano-genetic engineering could significantly gain farmers in waterless regions, which are most impacted by climate change [18, 23].

The ethical code of justice also describes for consideration how the values of nano-delivering genetic engineering in agriculture are distributed within societies. Even inside developed nations, there are significant discrepancies in the approach to nutritious food. As these technologies enhance the nutritional content or shelf life of several foods, it is important to ensure that these improved products are available to low-income groups and not exclusively available as premium options for wealthy consumers. Attending equity and accessibility establishments in nano-delivering genetic engineering needs a multifaceted approach. A solution is the improvement of public-private partnerships that prioritize equitable contact with these technologies. Many organizations could focus on designing open-source platforms, safeguarding that the fundamental tools and knowledge are available to a broader range of users [18, 19].

International assistance and technology transference programs have a crucial role in promoting equity. Developed nations and research institutions could launch proposals to share knowledge, offer training, and support the teaching of local capacity in nano-genetic engineering technologies in a smaller amount of developed regions. This view not only deals with immediate accessibility issues but also impacts long-term sustainability by allowing local communities to adapt and update with these technologies. Education and ability building are necessary elements of adopting equity and accessibility issues. Investments in agricultural training, specifically in developing countries, can bridge the data gap and enable more varied participation in the development and application of these tools. This includes not only practical training but also teaching on the ethical, social, and environmental consequences of nano-delivering genetic engineering [19, 20].

Lastly, constant ethical judgment and supervising of the impacts of these technologies on worldwide food systems and social equity are vital. This requires creating mechanisms for constant evaluation of how nano-genetic engineering is altering different sponsors, particularly helpless populations. These equity and accessibility fear proactively, the agricultural and food areas can work near to ensuring that the possible benefits of nano-delivering genetic engineering are shared approximately and contribute to a more just and sustainable global food system [25, 29].

2.3 Environmental and Ecological Ethics

As humanity searches for innovating and improvement, it is crucial to consider the environmental and ecological consequences of these accomplishments. Ethical stewardship of the planet requires a promise to sustainability and the minimization of damage to ecosystems. This implies exploring practices such as resource extraction, land-use alterations, and waste management to ensure they do not compromise ecological offset. Ethical frameworks should conduct decision-making practices to safeguard biodiversity and raise regenerative practices that restore and maintain the health of the earth [24, 30, 31].

One of the principal ethical considerations in this field is the principle of environmental stewardship. This principle speculates that humans have a moral acknowledgement to protect and preserve the natural ecosystem for recent and future generations. Nano-delivering genetic engineering in agriculture has the capacity to modify ecosystems and natural processes, raising doubts about the limit to which we should interfere in nature. While protagonists claim that these technologies could help environmental contests such as climate change adaptation or reduced pesticide use, detractors worry about unintentional consequences and the long-term effects of introducing engineered organisms into intricate ecological systems. The precautionary principle is a fundamental ethical framework when respecting the environmental implications of nano-genetic engineering. This principle informs that when an accomplishment has a presumed risk of causing damage to the environment or human health, in the non-appearance of scientific agreement. Agreed the newness and complexity of nano-delivering genetic engineering, there are many strangers regarding its long-term environmental influences [17, 30].

Biodiversity preservation is another critical ethical thought. Biodiversity is not simply intrinsically valuable but also vital for ecosystem resilience and the provision of ecosystem deals. Nano-genetic engineering in agriculture could affect biodiversity in several directions. On one hand, it might help conserve crop diversity by supporting the development of varieties resistant to climate change or pests. On the other hand, the prevalent adoption of genetically engineered crops could advance to monocultures and reduce agricultural biodiversity. There's also the hazard of gene issues from engineered crops to wild relatives, in theory modifying natural plant populations and ecosystems [32].

The perception of ecological integrity is central to environmental ethics and presents important questions for nano-delivering genetic engineering. This principle emphasizes the value of keeping the fullness and continuity of ecosystems. Leading genetically engineered organisms improved by nanotechnology into the environment could interrupt existing ecological relationships and food nets. Such as, crops persuaded to be pest-resistant may have unintentional effects on beneficial insects or soil microorganisms. Ethical considerations must value for the complex mutuality within ecosystems and the potential dropping effects of introducing novel organisms [33].

Another important environmental ethical thought is the potential influence on soil diversity. Soil is a compound living system that has a vital function in ecosystem performance and agricultural productivity. Nano-delivering genetic engineering techniques that modify plant-soil interactions or the introduction of engineered microorganisms could have far-reaching outcomes for soil ecology. The worries about the potential increase of engineered nanoparticles in soil and their consequences on soil microorganisms. Ethical frameworks must ponder the long-term sustainability of agricultural systems and their factor on this critical, often unseen, component of ecosystems [24, 33].

The principle of intergenerational justice is specifically relevant now for considering the environmental effects of nano-delivering genetic engineering. This assumption establishes that modern generations have an obligation and responsibility to maintain the environment and natural resources for the next generations. The long-term and possibly irreversible nature of genetic modifications increased by nanotechnology mentions questions about our right to take decisions that could necessarily affect ecosystems for generations to come. Ethical frameworks must balance potential short-term benefits against long-term environmental sustainability and the rights of future generations to inherit a healthy, functioning planet [31].

Another criticism is the probable for nano-genetic engineering to intensify or mitigate existing environmental problems. Once these technologies might decrease the requirement for chemical pesticides, they could also facilitate the intensive farming practices that have other adverse environmental impact. The ethical principle of holistic environmental managing invites for a comprehensive evaluation of how these technologies co-operate with broader agricultural systems and environmental challenges [30, 31] .

Transparency and public situation in environmental decision-making are solutions for ethical considerations. Given the complexity and possible far-reaching effects of nano-genetic engineering, there is an ethical domineering to ensure that diverse sponsors such as: environmental groups, indigenous communities, and future generations [18, 32].

Finally, the conservation and ecological ethical concerns surrounding nano-delivering genetic engineering in agriculture require a counteractive, holistic, and general approach. By the time these technologies provide a potential benefit for global challenges such as food security and climate change variation, their expansion and operation must be directed by robust ethical frameworks that arrange long-term environmental sustainability, biodiversity conservation, and the health of ecosystems [25, 29].

As we cross this complicated ethical landscape, ongoing investigation, monitoring, and adaptive management strategies will be decisive. Ethical decision-making in this area requires a responsibility to continuous knowledge and reassessment as new information happens to be available about the environmental factors of nano-genetic engineering. In the direction of environmental and ecological ethics in the advance and application of these powerful technologies, we can attempt to connect their potential benefits while preserving the health and integrity of the natural systems [34].

2.4 Cultural and Religious Sensitivities

Cultural and religious attitudes significantly impact on how societies recognize technology and its applications. It is fundamental to recognize and respect these viewpoints, which can depend widely between communities. Connecting with cultural leaders and integrating traditional knowledge into planning and implementation can facilitate adopting a sense of ownership and acceptance. Furthermore, avoiding action practices that may conflict with cultural or religious principles is not only an ethical obligation but can also augment the efficacy of initiatives by supporting the community's beliefs and practices [35]. The improvement of nano-delivering genetic engineering in agriculture and food systems not only improves scientific and environmental doubts but also pass across deeply convened cultural and religious beliefs. These technologies exhibit the potential to trial traditional worldviews, interrupt long-standing agricultural methods, and conflict with religious dietary laws and ethical standards. Perception and attending these cultural and religious thoughtfulness are crucial for the ethical development and application of nano-genetic engineering in global food systems [36–38].

The main ethical consideration in this context is the perception of "playing God" or delaying with nature's order, which is an involvement shared by many religious and cultural traditions. The capacity to manipulate genetic material at the nanoscale and make organisms that would not naturally take place can be perceived as violating the restrictions of human authority and interfering into the divine realm. For example, many Indigenous cultures viewpoint nature as sacred and interconnected, and the idea of essentially altering living organisms may be seen as a damage of natural law and traditional wisdom [35, 39].

In Abrahamic religions (Judaism, Christianity, and Islam), there are contrasting interpretations of the ethical suggestions of genetic engineering. Some faithful leaders and scholars discuss that humans have been given stewardship over nature and that guilty use of technology to improve food production and reduce experiencing is adequate or even persuaded. Others continue that genetic engineering, specifically when increased by nanotechnology, correspond to an unacceptable level of human intervention in God's creation. Buddhist and Hindu outlooks often emphasize the interconnection of all life and the value of preserving harmony with nature. The model of ahimsa (non-violence) in these civilizations can offer concerns about potential harm to organisms and ecosystems via genetic manipulation. However, versions change, with some claiming that if nano-genetic engineering can lower suffering and hunger, it might be associated with religious principles of empathy and overhaul for all beings [38, 40, 41].

Cultural feelings also appear around traditional agricultural systems and indigenous knowledge systems. Many villages have developed complicated agricultural techniques over generations, genuinely intertwined with cultural identity and traditional ecological understanding. The introduction of nano-delivering genetic engineering could be identified as an intimidation to these practices and the cultural heritage they symbolize. There are ethical worries about the potential loss of traditional crop varieties, farming techniques, and the linked cultural practices if genetically engineered crops become prevailing [42].

The dispute of food dominance is another critical ethical regard at the junction of culture and nano-genetic engineering. Many aboriginal and local communities claim their right to maintain control over their food systems, including the types of crops grown and the techniques used. The introduction of genetically engineered crops boosted by nanotechnology could be seen as a burden of external technologies that weaken local autonomy and decision-making developments [37].

Religious dietetic laws and restrictions exhibit another layer of ethical involvedness. For example, in Judaism and Islam, there are specific guidelines about what foods are contemplated kosher or halal. The use of nano-delivering genetic engineering increases questions about whether genetically modified organisms (GMOs) developed by nanotechnology obey these religious dietary requests. If genes from non-kosher or non-halal supplies are used in crop engineering, it could create conflicts with religious performance and personal beliefs. The perception of "purity" in food, which is important in many cultural and religious circumstances, can also be questioned by nano-genetic engineering. For some, the perception of consuming food that has been genetically modified at the nanoscale may feel artificial or impure, despite scientific guarantees of safety. This perception can lead to challenges with these technologies and create ethical questions about accepting diverse cultural values in food production and consumption [40].

Ethical factors also extend to the potential influence of nano-delivering genetic engineering on conventional seed-saving practices. Many cultures control long-standing traditions of saving and switching over seeds, which are often infused with cultural and spiritual magnitude. Genetically engineered seeds, especially if protected by patents, could interrupt these practices and the attendant cultural traditions [43]. The theory of informed consent is particularly appropriate when considering cultural and religious sensitivities. There is an ethical authoritative to ensure that communities are absolutely informed about the nature of nano-genetic engineering technologies, and their impact on local food systems and cultural practices. This adds in presenting information in culturally applicable ways and recognizing the right of communities to make decisions about whether to take on these technologies based on their own values and beliefs [44].

Another important ethical point is the potential for nano-delivering genetic engineering to exacerbate cultural homogenization in food systems. This is a significant ethical concern because as these technologies consent to the development of crops with higher traits, there's a risk that a small number of genetically engineered varieties could tower over the control of global agriculture. This could lead to a decline in the diversity of traditional crop varieties and associated cultural practices [41, 42].

The ethical implications of this potential homogenization are multifaceted:

1. **Loss of cultural legacy**: Many communities have acquired unique crop varieties over generations, which are fundamental to their cultural identity and traditional gastronomy. The prevalent adoption of genetically engineered crops could terrorize these culturally significant varieties.
2. **Impact on traditional knowledge**: As traditional crop varieties reduce, the associated information of cultivation, preparation, and use may also be lost, affecting the intergenerational transfer of cultural knowledge.
3. **Dietary shifts**: The dominance of certain genetically engineered crops could alter traditional regimes, potentially conflicting with cultural food inclinations and nutritional practices.

Many cultures and religions panorama life as holy or infused with intrinsic value beyond its practical aspects. The capacity to patent genetically engineered organisms induces ethical questions about the possession and commercialization of life forms, which may argument with cultural and religious philosophy about the purity of life and nature. Another key point argued is the potential for nano-delivering genetic engineering to craft or exacerbate power discrepancies in global food systems. There are ethical fears about large corporations governing these advanced technologies and potentially weakening the autonomy of local farmers and communities. This matches the concerns of justice, equity, and the right of communities to preserve control over their food production methods [31, 45].

Finally, the magnitude of comprehensive dialogue and ethical frameworks that respect cultural and religious diversity. It submits that the development and implementation of nano-delivering genetic engineering in agriculture should require consultation with diverse participants, including religious leaders, cultural representatives, and indigenous communities. This attempt intentions to ensure that these technologies are established and used in ways that are sensitive to various cultural and religious perspectives [24, 36].

2.5 Long-Term Effects on Food Security

The long-term associations of innovations on food security must be carefully estimated to ensure they impact positively to future generations. While certain technologies may increase productivity in the short term, it is essential to measure their sustainability and resilience in the face of climate change and other global challenges. Ethical considerations in food security involve highlighting practices that promote nutritional health, environmental sustainability, and economic sustainability for farmers. Participating with communities to understand their specific food needs and preferences will assistance in developing strategies that protect food systems and guarantee equitable access to nutritious food for all [36, 46].

The application of nano-delivering genetic engineering in agriculture and food systems presents in cooperation promising opportunities and substantial ethical challenges when considering its long-term effects on food security. As we cross this complex landscape, it's crucial to examine the potential impacts, both positive and negative, that these technologies may receive on our ability to provide sustainable, equitable, and sufficient food for an increasing global population [47].

One of the principal ethical considerations is the potential for nano-genetic engineering to enhance crop yields and strength. Supporters argue that these technologies could change crops with improved resistance to pests, diseases, and environmental stresses such as drought or salinity. This could indicate more constant and better food production, particularly in regions confronting challenging growing conditions due to climate change. The ethical obligation to address global hunger and malnutrition lends weight to the development of technologies that could enhancement food security [46, 47].

However, this potential benefit must be weighed against the ethical concern of creating dependency on high-tech solutions. There's a risk that farmers, particularly in developing countries, could become reliant on genetically engineered seeds and associated technologies restricted by a small number of companies. This presents questions about food sovereignty and the right of communities to maintain control over their food production systems. The ethical code of autonomy suggests that farmers should have the freedom to decide their agricultural practices without being persuaded by market forces or technological lock in [24]. Another crucial ethical consideration is the potential impact on biodiversity and genetic resources. While nano-delivering genetic engineering could help conserve certain crop varieties by making them more robust, there's also a risk of reducing overall agricultural biodiversity if a small number of engineered varieties dominate global agriculture. The statement of intergenerational justice requires us to deliberate how our actions today might affect the genetic resources available to future generations for food security [33].

The concept of sustainability is fundamental to ethical discussions about long-term food security. Nano-genetic engineering could possibly contribute to more sustainable agricultural practices by reduction the need for chemical inputs or enabling more efficient use of resources like water and nutrients. However, there are also alarms about the long-term ecological impacts of introducing engineered organisms into complicated ecosystems. The precautionary principle recommends that we should carefully evaluate and monitor these impacts to avoid involuntary consequences that could ultimately compromise food security [29].

The dispute of equity in approach to nano-genetic engineering technologies is another critical ethical understanding. While these technologies have the potential to increase food security, there's a risk that they could make existing inequalities in the global food system worse. Smallholder farmers in developing countries might be incapable of affording or accessing these advanced technologies, potentially expanding the productivity gap between large-scale industrial agriculture and small-scale farming. This raises questions of distributive justice and the ethical obligation to ensure that technological advancements benefit all segments of society, not just those with the resources to access them [34, 48].

The potential for nano-delivering genetic engineering to attend nutritional deficiencies is an important consideration for long-term food security. These technologies could be applied to develop biofortified crops with better nutritional profiles, potentially addressing issues of malnutrition and hidden hunger. However, this approach raises ethical questions about the most appropriate ways to address nutritional challenges. Some argue that a more holistic approach focusing on dietary diversity and traditional food systems might be more sustainable and culturally appropriate than relying on engineered solutions [49].

Another significant ethical consideration is the potential influence on traditional and indigenous agricultural knowledge systems. Many communities have developed advanced agricultural practices over generations, designed to local conditions and deeply linked with cultural identity. The introduction of nano-genetic engineering could potentially relocate these traditional practices, leading to a loss of valuable knowledge and cultural legacy. There's an ethical imperative to respect and conserve diverse agricultural traditions while exploring the potential benefits of new technologies [25, 29].

The stem of transparency and informed consent is crucial when considering the long-term effects of nano-delivering genetic engineering on food security. Consumers and farmers have a right to know how their food is made and what technologies are involved. This introduces ethical questions about labeling requirements for genetically engineered foods and the extent to which the difficulties of nano-genetic engineering can be effectively communicated to the public. The principle of autonomy suggests that individuals should be able to make informed preferences about the foods they consume and the agricultural practices they support [46]. The potential for involuntary ecological consequences is a significant ethical concern when considering long-term food security. While nano-genetic engineering might solve certain agricultural challenges, it could also create new ones. For instance, crops engineered to be pest-resistant might advance to the evolution of more resilient pests or the trouble of beneficial insect populations. The principle of environmental stewardship requires us to evaluate and mitigate these potential risks carefully to maintain the ecological balance necessary for sustainable food production [30].

An additional ethical concern is the potential for nano-genetic engineering to intensify power differences in the global food system. As these technologies become more advanced and concentrated in the hands of a few major corporations, there's a risk of increased corporate control over the food supply. This could further disregard small-scale farmers and undermine their ability to maintain food security for their communities. The ethical principle of justice calls for a more equitable distribution of power and decision-making in the food system. The issue of food safety is another critical long-term consideration. While recent risk assessment frameworks may consider genetically engineered crops enhanced by nanotechnology to be safe for human consumption, there are relates about potential long-term health impacts that may not be immediately evident. The precautionary principle recommends that we should go astray on the side of caution and conduct rigorous, long-term studies to ensure the safety of these technologies before extensive deployment [17, 47].

Ultimately, the ethical considerations around the long-term effects of nano-delivering genetic engineering on food security require a comprehensive, system-level approach. We must prudently weigh the potential benefits of these technologies with their potential risks and unintentional consequences, always prioritizing the principles of sustainability, equity, and the preservation of diverse agricultural knowledge and practices. Constant dialogue, inclusive decision-making, and adaptive management plans will be crucial in directing this multifaceted ethical landscape and safeguarding that the development and application of nano-genetic engineering technologies contribute to a more just, resilient, and food-secure future for all [32], while genetic alteration has been a projecting tool in agricultural biotechnology, developing technologies like nanotechnology may offer different or complementary methods to enhance crop traits and processing agricultural production systems. This perspective encourages a more universal view of agricultural improvement, contemplating several technological solutions to direct global food security and sustainability challenges.

3 Limitations of Nano-delivery in Genetic Engineering

Nano-delivery systems in genetic engineering present promising alternatives to conventional methods, offering new possibilities for enhancing gene delivery efficiency. However, several limitations still hinder their widespread application. Addressing these challenges is crucial to unlocking the full potential of nanotechnology in gene delivery.

3.1 *Technical and Scientific Barriers*

Plant genetic engineering has advanced significantly in recent years, contributing to improved crop quality and yield, enhanced secondary metabolite production in medicinal plants, and the development of crops suited for sustainable agriculture [50]. Although biolistic particle delivery and Agrobacterium-mediated delivery are widely established techniques for plant genetic transformation, they present notable drawbacks such as tissue damage, low efficiency, and limited host specificity. Electroporation and polymer-mediated delivery offer alternative approaches but are often constrained by species compatibility and require extensive tissue culture processes [51, 52]. Table 1 summarizes common delivery methods used in plant genetic transformation, detailing their associated adverse effects, target species, cargo types, and limitations.

Nanoparticles (NPs) are highly effective for intracellular biomolecule delivery due to their ability to cross membranes, protect cargo, and enable targeted release through chemical and physical modifications [58]. NP-mediated delivery addresses key limitations in current plant delivery techniques by enabling cell wall traversal through size control, optimizing charge and surface properties for diverse cargo transport, and providing broader applicability across plant species [59]. In recent decades, significant progress has been made in nanomaterial-based DNA delivery in living cells, opening new opportunities for delivering biomolecules in plant cells as well [60]. However, several technical and scientific challenges remain in optimizing these nano-delivery systems for effective and safe use [61].

To better understand the challenges associated with nano-delivery systems in genetic engineering, it is important to identify the primary technical and scientific barriers that hinder their efficiency and safety. Table 2 outlines these obstacles, highlighting key issues researchers face in improving nano-delivery systems for enhanced gene therapy applications.

While NPs can enhance plant growth and improve disease resistance, they also pose potential genetic risks by inducing various genetic changes in plants. These changes may affect crop quality, disrupt ecosystem balance, and pose risks to human health [68, 69]. Table 3 outlines key genetic risks, their effects, and possible consequences related to nanoparticle exposure in agricultural systems.

Nanoparticles provide promising benefits for agricultural productivity and plant disease management; however, they also present significant genetic and environmental risks [52, 76]. These concerns include potential genomic alterations in plants and the possibility of nanoparticles entering the food chain [77, 78]. Balancing these benefits with potential risks is essential to ensure the safe and sustainable use of nanotechnology in agriculture.

Table 1 Comparison of plant genetic transformation delivery methods

Type	Delivery method	Adverse effects of delivery	Target species/tissue	Cargo type and size	Limitations	References
Physical	Biolistic particle—mediated delivery	Scrambled and multiple integrations, damage to the target tissue, and specialized equipment required	Monocots, dicotyledons/ microspores and intact tissue	DNA, siRNA, miRNA, ribonucleo proteins	Cell damage, low integration efficiency, low expression levels	Sanford [53]
	Electroporation	Non-specific transport of material and damage to the target tissue	Green algae, monocots, dicotyledons, pollen grains, protoplasts and meristems	DNA, siRNA, miRNA proteins	Damage tissue, non-specific. Applicable only for protoplast. Limited to only a few species Limited to cargo-carrying capacity	Ozyigit [54]
Chemical	Polymer-mediated	High charge densities induce cytotoxicity	Monocots, dicotyledons, protoplasts	Nucleic acids (DNA, siRNA, mi RNA)	High inefficient regeneration for most species in transient studies and requires tissue culture	Cunningham et al. [52]
Biological	*Agrobacterium*-mediated delivery	Can lead to apoptosis and necrosis, random integration, high host specificity	Root, shoot apical meristem, leaf, flower, hypocotyl, cotyledon, pollen tube	siRNA, miRNA, DNA	Limited to DNA cargo, regeneration protocols required, poor gene transfer efficiency	Gelvin [55], Tzfira and Citovsky [56]
	Virus delivery	Integration of viruses	Callus and meristems, and cells	DNA, siRNA, miRNA	Limited DNA loading capability Restriction specifically for species, lack of selectivity, transient expression	Mahmood et al. [57]

Table 2 Technical and scientific barriers in nano-delivery for genetic engineering

Barrier type	Challenge	Description	References
Technical	Low transfection efficiency	Non-viral vectors, while safer, often show poor gene delivery efficiency, limiting their clinical success. Improving transfection methods remains a major focus of current research	Garcia-Guerra et al. [62], Mirón-Barroso et al [63]
	Intracellular barriers	NPs must overcome endosomal entrapment and enable efficient DNA translocation to the nucleus for effective gene expression. Understanding these processes is key to improving delivery success	Degors et al. [64], Durymanov and Reineke [65]
Scientific	Toxicity concerns	Nanomaterials may interact with biological systems in ways that induce toxicity, posing risks for clinical applications. Ensuring biocompatibility is critical for safe gene delivery	Garcia-Guerra et al. [62], Mirón-Barroso et al. [63], Yazdani et al. [66]
	Biological interactions	The complex and often unpredictable interactions between nanomaterials and biological systems can affect gene delivery stability, distribution, and efficacy. Efforts are ongoing to better understand and manage these effects	Degors et al. [64], Torres-Vanegas et al. [67]

Table 3 Genetic risks of nanoparticles in agriculture and their consequences

Genetic risk	Description	Potential consequences	References
Genomic alterations	NPs may cause genomic, metabolomic, and proteomic changes, potentially leading to genetic modifications in plants	Altered plant development, reduced crop quality, and compromised food safety	Ahmed et al. [70], Ma et al. [71], Paramo et al. [72]
Phytotoxicity	NPs accumulation in soil and plants can trigger physiological and molecular alterations	Impaired plant growth, reduced yield, and decreased nutritional value	Ma et al. [71], Yan and Chen [73]
Translocation and bioaccumulation	NPs may translocate within plant tissues and bioaccumulate, entering the food chain	Potential health risks for humans and animals due to nanoparticle ingestion	Pacheco and Buzea [74], Yan and Chen [73]
Mechanisms of genetic impact	NPs can deliver biomolecules such as nucleotides and proteins, influencing genetic expression	Possible induction of disease resistance genes or creation of transgenic plants	Demirer et al. [4], Dong et al. [75], Shivashakarappa et al. [58]

3.2 Regulatory and Policy Challenges

Nanoparticles in agriculture offer promising benefits, such as improved crop yields and reduced agrochemical use; however, they also present potential environmental and health risks [72, 79]. The regulation of nanotechnology in agriculture varies significantly worldwide, reflecting diverse approaches to safety, environmental impact, and consumer acceptance. This variability stems from differences in national priorities, regulatory frameworks, and levels of technological adoption [80].

The regulatory framework for the use of nanoparticles in agriculture is still evolving, highlighting the need for comprehensive risk assessments and well-defined policies [81, 82]. Current regulations remain insufficient to address the complex risks associated with nanoparticles, emphasizing the importance of adopting a transdisciplinary approach within regulatory frameworks [82]. Policymakers face the challenge of developing regulations that support nanotechnology advancement while safeguarding environmental and consumer well-being. Achieving this requires regulations grounded in robust scientific research [79, 83].

Countries have implemented various regulatory measures for nano-based agricultural products, focusing primarily on safety assessments to protect consumers and the environment [80]. The European Union (EU) is among the few regions with specific provisions for nanotechnology in agriculture, which are integrated into its existing legislative framework [84]. The EU has included nano-specific requirements in its regulations for the agricultural, feed, and food sectors, mandating safety assessments to ensure nano-enabled products do not pose undue risks to consumers or the environment [84, 85]. The EU's regulatory strategy for nanotechnology aligns with its approach to genetically modified organisms (GMOs), involving stringent modifications to existing frameworks to address environmental, health, and safety concerns. This strategy reflects the EU's broader precautionary regulatory approach, emphasizing a comprehensive and cautious integration of regulations [86].

In contrast, the regulatory framework in the U.S. remains underdeveloped in addressing the unique challenges posed by nanotechnology in agriculture [13]. This reflects a broader trend in U.S. policy, which places greater emphasis on economic interests and technological advancement over precautionary measures [87]. Although efforts to regulate nanotechnology in U.S. agriculture are ongoing, the focus remains on mitigating safety, environmental, and health risks [88]. Nonetheless, clearer guidelines are essential to effectively manage the long-term environmental impacts and ensure the safety of nano-enabled agricultural products [80].

India currently lacks specific regulations dedicated to nanotechnology in agriculture. The focus on nanoregulation is still emerging, with only a few countries worldwide having established comprehensive regulatory frameworks [89]. Despite this, the Indian government actively promotes the expansion of nanotechnology in agriculture by encouraging private sector investments and partnerships. However, challenges such as toxicity concerns, health risks, and regulatory inconsistencies persist as significant hurdles [90]. Similarly, in China, nanotechnology has been applied across various agricultural sectors, including crop production, livestock, and aquaculture [91]. However, despite these advancements, China's regulatory framework continues to evolve to keep pace with the rapid development of nanotechnology [13]. Comprehensive regulations are urgently needed to address the safety, environmental impact, and ethical considerations associated with nanotechnology applications in agriculture [92].

More extensive research is required across various regions to better understand the long-term environmental impacts of nanotechnology and develop evidence-based regulations [88]. The ambiguity in existing regulatory frameworks, coupled with potential risks linked to nanotechnology, highlights the need for continued investigation [85].

3.3 Cost and Scalability

The cost and scalability of nanoparticle production are critical factors that determine the feasibility of their commercial applications [93]. The production costs of nanoparticles can vary significantly depending on the materials and methods used [94].

Scaling up nanoparticle production from laboratory settings to field applications can be expensive. Ensuring that production costs remain affordable compared to traditional agricultural inputs is a significant challenge [95]. Using inexpensive technical-grade reagents has proven effective in developing cost-efficient preparation methods, such as urea-functionalized nanoparticles, which improve nitrogen use efficiency [96]. These innovative nanoparticles enhance nutrient uptake, reducing the overall quantity required, lowering costs for farmers, and promoting more sustainable agricultural practices [27, 97]. While some nanoparticle synthesis methods are cost-effective, others depend on expensive processes or materials, underscoring the need for controlled, safe, and economical synthesis techniques to reduce costs [79, 98]. Developing simple and scalable synthesis methods is crucial for advancing nanoparticle use in agriculture. Techniques such as in situ coating have shown promise in enabling the efficient production of nanohybrids that can be applied as fertilizers [99]. Nevertheless, large-scale industrial production remains challenging due to the complexity of synthesis processes and the need for further research to ensure safe and effective applications [100, 101].

Integrating nanotechnology with modern agricultural practices, such as speed breeding and CRISPR/Cas, presents a promising approach to improving scalability by enhancing efficiency and effectiveness in crop improvement programs [51, 102]. Nanotechnology's versatility in enhancing plant stress tolerance, nutrient delivery, and pest resistance offers broad applicability across diverse crops and agricultural systems. As these benefits become more evident, the scalability of nanoparticle-based solutions is expected to expand, contributing to more sustainable and efficient agricultural practices [15, 103]. NP-based delivery systems also offer a potential solution to the challenges of genetic engineering by enabling direct germline modifications in plant tissues [104]. However, integrating these systems requires substantial initial investments, posing a financial barrier for many stakeholders. Currently, the economic return on these investments is often insufficient to offset the high development and implementation costs, limiting widespread commercial adoption [105]. As technology advances and production scales up, costs are expected to decrease. Additionally, integrating nanotechnology with existing agricultural practices has the potential to enhance efficiency and reduce the need for inputs such as fertilizers and pesticides, ultimately contributing to long-term cost savings [15, 106].

While nano-genetic systems hold significant potential for transforming agriculture, their cost and scalability are currently limited by high initial investments, regulatory challenges, and technical barriers. To achieve sustainable agricultural practices, research and development efforts must focus on addressing these scientific gaps while ensuring that nanoparticle integration does not disrupt existing production systems [97, 107].

3.4 Biosafety Concerns

In genetic engineering, nano-delivery presents biosafety limitations, with one of the main concerns being the potential for horizontal gene transfer (HGT). HGT refers to the movement of genetic information between different species, influencing their genetic, physiological, and ecological traits. While this process can drive biotechnological innovation, it also poses severe adverse effects [61, 108]. The use of NPs to deliver modified genes offers advantages in producing transgenic plants by enabling the passive administration of biomolecules in a minimally invasive, species-independent manner [109]. However, it can also facilitate or inhibit genetic material transfer between organisms, potentially altering the genetic composition of non-target species and leading to unforeseen environmental consequences that are difficult to monitor and mitigate [110, 111].

One major risk associated with HGT is the unintended transfer of herbicide resistance genes to wild plant species or weeds, which could result in the emergence of "superweeds" that are difficult to control using conventional herbicides. This unintended gene flow may disrupt ecosystems and complicate agricultural practices [112]. Furthermore, NPs can interact with soil components, potentially increasing their toxicity and affecting microbial communities [113]. Genetic alterations in plants may lead to increased microbial density in soil by disrupting non-target microbial populations. The hybridization or horizontal gene transfer from transgenic plants to weeds or natural vegetative species can lead to herbicide tolerant weeds, thereby reducing agricultural productivity [114, 115]. Targeted gene delivery to organelles is crucial to preventing gene proliferation in weed relatives, given that NPs can penetrate the cell membrane and chloroplast envelope [116].

Another critical biosafety concern is the potential development of "super pests" due to unintended ecological effects. Nano-delivery systems designed to enhance pest resistance in crops may inadvertently exert selective pressure on pest populations, accelerating their adaptation and leading to pesticide resistant pests [117]. Additionally, continuous sublethal exposure to these agents may induce stress responses in insects, promoting detoxification and repairing mechanisms that contribute to resistance. This could undermine pest control strategies, increasing reliance on chemical pesticides and exacerbating environmental contamination [118]. Additionally, interactions between NPs and living organisms in agricultural ecosystems remain poorly understood, raising concerns about their long-term ecological impact and potential disruptions to natural food chain dynamics [60, 119]. Genetic interference methods have limitations, as non-target insect toxicity increases the risk of insect resistance following chronic exposure to RNA interference (RNAi) plants [120, 121].

It has been demonstrated that nanomaterials such as silver NPs, nano-alumina, and natural sphalerite NPs promote horizontal gene transfer of antibiotic resistance genes (ARGs) among bacteria. This occurs primarily due to oxidative stress induction, membrane damage, and genetic expression changes related to conjugative transfer, potentially enhancing ARG spread in the environment [122, 123].

Further research is imperative to understand the long-term effects and efficacy of nanomaterials in agricultural environments [119, 124]. Moreover, regulatory frameworks and guidelines for the use of nanotechnology in agriculture are not well established, posing a challenge for its adoption [76, 125–127]. Figure 1 shows

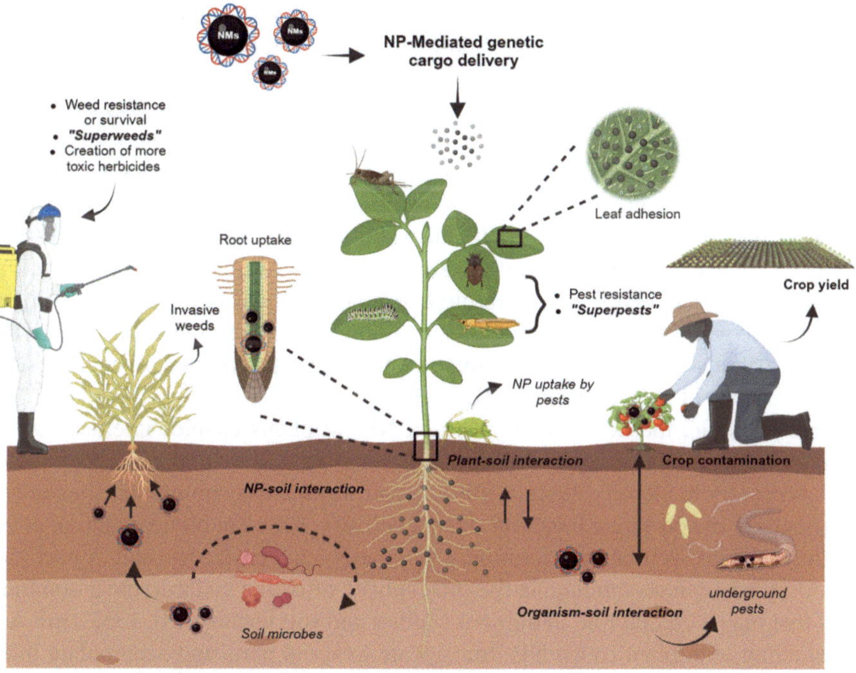

Fig. 1 Application of NP-mediated genetic cargo delivery in plants

examples of "superweeds" and "superpests" due to genetic engineering nano-delivery.

3.5 Lack of Long-Term Studies

Although nanomaterials show great promise for gene delivery in plants by overcoming the limitations of conventional methods, the long-term effects of their use remain unclear. Some studies using magnetic nanomaterials in cotton and maize pollen have demonstrated stable integration and inheritance of exogenous DNA. In a study involving Fe_3O_4 nanocomposites combined with MOS_2, their ability to inhibit the transfer of antibiotic resistance genes (ARGs) was demonstrated, offering potential strategies to manage risks that are crucial to mitigating the environmental impact of nanomaterials in agricultural settings [128]. However, further research is needed to identify integration sites, and border sequences, and to understand the underlying mechanisms of these processes [129].

The lack of comprehensive long-term studies on nano-delivery systems in genetic engineering poses risks to soil, plants, and human health [130]. A key uncertainty is how these NPs penetrate, absorb, and internalize in plants [15, 79, 131, 132]. Current methods for monitoring gene transfer facilitated by nanomaterials in transgenic plants are not highly sensitive, making it difficult to assess their environmental impact effectively. This has implications for regulatory agencies responsible for ensuring the safety of genetically modified organisms (GMOs) [133].

A major concern associated with these technologies is the potential accumulation of NPs in soil and plant tissues, which over time could lead to new types of environmental waste. The behavior of these materials in the environment, including their potential to contaminate ecosystems, is not yet fully understood [120, 134]. The use of nanomaterials in agriculture may introduce unintended ecological risks, such as the spread of ARGs in soil and water systems, facilitating the transfer of resistance genes among bacterial genera in wastewater treatment and agricultural runoff [120, 122]. This occurs through leaching, which permeates the soil and reaches the water supply, thereby exposing humans to these compounds. Given their high distribution and bioaccumulation potential, long-term exposure could result in chronic and acute health effects [70, 120].

For example, the delivery of spherical NPs in plants enables interactions with plant cells, leading to entry and absorption processes that are not comprehended yet. It is necessary to identify the transmission pathways and accumulation sites of NPs, particularly in non-target locations during delivery. Efficient biodegradation of NPs within cells is essential to determining their safety and environmental impact [50].

Also, NPs can accumulate in the rhizosphere, affecting soil microbial communities that play a crucial role in nutrient cycling and soil fertility [70, 135, 136]. After interacting with plant roots, NPs translocate to aerial parts and accumulate in cellular or subcellular organelles such as vacuoles and chloroplasts. Some studies have shown that mesoporous silica NPs, carbon nanotubes (CNTs), quantum dots, and metal/metal oxide NPs exhibit phytotoxicity through vascular blockage, oxidative stress, or structural DNA damage [52, 137]. For example, CuO, ZnO, and Ag NPs accumulate in the form of metallic ions [136, 138–141]. Several studies indicate that NPs influence crop yield, growth, and development, altering photosynthesis, inducing oxidative and antioxidant stress, and disrupting the nutritional content of edible crops [70, 134, 140, 141].

It is important to note that the toxicity of nanoformulations is influenced by their size, shape and chemical structure [142]. The development of agricultural and food-related nanoproducts that have not been thoroughly evaluated may threaten soil microbiota and enter the food chain through plants, ultimately affecting the health of living organisms via different mechanisms of action [143, 144]. The toxic effects of NPs on soil microflora disrupt enzymatic activities and microbial physiological processes. Once absorbed by microbial cells, NPs release ions that interact with DNA and other cellular biomolecules according to the uptake by roots and soil, leading to alterations in protein expression, membrane stability, and nutrient metabolism [144, 145].

While nanotechnology promises to revolutionize agriculture and genetic engineering, it is crucial to address the associated risks to establish guidelines that ensure the responsible use of nano-delivery in genetic engineering. This includes understanding nanotoxicity, managing occupational and environmental risks, and developing robust regulatory frameworks to ensure safe and sustainable application [146, 147]. Figure 2 shows the uptake of NPs into plant-soil systems allowing entry through the food chain.

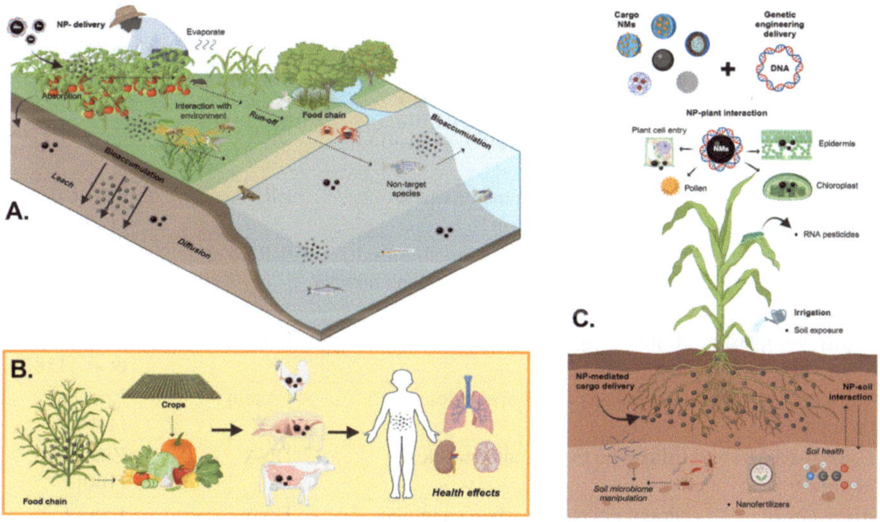

Fig. 2 Risks of nano-delivery systems in genetic engineering: (**a**) bioaccumulation in the environment, (**b**) toxic effects through the food chain and (**c**) interaction with soil components and cytotoxicity in plant cells

4 Case Studies and Examples

To date, the use of nanotechnology to deliver intracellular cargoes for plant genetic engineering is a reality. Carbon nanotubes, nanoparticles, and carbon quantum dots have been used to deliver genetic material into plant cells, enabling transient and stable genetic transformation (Table 4). Moreover, reports show that nanoparticles are also a feasible vehicle to distribute bigger molecules, such as proteins, into plant cells [163], which opens the possibility of nano-driven genome editing in plants. Technologies such as Zing Finger Nucleases (ZFNs), Transcription Activator-Like Effector Nucleases (TALENs), and Clustered Regularly Interspaced Short

Table 4 Nanotechnology applications in genetic engineering

Target crop/tissue	Nanomaterial	Cargo	References
Nicotiana benthamiana	Carbon nanotubes	siRNA	Demirer et al. [148]
Nicotiana benthamiana	Amine-functionalized mesoporous silica nanoparticles	dsRNA	Sangwan et al. [149]
Arabidopsis thaliana	Zeolitic imidazolate framework-8 NPs	Plasmid DNA	Yu et al. [150]
Oryza sativa	PEI functionalized carbon dots	Plasmid DNA	Wang et al. [151]
Nicotiana benthamiana *Solanum lycopersicum*	PEI functionalized carbon dots	siRNA	Schwartz et al. [152]
Solanum lycopersicum	Layered double hydroxide nanoparticles	dsRNA	Molesini et al. [153]
Arabidopsis thaliana	Virus like nanocarriers	Singlestranded DNA	Islam et al. [154]
Arabidopsis thaliana	PEI functionalized gold nanoparticles	siRNA	Qi et al. [155]
Nicotiana benthamiana	Mesoporous silica nanoparticles	siRNA	Cai et al. [156]
Nicotiana benthamiana	Chitosan quaternary ammonium salt, amine functionalized silica nano powder, and carbon quantum dots	dsRNA	Xu et al. [157]
Citrus protoplasts	Cationic lipid nanoparticles	DNA	Mahmoud et al. [158]
Tobacco protoplasts	Cationic lipid nanoparticles	Protein	Liu et al. [159]
Gossypium hirsutum	Magnetic Fe_3O_4 nanoparticles	DNA	Zhao et al. [160]
Arabidopsis thaliana mitochondria	Polymer coated carbon nanotubes (SWNT-PM)	DNA	Law et al. [161]
Solanum lycopersicum	Sheet-like clay nanoparticles deliver	RNA	Yong et al. [162]

Palindrome Repeats (CRISPR CAS9), which allow the modification of specific target genes, have already been delivered into cells using nanocarriers [164]. Therefore, it is reasonable to consider that the use of nanotechnology as a tool for plant biotechnology will only expand in the near future.

In this context, we can address significant ethical considerations around the use of nanobiotechnology in plant genetic engineering with hypothetical case studies.

4.1 Drought-Resistant Crops: The Maize Case

Since their advent in 1980, genetically engineered crops have been adopted in more than 75 countries [165]. Maize, one of the most cultivated crops worldwide, was, not surprisingly, one of the first crops to be genetically modified, with the introduction of its first transgenic line in 1996 [166]. Since then, considerable progress has been made in maize breeding, and recently, genetic editing technologies, such as CRISPR/CAS9, have been used to obtain drought-resistant lines [167]. Although genetic transformation is differentiated from genetic editing, mainly because the second does not insert foreign genes into the genome, the designation of Genetically Modified Organism (GMO) is still not homogeneous due to the differences in legislation around the globe. This disparity raises an initial concern, which attains equity and fairness. In addition, the use of nanotechnology for genetic transformation has no specific legislation, further complicating this issue.

The fact that some countries limit the use of genetic engineering is denying their populations access to potentially beneficial technologies. Divergent regulations can create trade barriers and economic disparities. Countries with strict GMO regulations may impose import restrictions on GMO products from countries with more lenient regulations, affecting global trade [168]. This can disadvantage farmers and industries in developing countries that might benefit from the use of nano-based genetic engineering but are unable to export their products to markets with stricter regulations. Moreover, differences among regulations can ultimately affect the level and quality of information provided to the final consumers, undermining public trust in science and regulatory bodies and raising significant concerns about consumer rights. The case of maize helps expose this issue, as in EU countries, GMO labeling is mandatory, allowing consumers to make informed choices, while in the US, labeling is not permitted [169]. This disparity fosters the misuse of labeling for conducting commercial campaigns that may lead to actual disinformation [170].

Another relevant discussion points winds up around economic inequality. The potential benefits of nano-enabled plant biotechnology, such as drought-resistant maize, may not be equally accessible to all farmers, particularly smallholders in drought-prone regions [171]. Most genetically engineered plant seeds, especially those based on the newest technologies, such as nanotechnology, are produced only by a few transnational enterprises in developed countries. This oligopolistic control over highly profitable genetic transformation technologies raises moral questions about whether these innovations truly serve the common good and benefit humanity

[168]. Small farmers in developing countries, who are often the most vulnerable to climate change and food insecurity, may find it difficult to access these advanced technologies due to high costs, intellectual property restrictions, and lack of infrastructure. This creates a paradox: While nano-enabled plant biotechnology has the potential to address critical challenges such as drought resistance and food security, its benefits may not reach those who need it most. This inequity highlights the need for international cooperation, equitable technology transfer mechanisms, and policies that ensure small farmers in drought-prone regions can access and benefit from these innovations. Without such measures, the promise of nano-enabled plant biotechnology risks exacerbating existing inequalities rather than alleviating them.

4.2 Pest Resistant Crops: The Cotton Case

Pests are another factor that significantly limits crop productivity. The application of nanotechnology in genetic engineering has opened new avenues for developing crops with enhanced traits, such as pest resistance [124]. Cotton, a crop of immense global importance, has been a prime candidate for such innovations, with recent advancements enabling the creation of pest-resistant varieties through precise gene-editing techniques [172]. These developments appear promising for reducing pesticide use and increasing agricultural productivity; they also raise significant ethical concerns related to their potential environmental impacts. The unintended consequences of implementing nano-enabled genetic engineering for breeding pest-resistant cotton could threaten biodiversity and alter ecosystems in ways that are not yet fully understood [173].

One of the most pressing environmental concerns is the potential loss of habitat for species that depend on pests for survival. Pest-resistant cotton, designed to suppress or eliminate specific insect populations [174], could inadvertently harm non-target organisms that rely on these pests as a food source or as part of their ecological niche [175]. For example, birds, bats, and beneficial insects that feed on cotton pests may face reduced food availability, leading to population declines and cascading effects in the ecosystem [176]. This disruption of food chains could result in biodiversity loss and the destabilization of ecosystems, raising ethical questions about the trade-offs between agricultural efficiency and environmental preservation.

Another significant environmental risk is the potential for gene flow from genetically edited cotton to wild relatives or non-target species [177]. If pest-resistant traits are transferred to wild cotton populations or other plants, it could lead to the emergence of "superweeds" or other unintended ecological consequences [178]. This risk is particularly concerning in regions where wild cotton relatives grow near cultivated fields, as the unintended spread of pest-resistant traits could have far-reaching and irreversible impacts on natural ecosystems.

Additionally, the widespread adoption of nano-enabled pest-resistant cotton could lead to the development of resistance in pest populations over time. Just as the overuse of chemical pesticides has led to the evolution of resistant pests, the continuous exposure of insect populations to pest-resistant cotton could drive the emergence of resistant strains [179]. This would not only undermine the effectiveness of the technology but could also require additional control measures, potentially leading to increased chemical pesticide use or the development of even more extensive genetic modifications. Such a scenario would create a cycle of escalating interventions, further exacerbating ethical dilemmas.

The potential for nano-enabled genetic editing to alter soil health and microbial communities is another area of concern. The introduction of genetically edited cotton plants into agricultural systems could affect soil microbiota, which play a critical role in nutrient cycling, plant health, and ecosystem stability [180]. Changes in microbial communities could have cascading effects on soil fertility, plant growth, and the broader ecosystem, potentially leading to long-term environmental degradation. These impacts highlight the need for comprehensive studies to assess the effects of nanotechnology-enabled genetic editing on soil ecosystems before widespread adoption [181].

Finally, the ethical implications of these environmental risks are compounded by the lack of robust regulatory frameworks to address them. Current regulations on genetically edited crops, particularly those developed using nanotechnology, are often inconsistent or inadequate, leaving gaps in the assessment and mitigation of environmental impacts [168]. Without stringent oversight and precautionary measures, the deployment of pest-resistant cotton could proceed without fully understanding or addressing its potential ecological consequences. This regulatory uncertainty stresses the need for a precautionary approach that prioritizes environmental protection and ethical responsibility.

4.3 Nutrient-Enriched Crops

A balanced diet requires an adequate combination of macronutrients and micronutrients. Nutritional deficiencies, defined as the insufficient intake of one or more nutrients, impair key physiological functions and increase the risk of diseases. In severe cases, they can lead to growth delays, cognitive development disorders, and increased susceptibility to infections. These effects can be mitigated through dietary strategies, supplementation, or biofortification of staple foods [182].

A recent study by Han analyzed the global prevalence of vitamin A, iron, and iodine deficiencies between 1990 and 2019 [183]. Although a downward trend was observed in most regions, critical challenges persist more than 2 billion people are still at risk of iodine deficiency, and vitamin A deficiency remains high in South Asia and Sub-Saharan Africa. Iron deficiency, although reduced, still affects vulnerable populations such as women and children in developing countries. However, the study did not address other relevant deficiencies (e.g., zinc, calcium, folate) due to a lack of reliable data.

Biofortification is a strategy aimed at increasing the density of vitamins and minerals in crops through various cultivation techniques, genetic engineering, and agronomic practices [184]. This alternative holds great potential for reaching the most vulnerable populations, those who do not have access to supplements or fortified foods, by providing a direct source of essential nutrients.

An illustrative case study is that of vitamin A biofortified maize. Through the HarvestPlus program, implemented by the Consultative Group on International Agricultural Research (CGIAR), conventional genetic improvement techniques have been used to achieve provitamin A carotenoid concentrations that cover up to half of the average requirement of the population [185]. Additionally, genetic sequences associated with higher concentrations of these carotenoids have been identified, which has allowed for an 18-fold increase in their content in certain crosses. Key loci, such as the crtRB1 gene, influencing β-carotene concentration and conversion in the grain, have also been determined. Consequently, phenotypic selection strategies and marker-assisted selection have been implemented to optimize these traits, significantly contributing to the improvement of maize nutritional quality [186].

The selection of maize varieties and their crossing can be a very time-consuming process. One of the most notable disadvantages of conventional breeding, compared to transgenic strategies, is its dependence on alleles already present in the species' genetic pool [187]. However, its main advantage lies in relying on the intrinsic properties of the crop, simplifying the process and leading to significantly lower regulatory requirements. Recently, certain quantitative trait loci (QTLs) have been identified that can be used for crop biofortification.

The identification of QTLs in wheat, through the analysis of 188 recombinant lines obtained from crossing a high-yielding bread wheat (HD2932) and a synthetic hexaploid wheat (Synthetic 46), led to the detection of nine genomic regions associated with grain protein content (GPC), iron (GFeC), zinc (GZnC), and thousand-kernel weight (TKW). Two stable and co-localized QTLs on chromosome 4B stand out, reinforcing their potential for marker-assisted selection (MAS). Additionally, in silico analysis revealed candidate genes involved in iron homeostasis, zinc transport, nutrient mobilization, and grain size regulation. These findings are crucial for biofortification strategies, as they allow the use of genetic engineering and nanotechnology to optimize crop nutritional quality [188].

On the other hand, a meta-analysis of 155 QTLs from 26 studies consolidated 57 MQTLs, significantly reducing confidence intervals and facilitating the precise localization of genomic regions linked to zinc content. These regions, enriched with genes regulating metal homeostasis, have revealed superior haplotypes that increase zinc in the grain by approximately 2 mg/kg. This knowledge supports its application in marker-assisted selection programs and its integration into genetic engineering and nanotechnology strategies for developing biofortified rice varieties, crucial for combating malnutrition [189].

The combination of nanotechnology and genetic engineering opens new possibilities in crop biofortification. Carbon nanotubes, for example, have proven to be an effective tool for delivering genetic material without integrating exogenous DNA into the plant genome. Unlike traditional techniques like Agrobacterium-mediated transformation, carbon nanotubes allow the efficient diffusion of plasmids and linear amplicons in various species, protecting genetic material from degradation and promoting transient protein expression. When combined with technologies like CRISPR-Cas9, this platform allows permanent genome modifications without the regulatory challenges faced by genetically modified organisms, paving the way for genetic engineering of crops with a direct impact on food security [109].

In addition to improving the concentration of essential nutrients such as iron, the application of iron nanoparticles (Fe) has shown promising results in crop biofortification. Studies in plants such as Cucurbita maxima and wheat have demonstrated that iron nanoparticles not only increase the concentration of this mineral but also improve nutrient translocation to other parts of the plant, enhancing its absorption. This technique has been particularly effective when applied foliar, allowing direct absorption through the leaves. Although these advances are promising, a greater understanding of the factors influencing the absorption and translocation of these nanoparticles is required to optimize their use on a larger agricultural scale [4].

- **Ethical issue**: Prioritization of such technologies for human benefit over environmental sustainability.

One of the main aspects to consider before implementing food fortification strategies and supplementation programs, whether through genetic improvement, variety selection, or nanotechnology, is the evaluation of existing interventions in each region. This helps to avoid program overlap and prevent over-supplementation. A clear example of this issue is reported by Tanumihardjo, who, in a study with 133 Zambian children, found that 59% were diagnosed with hypervitaminosis A and 16% showed biomarkers of toxicity. Similarly, in South Africa, 72.5% of 40 children with frequent liver consumption showed elevated biomarkers of toxicity [190]. These situations resulted from the combination of mandatory fortified foods and high coverage of supplementation in preschool-aged children, highlighting the importance of careful planning when implementing these nutritional strategies.

Another critical aspect to consider before implementing supplementation and crop biofortification strategies is the potential interaction between increased micronutrients and macronutrients in food and preexisting health conditions in the population. A clear example of this interaction is iron supplementation, which can promote the growth of intestinal pathogens and consequently increase the risk of infectious diseases, especially in infants. For this reason, iron supplementation should be strictly administered and only to individuals with confirmed deficiencies of this mineral [191].

Beyond the, it is essential to contextualize the strategies for developing nutrient-enriched crops from a holistic perspective. Although genetic advancements and current nanotechnology strategies are encouraging, they must be integrated into a comprehensive biofortification effort that not only includes technological improvement but also key aspects such as seed development and distribution infrastructure, social acceptance, dietary habits, and actual nutritional impact. Only through this multifaceted approach can sustainable and widespread benefits for population health be achieved [192].

5 The Role of Stakeholders and Strategies for Ethical and Sustainable Adoption

5.1 Role of Scientists and Innovators

While some organizations focus primarily on developing and promoting nanotechnology, others are dedicated to ensuring that the materials used are safe for both human health and the environment [69]. The advancement of nanotechnology in agriculture relies on the combined efforts of researchers, developers, regulatory bodies, industry experts, and consumers, each contributing in distinct ways to support innovation, address challenges, and promote sustainable practices (Fig. 3) [193, 194].

Researchers and developers are responsible for advancing scientific and technical research in nano-agrifoods. However, they face several challenges, including limited data availability, insufficient product oversight, and collaboration barriers, all of which can hinder progress in responsible innovation and the implementation of nanotechnology in agriculture [195]. Regulatory bodies contribute by addressing safety and regulatory concerns, ensuring that nanotechnology applications are safe for both the environment and human health. Their efforts are crucial for gaining consumer acceptance and improving the marketability of nano-enabled agricultural products [125, 146]. Industry experts help overcome the high initial production

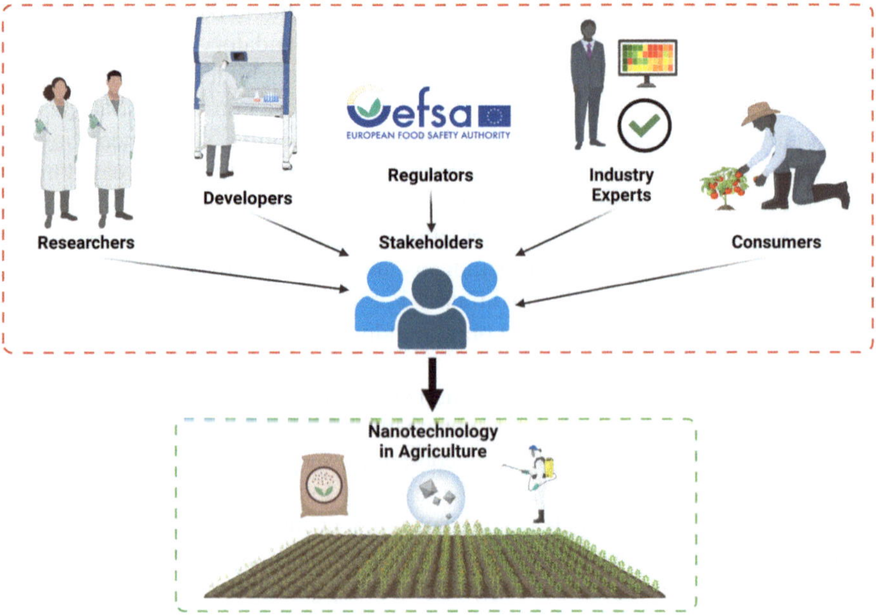

Fig. 3 Key stakeholders in the advancement of nanotechnology in agriculture

costs of nano-enabled agricultural products while ensuring their successful entry into the market. By translating research findings into practical field applications, they help address economic barriers that could otherwise slow the adoption of nanotechnology in agriculture [125, 196] Strengthening collaboration among these groups is essential to overcoming existing barriers and aligning efforts with the principles of responsible innovation. Open communication can enhance understanding of potential risks and support strategies to mitigate them effectively [146, 195].

Beyond technical and economic aspects, ethical considerations are also critical in the development and application of nanotechnology in agriculture. Balancing innovation with safety, regulation, and societal concerns is necessary, particularly given the potential toxicity and environmental impact of nanomaterials. Ensuring the safety of agricultural workers and ecosystems requires thorough evaluation and oversight [14, 126, 132]. Regulatory efforts are key to promoting responsible innovation (RI) in nano-agrifoods, not only by ensuring product safety and efficacy but also by addressing ethical, legal, and social implications (ELSI) [197]. Although nanotechnology offers transformative potential for agriculture, concerns persist regarding its ethical implications and the need for additional regulatory measures to reduce risks to human health and the environment [69].

Public perception and acceptance are equally important in determining the success of nanotechnology in agriculture [195]. Engaging consumers in dialogue and maintaining transparency about both the benefits and risks of this technology can help build trust and support [92]. The public often perceives nanotechnology as greater risks than experts do, with scientists emerging as the most trusted group. Policymakers should consider these differences in risk perception and trust when developing strategies to improve communication [198]. Efforts to improve public awareness and acceptance must include clear explanations about the safety and performance of nanomaterials. Transparent communication plays a central role in addressing public concerns and promoting informed decision-making regarding nanotechnology use in agriculture [88].

The success of nanotechnology in agriculture ultimately depends on a coordinated effort among all involved stakeholders. By addressing barriers, ensuring safety, and fostering collaboration, these collective efforts can help unlock nanotechnology's potential to revolutionize agriculture in a sustainable manner.

5.2 Policymakers and Regulators

Nanotechnology in agriculture offers significant potential for improving productivity and sustainability [110]. However, its integration into agricultural practices requires well-defined regulatory frameworks to ensure consumer and environmental safety [80]. Comprehensive risk assessments are essential to better understand the environmental and human health impacts of nanomaterials used in agriculture [132]. Research efforts must focus on evaluating the fate, transport, and bioavailability of nanoparticles in the environment, as well as their potential toxicity to plants, soil, and soil microbiomes [85]. Limited knowledge about the long-term effects of these materials highlights the need for further investigation to inform evidence-based regulations [13].

Globally, several countries are working to refine their regulatory frameworks to address the challenges posed by nanotechnology in agriculture. The European Union (EU) and Switzerland have introduced nano-specific provisions within existing legislation, setting an example for other regions where nanomaterials are regulated indirectly through industry guidance [80, 84]. The European Food Safety Authority (EFSA) has played an important role in reviewing nanomaterial applications in the agri-food sector to ensure compliance with safety standards [199]. In contrast, the U.S. regulatory framework for nanotechnology in agriculture remains underdeveloped, facing challenges in addressing the unique risks associated with these technologies [13]. Although progress has been made, current efforts in the U.S. primarily focus on mitigating safety, environmental, and health risks [88]. Across various regions, more comprehensive research is needed to improve understanding of nanotechnology's long-term environmental impacts and to develop regulations based on scientific evidence [13].

The current state of regulatory frameworks reflects ongoing efforts to address global challenges and establish future directions for nano-engineering in agriculture [81, 83]. Despite its promising applications, concerns persist regarding nanomaterial interactions with soil and plants, their behavior in the environment, and potential risks to agricultural workers [200]. Existing regulations remain insufficient to manage these complex risks, underscoring the need for a transdisciplinary approach that integrates diverse scientific fields within regulatory frameworks [201]. Policymakers are tasked with developing regulations that support the advancement of nanotechnology while ensuring environmental and consumer protection [79] or nano-based products to gain approval, they must demonstrate safety through rigorous testing and adherence to international safety standards [80]. Achieving this balance requires regulatory frameworks grounded in robust scientific research to ensure responsible innovation in agriculture.

5.3 Farmers and Consumer Participation

Nanomaterials are increasingly used in agriculture to enhance crop production and protection by improving the efficiency of inputs such as fertilizers and pesticides [202]. Their ability to enable the controlled release of agrochemicals helps reduce environmental pollution while maximizing the effectiveness of farming practices [203]. Despite these benefits, farmers and consumers have expressed concerns about the use of nanomaterials in agriculture, Table 5 presents some of the primary concerns associated with their application.

Although nanotechnology shows considerable promise for improving agricultural productivity and sustainability, uncertainties persist regarding its safety,

Table 5 Key concerns regarding nanomaterial use in agriculture

Concern	Description	Stakeholder	References
Safety and toxicity	Limited knowledge about adverse effects on humans, ecosystems, and agro-system component	Farmers and consumers	Iavicoli et al. [132], Kaphle et al. [204], Singh et al. [79]
Environmental impact	Uncertainty about the fate, mobility, and ecological toxicity of nanomaterials, as well as their long-term impact on soil quality and plant growth	Farmers and consumers	Saleem and Zaidi [205], Tripathi et al. [206]
Regulatory and management challenges	Need for clear regulations regarding exposure levels, physicochemical features affecting toxicity, and occupational safety practices	Farmers and regulatory institutions	Kaphle et al. [204], Singh et al. [207]
Uncertainties in application	Lack of understanding about the mechanisms of nanomaterial penetration and uptake in crops, raising concerns about their safety and effectiveness	Farmers	Saleem and Zaidi [205], Singh et al. [79]
Research and commercialization	Further studies are needed to understand nanomaterials' behavior in agricultural environments and their long-term effects	Researchers and companies	Saleem and Zaidi [205], Usman et al. [126]
Consumer acceptance	Public perception plays a crucial role in the adoption of these technologies; clear communication of benefits and safety is necessary	Consumers	Kah et al. [92], Parisi et al. [105]

environmental impact, and regulatory oversight. Addressing these issues through comprehensive research and policy development is essential to ensure the responsible integration of nanotechnology in agriculture.

6 Future Directions and Conclusions

Nano-delivery systems for crop nutrition represent a significant advancement in sustainable agriculture. By enhancing nutrient efficiency and reducing environmental impact, these systems offer a promising alternative to conventional fertilizers. Continued research and development in this field are essential to fully realize their potential and address existing challenges in agricultural practices.

The potential of nanomaterials in plant genetic manipulation raises intriguing questions about the definition of genetically engineered organisms, especially when high-efficiency protein delivery is achieved without DNA integration or using pathogens. This could significantly reshape the landscape of genetically modified organisms.

Long-term monitoring programs for nano-genetically modified crops are essential to evaluate their safety and efficacy, addressing concerns about environmental impacts and regulatory gaps [208]. To establish effective programs, it is important to consider:

1. *Comprehensive Risk Assessment:* Conduct thorough investigations into the potential toxic effects of nanoparticles on non-target organisms and the environment [209]. This includes assessing the long-term environmental impacts on soil, water, and overall environmental and human health to inform evidence-based regulations.
2. *Regulatory Frameworks:* Develop and implement suitable regulatory frameworks to handle nanotechnologies in agriculture, feed, and food, ensuring the safe use of nano-based products without endangering consumers and the environment [13].
3. *Indicator Species:* Identify relevant indicator species to monitor the long-term impacts of nano-genetically modified crops and the application of corresponding herbicides on farmland biodiversity [210].
4. *Advanced Detection Methods:* Employ nanosensors utilizing quantum dots and plasmonic nanoparticles for early and accurate detection of plant pathogens, which is crucial for timely disease management [209].
5. *Balanced Approach:* Prioritize safety, sustainability, and environmental stewardship in the development and application of nanotechnology in agriculture.

As a result, advancements in nano-based agriculture must go beyond technological innovation and include the creation of policies that balance the immense potential of these technologies with the establishment of frameworks to assess their safety and, where appropriate, regulate their application. This dual focus on progress and regulation will be crucial for the responsible development of nano-based agricultural practices.

References

1. Ahmar S, Mahmood T, Fiaz S, Mora-Poblete F, Shafique M, Chattha MS, Jung K-H (2021) Advantage of nanotechnology-based genome editing system and its application in crop improvement. Front Plant Sci 12:663849. https://doi.org/10.3389/fpls.2021.663849
2. Wu K, Xu C, Li T, Haijie, Gong J, Li X, Sun X, Hu X (2023) Application of nanotechnology in plant genetic engineering. Int J Mol Sci 24:14836. https://doi.org/10.3390/ijms241914836
3. Yan Y, Zhu X, Yu Y, Li C, Zhang Z, Wang F (2021) Nanotechnology strategies for plant genetic engineering. Adv Mater 34:e2106945. https://doi.org/10.1002/adma.202106945
4. Demirer G, Silva T, Jackson C, Thomas J, Ehrhardt D, Rhee S, Mortimer J, Landry M (2021) Nanotechnology to advance CRISPR–Cas genetic engineering of plants. Nat Nanotechnol 16:243–250. https://doi.org/10.1038/s41565-021-00854-y
5. Vats S, Kumawat S, Brar J, Kaur S, Yadav K, Magar S, Jadhav P, Salvi P, Sonah H, Sharma S, Deshmukh R (2022) Opportunity and challenges for nanotechnology application for genome editing in plants. Plant Nano Biol 1:100001. https://doi.org/10.1016/j.plana.2022.100001

6. Ghani M, Iqbal A, Ghani H, Bibi S, Wang Z, Pei R (2023) Recent advances in nanocomposite-based delivery systems for targeted CRISPR/Cas delivery and therapeutic genetic manipulation. J Mater Chem B 11(24):5251–5271. https://doi.org/10.1039/d2tb02610d
7. Sharma B, Chauhan I, Kumar G, Bhardwaj K, Tiwari RK (2024) Nanoparticle carriers: a new era of precise CRISPR/Cas9 gene editing. MicroRNA. https://doi.org/10.2174/0122115366319848241022092805
8. Kumar A, Choudhary A, Kaur H, Mehta S, Husen A (2021) Smart nanomaterial and nanocomposite with advanced agrochemical activities. Nanoscale Res Lett 16:156. https://doi.org/10.1186/s11671-021-03612-0
9. Rasheed A, Li H, Tahir M, Mahmood A, Nawaz M, Shah A, Aslam M, Negm S, Moustafa M, Hassan M, Wu Z (2022) The role of nanoparticles in plant biochemical, physiological, and molecular responses under drought stress: a review. Front Plant Sci 13. https://doi.org/10.3389/fpls.2022.976179
10. Shah M, Shahnaz T, Zehab Ud D, Masoodi J, Nazir S, Qurashi A, Ahmed GH (2024) Application of nanotechnology in the agricultural and food processing industries: a review. Sustain Mater Technol 39:e00809. https://doi.org/10.1016/j.susmat.2023.e00809
11. Zhao L, Zhou X, Kang Z, Peralta-Videa J, Zhu Y-G (2024) Nano-enabled seed treatment: a new and sustainable approach to engineering climate-resilient crops. Sci Total Environ 910:168640. https://doi.org/10.1016/j.scitotenv.2023.168640
12. Rahim U, Qaswar M, Uddin M, Giannini C, Herrera M, Rea G (2021) Nano-enable materials promoting sustainability and resilience in modern agriculture. Nanomaterials 11:2068. https://doi.org/10.3390/nano11082068
13. Santos PA, Biraku X, Nielsen E, Ozketen A, Ozketen AA, Hakki E (2025) Agricultural nanotechnology for a safe and sustainable future: current status, challenges, and beyond. J Sci Food Agric 105(6):3159–3169. https://doi.org/10.1002/jsfa.13922
14. Behl T, Kaur I, Sehgal A, Singh S, Sharma N, Bhatia S, Al-Harrasi A, Bungau S (2022) The dichotomy of nanotechnology as the cutting edge of agriculture: nano-farming as an asset versus nanotoxicity. Chemosphere 288:132533. https://doi.org/10.1016/J.CHEMOSPHERE.2021.132533
15. Ali S, Al-Tohamy R, Koutra E, Moawad M, Kornaros M, Mustafa A, Mahmoud Y, Badr A, Osman M, Elsamahy T, Jiao H, Sun J (2021) Nanobiotechnological advancements in agriculture and food industry: applications, nanotoxicity, and future perspectives. Sci Total Environ 792:148359. https://doi.org/10.1016/j.scitotenv.2021.148359
16. Buchner J, Menrad K, Decker T (2025) Public acceptance of green hydrogen production in Germany. Renew Sust Energ Rev 208:115057. https://doi.org/10.1016/j.rser.2024.115057
17. Germani F, Kellmeyer P, Wäscher S, Biller-Andorno N (2021) Engineering minds? Ethical considerations on biotechnological approaches to mental health, well-being, and human flourishing. Trends Biotechnol 39(11):1111–1113. https://doi.org/10.1016/j.tibtech.2021.04.007
18. Maryamchik E, Ikonomou L, Roxland BE, Grignon F, Levine BL, Grilley BJ (2025) International Society for Cell & Gene Therapy Expanded Access Working Group position paper: key considerations to support equitable and ethical expanded access to investigational cell- and gene-based interventions. Cytotherapy 27(6):671–677. https://doi.org/10.1016/j.jcyt.2025.01.016
19. White GB (2020) Chapter 39—Ethical considerations for biotechnology leaders. In: Shimasaki C (ed) Biotechnology entrepreneurship, 2nd edn. Academic, New York, pp 583–590. https://doi.org/10.1016/B978-0-12-815585-1.00039-5
20. D'Amato ME, Joly Y, Lynch V, Machado H, Scudder N, Zieger M (2024) Ethical considerations for forensic genetic frequency databases: first report conception and development. Forensic Sci Int Genet 71:103053. https://doi.org/10.1016/j.fsigen.2024.103053
21. Liu Y, Timo de Vries W, Zhang G, Cui X (2024) From tradition to smart: a comprehensive review of the evolution and prospects of land use planning tools. Heliyon 10(23):e40857. https://doi.org/10.1016/j.heliyon.2024.e40857

22. Zainab R, Hasnain M, Ali F, Abideen Z, Siddiqui ZS, Jamil F, Hussain M, Park Y-K (2024) Prospects and challenges of nanopesticides in advancing pest management for sustainable agricultural and environmental service. Environ Res 261:119722. https://doi.org/10.1016/j.envres.2024.119722
23. Boldt J, Orrù E (2022) Towards a unified list of ethical principles for emerging technologies. an analysis of four European reports on molecular biotechnology and artificial intelligence. Sust Futur 4:100086. https://doi.org/10.1016/j.sftr.2022.100086
24. Bearth A, Otten CD, Cohen AS (2024) Consumers' perceptions and acceptance of genome editing in agriculture: insights from The United States of America and Switzerland. Food Res Int 178:113982. https://doi.org/10.1016/j.foodres.2024.113982
25. Ahmad Z, Niyazi S, Firdoos A, Wang C, Manzoor MA, Ramakrishnan M, Upadhyay A, Ding Y (2024) Enhancing plant resilience: nanotech solutions for sustainable agriculture. Heliyon 10(23):e40735. https://doi.org/10.1016/j.heliyon.2024.e40735
26. Singh NB, Kumar B, Usman UL, Susan MABH (2024) Nano revolution: exploring the frontiers of nanomaterials in science, technology, and society. Nano-Struct Nano-Objects 39:101299. https://doi.org/10.1016/j.nanoso.2024.101299
27. Al-Mamun MR, Hasan MR, Ahommed MS, Bacchu MS, Ali MR, Khan MZH (2021) Nanofertilizers towards sustainable agriculture and environment. Environ Technol Innov 23:101658. https://doi.org/10.1016/J.ETI.2021.101658
28. Rovina K, Wen Xia FL (2025) Chapter 20—Nano-engineering approaches for food analysis and related biosensing applications. In: Ramamurthy SS, Bhaskar S, Reddy N (eds) Nano-engineering at functional interfaces for multi-disciplinary applications. Elsevier, Amsterdam, pp 491–510. https://doi.org/10.1016/B978-0-443-21691-6.00021-4
29. Sikiru S, Abioye KJ, Adedayo HB, Adebukola SY, Soleimani H, Anar M (2024) Technology projection in biofuel production using agricultural waste materials as a source of energy sustainability: a comprehensive review. Renew Sust Energ Rev 200:114535. https://doi.org/10.1016/j.rser.2024.114535
30. Knorr D, Augustin MA (2024) Food systems restoration. Sustain Food Technol 2(5):1365–1390. https://doi.org/10.1039/D4FB00108G
31. Xu H, Waheed A, Kuerban A, Muhammad M, Aili A (2025) Dynamic approaches to ecological restoration in China's mining regions: a scientific review. Ecol Eng 214:107577. https://doi.org/10.1016/j.ecoleng.2025.107577
32. Molahalli V, Shetty A, Sharma A, Bijapur K, Soman G, Hegde G (2023) Chapter 2—Risks and ethics of nanotechnology: an overview. In: Seena S, Rai A, Kumar S (eds) Nanoparticles and plant-microbe interactions. Academic, New York, pp 35–68. https://doi.org/10.1016/B978-0-323-90619-7.00014-X
33. Neukart F (2025) Toward the stars: technological, ethical, and sociopolitical dimensions of interstellar exploration. Space Policy 72:101676. https://doi.org/10.1016/j.spacepol.2024.101676
34. Kiran V, Harini K, Thirumalai A, Girigoswami K, Girigoswami A (2024) Nanotechnology's role in ensuring food safety and security. Biocatal Agric Biotechnol 58:103220. https://doi.org/10.1016/j.bcab.2024.103220
35. Raparelli E, Bajocco S, Mugnozza GS (2017) The perception of biotechnology in agroforestry: the opinion of undergraduates and researchers. Land Use Policy 66:364–373. https://doi.org/10.1016/j.landusepol.2017.05.015
36. Bedu-Addo K, Okofo LB, Ntiamoah A, Mensah H (2024) Pollution of water bodies and related impacts on aquatic ecosystems and ecosystem services: the case of Ghana's booming 'galamsey' industry. Heliyon 10(24):e40880. https://doi.org/10.1016/j.heliyon.2024.e40880
37. Gu MM, Huang CF, Lee C-KJ (2023) Investigating university students' digital citizenship development through the lens of digital literacy practice: a translingual and transemiotizing perspective. Linguist Educ 77:101226. https://doi.org/10.1016/j.linged.2023.101226
38. van Gevelt T, Zaman T, George F, Bennett MM, Fam SD, Kim JE (2020) End-user perceptions of success and failure: narratives from a natural laboratory of rural electrification

projects in Malaysian Borneo. Energy Sustain Dev 59:189–198. https://doi.org/10.1016/j.esd.2020.10.008
39. Wan N-F, Chen J, Jiang J-X, Li B (2017) A conceptual framework for ecosystem management based on tradeoff analysis. Ecol Indic 75:352–361. https://doi.org/10.1016/j.ecolind.2016.12.032
40. Appio FP, Hernandez CT, Platania F, Schiavone F (2025) AI narratives in fiction and media: exploring thematic parallels in public discourse. Technovation 142:103201. https://doi.org/10.1016/j.technovation.2025.103201
41. McConnell J, Tian Z, Yazan B (2023) Monolingual content-area teacher candidates' identity work in an online teacher education course. Linguist Educ 77:101228. https://doi.org/10.1016/j.linged.2023.101228
42. Sáenz-Hernández I, Petreñas C, Lapresta-Rey C, Ubalde J (2023) 'They speak Arabic to make teachers angry': high-school teachers' (de)legitimization of heritage languages in Catalonia. Linguist Educ 78:101232. https://doi.org/10.1016/j.linged.2023.101232
43. Laxman L, Ansari AH (2023) Chapter 4—Negotiating the safe use of biotechnology to conserve agrobiodiversity: issues and challenges impacting universal sustainability. In: Laxman L, Ansari AH (eds) Conservation policies for agricultural biodiversity. Woodhead Publishing, Sawston, pp 107–141. https://doi.org/10.1016/B978-0-12-821441-1.00004-X
44. Wahbeh H, Radin D, Yount G, Michael A Woodley of Menie, Sarraf MA, Karpuj MV (2022) Genetics of psychic ability—a pilot case-control exome sequencing study. Explore 18(3):264–271. https://doi.org/10.1016/j.explore.2021.02.014
45. Thoma N (2023) Language education policy and transnational and translingual social practices at schools. Commentary on the special issue. Linguist Educ 78:101200. https://doi.org/10.1016/j.linged.2023.101200
46. Jameel S (2023) Climate change, food systems and the Islamic perspective on alternative proteins. Trends Food Sci Technol 138:480–490. https://doi.org/10.1016/j.tifs.2023.06.028
47. Rashed MS, Abdelkarim EA, Elsamahy T, Sobhy M, El-Mesery HS, Salem A (2025) Advances in cell-based biosensors: transforming food flavor evaluation with novel approaches. Food Chem X 26:102336. https://doi.org/10.1016/j.fochx.2025.102336
48. Osmani Z, Wang L, Xiao W, Kulka M (2024) Nanomaterials as tools in plant transformation: a protoplast-centric perspective. Plant Nano Biol 10:100100. https://doi.org/10.1016/j.plana.2024.100100
49. Antu UB, Islam MS, Ahmed S, Arifuzzaman M, Saha S, Mitu PR, Sarkar AR, Mahiddin NA, Ismail Z, Ibrahim KA, Idris AM (2024) Emerging technologies for efficient water use in agriculture: a review of current trends and future directions. J Water Process Eng 68:106317. https://doi.org/10.1016/j.jwpe.2024.106317
50. Lv Z, Jiang R, Chen J, Chen W (2020) Nanoparticle-mediated gene transformation strategies for plant genetic engineering. Plant J 104:880–891
51. Ahmar S, Mahmood T, Fiaz S, Mora-Poblete F, Shafique MS, Chattha MS, Jung KH (2021) Advantage of nanotechnology-based genome editing system and its application in crop improvement. Front Plant Sci 12:663849. https://doi.org/10.3389/FPLS.2021.663849/PDF
52. Cunningham FJ, Goh NS, Demirer GS, Matos JL, Landry MP (2018) Nanoparticle-mediated delivery towards advancing plant genetic engineering. Trends Biotechnol 36(9):882–897. https://doi.org/10.1016/J.TIBTECH.2018.03.009
53. Sanford JC (1990) Biolistic plant transformation. Physiol Plant 79(1):206–209. https://doi.org/10.1111/J.1399-3054.1990.TB05888.X
54. Ozyigit II (2020) Gene transfer to plants by electroporation: methods and applications. Mol Biol Rep 47(4):3195–3210. https://doi.org/10.1007/S11033-020-05343-4
55. Gelvin SB (2017) Integration of agrobacterium T-DNA into the plant genome. Annu Rev Genet 51:195–217. https://doi.org/10.1146/ANNUREV-GENET-120215-035320/CITE/REFWORKS
56. Tzfira T, Citovsky V (2003) The agrobacterium-plant cell interaction. Taking biology lessons from a bug. Plant Physiol 133(3):943–947. https://doi.org/10.1104/PP.103.032821

57. Mahmood MA, Naqvi RZ, Rahman SU, Amin I, Mansoor S (2023) Plant virus-derived vectors for plant genome engineering. Viruses 15(2):531. https://doi.org/10.3390/V15020531
58. Shivashakarappa K, Marriboina S, Dumenyo K, Taheri A, Yadegari Z, Ghosh I, Roy AK, Kumar Kamaraj S (2025) Nanoparticle-mediated gene delivery techniques in plant systems. Front Nanotechnol 7:1516180. https://doi.org/10.3389/FNANO.2025.1516180
59. Mujtaba M, Wang D, Carvalho LB, Oliveira JL, Espirito Santo Pereira AD, Sharif R, Jogaiah S, Paidi MK, Wang L, Ali Q, Fraceto LF (2021) Nanocarrier-mediated delivery of miRNA, RNAi, and CRISPR-Cas for plant protection: current trends and future directions. ACS Agric Sci Technol 1(5):417–435. https://doi.org/10.1021/ACSAGSCITECH.1C00146/ASSET/IMAGES/LARGE/AS1C00146_0006.JPEG
60. Jat SK, Bhattacharya J, Sharma MK (2020) Nanomaterial based gene delivery: a promising method for plant genome engineering. J Mater Chem B 8:4165–4175
61. Niazian M, Molaahmad Nalousi A, Azadi P, Ma'mani L, Chandler SF (2021) Perspectives on new opportunities for nano-enabled strategies for gene delivery to plants using nanoporous materials. Planta 254:83
62. Garcia-Guerra A, Dunwell TL, Trigueros S (2018) Nano-scale gene delivery systems: current technology, obstacles, and future directions. Curr Med Chem 25(21):2448–2464. https://doi.org/10.2174/0929867325666180108100723
63. Mirón-Barroso S, Domènech EB, Trigueros S (2021) Nanotechnology-based strategies to overcome current barriers in gene delivery. Int J Mol Sci 22(16):8537. https://doi.org/10.3390/IJMS22168537
64. Degors IMS, Wang C, Rehman ZU, Zuhorn IS (2019) Carriers break barriers in drug delivery: endocytosis and endosomal escape of gene delivery vectors. Acc Chem Res 52(7):1750–1760. https://doi.org/10.1021/ACS.ACCOUNTS.9B00177/ASSET/IMAGES/LARGE/AR-2019-001776_0004.JPEG
65. Durymanov M, Reineke J (2018) Non-viral delivery of nucleic acids: insight into mechanisms of overcoming intracellular barriers. Front Pharmacol 9:398662. https://doi.org/10.3389/FPHAR.2018.00971/PDF
66. Yazdani S, Mozaffarian M, Pazuki G, Hadidi N, Villate-Beitia I, Zárate J, Puras G, Pedraz JL (2024) Carbon-based nanostructures as emerging materials for gene delivery applications. Pharmaceutics 16(2):288. https://doi.org/10.3390/PHARMACEUTICS16020288
67. Torres-Vanegas JD, Cruz JC, Reyes LH (2021) Delivery systems for nucleic acids and proteins: barriers, cell capture pathways and nanocarriers. Pharmaceutics 13(3):428. https://doi.org/10.3390/PHARMACEUTICS13030428
68. Kannan M, Bojan N, Swaminathan J, Zicarelli G, Hemalatha D, Zhang Y, Ramesh M, Faggio C (2023) Nanopesticides in agricultural pest management and their environmental risks: a review. Int J Environ Sci Technol 20:10507–10532
69. Larsson S, Jansson M, Boholm Å (2019) Expert stakeholders' perception of nanotechnology: risk, benefit, knowledge, and regulation. J Nanopart Res 21(3):1–17. https://doi.org/10.1007/S11051-019-4498-1/TABLES/8
70. Ahmed B, Rizvi A, Ali K, Lee J, Zaidi A, Khan MS, Musarrat J (2021) Nanoparticles in the soil–plant system: a review. Environ Chem Lett 19(2):1545–1609. https://doi.org/10.1007/S10311-020-01138-Y
71. Ma C, White JC, Zhao J, Zhao Q, Xing B (2018) Uptake of engineered nanoparticles by food crops: characterization, mechanisms, and implications. Annu Rev Food Sci Technol 9:129–153. https://doi.org/10.1146/ANNUREV-FOOD-030117-012657/CITE/REFWORKS
72. Paramo LA, Feregrino-Pérez AA, Guevara R, Mendoza S, Esquivel K (2020) Nanoparticles in agroindustry: applications, toxicity, challenges, and trends. Nanomaterials 10(9):1654. https://doi.org/10.3390/NANO10091654
73. Yan A, Chen Z (2019) Impacts of silver nanoparticles on plants: a focus on the phytotoxicity and underlying mechanism. Int J Mol Sci 20(5):1003. https://doi.org/10.3390/IJMS20051003
74. Pacheco I, Buzea C (2017) Nanoparticle interaction with plants. In: Nanoscience and plant–soil systems. Springer, Berlin, pp 323–355. https://doi.org/10.1007/978-3-319-46835-8_12

75. Dong BR, Jiang R, Chen J-F, Xiao Y, Lv Z-Y, Chen W-S (2023) Strategic nanoparticle-mediated plant disease resistance. Crit Rev Biotechnol 43(1):22–37. https://doi.org/10.1080/07388551.2021.2007842
76. Wu K, Xu C, Li T, Ma H, Gong J, Li X, Sun X, Hu X (2023) Application of nanotechnology in plant genetic engineering. Int J Mol Sci 24:14836
77. Ahmad R, Pranaw K, Khare SK (2019) Effect of nanomaterials and their possible implication on the plants. In: Plant biotechnology: progress in genomic era. Springer Singapore, Singapore, pp 213–229. https://doi.org/10.1007/978-981-13-8499-8_10
78. Haliloğlu K, Türkoğlu A, Balpınar Ö, Nadaroğlu H, Alaylı A, Poczai P (2022) Effects of zinc, copper and iron oxide nanoparticles on induced DNA methylation, genomic instability and LTR retrotransposon polymorphism in wheat (Triticum aestivum L.). Plants 11(17):2193. https://doi.org/10.3390/PLANTS11172193
79. Singh RP, Handa R, Manchanda G (2021) Nanoparticles in sustainable agriculture: an emerging opportunity. J Control Release 329:1234–1248
80. Kumari R, Suman K, Karmakar S, Mishra V, Lakra SG, Saurav GK, Mahto BK (2023) Regulation and safety measures for nanotechnology-based agri-products. Front Genome Ed 5:1200987. https://doi.org/10.3389/FGEED.2023.1200987/PDF
81. Jeevanandam J, Barhoum A, Chan YS, Dufresne A, Danquah MK (2018) Review on nanoparticles and nanostructured materials: history, sources, toxicity and regulations. Beilstein J Nanotechnol 9(1):1050. https://doi.org/10.3762/BJNANO.9.98
82. Shandilya N, Barreau MS, Suarez-Merino B, Porcari A, Pimponi D, Jensen KA, Fransman W, Franken R (2023) TRAAC framework to improve regulatory acceptance and wider usability of tools and methods for safe innovation and sustainability of manufactured nanomaterials. NanoImpact 30:100461. https://doi.org/10.1016/J.IMPACT.2023.100461
83. Boyes WK, Thornton BLM, Al-Abed SR, Andersen CP, Bouchard DC, Burgess RM, Hubal EAC, Ho KT, Hughes MF, Kitchin K, Reichman JR, Rogers KR, Ross JA, Rygiewicz PT, Scheckel KG, Thai SF, Zepp RG, Zucker RM (2017) A comprehensive framework for evaluating the environmental health and safety implications of engineered nanomaterials. Crit Rev Toxicol 47(9):767–810. https://doi.org/10.1080/10408444.2017.1328400
84. Amenta V, Aschberger K, Arena M, Bouwmeester H, Botelho Moniz F, Brandhoff P, Gottardo S, Marvin HJP, Mech A, Quiros Pesudo L, Rauscher H, Schoonjans R, Vettori MV, Weigel S, Peters RJ (2015) Regulatory aspects of nanotechnology in the agri/feed/food sector in EU and non-EU countries. Regul Toxicol Pharmacol 73(1):463–476. https://doi.org/10.1016/J.YRTPH.2015.06.016
85. Mishra S, Keswani C, Abhilash PC, Fraceto LF, Singh HB (2017) Integrated approach of agri-nanotechnology: challenges and future trends. Front Plant Sci 8:254477. https://doi.org/10.3389/FPLS.2017.00471/PDF
86. Justo-Hanani R, Dayan T (2016) Explaining transatlantic policy divergence: the role of domestic politics and policy styles in nanotechnology risk regulation. Global Environ Politics 16(1):79–98. https://doi.org/10.1162/GLEP_A_00337
87. Justo-Hanani R, Dayan T (2014) The role of the state in regulatory policy for nanomaterials risk: analyzing the expansion of state-centric rulemaking in EU and US chemicals policies. Res Policy 43(1):169–178. https://doi.org/10.1016/J.RESPOL.2013.06.008
88. He X, Deng H, Hwang H-m (2019) The current application of nanotechnology in food and agriculture. J Food Drug Anal 27(1):1–21. https://doi.org/10.1016/J.JFDA.2018.12.002
89. Mishra M, Dashora K, Srivastava A, Fasake VD, Nag RH (2019) Prospects, challenges and need for regulation of nanotechnology with special reference to India. Ecotoxicol Environ Saf 171:677–682. https://doi.org/10.1016/J.ECOENV.2018.12.085
90. Pandey G (2018) Challenges and future prospects of agri-nanotechnology for sustainable agriculture in India. Environ Technol Innov 11:299–307. https://doi.org/10.1016/J.ETI.2018.06.012
91. Huang S, Wang L, Liu L, Hou Y, Li L (2015) Nanotechnology in agriculture, livestock, and aquaculture in China. A review. Agron Sustain Dev 35(2):369–400. https://doi.org/10.1007/S13593-014-0274-X/TABLES/6

92. Kah M, Tufenkji N, White JC (2019) Nano-enabled strategies to enhance crop nutrition and protection. Nat Nanotechnol 14(6):532–540. https://doi.org/10.1038/s41565-019-0439-5
93. Liu P, Yang M, Hermanowicz SW, Huang Y (2022) Efficacy-associated cost analysis of copper-based nanopesticides for tomato disease control. ACS Agric Sci Technol 2(4):796–804. https://doi.org/10.1021/acsagscitech.2c00098
94. Abbati De Assis C, Greca LG, Ago M, Balakshin MY, Jameel H, Gonzalez R, Rojas OJ (2018) Techno-economic assessment, scalability, and applications of aerosol lignin micro- and nanoparticles. ACS Sustain Chem Eng 6(9):11853–11868. https://doi.org/10.1021/ACSSUSCHEMENG.8B02151/ASSET/IMAGES/LARGE/SC-2018-02151P_0011.JPEG
95. Pérez-de-Luque A (2017) Interaction of nanomaterials with plants: what do we need for real applications in agriculture? Front Environ Sci 5:259755. https://doi.org/10.3389/FENVS.2017.00012/BIBTEX
96. Carmona FJ, Dal Sasso G, Ramírez-Rodríguez GB, Pii Y, Delgado-López JM, Guagliardi A, Masciocchi N (2021) Urea-functionalized amorphous calcium phosphate nanofertilizers: optimizing the synthetic strategy towards environmental sustainability and manufacturing costs. Sci Rep 11(1):1–14. https://doi.org/10.1038/s41598-021-83048-9
97. Raliya R, Saharan V, Dimkpa C, Biswas P (2018) Nanofertilizer for precision and sustainable agriculture: current state and future perspectives. J Agric Food Chem 66(26):6487–6503. https://doi.org/10.1021/ACS.JAFC.7B02178/ASSET/IMAGES/MEDIUM/JF-2017-02178K_0010.GIF
98. Rai PK, Kumar V, Lee SS, Raza N, Kim KH, Ok YS, Tsang DCW (2018) Nanoparticle-plant interaction: implications in energy, environment, and agriculture. Environ Int 119:1–19. https://doi.org/10.1016/J.ENVINT.2018.06.012
99. Kottegoda N, Sandaruwan C, Priyadarshana G, Siriwardhana A, Rathnayake UA, Berugoda Arachchige DM, Kumarasinghe AR, Dahanayake D, Karunaratne V, Amaratunga GAJ (2017) Urea-hydroxyapatite nanohybrids for slow release of nitrogen. ACS Nano 11(2):1214–1221. https://doi.org/10.1021/ACSNANO.6B07781/SUPPL_FILE/NN6B07781_LIVESLIDES.MP4
100. Dimkpa CO, Bindraban PS (2018) Nanofertilizers: new products for the industry? J Agric Food Chem 66(26):6462–6473. https://doi.org/10.1021/ACS.JAFC.7B02150/ASSET/IMAGES/MEDIUM/JF-2017-02150Z_0004.GIF
101. Kalwani M, Chakdar H, Srivastava A, Pabbi S, Shukla P (2022) Effects of nanofertilizers on soil and plant-associated microbial communities: emerging trends and perspectives. Chemosphere 287:132107. https://doi.org/10.1016/J.CHEMOSPHERE.2021.132107
102. Wu H, Li Z (2022) Recent advances in nano-enabled agriculture for improving plant performance. Crop J 10(1):1–12. https://doi.org/10.1016/J.CJ.2021.06.002
103. Pereira ADES, Oliveira HC, Fraceto LF, Santaella C (2021) Nanotechnology potential in seed priming for sustainable agriculture. Nanomaterials 11(2):267. https://doi.org/10.3390/NANO11020267
104. Osmani Z, Kulka M (2025) Form and function: the factors that influence the efficacy of nanomaterials for gene transfer to plants. Molecules 30:446
105. Parisi C, Vigani M, Rodríguez-Cerezo E (2015) Agricultural nanotechnologies: what are the current possibilities? Nano Today 10(2):124–127. https://doi.org/10.1016/J.NANTOD.2014.09.009
106. Fraceto LF, Grillo R, de Medeiros GA, Scognamiglio V, Rea G, Bartolucci C (2016) Nanotechnology in agriculture: which innovation potential does it have? Front Environ Sci 4:186737. https://doi.org/10.3389/FENVS.2016.00020/BIBTEX
107. Liu R, Lal R (2015) Potentials of engineered nanoparticles as fertilizers for increasing agronomic productions. Sci Total Environ 514:131–139. https://doi.org/10.1016/J.SCITOTENV.2015.01.104
108. Emamalipour M, Seidi K, Zununi Vahed S, Jahanban-Esfahlan A, Jaymand M, Majdi H, Amoozgar Z, Chitkushev LT, Javaheri T, Jahanban-Esfahlan R, Zare P (2020) Horizontal gene transfer: from evolutionary flexibility to disease progression. Front Cell Dev Biol 8:229

109. Demirer GS, Zhang H, Matos JL, Goh NS, Cunningham FJ, Sung Y, Chang R, Aditham AJ, Chio L, Cho M-J, Staskawicz B, Landry MP (2019) High aspect ratio nanomaterials enable delivery of functional genetic material without DNA integration in mature plants. Nat Nanotechnol 14(5):456–464. https://doi.org/10.1038/s41565-019-0382-5
110. Panpatte DG, Jhala YK (2019) Nanotechnology for agriculture: advances for sustainable agriculture. Springer Singapore, Singapore. https://doi.org/10.1007/978-981-32-9370-0
111. Xin X, Judy JD, Sumerlin BB, He Z (2020) Nano-enabled agriculture: from nanoparticles to smart nanodelivery systems. Environ Chem 17:413–425
112. Grillo R, Mattos BD, Antunes DR, Forini MML, Monikh FA, Rojas OJ (2021) Foliage adhesion and interactions with particulate delivery systems for plant nanobionics and intelligent agriculture. Nano Today 37(40)
113. Guleria P, Kumar V (2018) Impact of recombinant DNA technology and nanotechnology on agriculture. In: Sustainable agriculture reviews. Springer, Cham, pp 271–292. https://doi.org/10.1007/978-3-319-98914-3_11
114. Bain C, Selfa T, Dandachi T, Velardi S (2017) 'Superweeds' or 'survivors'? Framing the problem of glyphosate resistant weeds and genetically engineered crops. J Rural Stud 51:211–221. https://doi.org/10.1016/j.jrurstud.2017.03.003
115. Lichtfouse E (2019) Sustainable agriculture reviews 32. Scholars Portal, Toronta
116. Vega-Vásquez P, Mosier NS, Irudayaraj J (2020) Nanoscale drug delivery systems: from medicine to agriculture. Front Bioeng Biotechnol 8:79
117. Saakre M, Jaiswal S, Rathinam M, Raman KV, Tilgam J, Paul K, Sreevathsa R, Pattanayak D (2024) Host-delivered RNA interference for durable pest resistance in plants: advanced methods, challenges, and applications. Mol Biotechnol 66:1786–1805
118. Halder K, Chaudhuri A, Abdin MZ, Majee M, Datta A (2022) RNA interference for improving disease resistance in plants and its relevance in this clustered regularly interspaced short palindromic repeats-dominated era in terms of dsRNA-based biopesticides. Front Plant Sci 13:885128
119. Law SSY, Miyamoto T, Numata K (2023) Organelle-targeted gene delivery in plants by nanomaterials. Chem Commun 59(47):7166–7181. https://doi.org/10.1039/d3cc00962a
120. Chaud M, Souto EB, Zielinska A, Severino P, Batain F, Oliveira-Junior J, Alves T (2021) Nanopesticides in agriculture: benefits and challenge in agricultural productivity, toxicological risks to human health and environment. Toxics 9:131
121. Nezhmetdinova FT, Guryleva ME, Sharypova NK, Zinurova RI, Tuzikov AR (2020) Risks of modern biotechnologies and legal aspects of their implementation in agriculture. BIO Web Conf 17:00227
122. Li G, Chen X, Yin H, Wang W, Wong PK, An T (2020) Natural sphalerite nanoparticles can accelerate horizontal transfer of plasmid-mediated antibiotic-resistance genes. Environ Int 136:105497. https://doi.org/10.1016/j.envint.2020.105497
123. Lu J, Wang Y, Jin M, Yuan Z, Bond P, Guo J (2020) Both silver ions and silver nanoparticles facilitate the horizontal transfer of plasmid-mediated antibiotic resistance genes. Water Res 169:115229. https://doi.org/10.1016/J.WATRES.2019.115229
124. Kumar S, Nehra M, Dilbaghi N, Marrazza G, Tuteja SK, Kim K-H (2020) Nanovehicles for plant modifications towards pest- and disease-resistance traits. Trends Plant Sci 25(2):198–212. https://doi.org/10.1016/j.tplants.2019.10.007
125. Acharya A, Pal PK (2020) Agriculture nanotechnology: translating research outcome to field applications by influencing environmental sustainability. NanoImpact 19:100232. https://doi.org/10.1016/J.IMPACT.2020.100232
126. Usman M, Farooq M, Wakeel A, Nawaz A, Cheema SA, ur Rehman H, Ashraf I, Sanaullah M (2020) Nanotechnology in agriculture: current status, challenges and future opportunities. Sci Total Environ 721:137778. https://doi.org/10.1016/J.SCITOTENV.2020.137778
127. Wang P, Zhao FJ, Kopittke PM (2019) Engineering crops without genome integration using nanotechnology. Trends Plant Sci 24:574–577

128. Wang H, Qi H, Gong S, Huang Z, Meng C, Zhang Y, Chen X, Jiao X (2020) Fe3O4 composited with MoS2 blocks horizontal gene transfer. Colloids Surf B Biointerfaces 185:110569. https://doi.org/10.1016/j.colsurfb.2019.110569
129. Zhi H, Zhou S, Pan W, Shang Y, Zeng Z, Zhang H (2022) The promising nanovectors for gene delivery in plant genome engineering. Int J Mol Sci 23:8501
130. Mukherjee A, Majumdar S, Servin AD, Pagano L, Dhankher OP, White JC (2016) Carbon nanomaterials in agriculture: a critical review. Front Plant Sci 7:172
131. El-Ramady H, Abdalla N, Sári D, Ferroudj A, Muthu A, Prokisch J, Fawzy ZF, Brevik EC, Solberg S (2023) Nanofarming: promising solutions for the future of the global agricultural industry. Agronomy 13:1600
132. Iavicoli I, Leso V, Beezhold DH, Shvedova AA (2017) Nanotechnology in agriculture: opportunities, toxicological implications, and occupational risks. Toxicol Appl Pharmacol 329:96–111. https://doi.org/10.1016/J.TAAP.2017.05.025
133. Philips JG, Martin-Avila E, Robold AV (2022) Horizontal gene transfer from genetically modified plants—regulatory considerations. Front Bioeng Biotechnol 10:971402
134. Khan ZS, Rizwan M, Hafeez M, Ali S, Javed MR, Adrees M (2019) The accumulation of cadmium in wheat (Triticum aestivum) as influenced by zinc oxide nanoparticles and soil moisture conditions. Environ Sci Pollut Res 26(19):19859–19870. https://doi.org/10.1007/s11356-019-05333-5
135. Djibril Sekou K, Patel H (2023) A review on the interaction between nanoparticles and toxic metals in soil: meta-analysis of their effects on soil, plants and human health. Soil Sediment Contam Int J 32(4):417–447. https://doi.org/10.1080/15320383.2022.2096564
136. You T, Liu D, Chen J, Yang Z, Dou R, Gao X, Wang L (2017) Effects of metal oxide nanoparticles on soil enzyme activities and bacterial communities in two different soil types. J Soils Sediments 18:211–221. https://doi.org/10.1007/s11368-017-1716-2
137. Tripathi DK, Shweta, Singh S, Singh S, Pandey R, Singh VP, Sharma NC, Prasad SM, Dubey NK, Chauhan DK (2017) An overview on manufactured nanoparticles in plants: uptake, translocation, accumulation and phytotoxicity. Plant Physiol Biochem 110:2–12. https://doi.org/10.1016/J.PLAPHY.2016.07.030
138. Ge Y, Schimel JP, Holden PA (2011) Evidence for negative effects of TiO2 and ZnO nanoparticles on soil bacterial communities. Environ Sci Technol 45(4):1659–1664. https://doi.org/10.1021/ES103040T/SUPPL_FILE/ES103040T_SI_001.PDF
139. Grün AL, Straskraba S, Schulz S, Schloter M, Emmerling C (2018) Long-term effects of environmentally relevant concentrations of silver nanoparticles on microbial biomass, enzyme activity, and functional genes involved in the nitrogen cycle of loamy soil. J Environ Sci 69:12–22. https://doi.org/10.1016/J.JES.2018.04.013
140. Rajput V, Minkina T, Mazarji M, Shende S, Sushkova S, Mandzhieva S, Burachevskaya M, Chaplygin V, Singh A, Jatav H (2020) Accumulation of nanoparticles in the soil-plant systems and their effects on human health. Ann Agric Sci 65(2):137–143. https://doi.org/10.1016/J.AOAS.2020.08.001
141. Rajput V, Minkina T, Sushkova S, Tsitsuashvili V, Mandzhieva S, Gorovtsov A, Nevidomskyaya D, Gromakova N (2018) Effect of nanoparticles on crops and soil microbial communities. J Soils Sediments 18:2179–2187. https://doi.org/10.1007/s11368-017-1793-2
142. Shrivastava M, Srivastav A, Gandhi S, Rao S, Roychoudhury A, Kumar A, Singhal RK, Jha SK, Singh SD (2019) Monitoring of engineered nanoparticles in soil-plant system: a review. Environ Nanotechnol Monit Manag 11:100218. https://doi.org/10.1016/J.ENMM.2019.100218
143. Shukla PK, Misra P, Kole C (2016) Uptake, translocation, accumulation, transformation, and generational transmission of nanoparticles in plants. In: Plant nanotechnology. Springer International Publishing, Berlin, pp 183–218. https://doi.org/10.1007/978-3-319-42154-4_8
144. Zhang P, Guo Z, Zhang Z, Fu H, White JC, Lynch Zhang IP, Guo Z, Lynch I, Zhang Z, Fu H, White JC (2020) Nanomaterial transformation in the soil–plant system: implications for food safety and application in agriculture. Small 16(21):2000705. https://doi.org/10.1002/SMLL.202000705

145. Achari GA, Kowshik M (2018) Recent developments on nanotechnology in agriculture: plant mineral nutrition, health, and interactions with soil microflora. J Agric Food Chem 66:8647–8661
146. Hofmann T, Lowry GV, Ghoshal S, Tufenkji N, Brambilla D, Dutcher JR, Gilbertson LM, Giraldo JP, Kinsella JM, Landry MP, Lovell W, Naccache R, Paret M, Pedersen JA, Unrine JM, White JC, Wilkinson KJ (2020) Technology readiness and overcoming barriers to sustainably implement nanotechnology-enabled plant agriculture. Nat Food 1(7):416–425. https://doi.org/10.1038/s43016-020-0110-1
147. Li M, Sun X, Yin M, Shen J, Yan S (2023) Recent advances in nanoparticle-mediated co-delivery system: a promising strategy in medical and agricultural field. Int J Mol Sci 24:5121
148. Demirer GS, Zhang H, Goh NS, Pinals RL, Chang R, Landry MP (2020) Carbon nanocarriers deliver siRNA to intact plant cells for efficient gene knockdown. Sci Adv 6(26):eaaz0495. https://doi.org/10.1126/sciadv.aaz0495
149. Sangwan A, Gupta D, Singh OW, Roy A, Mukherjee SK, Mandal B, Singh N (2023) Size variations of mesoporous silica nanoparticle control uptake efficiency and delivery of AC2-derived dsRNA for protection against tomato leaf curl New Delhi virus. Plant Cell Rep 42(10):1571–1587. https://doi.org/10.1007/s00299-023-03048-z
150. Yu P, Zheng X, Alimi LO, Al-Babili S, Khashab NM (2024) Metal–organic framework-mediated delivery of nucleic acid across intact plant cells. ACS Appl Mater Interfaces 16(15):18245–18251. https://doi.org/10.1021/acsami.3c19571
151. Wang ZP, Zhang ZB, Zheng DY, Zhang TT, Li XL, Zhang C, Yu R, Wei JH, Wu ZY (2022) Efficient and genotype independent maize transformation using pollen transfected by DNA-coated magnetic nanoparticles. J Integr Plant Biol 64(6):1145–1156. https://doi.org/10.1111/jipb.13263
152. Schwartz SH, Hendrix B, Hoffer P, Sanders RA, Zheng W (2020) Carbon dots for efficient small interfering RNA delivery and gene silencing in plants. Plant Physiol 184(2):647–657. https://doi.org/10.1104/pp.20.00733
153. Molesini B, Pennisi F, Cressoni C, Vitulo N, Dusi V, Speghini A, Pandolfini T (2022) Nanovector-mediated exogenous delivery of dsRNA induces silencing of target genes in very young tomato flower buds. Nanoscale Adv 4(21):4542–4553. https://doi.org/10.1039/D2NA00478J
154. Islam MR, Youngblood M, Kim H-I, González-Gamboa I, Monroy-Borrego AG, Caparco AA, Lowry GV, Steinmetz NF, Giraldo JP (2024) DNA delivery by virus-like nanocarriers in plant cells. Nano Lett 24(26):7833–7842. https://doi.org/10.1021/acs.nanolett.3c04735
155. Qi J, Li Y, Yao X, Li G, Xu W, Chen L, Xie Z, Gu J, Wu H, Li Z (2024) Rational design of ROS scavenging and fluorescent gold nanoparticles to deliver siRNA to improve plant resistance to Pseudomonas syringae. J Nanobiotechnol 22(1):446. https://doi.org/10.1186/s12951-024-02733-9
156. Cai Y, Liu Z, Wang H, Meng H, Cao Y (2024) Mesoporous silica nanoparticles mediate SiRNA delivery for long-term multi-gene silencing in intact plants. Adv Sci (Weinh) 11(9):e2301358. https://doi.org/10.1002/advs.202301358
157. Xu X, Yu T, Zhang D, Song H, Huang K, Wang Y, Shen L, Li Y, Wang F, Zhang S, Jiao Y, Yang J (2023) Evaluation of the anti-viral efficacy of three different dsRNA nanoparticles against potato virus Y using various delivery methods. Ecotoxicol Environ Saf 255:114775. https://doi.org/10.1016/j.ecoenv.2023.114775
158. Mahmoud LM, Kaur P, Stanton D, Grosser JW, Dutt M (2022) A cationic lipid mediated CRISPR/Cas9 technique for the production of stable genome edited citrus plants. Plant Methods 18(1):33. https://doi.org/10.1186/s13007-022-00870-6
159. Liu W, Rudis MR, Cheplick MH, Millwood RJ, Yang JP, Ondzighi-Assoume CA, Montgomery GA, Burris KP, Mazarei M, Chesnut JD, Stewart CN Jr (2020) Lipofection-mediated genome editing using DNA-free delivery of the Cas9/gRNA ribonucleoprotein into plant cells. Plant Cell Rep 39(2):245–257. https://doi.org/10.1007/s00299-019-02488-w

160. Zhao X, Meng Z, Wang Y, Chen W, Sun C, Cui B, Cui J, Yu M, Zeng Z, Guo S, Luo D, Cheng JQ, Zhang R, Cui H (2017) Pollen magnetofection for genetic modification with magnetic nanoparticles as gene carriers. Nat Plants 3(12):956–964. https://doi.org/10.1038/s41477-017-0063-z
161. Law SSY, Liou G, Nagai Y, Giménez-Dejoz J, Tateishi A, Tsuchiya K, Kodama Y, Fujigaya T, Numata K (2022) Polymer-coated carbon nanotube hybrids with functional peptides for gene delivery into plant mitochondria. Nat Commun 13(1):2417. https://doi.org/10.1038/s41467-022-30185-y
162. Yong J, Zhang R, Bi S, Li P, Sun L, Mitter N, Carroll BJ, Xu ZP (2021) Sheet-like clay nanoparticles deliver RNA into developing pollen to efficiently silence a target gene. Plant Physiol 187(2):886–899. https://doi.org/10.1093/plphys/kiab303
163. Wang JW, Cunningham FJ, Goh NS, Boozarpour NN, Pham M, Landry MP (2021) Nanoparticles for protein delivery in planta. Curr Opin Plant Biol 60:102052. https://doi.org/10.1016/j.pbi.2021.102052
164. Wei T, Cheng Q, Min Y-L, Olson EN, Siegwart DJ (2020) Systemic nanoparticle delivery of CRISPR-Cas9 ribonucleoproteins for effective tissue specific genome editing. Nat Commun 11(1):3232. https://doi.org/10.1038/s41467-020-17029-3
165. Thomson JA (2002) Research needs to improve agricultural productivity and food quality, with emphasis on biotechnology. J Nutr 132(11):3441S–3442S. https://doi.org/10.1093/jn/132.11.3441S
166. Deng P, Zhou X, Zhou P, Du Z, Hou H, Yang D, Tan J, Wu X, Zhang J, Yang Y, Liu J, Liu G, Li Y, Liu J, Yu L, Fang S, Yang X (2008) Edible safety requirements and assessment standards for agricultural genetically modified organisms. Food Chem Toxicol 46(5):1414–1436. https://doi.org/10.1016/j.fct.2008.01.007
167. Li X-D, Gao Y-Q, Wu W-H, Chen L-M, Wang Y (2022) Two calcium-dependent protein kinases enhance maize drought tolerance by activating anion channel ZmSLAC1 in guard cells. Plant Biotechnol J 20(1):143–157. https://doi.org/10.1111/pbi.13701
168. Jiang L (2020) Commercialization of the gene-edited crop and morality: challenges from the liberal patent law and the strict GMO law in the EU. New Genet Soc 39(2):191–218. https://doi.org/10.1080/14636778.2019.1686968
169. Guehlstorf NP, Hallstrom LK (2005) The role of culture in risk regulations: a comparative case study of genetically modified corn in The United States of America and European Union. Environ Sci Pol 8(4):327–342. https://doi.org/10.1016/j.envsci.2005.04.007
170. Ryan CD, Elizabeth H, Samantha G, Schaul AJ, Swarthout JT (2024) Exploring the GMO narrative through labeling: strategies, products, and politics. GM Crops Food 15(1):51–66. https://doi.org/10.1080/21645698.2024.2318027
171. Tambo JA, Abdoulaye T (2012) Climate change and agricultural technology adoption: the case of drought tolerant maize in rural Nigeria. Mitig Adapt Strateg Glob Chang 17(3):277–292. https://doi.org/10.1007/s11027-011-9325-7
172. Juturu VN, Mekala GK, Kirti PB (2015) Current status of tissue culture and genetic transformation research in cotton (Gossypium spp.). Plant Cell Tissue Organ Cult 120(3):813–839. https://doi.org/10.1007/s11240-014-0640-z
173. Kolseth A-K, D'Hertefeldt T, Emmerich M, Forabosco F, Marklund S, Cheeke TE, Hallin S, Weih M (2015) Influence of genetically modified organisms on agro-ecosystem processes. Agric Ecosyst Environ 214:96–106. https://doi.org/10.1016/j.agee.2015.08.021
174. Tabashnik BE, Liesner LR, Ellsworth PC, Unnithan GC, Fabrick JA, Naranjo SE, Li X, Dennehy TJ, Antilla L, Staten RT, Carrière Y (2021) Transgenic cotton and sterile insect releases synergize eradication of pink bollworm a century after it invaded the United States. Proc Natl Acad Sci U S A 118(1):e2019115118. https://doi.org/10.1073/pnas.2019115118
175. Lundgren JG, Gassmann AJ, Bernal J, Duan JJ, Ruberson J (2009) Ecological compatibility of GM crops and biological control. Crop Prot 28(12):1017–1030. https://doi.org/10.1016/j.cropro.2009.06.001

176. Yang LH, Gratton C (2014) Insects as drivers of ecosystem processes. Curr Opin Insect Sci 2:26–32. https://doi.org/10.1016/j.cois.2014.06.004
177. Vázquez-Barrios V, Boege K, Sosa-Fuentes TG, Rojas P, Wegier A (2021) Ongoing ecological and evolutionary consequences by the presence of transgenes in a wild cotton population. Sci Rep 11(1):1959. https://doi.org/10.1038/s41598-021-81567-z
178. Ellstrand NC, Meirmans P, Rong J, Bartsch D, Ghosh A, de Jong TJ, Haccou P, Lu B-R, Snow AA, Neal Stewart C, Strasburg JL, van Tienderen PH, Vrieling K, Hooftman D (2013) Introgression of crop alleles into wild or weedy populations. Annu Rev Ecol Evol Syst 44:325–345. https://doi.org/10.1146/annurev-ecolsys-110512-135840
179. Tabashnik BE, Brévault T, Carrière Y (2013) Insect resistance to Bt crops: lessons from the first billion acres. Nat Biotechnol 31(6):510–521. https://doi.org/10.1038/nbt.2597
180. Bakker PAHM, Berendsen RL, Van Pelt JA, Vismans G, Yu K, Li E, Van Bentum S, Poppeliers SWM, Sanchez Gil JJ, Zhang H, Goossens P, Stringlis IA, Song Y, de Jonge R, Pieterse CMJ (2020) The soil-borne identity and microbiome-assisted agriculture: looking back to the future. Mol Plant 13(10):1394–1401. https://doi.org/10.1016/j.molp.2020.09.017
181. Khan ST (2020) Interaction of engineered nanomaterials with soil microbiome and plants: their impact on plant and soil health. In: Hayat S, Pichtel J, Faizan M, Fariduddin Q (eds) Sustainable agriculture reviews 41: nanotechnology for plant growth and development. Springer International Publishing, Berlin, pp 181–199. https://doi.org/10.1007/978-3-030-33996-8_10
182. Murray MT, Nowicki J (2020) 136—Immune system support. In: Pizzorno JE, Murray MT (eds) Textbook of natural medicine, 5th edn. Churchill Livingstone, London, pp 1009–1016.e1002. https://doi.org/10.1016/B978-0-323-43044-9.00136-9
183. Han X, Ding S, Lu J, Li Y (2022) Global, regional, and national burdens of common micronutrient deficiencies from 1990 to 2019: a secondary trend analysis based on the global burden of disease 2019 study. eClinicalMedicine 44:101299. https://doi.org/10.1016/j.eclinm.2022.101299
184. Galani YJH, Orfila C, Gong YY (2022) A review of micronutrient deficiencies and analysis of maize contribution to nutrient requirements of women and children in Eastern and Southern Africa. Crit Rev Food Sci Nutr 62(6):1568–1591. https://doi.org/10.1080/10408398.2020.1844636
185. De Moura FF, Alexander M, Boy E (2015) Retention of provitamin A carotenoids in staple crops targeted for biofortification in Africa: cassava, maize and sweet potato. Crit Rev Food Sci Nutr 55(9):1246–1269. https://doi.org/10.1080/10408398.2012.724477
186. Manjeru P, Angeline VB, Labuschagne M (2019) The development and release of maize fortified with provitamin A carotenoids in developing countries. Crit Rev Food Sci Nutr 59(8):1284–1293. https://doi.org/10.1080/10408398.2017.1402751
187. Zhu C, Naqvi S, Gomez-Galera S, Pelacho AM, Capell T, Christou P (2007) Transgenic strategies for the nutritional enhancement of plants. Trends Plant Sci 12(12):548–555. https://doi.org/10.1016/j.tplants.2007.09.007
188. Jadon V, Sharma S, Krishna H, Krishnappa G, Gajghate R, Devate NB, Panda KK, Jain N, Singh PK, Singh GP (2023) Molecular mapping of biofortification traits in bread wheat (Triticum aestivum L.) using a high-density SNP based linkage map. Genes 14(1):221. https://www.mdpi.com/2073-4425/14/1/221
189. Joshi G, Soe YP, Palanog A, Hore TK, Nha CT, Calayugan MI, Inabangan-Asilo MA, Amparado A, Pandey ID, Cruz PCS, Hernandez JE, Swamy BPM (2023) Meta-QTLs and haplotypes for efficient zinc biofortification of rice. Plant Genome 16(4):e20315. https://doi.org/10.1002/tpg2.20315
190. Tanumihardjo SA, Kaliwile C, Boy E, Dhansay MA, van Stuijvenberg ME (2019) Overlapping vitamin A interventions in the United States, Guatemala, Zambia, and South Africa: case studies. Ann N Y Acad Sci 1446(1):102–116. https://doi.org/10.1111/nyas.13965
191. Mwangi MN, Prentice AM, Verhoef H (2017) Safety and benefits of antenatal oral iron supplementation in low-income countries: a review. Br J Haematol 177(6):884–895. https://doi.org/10.1111/bjh.14584

192. Martínez-Chávez LA, Hernández-Ramírez MY, Feregrino-Pérez AA, Esquivel Escalante K (2024) Cutting-edge strategies to enhance bioactive compound production in plants: potential value of integration of elicitation, metabolic engineering, and green nanotechnology. Agronomy 14(12):2822. https://www.mdpi.com/2073-4395/14/12/2822
193. Gamage A, Gangahagedara R, Subasinghe S, Gamage J, Guruge C, Senaratne S, Randika T, Rathnayake C, Hameed Z, Madhujith T, Merah O (2024) Advancing sustainability: the impact of emerging technologies in agriculture. Curr Plant Biol 40:100420. https://doi.org/10.1016/J.CPB.2024.100420
194. Grieger KD, Merck AW, Cuchiara M, Binder AR, Kokotovich A, Cummings CL, Kuzma J (2021) Responsible innovation of nano-agrifoods: insights and views from U.S. stakeholders. NanoImpact 24:100365. https://doi.org/10.1016/J.IMPACT.2021.100365
195. Cummings CL, Kuzma J, Kokotovich A, Glas D, Grieger K (2021) Barriers to responsible innovation of nanotechnology applications in food and agriculture: a study of US experts and developers. NanoImpact 23:100326. https://doi.org/10.1016/J.IMPACT.2021.100326
196. Lowry GV, Avellan A, Gilbertson LM (2019) Opportunities and challenges for nanotechnology in the agri-tech revolution. Nat Nanotechnol 14(6):517–522. https://doi.org/10.1038/s41565-019-0461-7
197. Merck AW, Grieger KD, Cuchiara M, Kuzma J (2022) What role does regulation play in responsible innovation of nanotechnology in food and agriculture? Insights and framings from U.S. Stakeholders. Bull Sci Technol Soc 42(3):85–103. https://doi.org/10.1177/02704676221102066
198. Capon A, Gillespie J, Rolfe M, Smith W (2015) Perceptions of risk from nanotechnologies and trust in stakeholders: a cross sectional study of public, academic, government and business attitudes. BMC Public Health 15(1):1–13. https://doi.org/10.1186/S12889-015-1795-1/TABLES/6
199. Peters RJB, Bouwmeester H, Gottardo S, Amenta V, Arena M, Brandhoff P, Marvin HJP, Mech A, Moniz FB, Pesudo LQ, Rauscher H, Schoonjans R, Undas AK, Vettori MV, Weigel S, Aschberger K (2016) Nanomaterials for products and application in agriculture, feed and food. Trends Food Sci Technol 54:155–164. https://doi.org/10.1016/J.TIFS.2016.06.008
200. Prasad R, Bhattacharyya A, Nguyen QD (2017) Nanotechnology in sustainable agriculture: recent developments, challenges, and perspectives. Front Microbiol 8:1014. https://doi.org/10.3389/FMICB.2017.01014/BIBTEX
201. Lombi E, Donner E, Dusinska M, Wickson F (2019) A One Health approach to managing the applications and implications of nanotechnologies in agriculture. Nat Nanotechnol 14(6):523–531. https://doi.org/10.1038/s41565-019-0460-8
202. Chhipa H (2017) Nanofertilizers and nanopesticides for agriculture. Environ Chem Lett 15(1):15–22. https://doi.org/10.1007/S10311-016-0600-4/TABLES/2
203. Shukla K, Mishra V, Singh J, Varshney V, Verma R, Srivastava S (2024) Nanotechnology in sustainable agriculture: a double-edged sword. J Sci Food Agric 104(10):5675–5688. https://doi.org/10.1002/JSFA.13342
204. Kaphle A, Navya PN, Umapathi A, Daima HK (2017) Nanomaterials for agriculture, food and environment: applications, toxicity and regulation. Environ Chem Lett 16(1):43–58. https://doi.org/10.1007/S10311-017-0662-Y
205. Saleem H, Zaidi SJ (2020) Recent developments in the application of nanomaterials in agroecosystems. Nanomaterials 10(12):2411. https://doi.org/10.3390/NANO10122411
206. Tripathi S, Mahra S, Victoria J, Tiwari K, Rana S, Tripathi DK, Sharma S, Sahi S (2023) Recent advances and perspectives of nanomaterials in agricultural management and associated environmental risk: a review. Nanomaterials 13(10):1604. https://doi.org/10.3390/NANO13101604
207. Singh H, Sharma A, Bhardwaj SK, Arya SK, Bhardwaj N, Khatri M (2021) Recent advances in the applications of nano-agrochemicals for sustainable agricultural development. Environ Sci: Processes Impacts 23(2):213–239. https://doi.org/10.1039/D0EM00404A

208. Wang J, Wu H, Wang Y, Ye W, Kong X, Yin Z (2024) Small particles, big effects: how nanoparticles can enhance plant growth in favorable and harsh conditions. J Integr Plant Biol 66(7):1274–1294. https://doi.org/10.1111/jipb.13652
209. Nizamani M, Hughes A, Zhang H, Wang Y (2024) Revolutionizing agriculture with nanotechnology: innovative approaches in fungal disease management and plant health monitoring. Sci Total Environ 928:172473. https://doi.org/10.1016/j.scitotenv.2024.172473
210. Hilbeck A, Meier M, Benzler A (2008) Identifying indicator species for post-release monitoring of genetically modified, herbicide resistant crops. Euphytica 164:903–912. https://doi.org/10.1007/s10681-008-9666-9

MIX
Papier aus verantwortungsvollen Quellen
Paper from responsible sources
FSC® C105338

FSC
www.fsc.org

If you have any concerns about our products,
you can contact us on
ProductSafety@springernature.com

In case Publisher is established outside the EU,
the EU authorized representative is:
Springer Nature Customer Service Center GmbH
Europaplatz 3, 69115 Heidelberg, Germany

Printed by Libri Plureos GmbH
in Hamburg, Germany